Energy Efficiency in Motor Driven Systems

Springer
Berlin
Heidelberg
New York
Hong Kong
London
Milan
Paris
Tokyo

Francesco Parasiliti
Paolo Bertoldi
Editors

Energy Efficiency in Motor Driven Systems

With 308 Figures
and 80 Tables

Springer

Professor Francesco Parasiliti
Department of Electrical Engineering
University of L'Aquila
Poggio di Roio
I-67040 L'Aquila, Italy
e-mail: rock@ing.univaq.it

Dr. Paolo Bertoldi
European Commission
DG Joint Research Centre
Via Fermi 1
I-21020 Ispra, Varese, Italy
e-mail: paolo.bertoldi@cec.eu.int

ISBN 3-540-00666-4 Springer-Verlag Berlin Heidelberg New York

Bibliographic information published by Die Deutsche Bibliothek
Die Deutsche Bibliothek lists this publication in the Deutsche Nationalbibliografie; detailed bibliographic data available in the internet at *http.//dnb.ddb.de*

Springer-Verlag Berlin Heidelberg New York
a member of BertelsmannSpringer Science + Business Media GmbH

http://www.springer.de

Cover design: Erich Kirchner, Heidelberg

SPIN 10917442 42/3130-5 4 3 2 1 0 – Printed on acid-free paper

Preface

The recent World Summit on Sustainable Development in Johannesburg concluded that changing unsustainable patterns of energy use is a key area for global action to ensure the survival of our planet. At the same time, the Summit also highlighted that nearly one third of mankind does not yet have access to electricity and basic energy services, while another third has only poor or unreliable access. Energy efficiency improvements in motor systems can play a key role in assuring a sustainable energy future and socio-economic development, and at the same time mitigate climate change. Energy efficiency measures related to motor systems are among the most cost-effective CO_2 emission reduction actions, and offer the best opportunity to increase the security and reliability of energy supply.

However market, policy, trade and information barriers impede the further penetration of energy efficient motor systems, resulting in a missed opportunity for climate change mitigation and socio-economic development.

Electric motor driven systems account for the greatest part of industrial electricity consumption in all countries. Numerous studies on individual component (motors, pumps, compressors, fans) and on the consumption characterisation have shown the considerable potential for improvement of energy efficiency of these systems, and have recommended suitable policy actions. The policy actions so far implemented cover individual components, in particular motors. For motors most of the developed countries have adopted mandatory or voluntary efficiency requirements, classification systems and motor selection database. Other policy initiatives cover end-use equipment such as pumps, compressors, and fans. These initiatives tend to be of a voluntary nature and they include: information dissemination, best practice, voluntary agreement, audit schemes, and financial and fiscal incentives. More recently the attention of policy makers and programme designers has moved to the "systems" and to the numerous possibilities for improving efficiency and save energy in the systems design, operation and maintenance.

The international community of stakeholders dealing with motor systems (including manufacturers, system designers, utilities, governments, international organisations and agencies, academia and experts) have already gathered twice at the **International Conference on Energy Efficiency in Motor Driven Systems (EEMODS)** (Lisbon 1996 and London 1999) to discuss the progress achieved in technologies, programmes and policies, and the strategies to be implemented to further this progress.

The EEMODS conferences have been very successful in attracting an international audience, representing a wide variety of stakeholders involved in policy implementation and development, and manufacturing and promotion of energy efficient motor systems. The EEMODS conference has established itself as an influential and recognised international event where participants can discuss the latest developments and build international partnerships among stakeholders.

The third EEMODS conference was held in Treviso (Italy), September 18-20 2002, and was organised by the European Commission JRC and University of L'Aquila. 250 participants from 35 countries (many from developing countries) attended the conference. Key representatives of the manufacturing industry (motors, pumps, compressors) and trade organizations also attended the conference.

In the plenary sessions the main EU and US trade associations for motor systems components gave an overview of the recent developments. In particular the European Commission introduced the new Motor Challenge programme; Motor Challenge will constitute the new frame to include the various actions identified in the sectorial studies. The recent developments both in the US under the Best Practice programme, and in developing countries were also presented.

79 papers were presented in the 3 concurrent sessions covering the following topics:

Compressed Air
Electrical Motors (technologies, policies, and test methods)
Fans and Fan Systems
Management Issues
Motor System Audit and Programs
Policies and International Issues
Power Electronics and Electrical Drives
Pumps and Pump Systems

This book contains the key presentations made in the plenary sessions, as well as the papers presented in the parallel sessions. It is hoped that its availability will enable a large audience to benefits from the presentations made in EEMODS'02. Potential readers who may benefit from this book include researchers, engineers, policymakers, energy agencies, electric utilities, and all those who can influence the design, selection, application and operation of electrical motor driven systems.

EEMODS'02 had as its main sponsor the INTERNATIONAL COPPER ASSOCIATION.

The Organising Committee also wish to acknowledge the following organisations that helped by sponsoring the event: ABB INDUSTRIA, ATLAS COPCO, FERRI ELETTROFORNITURE, AEG-LAFERT, CONTROL TECHNIQUES, EARP-ELETTRO ADDA-REEL, EUROVENT-CECOMAF, SWISS FEDERAL OFFICE OF ENERGY, MARELLI MOTORI, TEXAS INSTRUMENTS, CARISPAQ and the UNIVERSITY of L'AQUILA. The support of all the sponsors and their generous funding were very important in ensuring the success of EEMODS'02, and are greatly appreciated.

The Editors

Francesco Parasiliti
Paolo Bertoldi

Contents

Fans and Fan Systems

Management Issues

Motor System Audit and Programs

Policies and International Issues

Power Electronics and Electrical Drives

Pumps and Pumping Systems

European Compressed Air Industry Energy Saving Strategy

Harry David Craig

Representing:

PNEUROP, Diamant Building, Boulevard A Reyers, 80, B-1030 Brussels, Belgium
Ph. +32 2 706 8230 - Fax +32 2 706 8250 - E-mail secretariat@pneurop.org, and

CompAir UK Limited, Hughenden Avenue, High Wycombe, Bucks HP13 5SF, UK.
Ph. +44 1494 605300 - Fax +44 1494 462624 - E-mail sales@compair.com

Introduction

The compressed air industry has been conscious of the need to have energy utilization as a key criterion when creating products and when designing and installing systems.

This is not a recent concern but focus on the need has increased with Kyoto and other energy saving initiatives. A difficulty with compressed air systems is the wide range of organisations that could be involved in the compressors, ancillaries and in the installation of systems as well as its optimisation and ongoing maintenance.

Pioneering work in this field has been done as part of the US Compressed Air Challenge, which was launched in 1997 and has had significant achievements, particularly in the fields of education and provision of excellent publications.

Pneurop is the organisation of European manufacturers of compressed air equipment.

It was founded in 1960 and its membership now comprises Trade Associations of Austria, Belgium, Great Britain, Finland, France, Germany, Italy, Sweden and Switzerland. It has a record of concern about energy efficiency and the safe use of compressed air.

Pneurop

1. Provides a formal focal point and funnel for diverse manufacturers' interests in compressed air, vacuum and pneumatics as well as allied equipment
2. Provides access to European legislators and to detailed guidance on interpretation and understanding of Directives affecting the Compressed Air Industry.

3. Develops draft standards and works on the updating of existing ones providing the results to CEN and ISO
4. It is organised into seven committees handling:

- Compressors
- Tools
- Vacuum Technology
- Pressure Equipment
- Commercial Matters
- Air Treatment
- Process Compressors

The industry, largely through Pneurop, has developed ISO/CEN standards for performance measurement (ISO1217), noise emission (ISO2151), Air Purity classes (ISO 8573) and Safety (ISO5388) and CEN (1012-1)

Pneurop has always realised the importance of energy efficiency and welcomed the invitation to participate in the EC sponsored "Compressed Air Systems" study of 1997. Probably its main contribution to this was to be able to emphasise and have included in this work a global system approach to energy saving.

Pneurop contributed to the 2001 "Compressed Air Systems in the European Union" Report and was asked as a result of this to contribute to the "Motor Challenge Programme" which will be presented and discussed in a later session of this Conference

Pneurop set up its own "Energy" Working Group three years ago so that it had a team to consider and comment on key documents, in particular the Motor Challenge Programme" Compressed Air" module.

The importance of a system approach was clearly identified in the "Compressed Air System in the European Union" report which ranked the energy saving potential of various simple and sophisticated possibilities of which the most important in most systems was "elimination of leaks".

However, a systems approach needs measurement to establish "What exists" and the "difference" after changes are made and this necessitates a full understanding of the end users' requirements in terms of volume, pressure & air quality, which, in turn, raises the question of compressed air education. It is not possible to audit a system and draw sound conclusions without a good understanding of different types of machines and ancillaries and the ways in which they can be matched.

Good publications on methods of reducing energy consumption have been produced, as have excellent case studies, but what is vital is to preach to the "unconverted" not the "converted".

Virtually all those in this room today are converted to the importance of energy saving, people in the industry in general are not as they have other more pressing matters on which to concentrate and they will not become energy saving enthusiasts without

- information
- education & training

- access to Energy Saving Case studies
- Best Practice manuals

In this the US Compressed Air challenge have done excellent work with, to date, some 4,000 trained as qualified instructors, both end users and suppliers.

The audience is crucial – training for energy saving enthusiasm and competence must be for a wide variety of skills and responsibilities not just for one group of people, for example maintenance engineers.

Company management must see real benefit from Energy Saving Programmes if they are to take these up with enthusiasm. In the case of compressed air studies (including those mentioned above) show that to end-users the most important features of a compressed air system are

- reliability of supply
- quality of air and its compatibility with process requirements
- safety
- cost.

As a result of this it is essential that energy initiatives must be "sold" on the basis of enhancement of company productivity, not on energy saving alone. Also, of course, on the basis of worthwhile saving in relation to the total cost of system enhancement, which will almost certainly include both capital and revenue expense. Payback on investment will almost certainly have to be less than two years to compete with other projects.

A system approach sounds logical and attractive but it does mean that many different aspects need to be considered, and discipline becomes crucial in that, unless each stage of system enhancement or establishment are considered, energy saving gains may not be achieved or maintained. Discipline is required in:

- Selection and purchasing of system elements (both crucial for low LCC)
- Use, including production, distribution and end use devices
- Control both locally and overall
- Maintenance

Unless there is discipline in the approach to each aspect it is not possible to guarantee that the maximum possible savings have been obtained: in fact, in practice, it is unlikely that they ever will be because some system modifications – for example the addition of sophisticated controls - may not be economically viable. The other big challenge is the way the system is actually used, whatever management intends may not be possible because of , for example, limited orders, difficulties with suppliers or failure of significant items of production plant. But, unless the best possible saving is aimed for, the result is likely to be disappointing.

The compressed air industry has developed energy efficient products, for example variable speed compressors and several highly efficient items of downstream equipment, and it will continue to do so as it is committed to energy saving, and individual companies must do so to remain competitive with industry leaders.

The industry has a record of eco-efficiency, to which Pneurop has contributed by its development of standards, which enable gains to be quantified precisely. There have been significant gains in:

- Specific energy performance
- Low noise
- Air purity
- Disciplined disposal practices
- Consideration of LCC

I am aware that I have spoken more about beliefs than strategies; it will be interesting to hear how many of these come up in the following sessions of this conference: many of these must be common regardless of product type.

Finally two points:

1. What incentives are needed to convince the management of companies that energy savings are worthwhile for them - I am sure some are needed,

2. A reminder of our beliefs:

   - System Approach
   - Measurement
   - Education and Training
   - Sell Productivity
   - Save and Follow-up
   - Need for Incentives

Recent Compressed Air Energy Saving Products

Peter Seroczynski[1], Harry David Craig[2]

[1] Ingersoll-Rand European Sales Ltd representing PNEUROP, Diamant Building, 80, Boulevard A Reyer, B-1030 Brussels, Belgium

[2] Compair UK Ltd representing PNEUROP

Introduction

This paper presented by PNEUROP, the European Association of Compressed Air Vacuum and Allied Equipment Manufacturers, demonstrates that our industry has and will continue to develop energy saving products to complement developments in motors and their control systems.

The 2001 report "Compressed Air Systems in the European Union" showed that substantial savings of the order of 30% were possible, but for this level to be attained many aspects needed to be reviewed and possibly enhanced or replaced. The report subsequent work on the Motor Challenge Programme Compressed Air System module considers systems in three parts:

- Compressed air production
- Distribution
- End use devices

This paper identifies recent developments in each of these three parts.

Compressed Air Production

Improved Efficiency in Rotary Screw Compressors

The predominant type of compressor used in general industry to produce compressed air is the Rotary Screw lubricated compressor. Over the years much development work has been done to improve the profile of the rotor lobes and the ratio of the male to female lobes to improve both sealing and the air flow through the rotors. On these single stage compressors most of the improvements in efficiency have been attained and future developments are expected to give small gains in comparison with those in the past, which were in the magnitude of 15%.

A further means of improving compression efficiency is to use a two-stage screw compressor to compress the air to the same pressure as the single stage version.

These are normally larger compressors used where a high continuous usage of compressed air is required. These can give savings over a single stage equivalent in the same application of 15%.

Speed Controlled Screw Compressors

As stated previously we have made most of the gains possible in the screw air end technology. A further area where efficiency gains have and are being made is in the use of variable speed motors and controllers which match the speed and output of the compressor to the system demand. These controls can be used on all types of compressors, oil lubricated single and two stage screw, oil free single and two stage screw, and oil free lobe compressors. Over a constant speed compressor gains of 25% in power reduction can be seen depending on the duty requirement of the air system.

These speed-controlled compressors are ideal where there is a base load requirement for the air system with fluctuations during the manufacturing shifts. This allows for the selection of a base load compressor and a speed-controlled compressor to cover the fluctuating requirements.

Water Injected Screw Compressors

Water injected screw compressors are not so common in the market although the concept is well understood in the industry. The normal oil lubricated screw compressor uses the oil to seal the gaps between the rotors and to lubricate the rotors and to dissipate the heat created during the compression. This lubricant has to be separated from the compressed air prior to the air entering the system. On a regular basis this lubricant has to be replaced and the old disposed of. With the water injected screw the water replaces the lubricant so avoids the separation system and the disposal of the lubricant thus making it environmentally friendly. Some form of water treatment may be required such as ionisation.

Electronic Control Systems

Electronic Control Systems must be understood, and that in the majority of air systems the demand for air is very dynamic and the load requirement will constantly change throughout the day. Different days will produce different demand patterns depending on the manufacturing and maintenance processes being employed.

Where multiple compressor installations are used, it is essential that some form of sequencing control system be employed to run only the compressors that are required to meet the air system demand at that time. Failure to install such a system can result in wasted energy costs of approximately Euro 6000 (depending on energy costs) for an installation of six 75 kw screw compressors working 8000 hrs per year. Obviously if larger or more compressors are being used then the wasted energy will be greater.

System controllers used today are microprocessor devices that can control normally between five and fifteen compressors and usually work with the microprocessor controller installed on the compressor. If the compressor does not have this type of controller then an interface can be fitted to the compressor to activate its pressure switch control.

The system controller will normally be set for a system pressure and will have a system pressure band to work within. These are set to suit the system requirements.

There are many sequences of control which can be used, but the main ones are the "Timed Mode" and the "Event Mode".

The Timed Mode is normally used where all the compressors have the same power and capacity and each compressor will be started and stopped in sequence as the system pressure demands. After a pre-determined running period the compressor sequence will change ensuring that the running hours on each compressor are similar, thus giving better control of the maintenance requirements.

The Event Mode is where there are several compressors of varying capacity and the plant experiences varying load conditions. Examples include multi-shift operations or operations that use equipment with high air requirements on certain days or at certain times. Energy efficiency can be maximised by programming the compressors that most closely match the air demand. Therefore the system controller programmes the event by day and time to ensure the correct sequence of compressors are operating to meet the demand.

Condition Monitoring

For a process operator or a manufacturer the main criteria is that the air supply will maintain the process being undertaken. This means the compressor and ancillary equipment are operating at the correct pressure and any evidence of impending failure is detected and corrected when the demand is not critical.

This is where condition monitoring is used to detect if the compressor and ancillary equipment are operating correctly. It is normally done by measuring temperature, pressure and vibrations at critical points on the equipment. The monitoring can be completed manually at the compressor, or be done centrally within the plant or remotely from the plant in a location set up by the manufacturer or service provider. This should therefore ensure that the equipment is serviced at the correct intervals and that costly breakdowns and production stoppage are prevented.

Condition monitoring can also be applied to the whole air system to see where problems are arising.

Distribution

This is the systems and components that take the compressed air from the compressor house to the point of use. Many different items and components can be used in the distribution system, and in Europe they will now be covered by the Pressure Equipment Directive. For this paper we will only discuss certain components which comprise part of the distribution system.

Low Friction Pipe Systems

The normal pipe systems have been made from Schedule 40 pipe which is normally welded in situ to meet the air systems layout. This pipe is reliable but has a relatively high pressure drop and can corrode, which can mean particles of rust in the air close to the point of use. To overcome this galvanised or stainless steel piping can be used. However, there is another alternative which is low friction anodised alumimium extrusion, which has very low pressure drops and therefore smaller sized piping can be selected for a given air flow. The piping is modular in design and can be assembled on site. Being modular simplifies the modifications or additions to the system, and being anodised means no corrosion will occur in the piping.

Electronic No Loss Drain Traps

The drain valves previously and currently sold are the float type where the condensate level rises in the drain and opens a valve which allows the condensate to escape. Unfortunately this type of valve also allows compressed air to escape in normal operation and should the valve stick open, as they are prone to do, then you have a constant loss of air. As there can be several drain traps fitted in the system then this can be a large loss of energy. For one small trap this can cost Euro 25 per week in lost energy.

To overcome this electronic no loss drain traps are used. These work with a liquid level indicator with a high and low set point, or with a high set point and timer. These only allow a certain amount of condensate to escape before closing so there is always some condensate in the trap. In this way no compressed air is allowed to escape as the valve is closed before this can happen. By fitting these valves in a new installation or up-grading an existing system considerable energy costs can be eliminated.

Air System Audits to Optimise Operating Performance of the Manufacturing Process

Most of the energy losses in the total compressed air system are as a result of the inefficiencies in air distribution system. There are many areas to consider to find these inefficiencies and we will only cover these briefly.

They are:

1. Air leaks
2. Miscellaneous use which is unnecessary use of compressed air
3. Artificial demand using too high a pressure to compensate for system problems
4. Open blowing for production
5. Open blowing for drainage
6. Worn or enlarged orifices and nozzles
7. Where applicable dryer purge air

To find out what is wrong with the system requires a physical audit from someone knowledgeable in compressed air systems and the operation of end use devices. This can be accomplished by a plant visit to identify the obvious problems, but for a complex system some monitoring is required over a period of at least one working week to identify the loading patterns. These services are available from the Member companies of PNEUROP.

The savings that can be achieved, as previously stated, are in the region of 30% of the installed power. For a plant with 500 kW of installed power and electrical cost of Euro .05 with 7000 hours of annual operation the electrical cost will be Euro 190,000, so the savings can be large with a quick return on the Capital invested.

End Use Devices

Many applications for compressed air are an expensive means of operation, such as blowing - the generation of vacuum by a venturi tube. This can be replaced by a low pressure blower or a vacuum pump respectively, which are inexpensive to purchase and use far less energy for the application. On some end user devices the use of worn orifices and nozzles can cause wasted energy; as can filtration fitted to these devices as they are rarely maintained and result in high pressure drops across the filter element. If end user devices are modified to suit the production, or the cycle times are changed then the system response has to be checked to ensure it is compatible.

Pneumatic Tools

Pneumatic Tools are still one of the large users of compressed air in assembly plant operations. These tools are being developed to operate more quickly and, where required in the case of impact wrenches, more accurately in meeting the torque requirements. This will therefore reduce the air requirements of these tools.

Blow guns have also been developed to reduce the amount of air required by using a venturi arrangement to draw atmospheric air in to increase the volume of air used in the blowing process. In other developments pneumatic tools are being manufactured to be lighter and more ergonomic to make them easier to operate and reduce operator fatigue thus increasing productivity.

Compressed Air Optimization in Switzerland

Rolf Gloor[1], Christian Bachmann[2]

[1] Gloor Engineering, CH-7434 Sufers

[2] pcb Pressebüro, CH-8501 Frauenfeld

Estimation of Energy-Saving Potential

In 2000, on behalf of the Swiss Federal Office of Energy, a study on the possibilities of energy saving in compressed air installations in Switzerland has been undertaken. As a result, energy-saving potentials of 5 to 50 % have been calculated, depending on the type of industries and applications.

Background and goals

The energy efficiency of pneumatic processes is low. But little is known about the amount of energy that could be saved by optimizing these processes. The goals of this study were to estimate the total energy consumption of compressed air installations in Switzerland, to calculate the optimizing potential, to identify energy-saving measures and to analyze the market situation.

Methods and results

The total energy consumption of compressed air installations in Switzerland was estimated in two different ways: by market survey in suppliers of compressed air units and by the percentage of energy consumption for the generation of compressed air in some important industries. Both methods lead to corresponding estimates of 150,000 compressed air units installed in Switzerland, consuming 750 GWh of power annually, which is 1.5% of national power consumption and up to 25% of industrial power consumption.

Conclusions

Saving potentials vary with different types of applications between 5% and 50%. The nation-wide saving potential is, in theory, 300 GWh annually. As an optimistic, but reasonable assumption, 100 GWh could be saved annually by optimizing compressed air installations.

Table 1. : Statistics of compressed air installations in Switzerland

Switzerland 95/98	**General**	**Installations of compressed air**				**Total comp. air**	
NOGA - nomenclature	*GWh/a*	*<3 kW*	*3–15*	*18-90*	*>90 kW*	*GWh/a*	*part*
Agriculture	1'000	5'000	200			2	0.2%
Fishery	2	300				0	1.5%
Energy production	1		10			0	5.0%
Quarry	80	100	250	150	5	10	12.5%
Food industry	378	1'000	600	300	30	26	6.8%
Textile industry	533	1'000	300	300	10	19	3.6%
Leather goods	40	200	70	20		1	3.4%
Wood industry	400	4'000	2'500	300	10	30	7.6%
Paper industry	1'243	2'000	800	500	20	34	2.8%
Petrochemicals	30			10		1	1.7%
Chemical industry	2'255	500	400	500	200	77	3.4%
Plastics processing	500	200	400	300	50	30	5.9%
Minerals	423	500	500	200	40	23	5.3%
Metal industry	3'700	3'000	2'800	1'200	100	99	2.7%
Machine industry	2'622	500	2'500	1'100	200	118	4.5%
Apparatus manufacture	2'400	1'500	3'000	1'000	100	90	3.8%
Construction of vehicles	200	300	200	100	5	7	3.6%
Production	300	2'000	2'000	400	5	31	10.5%
Supply	300	500	500	100	5	9	2.9%
Construction	1'000	20'000	3'000	300	10	35	3.5%
Trade	3'000	10'000	4'000	500		46	1.5%
Gastronomy	2'500					0	0.0%
Traffic	4'000	1'000	4'500	100		28	0.7%
Financial services	1'000					0	0.0%
Business consulting	1'800					0	0.0%
Government	600	1'000	200	10		2	0.3%
Education	500	400	20	10		1	0.1%
Health	3'500	3'000	750	100		9	0.3%
Services	700	2'000	500	500	10	30	4.3%
Homes	15'000	50'000				5	0.0%
Total units		**110'000**	**30'000**	**8'000**	**800**		
Energy per unit (MWh/a)		0.1	5	50	250		
Total energy (GWh/a)	**50'007**	**11**	**150**	**400**	**200**	**761**	**1.5%**

Possible Measures to Save Energy

The main goal of the study described above was to identify measures to be taken on various levels, from optimizing the installations up to influencing the market mechanisms.

Economic context

The largest 10,000 units use approximately 80% of the power consumed by all of the 150,000 compressed air installations in Switzerland. Annual energy costs of these larger units exceed 5000 Swiss Francs.

Components of compressed air installations

Energy-saving measures must begin with consumers, because the whole system depends on their need of compressed air. As a second step, the distribution is to be optimized, mainly by replacing leaky elements. The saving potential of compressors, however, is limited to 10% in most cases.

Possibilities to influence the market

Main players in the compressed air market are the users in a considerable number of industries, the suppliers of compressors and of fittings, the engineers and fitters, the manufacturers and traders of air-driven tools and of machines using compressed air. In most cases, a plumber installs the pressure lines during the construction of a building. In simple cases, the engineers planning compressed air systems often use standard configurations. In more complex installations, they use to take the advice of compressor suppliers.

There is, in brief, a number of possibilities to influence the market players: Centers of competence, documentation, check lists, labels, setting examples, contracting, subventions of energy-efficient products.

The feasibility of energy-saving measures has been shown in two industrial companies: a kitchen manufacture and a weaving mill.

Optimizing Compressed Air in a Kitchen Manufacture

This example shows a substantial optimizing potential. However, this potential is too small to justify the consultation of external expertise. But the appropriate measures are so simple that they can be taken easily by installation owners, best with some assistance of maintenance men.

Description of the facility

Schneebeli AG is a kitchen manufacturer with 22 employees in Ottenbach, Switzerland. Main pressure consumers are a CNC working center, an edge gluer and a plate saw with pneumatic cylinders, an air-cleaned band grinding machine, various working machines and presses, and 12 working benches equipped with air-driven hand equipment and cleaning guns. The elements of the pressure supply are a pressure line network, a refrigeration dryer, an ultra filter, and 4 pressure containers. The pressure generators are mainly a screw compressor of 5.5 kW, and a piston compressor running only in marginal hours.

Assessment of status quo

As a first step, the following values have been measured daily and added up weekly:

total power consumption of the enterprise (2500 kWh/w);

working hours of the screw compressor (50 h/w) and the piston compressor (2 h/w);

power consumption of the screw compressor (10% of total);

power consumption of the refrigeration dryer (0.5% of total).

Further assessment led to the conclusion that there is only limited saving potential in pressure consumers and in the pressure line network. Therefore, we focused our efforts on the compressor. Its running time of 50 hours a week was indicating a considerable saving potential.

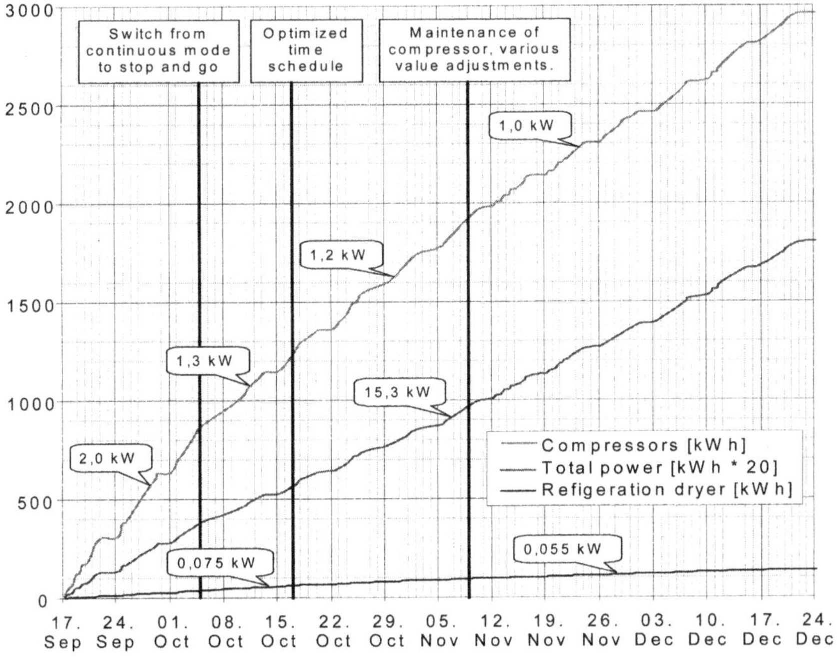

Fig. 1. Power consumption of the enterprise, of the screw compressor and of the refrigeration dryer from September to December 2001.

Optimization and results

Changing the operating mode of the compressor from continuous to stop and go lowered the power consumption by 35%.

Optimizing the time schedules of piston compressor, screw compressor and refrigeration dryer by limiting the availability of pressure to the main working hours lowered the power consumption by 8%.

Reducing the after-running time and pressure limits of the screw compressor lowered the power consumption by 17%.

By all these measures, the power consumption of the compressed air unit has been reduced by a total of 30% without any investment in material.

Conclusions

By the optimizing measures described above, annual cost savings are about 1000 Swiss Francs. This amount is considerable, but by far too small to justify an external consulting. It is necessary to give installation owners a simple guide, so that they know how to optimize the compressed air units themselves. The best moment to do this is a planned maintenance, in collaboration with the maintenence man. An article in trade journals will communicate this message.

Optimizing Compressed Air in a Weaving Mill

This example shows a very high energy optimizing and cost saving potential.

Description of the facility

Jenny Fabrics is a manufacture of cotton tissue in Ziegelbrücke, Switzerland, operating 120 gripper looms in a 6 days, 24 hours schedule, resulting in a production time of 6500 hours annually. Each gripper loom needs only small amounts of compressed air for thread woof and for cleaning.

Assessment of status quo

Vibrations cause huge leakage losses in the compressed air supplies of a large number of looms. The losses exceed the consumption of the machinery by a factor of three. The leaks have been detected in an acoustic search with pressure supply turned on and looms turned off. This test cannot be done during working hours because of the loud noise.

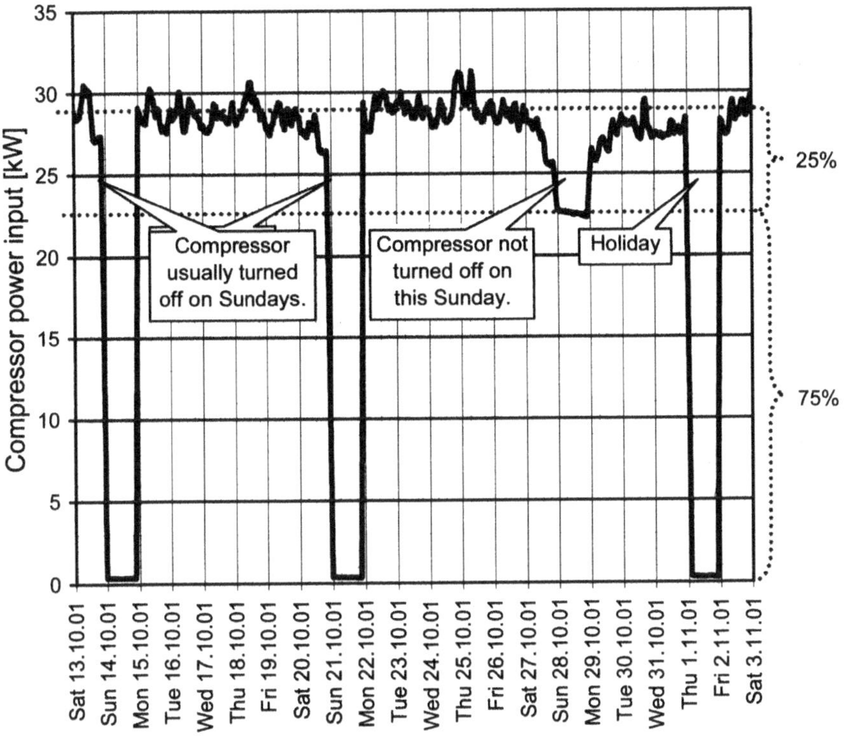

Fig. 2. Power consumption log of the compressor. Running the compressor on one Sunday showed high power input caused by leaks.

Optimization and results

By repairing the leaky fittings, 150,000 kWh of electric power and costs of over 10,000 Swiss Francs have been saved annually. An investment of 3000 Swiss Francs in material and in 30 hours of manpower was needed. The payback time is only a few months.

Conclusions

This example shows that permanent vibrations in textile machinery may lead to hidden leakages of pressure supply, causing high losses of energy and cost. An article in trade journals of the textile industry will communicate this message.

Developments in Cas Energy Management

Ludo Van Nederkassel

Engineering Manager, Industrial Air Division Atlas Copco Airpower NV,
P.O Box 103 , B 2610 Wilrijk, Belgium

Atlas Copco

Atlas Copco is a Swedish based international company, employing over 26.000 people worldwide.

32% of its business is coming from the very much energy related Compressor Technique business area, covering a wide variety of products and services in the field of gas and process compressors, industrial and portable compressors as well as diesel engine driven electricity generators.

The topics covered below relate to electric motor driven industrial air compressors and to their related ancillary equipment.

1. Compressors and Air Treatment Equipment

When looking to the build up of an industrial compressor package, there are many opportunities to loose efficiency, starting from drive inefficiencies, suction losses at the air intake, to pressure drops over the oil separator filter, coolers, the refrigerant dryer heat exchangers, condense water separators and the outlet valve before the compressed air can leave the compressor package.

Depending on the size and the design of the components used, only 68...83 % of the electricity consumed, is effectively used to deliver compressed air at the outlet.

No need to say, that improving the efficiency of the compressor element itself is a prime concern to the manufacturers, but progress is difficult and has limited potential for spectacular results.

Efficiency in cooling , flow control and air drying are a few examples where progress is possible.

1.1 Cooling

Compressors require extensive cooling , mostly by forced ventilation through air cooled heat exchangers. Design of sufficiently large and efficient coolers and the use of optimized fans have allowed to reduce the ventilation to 2..5 % of the installed compressor power . Vibro-acoustic optimization of the compressor parts has allowed to reach low noise levels with open package structures offering low ventilation restriction.

A cooling system is designed for worst case continuous full load conditions and normally is in use as long as the compressor runs. Lower ambient temperatures and part load conditions offer possibilities to reduce the ventilation requirement, which can be achieved by temperature controlled speed variation or start-stop regulation of the fan.

1.2 Flow Control

Flow control is probably the most extensively pursued energy saving possibility, simply because all common flow control solutions are far from efficient.

Compressors are designed for best efficiency at continuous full load operation. However very few compressors are effectively used under these conditions. To have an understanding of the real situation in industry, Atlas Copco logged the compressor loads for a full week in hundreds of installations.

For compressor installations below 100 kW, 3 typical air demand profiles were identified.

Profile NR1 is typical for a plant operating 24 hrs /day, with a peak demand during day shifts, a dip for lunch, and with a constant low air demand during the week-end, often showing the leakages in the installation. This type of air demand profile was found in 64 % of the investigated installations.

Profile NR 2 is typical for a plant operating 5 days/week in 2 shifts, with no activity during night or weekend. Air demand fluctuates very erratically. It was found in 28 % of the installations.

Profile NR 3 is typical for a plant with fixed air demand, found in 8 % of the installations visited.

It is evident that with so many installations with very fluctuating air demand, variable speed flow control is to be considered.

Positive displacement air compressors such as the rotary screw type, have a basically constant torque characteristic over speed, at constant pressure. The compressor efficiency, expressed as specific power consumption in $kW/m^3/min$ or Joule/liter, in a regulating speed range of 100 % down to 20..25 % should be optimized at 50..80 % of the max speed.

Comparing the efficiency of a VSD compressor with a constant speed compressor optimized for full load operation shows that the latter one is less efficient for a VSD application.

The power over speed characteristic shows the almost linear variation of power consumed with compressor flow. For flows, smaller then the minimum allowable operating speed, start stop control is possible thanks to the practically unlimited number of starts of the electric motor with an electronic frequency converter.

Energy saving from VSD control, not only is the result of an almost constant specific power consumption over the entire control band, but also from the very precise and narrow pressure band control , allowing to operate at lower average working pressure. Also the total elimination of any blow off of compressed air helps to save energy.

To effectively evaluate the advantage of a VSD controlled compressor versus a load-unload controlled constant speed compressor, a comparative test was performed by Laborelec, the laboratory of the electrical power supply utility Electrabel in Belgium. Both compressors controlled the pressure in a 1000 liter air receiver subject to the 3 different air demand profiles as described before.

The result was an average saving of 30 % over the 3 demand profiles, when weighted for the occurrence of the 3 profiles the saving was over 35 % .

The conclusion of these investigations was that 88 % of the installations below 100 kW show important fluctuations in demand and that in 70 % of these installations, load varies between 40 and 80 %, offering substantial potential for VSD energy savings. A measurement on site is the first step.

1.3 Air Dryers

Common compressed air dryers are either of the refrigerant type, or of the adsorption type when dew points below freezing are required.

Refrigerant dryers normally include air to air heat exchangers recovering 70 % of the generated cold.

Additionally, for larger dryers, VSD control of the refrigerant compressor is justified by the energy saving at lower then full load or worst case conditions.

Adsorption dryers, needed for dew points below 0°C, either require important purge losses for regeneration in the so called heat less dryers or require heat for regeneration. Dewpoint is timer controlled regeneration and /or the use of heat of compression can substantially reduce, even eliminate completely the need for regeneration energy.

Another form of energy cost is resulting from pressure drops. Granulate bed adsorption dryers have to be equipped with a filter before and a filter after the dryer, with a resulting total pressure drop of typically 0.5 bar over the entire dryer installation, representing an additional 3 % energy cost.

One such example is Atlas Copco's combination of an oil free Z compressor with a corresponding MD dryer.

The core of the dryer is a rotating drum, impregnated with a moisture absorbing desiccant. After saturation of the drying section, the drum rotates into a regeneration sector, where hot unsaturated air is used to remove the water from the desiccant.

The major part of the compressed air is passing through an after cooler, a watertrap for condensate removal, an ejector and the drying section of the dryer rotor. The remaining part of the compressed air, hot and unsaturated, is taken from before the after cooler, directly at the outlet of the compressor final compression stage. After passing through the regeneration section of the dryer, this air, now saturated, passes through a regeneration cooler where water is separated by condensation. After removal of the condensate water, the regeneration air is mixed again with the mainstream air in the ejector, delivering the driving force for the regeneration flow.

Power consumption is limited to the power for the fan of the regeneration cooler, typically 1 % of the compressor power, and over-all pressure drop is no more then 0.25 bar thanks to the low resistance drying rotor and absence of pre or after filtration requirement.

2. Compressed Air System Analysis and Control

2.1 CAS Analysis

A compressor status registration system has been developed, allowing to measure the load condition of up to four compressors. Expansion of this tool to measure also flow, pressure, dew point or power consumption at various points in the compressed air system is under development.

The results are logged for a full normal work week.

After this the data can be downloaded in a PC for printing and analyzing the results. A simulation tool, in combination with an air system configurator allows to evaluate the potential for air system improvements.

2.2 CAS Control & Monitoring

In a complete air system, several compressors, dryers, filters, receivers, piping and valves form a complex set up with many components often without an over-all control, operating in off optimum conditions.

To satisfy the need for central control of a complete air system Atlas Copco has developed various solutions.

The basis is the newest generation of compressor controllers, easily adaptable for one of up to 4 controllers to act as a master controller over the others. Multiple compressor control with pressure band regulation and automatic sequencing is a possibility.

By including an additional separate central control unit to the system bus, a virtually unlimited number of compressors and ancillary equipments can be controlled and monitored.

A communication port is available to interface to industrial PLC busses like Profibus or Modbus allowing integration of the CAS in the plant process control.

A second communication port is available for interfacing to an Ethernet bus , from where communication to the external world is open.

Multiple compressor control permits to operate the system in the most economic way.

Obviously first priority remains to maintain the pressure within the required pressure band. Adjusting the pressure band requirements to the plant process needs potentially allows to lower system pressure during certain periods of the day or the week.

The second priority then can be system optimization e.g. for lowest possible energy, or for equal load distribution, or less common for enforced sequences.

Additionally, continuous monitoring of the system allows to optimize maintenance, even to detect air leakage.

One example is a load distribution between 2 VSD compressors, in order for both of them to operate as much as possible within the band of best operating efficiency.

In an installation with 2 VSD compressors of 160 kW each , this resulted in a more precise pressure control and an energy saving of 1.1 %.

A second example is one with 2 turbo compressors. Turbo compressors allow flow regulation by means of inlet guide vanes in a reasonably effective way, but only in the so-called turndown zone. For further flow reduction the very inefficient blow-off or load/unload control is needed.

A central control , in an installation with 2 turbo compressors of 530 kW each, replacing a combination of full load operation of one compressor with turndown and blow off control on the other one, buy a combination of both compressors turndown controlled without blow off of any of them, resulted in an energy saving 880,000 kWh/yr or 12 % on the electricity bill.

3. The WorkPlace Air System Concept

Not all compressed air systems receive the attention, or can carry the investment of sophisticated central controls.

A typical medium size industrial CAS, has grown over time as the business was growing. Old, less efficient equipment is kept in operation and the system has become very complex. The initial piping is not updated to cope with the increasing flows, excessive pressure drops and air leakage on old fittings are common. And on top of that, the CAS not being related to a specific production process, receives no management attention.

The WorkPlace Air System concept can offer an answer to this situation.

Integration of all compressed air equipment in one package, including a low noise, probably VSD compressor, a dryer, compressed air filter(s), electronic drains, condensate treatment and possibly even heat recovery equipment, ensures proper matching of all system components and best over-all efficiency.

Ultimately integration of the compressed air package in the production process eliminates distribution pressure losses and leakage, allows to shut down the CAS when the production process stops and also the CAS will receive management attention.

4. Conclusion

Compressed air systems are a major consumer of electricity in industry. Energy consumption accounts for 60..80 % of the total life cycle cost of a compressor.

Electric motor drive efficiency improvements are contributing to the reduction of the electricity bill, but optimization of the compressed air system is believed to have much higher potential for energy savings.

A few examples were given. Definitely they are not applicable to all installations, and they are not the only possibilities for compressed air system optimization, but we believe variable speed drives, developments in air drying equipment, system monitoring and control and integration of the compressor in the production process will play a major role in the future.

Hybrid Permanent Magnet Motors Allow for the Development of High Efficiency Screw Compressors

Roger Cook

Ingersoll-Rand Co. Ltd.

Synopsis

Air compressors that use variable speed drives to improve overall compressor performance and efficiency have been commercially available since the early 1990`s.

With few exceptions these products have been adapted from standard compressor package designs with the addition of a variable frequency inverter and upgraded main drive motor.

The justification for installing energy efficient, variable speed drive compressors within a compressed air system is now fully understood and widely accepted. As such this particular topic will not form part of this paper.

This paper will however demonstrate how the collaborative development of a new electric motor technology and a technically advanced compressor package allowed for the development of a fully integrated, innovative and efficient range of rotary screw air compressors by Ingersoll-Rand.

Introduction

The concept of developing an advanced variable speed compressor started with the design, construction and evaluation of a prototype switched reluctance drive machine as early as 1989

To achieve the ideal variable speed compressor it was possible from this exercise to demonstrate that:

- Peak load motor efficiencies needed to be high to offset any inverter drive efficiency loss
- Part load motor efficiencies needed to be maximised across a broad speed range to best match the variable load conditions typically experienced with this type of compressor
- Power losses could be reduced or even eliminated in some cases

- Maintenance requirements could be significantly reduced
- Significant component reduction could be achieved by using innovative design techniques
- Assembly times could be improved for enhanced manufacturing efficiency

If all of these features and benefits could be realised then it was argued that significant efficiency improvements for both the customer and the manufacturer would result.

The key to success would undoubtedly be linked to the availability of the ideal drive technology, be it an existing technology or a totally new concept.

Drive Technology Selection and Development

Working with some of the largest motor manufacturers in the world an extensive program was set up to fully evaluate all the principal motor technologies. Induction motor, switched reluctance, brush-less DC, and permanent magnet motors were all evaluated in great detail. Their principle characteristics, features and benefits were compared with the specific requirements established for the ideal compressor drive. No single technology proved capable of meeting all the requirements for the proposed new machine.

All were too large limiting the possibilities for successful integration and component reduction. Motor performances at peak load could be improved however no technology was capable of combining the peak load performance with the desired part load characteristics to achieve the ideal variable speed compressor.

The breakthrough was made when a new synchronous motor technology being developed by **Leroy Somer** called **Hybrid Permanent Magnet Motor** was introduced to **Ingersoll-Rand**.

HPM motor technology through a collaborative development effort between Leroy Somer and Ingersoll-Rand offered unparalleled opportunities for the development of a technically advanced, directly driven, air compressor for both oil free and oil flooded application.

The major benefits offered by HPM Technology for compressor design included:

- High peak efficiencies and excellent power factors
- High performance values maintained across a very broad speed range
- Capable of sustained high speed operation
- Unlimited starting capability for the motor and drive
- Constant torque and high torque capability
- Very low torque ripple through innovative rotor design
- Extremely compact "pancake" style motor for any given power output
- Uses a variable frequency inverter (with special software)
- Capable of cantilever mounting for direct compressor drive
- Opportunities for significant drive line component reduction

- Motor designed without bearings for minimal maintenance and maximum reliability
- Salient pole stator construction for increased reliability and individual coil replacement (if required)
- Simple stator or rotor removal for exchange or replacement (if required)

Motor and Airend Design

The HPM motors were designed specifically for air compressor application taking 2 years to fully develop and refine. The airends (compression modules) are essentially standard Ingersoll-Rand rotary screw airends modified to operate efficiently and reliably with the directly mounted HPM motor.

Fig. 1. Typical HPM motor cutaway and airend / motor module

The HPM motor is a synchronous 8 or 12 pole motor design with salient pole stator construction. The precision balanced rotor assembly (including permanent magnets) is cantilever mounted directly onto the airend input shaft. Hall effect sensors are used for speed and rotor positioning purposes and thermal protection is provided by means of intimate contact thermistors. Enclosure construction is IP23, drip proof and provision is made for a simple four bolt mounting arrangement. (see Fig.1.)

Power transmission inefficiencies are eliminated, there are no expensive gears or couplings, there are no motor bearings or seals requiring maintenance, components are significantly reduced and the assembly process is both efficient and rapid.

These motors and assemblies are a significant advance in the development of ultra compact, efficient, reliable and easy to service air compression modules.

Motor Performance

Collaboration with Leroy Somer into all aspects of the motor design for the Nirvana compressor has resulted in exemplary motor performance. Efficiency levels are very high and are maintained across a very broad speed range making the motors ideal for variable speed compressor application.

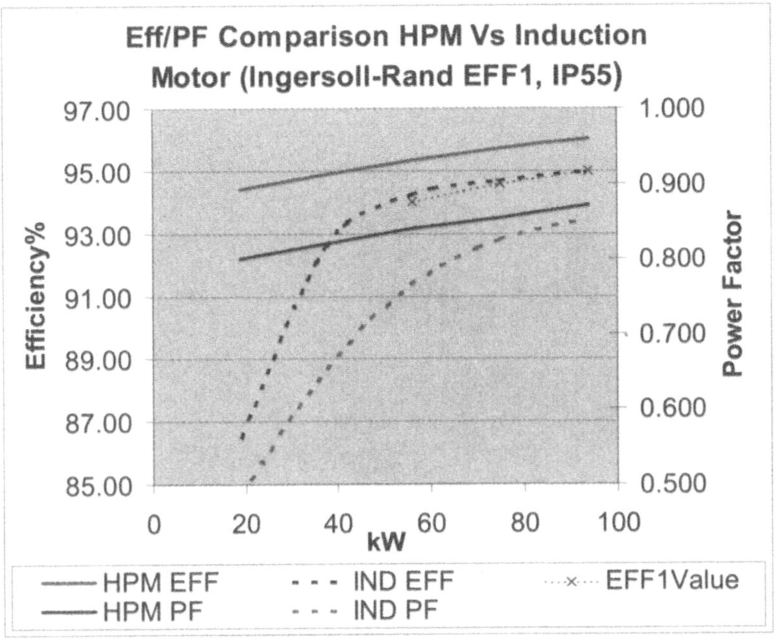

Fig. 2. HPM motor / IR induction motor comparison (typical)

Fig.2. Shows typical PF and Efficiency for the HPM motor and induction motor. The HPM motors efficiency comfortably exceeds EFF1 requirements also peak efficiencies and power factors are better than the comparable Ingersoll-Rand, EFF1 compliant induction motor. Of crucial importance however for a variable speed compressor is the fact that the performance is dramatically improved for part load operating conditions across a very broad speed range

All other operational characteristics for the HPM motor meet or exceed the demanding Nirvana specification including high thermal reserve and low vibration levels. All HPM motors have less than 1% torque ripple across their entire speed range and the requirement for motor maintenance is minimal. Lubrication is not required, motor stop/starts are unlimited and the compressor can also be started on full load.

The Nirvana Compressor Package

Both the HPM motor drive technology and the compressor itself are very new to the air compressor industry. In actual fact there are no existing components of any significance used in this totally new machine.

The challenge from day one of the Nirvana project was to significantly enhance all aspects of efficiency, the new HPM motor and drive may have been key to the products overall success however innovative design of an efficient compressor package was also essential. Designs that helped with the introduction of enhanced manufacturing techniques were encouraged. Considerable design effort was focused on eradicating power consuming components or at least in minimising their detrimental effects whenever possible.

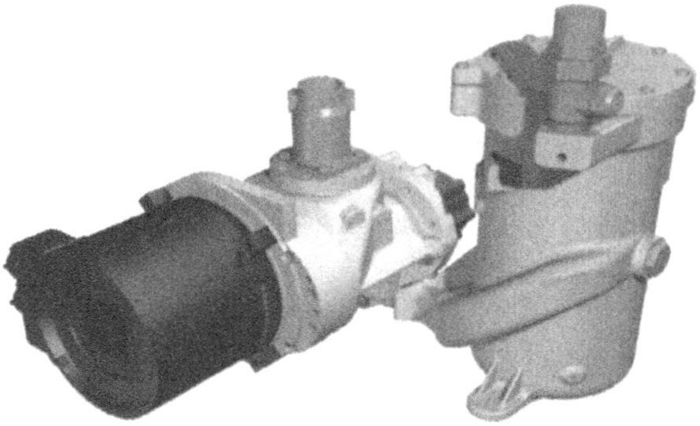

Fig. 3. Typical airend/motor and separator assembly.

The separator vessel for example was designed as a cast component inclusive of many features that would normally be added during the final assembly stage. Careful design of the separation system reduced power absorbing pressure drops and improved fluid separation performance. Many potential leak paths were eliminated and servicing operations were made more efficient with the easy to remove vessel cover that does not disturb pipe work or fluid connections. (Fig.3.)

Other benefits of the compressor package that will contribute to long term efficient operation include a patented anti condensation system for low ambient or high humidity operation, centrifugal blower main cooling technology for high ambient conditions and single point servicing access for routine maintenance. Major maintenance requirements such as motor lubrication have been removed completely and the remaining service intervals extended to set new standards in the efficient operation of today's rotary screw air compressor.

Conclusion

The final compressor package is a fully assembled, comprehensively equipped, low maintenance air compressor that uses the latest state of the art, motor and drive technology. The collaborative development effort between Leroy Somer, as the electric motor and drive manufacturer with Ingersoll-Rand as the air compressor manufacturer enabled Hybrid Permanent Magnet Motor Technology to be fully developed to ideally match the requirements for this new range of air compressors.

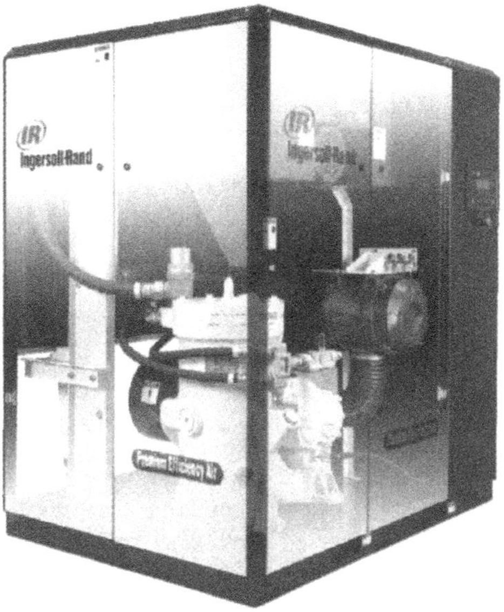

Fig. 4. The Nirvana Compressor Package

The high level of integration and very efficient operational characteristics of this new compressor would not have been possible to the same degree without the early identification and successful development of the chosen variable speed drive technology, **Hybrid Permanent Magnet Motor.**

Oil flooded single stage compressors are already in production, continued development will soon see the introduction of HPM variable speed two stage and oil free compressors. Ongoing development will see the introduction of new drives and motors that will continue the effort to raise efficiency levels and product specifications to a previously unattainable level.

This Nirvana Compressor sets very high standards for the efficient operation and manufacture of variable speed drive, air compressors. Additional benefits associated with long term reliability, low maintenance and ease of servicing complement the final package, altogether adding a new dimension to the meaning of **Compressed Air Energy Efficiency.**

A Global Approach to Energy Savings in Compressed Air

Kenneth Pauwels

Product Manager Energy Consulting, Electrabel, Regentlaan 8, 1000 Brussels

Introduction

Compressed air is one of the most expensive energy vectors in industry, and at the same time, one of the most commonly used. Considering the huge compressed air demand in industry, an efficient use of compressed air results into a large saving.

Electrabel, Belgium's leading electricity producer and supplier, has a lot of experience in executing compressed air audits in Belgian industry.

This paper discusses possible energy savings in compressed air, and highlights the importance of a global approach.

A global approach is required

In order to maximize energy savings in compressed air, a global approach is required: optimizations should be done at production level, as well as distribution and end-user level (fig. 1). Furthermore, optimizations cannot be seen separately. The case studies beneath, taken from Belgian industry, will show that a global optimization program can result in savings up to 15 %, or sometimes even more.

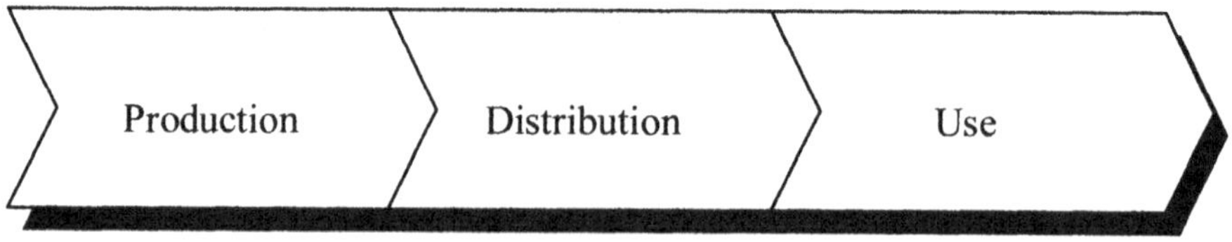

Fig. 1. A global approach

Case studies

Introduction

To obtain a maximum result, the following steps have to be executed chronologically:

a) A study of the end users of compressed air

The delivered air has to be used in an efficient way. If not, all necessary measures (pressure reduction, avoiding leakages, using another energy vector, ...) have to be taken.

b) A study of the distribution grid

Once it is sure that the air is used in an efficient way, there has to be controlled if the distribution is done properly. Possible measures are: fixing leakages, changing the concept of the grid, adapting piping sections, using other materials with less pressure losses,

c) A study of the compressed air production

After executing points a and b, it is sure now that the compressed air delivered to the grid is used efficiently. This means that the compressed air demand profile cannot be improved. The only remaining thing is to make sure that the production of this compressed air is done in the most efficient way.

All case studies beneath are taken from our vast experience in executing compressed air audits in Belgian industry. A classification into production, distribution and end user savings is made.

Energy savings at production level

Two issues are always coming back here: regulation of individual compressors and regulation of a group of compressors.

Regulation of individual compressors

The most currently used regulation principle in industry is the load/unload regulation (fig. 2). As the pressure becomes too high (a), the air inlet valve is closed. When the pressure becomes too low (b), the air inlet valve is opened again.

A more energy efficient way of regulation is frequency regulation: the amount of produced air is constantly varying by electronically adapting the speed of the compressor drive.

Depending on the load profile, energy savings from 15 up to 35 % were found in Belgian industry.

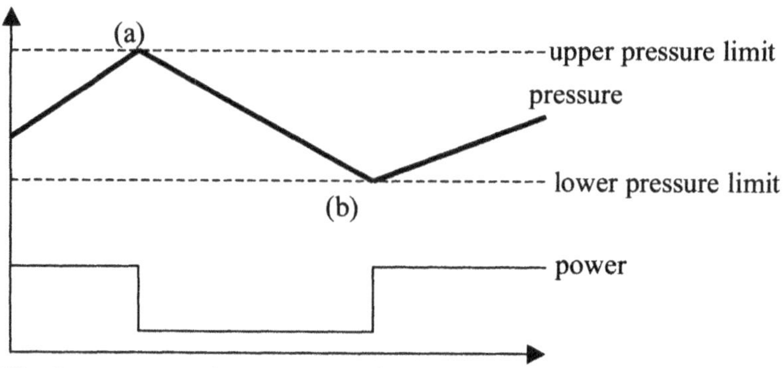

Fig. 2. pressure and power curve for load/unload regulation

Regulation of a group of compressors

In industry, the most frequently used regulation system for a group of compressors, is the so-called cascade system: each compressor is equipped with a load/unload regulation system and each compressor has its own pressure limits.

However, there exists a more efficient way of regulation. Here, a PLC decides which compressor has to switch on as a function of the pressure variation. The pressure is measured by means of a pressure sensor.

The result of this improved regulation system is a very narrow pressure band (fig. 3).

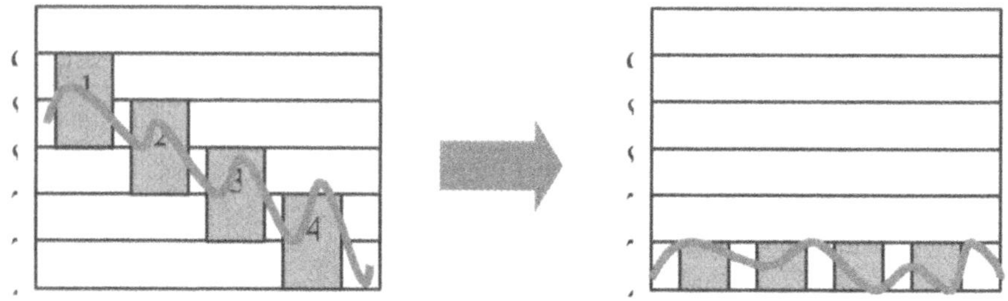

Fig. 3. pressure bands with cascade (left) and intelligent regulation (right)

Energy savings on distribution level

Important parameters that have their influence on the efficiency of compressed air distribution, are:

- concept of the grid
- type of dryer
- leakage losses

The first two are shortly discussed, the third is discussed under 3.4.1.

Concept of the grid

Some factors that influence energy consumption:

- piping diameter
- structure of the grid (antenna – ring)
- volume of buffer

Case:
A customer has a 400m 2" pipe transporting 6 Nm^3/h of compressed air at 6.4 bar. After replacing this pipe by a 2.5" pipe, the pressure loss was reduced with 0.3 bar.

Type of dryer

The pressure dewpoint is a very important parameter for compressed air. This temperature is defined as the temperature beneath which the humidity starts to condense.

Under all conditions, water should be avoided in compressed air.

Water causes:

- corrosion problems ==> increased maintenance
- increased load losses
- freezing of the piping during winter periods

Measures that were adviced in Belgian industry:

- use of corrosion resistant piping materials
- use of another type of dryer (e.g. an absorption dryer with a pressure dewpoint of – 20°C instead of a cooling dryer with a pressure dewpoint of 4°C)

Energy savings at end user level

Three important aspects must always be looked at:

- the volume balance
- pressure needs
- possible alternatives

Volume balance

A volume balance expresses the equilibrium between production on the one hand, and consumption + leakages on the other hand.

Some reasons to establish a volume balance are:

- to know the compressed air cost per division
- to determine where most of the savings will be found
- to determine the amount of leakages

A volume balance is practically established by:

- measuring of compressed air production
- measuring of all compressed air consumers

By establishing a volume balance for some Belgian customers, the following things were discovered:

- in one case, the leak percentage was found to be 34 %
- another case revealed that 75 % of the compressed air was used in one division
- all volume balances permitted to allocate the exact cost of compressed air to each division

Pressure needs

The aim is to reduce the pressure to the minimum required level.
For a 7 bar grid, a pressure reduction of 1 bar equals an energy saving of 6 %.

The following steps should be executed:

- determine the minimum required pressure level
- put the general production pressure to a minimum
- use distribution grids with different pressures
- pressure reduction during nights and/or weekends
- local pressure reduction by means of expansion valves

Possible alternatives

As compressed air is an expensive energy vector, there should always be looked for possible alternative energy sources.
Some case from Belgian industry:

Case 1
A customer uses compressed air at 7 bar to dry textile. Tests with an air knife system revealed that an overpressure of 400 mbar was sufficient to dry the textile.
In this way, the customer attained a yearly saving of 50.000 €.

Case 2
A customer uses a lot of hand tools driven by compressed air. Replacing all these hand tools with electric driven machinery, resulted in a yearly saving of 20 %.

Conclusion

Compressed air systems have a high potential to save energy. However, a global approach is required to maximize the savings.

The Compressed Air Challenge: Making a Difference for US Industry

Aimee McKane[1], Bruce Medaris[2]

[1] Lawrence Berkeley National Laboratory

[2] Compressed Air Challenge

Abstract

In 1997, the Compressed Air Challenge® (CAC) was developed as an outgrowth of the US Department of Energy's (USDOE) Motor Challenge program. The Challenge began as a voluntary collaboration managed by a sponsor, the Energy Center of Wisconsin, that included 15 sponsors from equipment manufacturers and distributors' associations, utilities, state research and development agencies, energy efficiency organizations, and the USDOE. The mission of the CAC is to develop and provide resources that educate industry on the opportunities and benefits attainable through compressed air system optimization.

In 2000, the CAC became an independent not-for-profit organization that continues a development and deployment model based on shared interests and shared costs among public, private, and not-for-profit organizations that serve industrial customers. Since the first CAC training session in 1999, approximately 3900 people have been trained by CAC qualified instructors- both end users and suppliers. Using the CAC cost-share model has allowed sponsors to offer a very high-quality training product at a relatively low cost. For instance, USDOE typically spends $1 for each $10 spent to offer a CAC training session.

This paper will summarize the findings of two recent independent evaluations were conducted by USDOE to assess the impact of CAC training and the market for compressed air efficiency services. One significant finding is that while approximately 76% of participants in the CAC training took specific actions within the 12-month period following training to improve the efficiency of their compressed air system, less than half of those randomly selected from the untrained population had taken any action at all in the past two years, including fixing leaks. Energy savings from improvements as a direct result of CAC training are conservatively estimated at $12.1 million per year. The market assessment also offers the first documented evidence that the CAC is beginning to have an impact on the compressed air market in promoting and supporting a shift from a component-based to a system-based approach.

In 2002, the CAC sponsors have renewed their focus on training. Other projects recently undertaken by the CAC include: industry-driven enhancements to AIRMaster+ software; coordination with USDOE to offer training for industry professionals to become Qualified AIRMaster+ Specialists; coordination with Iowa State University on the first graduate engineering course on compressed air systems; publication of a Best Practices manual; and an Operators Certification program.

Compressed Air System Opportunities

Optimization of compressed air systems represents one of the largest non-process, industrial energy efficiency opportunities, with improvements of 20-50% readily achievable through the introduction of a best-practices approach. Lack of information has been a primary barrier to realizing substantial improvements in the efficiency, reliability, and productivity of industrial compressed air systems. Compressed air systems in U.S. manufacturing account for $4.5 billion U.S. ($4.2 billion Euros) per year of energy costs and 21 MMTCE of total U.S. carbon emissions (1-2 percent of total U.S. emissions).

Compressed air is industry's "fourth utility"; it is central to production for many industries, including: chemicals, textiles, general manufacturing, plastics, mining, glass, pulp and paper, shipbuilding, furniture, automobile and aircraft manufacturing, iron and steel-making, and petroleum refining. Compressed air is used extensively as a source of power for tools and equipment as well as in industrial processes for pressurizing, atomizing, agitating, and mixing applications.

Compressed air is the most expensive utility; a single compressor delivering 500 standard cubic feet per minute (scfm) 24-hrs per day can cost $100,000 U.S. ($93,000 Euros) per year to operate. If the company using the compressor has a 5% net profit ratio, the cost of operating the compressor will be the equivalent of $2 million U.S. ($1.85 million Euros) in production. In spite of its cost, this level of waste in a manufacturing facility is not uncommon, due to poor system operation coupled with a perception by production staff that compressed air is "free". This is largely because compressed air is already present in distribution piping in the plant and the user is neither knowledgeable of nor responsible for the electricity bill associated with its use.

An example of a market opportunity for improving compressed air systems is illustrated by the findings of the Assessment of the Market for Compressed Air Efficiency Services, completed for USDOE, in cooperation with the Compressed Air Challenge in 2001. Phone interviews of 222 industrial compressed air users revealed the following:

- Only 9% identified controlling energy costs as a primary concern
- 71% identified **consistent, reliable compressed air supply** as a principle objective of system management

- 35% had experienced **unscheduled shutdowns** during the previous 12 months- 60% of these shutdowns were for 2 days or more
- 30% had service contracts- had no effect on incidence of shutdowns
- 75% of system operators had **no formal training** in compressed air system efficiency
- 57% had **taken no action** (include leak repair) to improve compressed air system efficiency in past 2 years

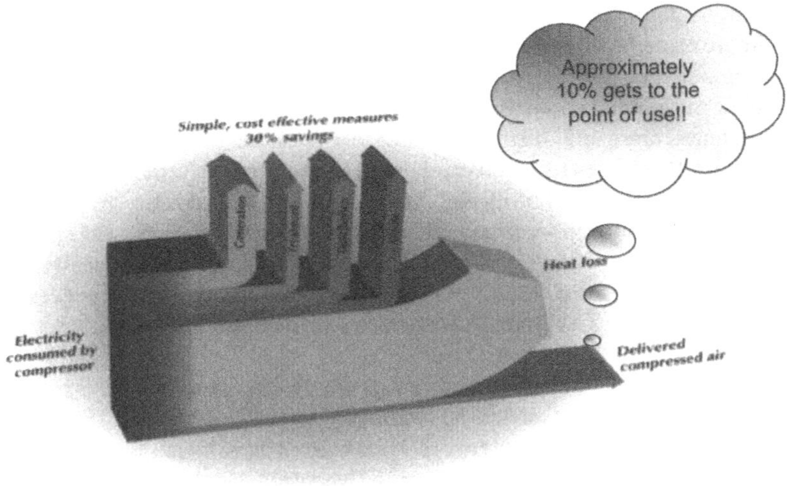

Fig. 1. 60-80% of the input energy to a compressed air system is converted to heat

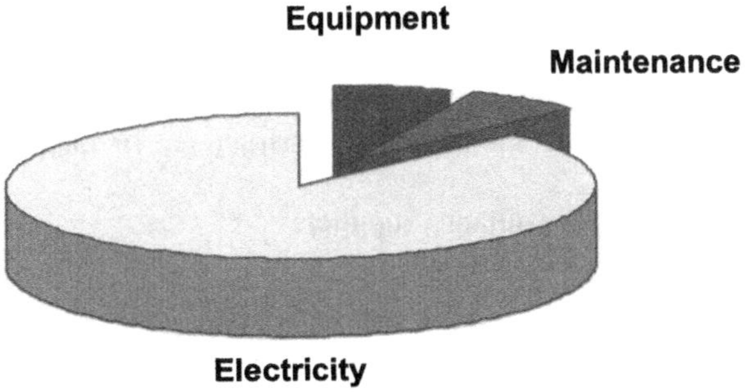

Fig. 2. The cost of operation far exceeds the purchase cost of a compressor

Overall, the Assessment findings indicated that compressed air users are not currently having their needs met by existing market delivery mechanisms, are unaware of the benefits of improving system efficiency, and are reluctant to purchase compressed air system efficiency services because they don't understand why they are beneficial. In short, there is still much work to be done to educate compressed air users.

The Compressed Air Challenge

The Compressed Air Challenge® (CAC) was formed in 1997 and is an outgrowth of work on industrial partnerships begun under the US Department of Energy's (USDOE) Motor Challenge program. The themes which ultimately led to this project were first identified in the April 1995 Roundtable on Market Transformation Strategies for Industrial Motor Systems breakout session on compressed air systems. A general point of consensus at the Roundtable was that the major improvement opportunities are in the compressed air system, not the individual components.

The CAC is a voluntary collaboration that includes participation by USDOE, equipment manufacturers and distributors and their associations, facility operating personnel, consultants, energy research and development agencies, energy efficiency organizations, and utilities. In all, the CAC counts 15 separate organizations as sponsoring members, all of which contribute both funding and time to the collaborative efforts. Since 1997, the CAC has been successful in raising and managing a pooled resource of product development funds in excess of $1 million. In 2000, the CAC was incorporated as a separate not-for-profit with an executive director.

The mission of the CAC is to increase awareness of the benefits of the systems approach to compressed air systems through training and education. Training is delivered in cooperation with the sponsors, with many sponsors hosting training at a discounted cost for their constituent groups. Organizations that are interested in hosting training but are not currently sponsors may host a CAC training in cooperation with a CAC sponsor, several of whom have a national scope.

The CAC has already accomplished the following:

- Published & distributed more than 4000 copies of Compressed Air Sourcebook
- Developed 1-Day Training "Fundamentals of Compressed Air Systems" & 2-day "Advanced Management of Compressed Air Systems"
- Qualified pool of 22 qualified CAC Fundamentals instructors; 10 qualified Advanced Instructors
- Trained more than 4000 end users, consultants, suppliers
- Website, more than 20 case studies, new projects

To accomplish its educational mission, the CAC must work with both the supply and use side of the market to shift the focus from individual components to a systems services approach. The strength of the CAC training is its emphasis on balancing system supply and demand.

Compressed air systems are especially dynamic- changes in demand over time can have dramatic effect on the volume of air required and the ability to sustain a stable system pressure. Taking a "systems approach" means using controls, storage, and demand management to design a system that meets peak requirements but also operates efficiently at part loads. Failure to address both situations leads to less than optimal performance, which can not only affect energy efficiency and cost of operation, but also may reduce productivity and reliability.

Recent CAC Activities

Not everyone that the CAC would like to reach with the systems message can participate in a one- or two-day training workshop. In today's economic climate, it is particularly difficult for plant personnel to take time away from their principal duties to go to an offsite training session. In recognition of this situation, the CAC has worked to developed a more complete portfolio of information to complement the two types of training.

The first effort involved developing an ongoing relationship with a magazine widely read by plant engineers and maintenance supervisors. Beginning in the last quarter of 2000, articles on different aspects of compressed air system efficiency, written by CAC Instructors, have been appearing every other month in Plant Services magazine. These articles are specifically designed to promote the CAC system message included in the training by providing a more in-depth view of one system-related topic per issue. A total of 12 articles have been published to date and have been read by hundreds of thousands of plant engineers and maintenance supervisors.

This year, the CAC will be publishing a Best Practices Manual, which is designed as a solutions-oriented desk reference for plant engineers responsible for managing compressed air systems. The purpose of the document is twofold, to provide systems-oriented guidance to plant engineers who are unable to participate in training and to serve as supplemental information for those who have already received the training.

The CAC has also cooperated with Iowa State University to offer the first US graduate-level course on compressed air systems for engineering students in Spring 2002. The initial course offering, which was oversubscribed in its first semester, has been evaluated and incorporated into the standard course offerings of the University starting in January 2003. The January course is also oversubscribed, with 75 students registering. USDOE has recently awarded a grant to the University to support development of a textbook by Professor Michael Pate.

In response to the results of the Assessment of the Market for Compressed Air Efficiency Services, the CAC developed a consensus document that provides guidance to purchasers of compressed air system services. The same document also provides a working market definition for system analysis services. The purpose is to give both the supply and demand side of the market a common vocabulary and set of working definitions when selling and purchasing compressed air energy efficiency system services. The objective of doing this is to create an educated customer base that understands the value of these services and is willing to purchase them. This document *-Guidelines for Selecting a Compressed Air System Service Provider-* includes criteria for obtaining quality energy efficiency services and also defines three levels of compressed air system analysis as follows:

- Walkthrough evaluation
- System Assessment
- System Audit

The CAC is working with its sponsors to publicize and encourage customers to use these *Guidelines* when obtaining energy efficiency services. These documents are available on the CAC website at www.compressedairchallenge.org .

Allied Partner Qualified Specialist Training

To substantially increase the use of its software tools with and by industrial end users, USDOE worked with Allied Partner industrial trade associations and their member companies in developing a Qualified Specialist training and recognition program. This approach is designed to build use of the assessment software into existing interactions between industrial customers and their suppliers. The Allied Partners, who are already highly skilled in their respective area of specialty (in this case, pumping systems) also become skilled, qualified users of the assessment software. There are several advantages to this approach. First, it addresses the two major barriers to increasing effective use of the software- the time required to understand the tool and technical expertise in systems required to make full use of the software's capabilities. The goal is to increase the number of professionals qualified to use each software tool from <5 to >50 and greater in a relatively short period (two years).

In addition, the process of developing the delivery program also encourages industry suppliers to provide detailed technical input to fine-tune the software, while building ownership in the final product. The long-term goal of USDOE is to develop a well-defined Qualified Specialist program element for each type of system assessment software and to seek out transfer of long term stewardship of Qualified Specialist programs to key industry associations. By involving these associations in the early stages of developing the software and the corresponding Qualified Specialist delivery mechanism, it is hoped that this transition can gradually evolve over a period of approximately five years from each Qualified Specialist program launch.

In order to engage industrial suppliers and consultants in the Qualified Specialist effort and sustain their interest in participating, USDOE offers the following benefits:

- Recognition as a skilled user of the software (PSAT, AIRMaster+, PHAST, etc) each Specialist gets a signed, numbered certificate;
- Listing on the DOE BestPractices Website & access to quantities of software;
- DOE refers callers and trainees interested in software to the website list of Specialists;
- May also become qualified to offer DOE software training classes to customers;
- Invitation to provide referrals for possible DOE case studies on energy-efficiency projects identified by using software;
- Invitation to participate in review of proposed changes to software; and
- Ongoing technical support.

The Qualified AIRMaster+ Specialist training is being offered in cooperation with the CAC. The training includes 2 days of classroom training on AIRMaster+ and measurement techniques, a practical exam on measurement techniques, a take home exam using AIRMaster+, and a 4-1/2 hour written qualifying exam. As with the Qualified Pump System Specialist exam, the AIRMaster+ Specialist exam is open book and designed to test the individual's ability to apply systems knowledge as well as use the software. To date, five classes have yielded 41 Qualified AIRMaster+ Specialists. These Specialists are being used as the primary avenue for delivering AIRMaster+ training to industrial end users through an awareness training module.

Program Results

In late 2001, the effectiveness of the CAC training was evaluated. This evaluation assessment, conducted by Xenergy for USDOE in cooperation with Lawrence Berkeley National Laboratory and Oak Ridge National Laboratory and the Compressed Air Challenge, drew a representative sample from the population of individuals that were trained by the CAC as of May 2001, which included 3029 individuals attending Fundamentals of Compressed Air Systems and 925 individuals attending Advanced Management of Compressed Air Systems. Phone surveys were conducted of 200 participants (100 end users +100 vendors) to assess what their reaction to the training and, more importantly, whether they had made any changes in the operation and maintenance of their compressed air systems as the result of the training or, in the case of vendors, any changes in the services offered to customers.

The draft edition of the *Compressed Air Challenge Training Program Evaluation* places the estimated annual savings from participants in the training is about US$12 million. Participants generally found the sessions to be both useful and of high quality. An impressive 76% of customers participating in CAC systems training reported that they had made significant capital or operating improvements to their compressed air system since attending the training. Using conservative estimates, participants saved 8% of compressed air system energy on average as the direct result of the training. In addition, end users who implemented compressed air system efficiency measures experienced significant non-energy benefits, including: reduced downtime, reduced system moisture and contamination, and more consistent system pressure. Some quotes from participants included:

> "As the result of the improvements, we saved time and money in all aspects of production""We gained sufficient air capacity to make quality products that we were previously incapable of producing"

The influence of the CAC on changing market interactions from a focus on components to a focus on system services was also assessed. The Evaluation found that 52% of vendors who participated in the training began to offer new energy efficiency services that included:

- Analysis of system efficiency
- Measurement of system performance (flow/power/pressure)
- Ultrasonic leak detection

Approximately 85% of vendors have used the training materials or information when evaluating customer's systems, with 58% using this information frequently. Although not addressed in the study, most major compressor manufacturers in the US are now selling an integrated line of system equipment rather than just components.

Next Steps

The CAC has much work left to do to educate the end use industrial customer on the benefits of taking a systems approach. New initiatives under consideration include 1) the development of an Operator Certification Program that recognizes maintenance staff who a skilled in taking a systems approach. and 2) a web-based version of the Fundamentals of Compressed Air Systems for individuals who are unable to participate in offsite training programs.

The CAC seeks to continue to host and promote CAC training, broaden awareness of the systems message through widespread adoption of the *Guidelines for Selecting a Services Provider and Levels of Analysis* by industrial end users, and promote greater use of the CAC website www.compressedairchallenge.org . In addition, the CAC is seeking additional strategic alliances to strengthen the organization as well as making the training available for licensing to use outside of the US.

References

[DOE] US Department of Energy. 2001. Assessment of the Market for Compressed Air Efficiency Services, Office of Industrial Technologies, June 2001, Washington, DC.

[DOE] US Department of Energy. 2002. Compressed Air Challenge Training Program Evaluation, Anticipated publication date December 2002, Washington, DC.

McKane, Aimee T, Chris Cockrill,, Vestal Tutterow, and Anthony Radspieler. 2002. Public/Private Sector Cooperation to Promote Industrial Energy Efficiency: Allied Partners and the US Department of Energy. In proceedings of Energy Efficiency in Motor Driven Systems, 3rd International Conference, Treviso, Italy, September 2002,

McKane, Aimee T, Chris Cockrill, and Vestal Tutterow. 2001. Making Industrial Energy Efficiency Mainstream and Profitable: Where Public Benefit and Private Interests Intersect. In Proceedings of the 2001 ACEEE Summer Study on Energy Efficiency for Industry, Tarrytown, NY, July 2001

McKane, Aimee T., 1999 The US Compressed Air Challenge, proceedings of Energy Efficiency in Motor Driven Systems, 2nd International Conference, 10-22 September 1999, London, UK

On Site Efficiency Measurements for Electrical Motors is Possible and Easier than Thought

Alfredo Muñoz, Pedro Maldonado

Energy Research Programme, University of Chile, Echaurren 750
email: amunoz@prien.tie.cl, pmaldona@prien.tie.cl

Abstract

The replacement of existing electrical motors by efficient ones is always a matter of controversy, because the efficiency of installed motors is usually unknown. On site, it is extremely complex to measure the variables required by the testing standards.

In order to estimate on site the electrical efficiency of an installed motor, three basic methods, not considered by the testing standards, have been proposed to Chile's main copper company, which is committed to promoting and implementing energy efficiency practices and technologies. It should be mentioned that this company consumes around 5,000 GWh per year, with motors consuming around 70% of the total. These methods provide results precise enough to ensure that adequate evaluations can be carried out to justify the replacement of an installed electrical motor.

The method considers separate evaluations of the no load losses, the stator losses, the rotor losses and the stray losses. For example, iron loss calculations are based on the measurement of input power for delta and star stator connections. The rotor losses are measured using the on load power input and the speed measurement. Finally a dynamic measurement of the efficiency is presented.

Introduction

The efficiency measurement of a machine, defined as the ratio between the consumed electric power and the mechanical output should be carried out with great care [1]. The norms establish a complex routine measurement, since it is necessary to have a precision dynamometer (0,25% of the nominal torque of the machine in test). The complex of this measurement has lead to make different efforts of diverse institutions (c.f. IEEE, IEC) to develop indirect methods of measurement of efficiency, especially those that consist on measuring the losses of the motor separately. In this scenery in site measurement of installed motors is difficult [2] and no special standards have been developed to solve this problem.

This paper intends to propose approximate methods for on site measurement of each one of the main losses of a motor, considering that the torque [3] and the speed [4] are not simple of measuring outside of the laboratory.

The Motor Stator Losses

The norms specify the measurement of the resistance Rt of the stator to the temperature t_t (temperature of the stator during the test). This measurement is easy to carry out. For the calculation of the losses of the motor the norm suggests to measure the resistance between two phases of the motor, independently that the motor is in delta or in wye connection, and to calculate the losses in the copper of the stator P_{Cu} assuming that the motor is symmetrical. It is better to measure the resistances separately and to use an expression that considers the asymmetry of each resistance:

$$P_{Cu} = 1{,}5 \times R_t \times I^2 \Rightarrow P_{Cu} = 0{,}5 \times (R_{t1} + R_{t2} + R_{t3}) \times I^2$$

No Load Losses.

According to norms, to determine the no load losses in a motor the current, voltage, input power and temperature should be registered, varying the voltage from 125% of the nominal voltage until the point where the voltage reduction produces an increment of the current. From these values it is possible to determine, the losses in the iron to nominal voltage and the ventilation and friction losses.

Practically, it is difficult to carry out this test to diverse feeding voltages; for this reason, in simplified form, it is possible to carry out one measurement with the motor connected in delta and another measurement with the motor connected in star. From this measurements, it is possible not only to determine the total no load losses but also to separate those losses results corresponding to the iron and those corresponding to ventilation and friction, which is, from the user's point of view, very important, because it allows him to diagnose the state of the iron (from the point of view of the efficiency) of a motor. The table 1 shows the results of the on site measurements carried out to a motor, first connected in star and then in delta.

Table 1. Measured and calculated data in a delta connected motor and in a wye connected motor

	Delta Motor	Wye motor
Voltage [V]	575,00	575,00
No load power (measured) [W]	720,00	282,99
Stator copper losses [W]	125,60	34,90
Friction and windage losses [W]	75,00	75,00
Core losses [W]	519,40	173,09

Rotor Losses

To measure the losses in the rotor it is only required, according to norms, of a precise measurement of the speed (with an error lower than 0,1%) and of the frequency of the supply voltage (with an error lower than 0,05%). It is also required to have obtained the losses figures previously in the stator and in the iron. The following equations allow us to determine the slip and the losses in the rotor:

$$s = \frac{\left\{2\pi f_{red} - \frac{p}{2} \times \frac{2\pi}{60} \times N\right\}}{2\pi f_{red}}$$

$$P_{rotor} = (P_{cons} - P_{Cu} - P_{fe}) \times s$$

s	= slip
f_{red}	= supply frequency [Hz]
p	= poles number
N	= motor speed [rpm]
P_{rotor}	= rotor losses [W]
P_{cons}	= input power [W]
P_{fe}	= iron losses [W]

To determine the rotor losses to different load degrees it is necessary, according to norms, to repeat the measurement for each selected load. Nevertheless, it is possible to approach these losses using the registrations of the current and the power of the existent motor which, surely, corresponds to the diverse load degrees in which the motor truly works. Thus, starting from the measurement according to norms of the losses of the rotor, it is possible to calculate:

$$k_{rotor} = \frac{P_{rotor}}{\left(I_{reg}^2 - I_0^2\right)}$$

I_{reg}	: Measured current related to P_{rotor} according to the norm.
I_o	: No load motor current.

Assuming that k_{rotor} is a constant it is possible to calculate the losses of the rotor for other values of the stator current using:

$$P_{rotor}(I) = k_{rotor} \times \left(I^2 - I_0^2\right)$$

Stray Load Losses

Unfortunately, all the methods to measure the stray load losses require the installation of the motor in a torquemeter, what is impracticable for on site measurements, since motors of the most diverse powers, voltages and speeds exist in the industrial system. As a value for defect, it is suggested to use the value of the stray load losses guaranteed by the manufacturer. If the manufacturer only gives the value of these losses to nominal power, it is a good approach to calculate the efficiency to other load degrees supposing the following equation:

$$P_{paras}(I) = \frac{P_{parasnom}}{I_{nom}^2 - I_o^2}\left(I^2 - I_o^2\right)$$

$P_{parasnom}$ =Nominal stray load losses.
I_{nom} =Nominal stator current

On Site-Measured Efficiency

With the proposed method it is possible to elaborate a chart like the one that comes next. The on site-measurement has good precision, it is easy to get, it allows us to visualize what happens to the motor to any load degree and, what is more important, allows us to predict the economic convenience of substituting the motor in analysis for another more efficient.

Table 2. On site motor efficiency measurement for different load charges

Load charge %	99%	76%	51%	26%
Stator current [A]	11,61	9,66	8,03	6,81
No load current [A]	6,32	6,32	6,32	6,32
Input power [kW]	8,88	6,78	4,73	2,71
Iron losses [kW]	0,519	0,519	0,519	0,519
Friction and windage losses [kW]	0,075	0,075	0,075	0,075
Stator copper losses [kW]	0,448	0,310	0,214	0,154
Rotor losses [kW]	0,292	0,164	0,076	0,020
Stray load losses [kW]	0,134	0,075	0,035	0,009
Mechanical power output [kW]	7,412	5,636	3,812	1,933
Total losses [kW]	1,468	1,144	0,918	0,777
Efficiency [%]	83,47	83,13	80,59	71,33

Efficiency During the Motor Starting Process.

Another on site problem is to measure the motor efficiency when there isn't the possibility to connect a mechanical load in the axis. For this case, it is proposed in this paper, to measure the efficiency during the starting transient of the motor. During that transitory it is easy to measure the input power consumed by the motor, that is to say:

$$P_{cons} = P_{Cu} + P_{vacío} + P_{rotor} + P_{mec}$$

On the other hand, the input power has the following relationship with the mechanical power in the axis of the motor:

$$P_{mec} = \frac{1-s}{s} P_{rotor} \Rightarrow P_{cons} = P_{Cu} + P_{vacío} + \frac{1}{1-s} P_{mec}$$

The efficiency, that we will designate with the name of dynamic efficiency, can be calculate by means of:

$$\eta_{dinam} = \frac{P_{mec}}{P_{cons}} = \frac{(P_{cons} - P_{Cu} - P_{vacío}) \times (1-s)}{P_{cons}}$$

Each term of this last expression can be measured with the no load motor during the motor starting. The following figure shows a sample of an experimental registration of the motor input power, the speed and the dynamic efficiency. The efficiency dynamic measure is, approximately, similar to the efficiency of the motor, for values of the speed around the synchronous speed. The maximum dynamic efficiency is a useful parameter of identification of the characteristics of the motor, since it is experimentally repeatable and representative of the losses of the motor. It is also possible to calculate from these records the efficiency in function of the mechanical output power of the motor.

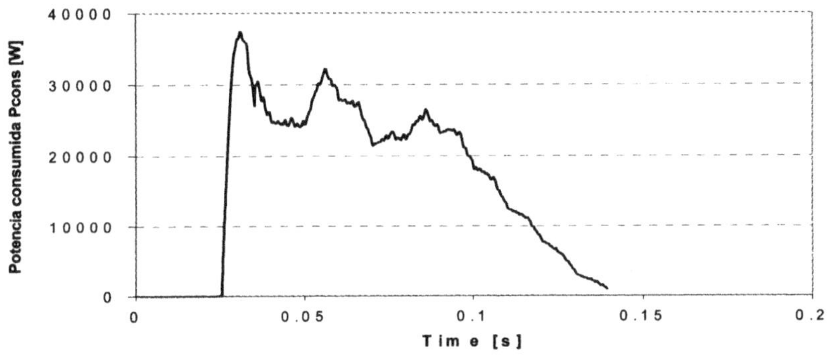

Fig. 1. Input power during an induction motor starting

Conclusions

In a great number of industrial and mining applications a test facility to determine efficiency is not used due to the complexity for getting an exact measurement. Standard measurements require at least to energize the motor at different voltages and to employ a calibrated dynamometer. In this paper an analysis of the induction efficiency as a result of on site measurements is carried out and a dynamic efficiency is calculated from the starting process of a no load induction motor.

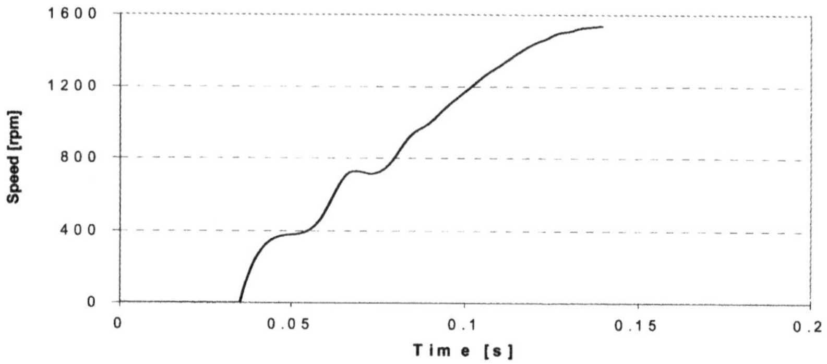

Fig. 2. Starting speed in a no load induction motor

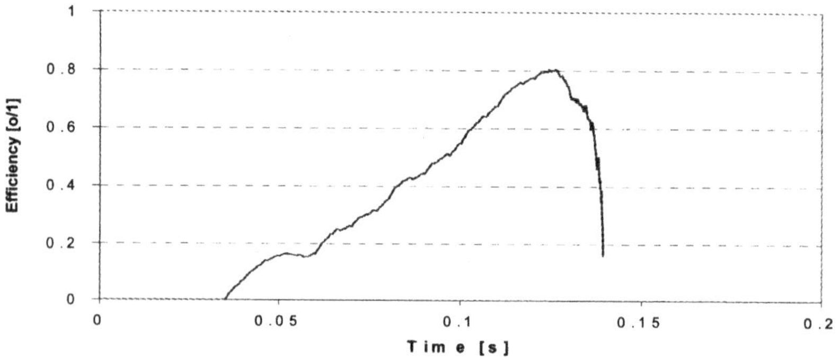

Fig. 3. Dynamic efficiency of an induction motor

References

[1] Nailen, R. : "Can field test prove motor efficiency?, IEEE Transactions on Industry Applications, Vol. 25, No 3, May-June, 1989, pp. 391-396.

[2] A. K. Wallace, E. J. Wiedenbrüg: "Motor efficiency determination: From testing laboratory to Plant Installation", Pulp and paper Conference, June 1999, p. 6.

[3] J. Hsu, P. Sorenson, "Field assessment of induction motor efficiency through air gape torque", IEEE Transactions on Energy Conversion, Vol. 11, No. 3, Sept., 1996, pp. 489-494.

[4] K. D. Hurst, Th. G. Habetler, "Sensorless speed measurement using current harmonic spectral estimation in induction machine drives", IEEE Transactions on Power Electronics, Vol. 11, No 1, January, 1996, pp. 66-73.

Determining the Efficiency of Electric Motors –

Does the Standard Draft IEC 61972 Provide Improvements for the European Market?

Herbert Auinger Eckehard Bunzel, Karin Friedrich

Dr. techn. Herbert Auinger
Siemens AG A&D, Erlangen

Dr.-Ing. habil. Eckehard Bunzel, Dipl.-Ing. Karin Friedrich
VEM motors GmbH, Wernigerode

Contribution of the ZVEI, German Manufacturers Association of Electric and Electronic Industries – Low Voltage Group, Frankfurt

General

Nowadays, the reduction of energy consumption, and, as a result, the protection of our environment is one of the most important concerns of reasonable technical activities. In the world wide competition, the numerical value of the nominal efficiency is therefore one of the essential features of a motor. Therefore, the determination of the efficiency has to be defined clearly and universally. This includes the reproducibility and tolerances of such procedures.

In Europe, it was in the past taken for granted that IEC 60034-2 was used for efficiency determination. The North American market had established the IEEE 112 for the same purpose. To give the electric motor industries and the user of electric motors some rules for the future reasonable design and use of motors, two different paths were trodden in the North American market and in Europe. With the US Energy Policy and Conservation Act (EPCA or EPAct) and the CSA Standard C390 legal rules were set, whereas the European Union has adopted the Voluntary Agreement between EU and CEMEP, the European Committee of Manufacturers of Electrical Machines and Power Electronics. So, the efficiency classification into three classes eff1, eff2 and eff3 is generally well-known.

It is understood, that the tendency will always go up to motors with higher efficiencies, but for motors with very short operational periods, motors in eff3 classification have still their right to exist, it would really be a waste of energy to raise their efficiencies if they are for example operated some minutes a day.

Standardisation

In preparation of the introduction of the Voluntary Agreement, the European Commission gave in December 1996 to CEN and CENELEC, the mandate M/244 to provide standards for test methods and for the measurement, calculation and designation of efficiency of general purpose low voltage three-phase induction motors, and in this context, also for excellent reproducibility of the measurements and for statements of the partial load behaviour of motors. The new standard should avoid the main drawbacks of the existing IEC 60034-2, i.e.

- the overall load-dependent additional losses taken with 0,5 % of the input into account,
- the stringent correlation of the thermal class reference temperature to the I^2R losses,
- that the measurement inaccuracies need only a small portion of the overall tolerance, even with Class ≤0,5 instruments. This leads in practice to the designation of "dressed-up" nominal values. Furthermore, to have the different tolerances for smaller motors up to 50 kW of −15 % $(1 - \eta)$ and of −10 % $(1 - \eta)$ for larger ones, as specified in IEC 60034-1. A way should be found to come to a uniform tolerance range for the whole affected output range.

An additional target of the above activities should be the introduction of a globally approved procedure for the efficiency determination and designation, to make the efficiency data better comparable. For this purpose, several Drafts of a new IEC 61972 were proposed to meet this objective.

The current Draft IEC 61972

This Draft offers two Methods to determine the motor efficiency, Method 1 is based on a direct measurement procedure, and Method 2 as an indirectly accomplished efficiency determination. According to this Draft, efficiencies of motors with outputs ≤ 150 kW are to be determined in accordance with Method 1 of the prepared new IEC 61972, the fourth draft of which is still in discussion.

This includes, incidentally, for a motor manufacturer, the product schedule of which covers an output range for both procedures, the fact to determine the efficiency of his motors in two different ways. And, self-understanding, this does not only cover the testing for research and new design purposes, but also the repeated type test procedure.

Some error and reproducibility considerations

The existing IEC 60034-2 and Method 2 of the IEC Draft 61972 both determine the motor efficiency in an indirect way. They differ in the value of taking the addi-

tional load dependent losses into account, for IEC 60034-2 with the fixed value of 0,5 % of the input power and, for Method 2 of IEC Draft 61972 with the assigned allowances of 2,5 % for 1 kW output down to 0,5 % of the input power for 10 MW.
So, the magnitude of the maximum uncertainty range of the indirectly determined efficiency is dependent on the random errors in the measured values and the efficiency itself. Both procedures are distinguished by an excellent reproducibility, independently of the efficiency range.

The accuracy of efficiency determination by Method 1 of the Draft IEC 61972 is finally limited through the direct measuring of the mechanical output P_2 and the electrical input power P_1. There are a lot of independent publications to verify the general statement that, from the basic principle the accuracy of Method 1 is limited to ±0,5 %. This uncertainty of ±0,5 % is therefore specified as the typical accuracy of the actual version of the IEC 61972, both for the efficiency and the portion of additional load losses. Finally, this results in the following situation, if Method 1 is applied for a 1,5 kW standard motor and a 160 kW high-efficiency motor, as seen in Fig. 1. Is this situation really an improvement? Therefore, the application of this method should be limited up to approximately 90 % efficiency.

	1,5 kW Standard motor $\eta_N = 76$ %	160 kW HEM $\eta_N = 96{,}5$ %
Permissible minimum value η_{min} at 15 % (1-η_N) tolerance	72.4 %	96 %
Required measured value $\geq \eta_{min}$ + 0.5% to cover measuring uncertainty	72.9 %	96.5 %
"Tolerance" available for production and material spread for "rounding up" η_N	+ 3.1%	0

Fig. 1. Results of tolerance definition for use of IEC 61972, Method 1

Here it is obvious, that we have a not acceptable and unbalanced situation.

If we compare the gap between uncertainty of measurement, respectively their reproducibility, with the allowed efficiency tolerances of IEC 60034-2 and Method 2 of Draft IEC 61972, the inaccuracy of the measurements requires only a small and nearly constant portion, even with less accurate measurement instruments of Class 0,5 (see Fig. 2). The remainder of the tolerance can be used for its basic purpose, i.e. the small spreads in material properties and in motor manufacturing processes.

Practical procedure according to Method 1

Measurement instruments with an accuracy class of 0,2 are required for the torque measurement. Formally, the modern torque measuring shafts designed according to the latest state of wire strain gauge technologies, fulfil the requirements. But, it is always understood, that the accuracy is based on the full scale data. In determining the efficiency according to IEC 61972 Draft, the torque must be measured within the output range of 25 up to 150 % of the nominal output. If, for instance, the torque measuring shaft is suitable for measuring the highest point of 150 % of the nominal output, then the lowest point of 25 % of the nominal output makes at least 16 % of the full scale of the torque measuring shaft, and I think, nobody would guarantee for an accuracy of 0,2 at this point.

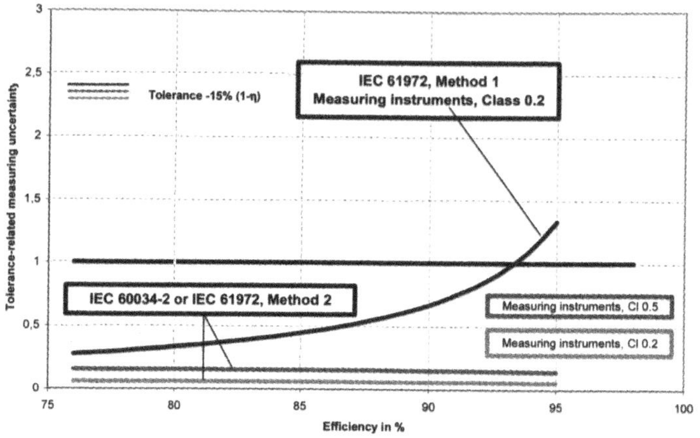

Fig. 2. Measuring uncertainty or reproducibility in relation to permissible tolerance

Procedural influences

In addition to the output measuring errors when measuring input and output, the following items can also have a decisive influence on the results, particularly on the determination of the additional losses:

- Operational conditions and design of the bearings, i.e. amount of grease filling, bearing sealing. And if, as done in some other rules for the efficiency determination, the option is offered to test the motors without this bearing sealing, then the test results may be physically correct, but the motor itself has in operation never such an efficiency, and the intended reduced energy consumption is never realised.
- Thermal conditions (the ambient temperature, thermal equilibrium)
- The measurement of the winding temperature (by thermocouples, extrapolation of resistance measurements after de-energising).

The constant reference temperature (see Draft IEC 61972) for winding losses at full, 5/4 and 6/4 load results principally to excessively high values for the additional losses, because the real copper losses in the winding are higher than they are taken into account. This error increases with decreasing thermal time constant of the motor and increasing duration of the measurement. This can result, for smaller motors, for example for an 2-pole motor with 0,75 kW, in a determination of additional losses being by 30 up 50 % too high.

Discrimination of eff 1 motors when Method 2 is used

On the market, motors in all 3 efficiency classes are represented. The increased amount of active material and/or use of higher-grade electrical steel are required for the high efficiencies of eff1-motors. As a rule, these motors have thermal reserves due to the significantly lower thermal and magnetic stresses. For this reason, using the standard optimisation for both motor efficiency classes, the copper and core losses as well as the additional losses are lower than for the highly utilised eff3-motors.

If the Draft IEC 61972 should come into force, then the situation for high-efficiency motors is changed to the worse, but the eff3-motors profit from, as seen from Fig. 3.

	Effl motor	Eff3 motor
	Current state based on Method 1	
V_{Fe}	No-load → load (ΔU = 3,5 %) 127 W → 116 W (-9 %)	No-load → load (ΔU = 5 %) 195 W → 167 W (-14 %)
V_{Fr+W}, V_{Cu}, V_{Rot}	31 W + 298 W + 136 W = 465 W	43 W + 428 W +197 W = 670 W
V_{add}	64 W = 1,04 % · P_1	96 W = 1,49 % · P_1
Total loss	645 W	933 W
P_1	6145 W	6433 W
Efficiency	89,5 %	85,5 %
	Application of Method 2	
$V_{Fe\,load}$	116 W	167 W
V_{add} (2,13 % of P_1)	132 W	138 W
Remaining loss	465 W	670 W
Total loss	713 W +10, 5 % compared with Method 1	975 W +4,5 % compared with Method 1
P_1	6213 W	6475 W
Efficiency	88,52 % = 0,98 % less	84,94 % = 0,55 % less
	Differences to current IEC 60034-2	
$V_{Fe\,no\text{-}load}$	127 W	195 W
V_{add} = 0,5 % P_1	31 W	32 W
Remaining loss	465 W	670 W
Total loss	623 W = -3,4 % compared with Method 1	897 W = -3,8 % compared with Method 1
P_1	6123 W	6397 W
Efficiency	89,83 % = +0,33 % compared with Method 1	85,98 % = +0,48 % compared with Method 1

Fig. 3. Application of different determination methods for effl and eff3 motors, P2= 5,5 kW in cross comparison

Conclusions

The principally correct results from Draft IEC 61972 Method 1 are obtained through the significantly higher expenditure required and the relatively poor reproducibility of ±0,5 percent, both for the efficiency and for the additional losses.

By comparison, the existing IEC 60034-2 as well as Draft IEC 61972, Method 2 are distinguished of significantly better reproducibility – even though less precise measuring instruments in class 0,5 are used. The principal errors present there are the overall consideration of the additional losses, which are, from the experience, too low in IEC 60034-2 and too high in Draft IEC 61972, Method 2.

In discussing the physically correct efficiency, it should not be neglected, that this depends on a series of influencing factors related to the real motor operation. The magnitude of operating voltage alone can increase or decrease the motor efficiency by several percent points, as discussed from Dr. Auinger at the previous EEMODS Conference in London 1999. As known, the low-voltage standard motors are designed not for a fixed design voltage, but rather for the European voltage range of 380...420 V. To this figure, the usual line voltage tolerances as well as the line harmonics and unbalances must be added. There is no information given on this basis relationship with the efficiency.

Here, it must be allowed to state clearly, that all these consideration are not at all a field of theoretical discussions, for motor manufacturers it could be vitally important, to have, with respect to legal rules for cases of warranty and liability, measurement procedures with excellent reproducibility.

On the other hand, very intensive efforts are made to measure the additional losses "precisely" or "accurately". However, the additional losses are not at all either constant values, but rather are subject to an ageing effect and can, on the other hand, depend on the motor being connected in star or delta circuitry. Moreover, their determination according to Method 1 of the Draft IEC 61972 is relatively inaccurate. With high-efficiency motors with additional losses of (0,5...1,0) % of the input power, the bandwidth for the measuring uncertainty is in the same magnitude as the measured value itself.

The information provided by a measured result of efficiency = 94,5 % means at the end of the day, that the "correct amount" will be between 94 and 95 %. This range applies also for the additional losses. As an example, a result of an amount of additional loss of 0,8 % of the input power means, that the additional losses really are in the range between 0,3 % and 1,3 % of the input power. This circumstances lead to the conclusion, that Method 1 is technically and economically practicable only for motor efficiencies lower than 92 %.

In terms of practical application, Draft IEC 61792 does not provide any overall improvement; the previously establish state with IEC 60034-2 with its familiar weak points was replaced by an other unsatisfactory method.

The drawbacks can be summarised as follows:

- For efficiencies in the range above 90 %, the not insignificant additional expenditure that must be spent by the manufacturer for investments and procedures, is not compensated by any recognisable benefit for the customer, and as an result, the customer would never pay for.

- The increased expenditure required for the "more precise determination method" would only appear justified when we could be sure that the achievable improvements are not lost again in the course of measuring inaccuracy and poor reproducibility and the existing error margins inside of the specified tolerances of the efficiency.
- It introduces new difficulties, to implement a new IEC 61972 with strong similarities to IEEE 112 without entailing the other limitations, for example the definition of relatively coarsely classified nominal efficiencies and their related minimal efficiencies.

My company, and also the German Association of the Electric and Electronic Industries, think, that since several years so much research work was done on the field of additional losses, that it should be enough data to come away from both the limits of overall estimation with fixed 0,5 % and depending on the output up to 2,5 % of the input power. From my knowledge, most well-known motor manufacturers should never exceed the range of 1,0 and 1,5 % for motors up to 150 kW. It could be a reasonable way to initiate an investigation in an independent scientific institution to come to a new internationally acceptable determination procedure for the efficiency.

As long as no conclusive overall concept within the sense of the EU mandate M/244 for determination and designation of motor efficiencies is presented, that takes all points of view into account, i.e. good reproducibility, practice-related tolerances, partial load conditions and reasonable expenditures for the motor testing, the status-quo should remain on the basis of the existing IEC 60034-2.

And, if I am allowed to speak for our potential customers and OEMs, we have to provide a reliable base for their comparisons between different motor manufacturers. And, on the existing basis of IEC 60034-2 the customer has, always having in mind the defects in additional load loss assessment, the necessary tool for the differentiation between the different brands.

References

[1] Auinger, H., Bunzel, E.: Wirkungsgradermittlung von Elektromotoren – Bringt der Normentwurf IEC 61972/CDV Verbesserungen für den europäischen Markt?, Antriebstechnik 40 (2001), S. 65 – 70

[2] Künzel, E., Auinger, H.: Vergleich genormter Verfahren zur Wirkungsgradbestimmung von Käfigläufer-Ausynchronnmaschinen, Elektrie 54 (2000), S. 284 – 302

[3] Auinger, H.: Wirkungsgrad von Elektromotoren unter realen Einsatzbedingungen, Elektrie 53 (1999), S. 74 - 77

[4] Bertoldi, P., de Almeida, A., Falkner, H.: Energy Efficiency Improvements in Electric Motors and Drives, Springer 2000

Comparison Between Nameplate Efficiency and Actual Measured Efficiency for Three-Phase Induction Motors

Pierre Angers

LTE-Hydro-Québec Research Institute, 600, ave. de la Montagne,
Shawinigan (Québec), Canada,
Ph.: (819) 539-1400 ext: 1427, Fax: (819) 539-1539,
e-mail: angers.pierre@ltee.ireq.ca

Introduction

In Canada, all regulated energy-using products imported into the country or shipped between provinces must carry an energy efficiency verification mark from a recognized certification organization. An energy efficiency verification mark indicates only that the energy performance of the product has been verified; it is not a safety certification mark. The verification mark must be put on the exterior of the product. To authorize an energy efficiency verification mark, a certification organization must be accredited by the Standards Council of Canada (SCC) and must administer an acceptable energy performance verification program for the product. Canadian Standards Association (CSA) International is one of the organizations accredited by SCC to certify electrical and electronic products and administrates an energy performance verification program. CSA works with Hydro-Quebec Research Institute (LTE), an independent laboratory who owns a facility accredited IEC / ISO Guide 17025 by SCC for testing motors. The laboratory is also accredited by CSA International for the same testing.

The Energy Efficiency Regulations requirements for electric motors in Canada came into effect in 1995, with amendments in 1997. The Regulations are administered by Natural Resources Canada (NRCan), a federal government department. These regulations apply to integral horsepower electric motors from 1 to 200 hp (0.746 to 150 kW), and to motors that are incorporated into other products, whether or not the other product is an energy consuming product.

The Regulations specify that this is the responsibility of the "dealer" (a dealer could be a manufacturer) to establish the minimum energy performance levels of the motors and he must ensure the motors meet the energy efficiency standard quoted on their nameplate. But how is the nameplate efficiency compared to the measured efficiency ?

Standards

There are few standard methods actually in use around the world to measure motor energy efficiency. In Canada, the Standard specified in the Energy Efficiency Regulations is: "Method for determining energy performance of three-phase induction motors, Method 1" CSA C90-93. The revision of this Standard (now CSA1 C90-98) is actually in use. The revision brought no change in the procedure to test the motors but now includes a table for efficiency requirements for IEC motors now covered by the Regulations and the addition of the 175 hp (130 kW) motor size. This Standard is equivalent to the well recognized Standard IEEE 112-1996, Method B: "Test Procedure for Polyphase Induction Motors and Generators" currently in use in the USA and some parts of the world. In addition, the upcoming IEC Standard 61972: "Method For Determining Losses and Efficiency of Three-Phase Cage Induction Motors", Direct Method will also be equivalent to the North American Standards so one will soon deal with "comparable" efficiency results throughout the world.

Accuracy

The accuracy of the measured motor's efficiency at the LTE testing facility according to CSA390 has been evaluated to ± 0.2 percentage point and the repeatability to ± 0.1 percentage point. These values are based on the instrumentation accuracy, calibration, error computation, comparative and repetitive tests on control motors over a period of several years.

Finally, for confidentiality purposes, letters have been substituted for real manufacturer's names in the paper.

Comparison of Efficiency and Limitations

Efficiency results concern only the motors tested. In most cases, the motors selected represent a sample of one (1) unit of one (1) model.

However, each motor nominal nameplate efficiency had a minimum associated (allowance for variation of 20 % of losses) for variations in materials, manufacturing processes and tests results and measured actual full-load efficiency had to be not less than this minimum.

Test Setup

All testing was conducted at the LTE-Hydro-Québec motor testing facility. This facility includes three (3) test setups for characterization of induction motors in the range of 1-500 hp.

The testing facility is characterized by a 500 kVA, 3-phase variable transformer capable of applying voltages in the range 0-4160 volts with each phase independently controlled for any application (near perfect 3-phase balanced voltage). Also part of the testing facility are three (3) computer controlled adjustable brake loads coupled with instruments for torque, speed, temperatures and all electrical measurements required for determination of motor's efficiency with high accuracy.

Results

A total of 258 motors have been tested during this period. Table 1 shows the motor quantity distribution according to the number of poles.

Table 1.

Number of poles	Number of motors
2	26
4	211
6	19
8	2

The motors were in the range of 1-500 hp (0.75 – 373 kW) and mainly low voltage (575 V), Totally Enclosed Fan Cooled (TEFC). For the purpose of clarity and to facilitate the comparison of the results, only the 4-pole motors in the range of 1-200 hp (0.75 – 149 kW) will be presented in the following figures, over the period 1994 to 2002 (182 motors).

Figure 1 and 2 present the comparison between the measured efficiency and the one found on the nameplate of the motors at 100 % load.

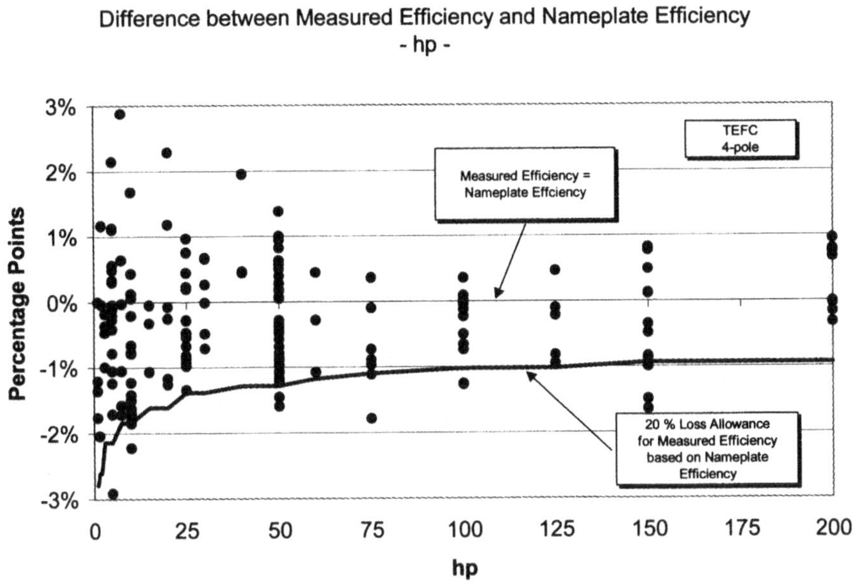

Fig. 1.

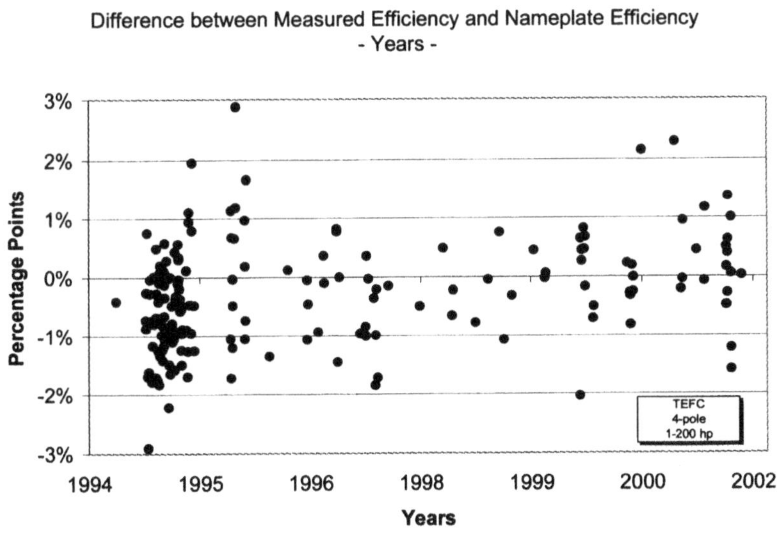

Fig. 2.

The 0 % line on the figure indicates that the measured efficiency equals the nameplate efficiency and the red curve on figure 1 gives the allowance for the 20 % loss as mentioned earlier. It can be seen that 174 of the 182 motors had a measured efficiency lower than their nameplate but higher than the minimum considering the 20 % allowance. *So one should be confident when reading a nameplate that the "real" efficiency should be at least equal to the minimum related to the 20 % loss allowance.*

Figure 3 presents the efficiency comparison of 147 motors from 13 manufacturers. This figure demonstrates that from one manufacturer to another, the results were significantly different. If we consider Manufacturers B, E, H, K, L and M, almost all motors obtained a measured efficiency lower than the nameplate efficiency. In contrast, Manufacturers C, F and I had measured efficiencies higher than nameplate efficiencies.

Finally, figures 4 and 5 compare the measured efficiency and related losses on motors of the same size from different manufacturers. It can be seen that all motors had their efficiency well beyond the minimum allowable. In addition, in the computation of each total losses, 9 of the 12 motors losses were within ± 10 %. *This is an example that manufacturers have a good control over their process and are capable of specifying the nameplate efficiency with good accuracy.*

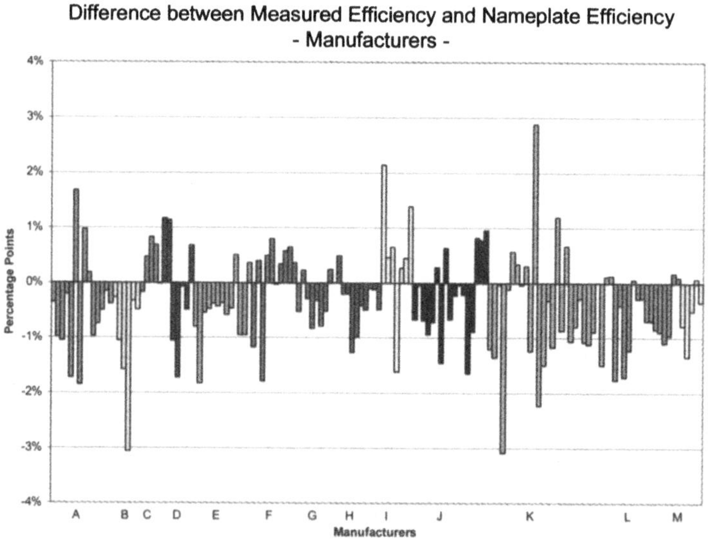

Fig. 3.

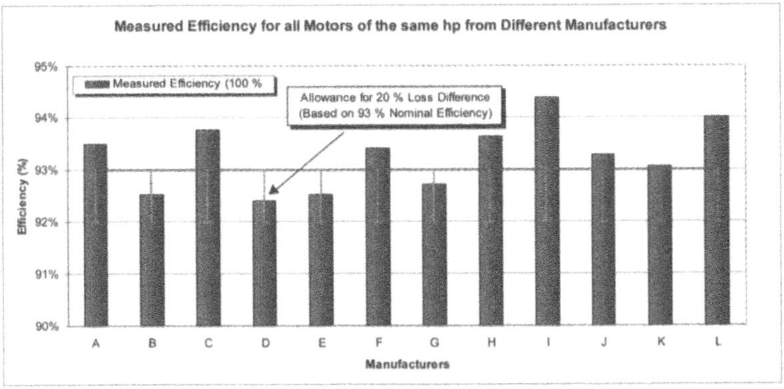

Fig. 4.

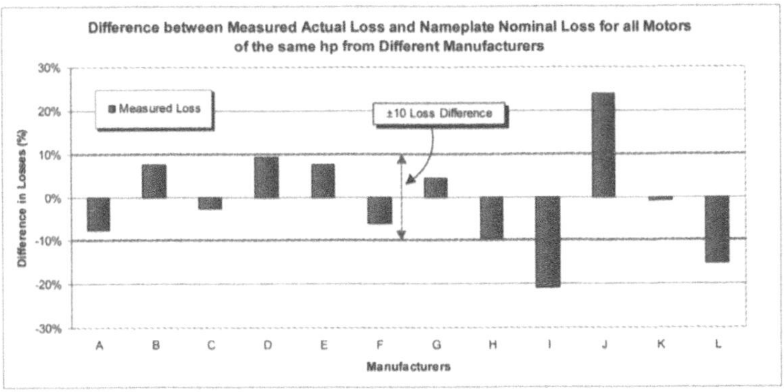

Fig. 5.

Conclusion

The comparison between nameplate and measured efficiency on 182 motors (TEFC, 4-pole, 1-200 hp) has shown that most motors had a measured efficiency lower than their nameplate but higher than the minimum considering the 20 % allowance. So one should be confident when reading a nameplate that the "real" efficiency should be at least equal to the minimum related to the 20 % loss allowance. Moreover, a comparison of groups of motors from different manufacturers demonstrated that certain manufacturers had most of their motors with measured efficiency lower than the nameplate and while the opposite situation was true for others. Finally, a comparison of the same size of motor from different manufacturers indicated that for 9 of the 12 motors, there was only a small discrepancy between measured and nameplate efficiencies, one example that manufacturers are capable of specifying the nameplate efficiency with good accuracy.

Optimal Design and Efficiency Test Method of Induction Motors Fed by Frequency Converters

Atanasi Jornet[1], Angel Orille[2], Albert Pérez[2], and Salvador Jareño[2]

[1] AEG Fábrica de Motores, S.A.Head of Test Laboratory
Carretera Castellar, 225, E-08226 Terrassa, Barcelona
Telephone: +0034937398493, Fax: +00347398488
E-mail: atanasi.jornet@lafert.com

[2] Universitat Politécnica de Catalunya, Department Electrical Engineering,
Campus.Colon St. 1, E-08222 Terrassa, Barcelona
Telephone: +0034934011982, Fax:+00347398256
E-mail: orille@ee.upc.es

Abstract

The different international energy policies recommend the use of electrical drives with variable speed. With these variable speed drives (e.g. centrifugal pumps, fans, etc.) a higher efficiency of the process is achieved, thus allowing for energy saving.

Efficiency of electrical motors has been improved for normal sinusoidal supply, without taking into account the negative effects when feeding the motors by means of frequency converter.

The frequency converters produce first a high amount of hysteresis and eddy losses in both stator and rotor iron and secondly a temperature increases of the rotor because of the current distribution in its rotor slots.

The conventional calculation using analytical tools could not calculate precisely the required parameters in order to obtain an optimal model to build a prototype that its properties confirm the calculated values with the model.

With a finite element method (FEM) application for magnetic field and heat transfer could be elaborated very precisely the required elements to design a new prototype and it is also a tool to prove the already existing motors for this application.

The presently applied standards to determine efficiency are not suitable to recognise this improvement. Applying high-tech tools we can define test method to measure the actual efficiency of this type of motors. This allows us also to create energy efficient electrical drives according to the recommendations of the last EEMODS conference held in London in 1999 with the support of the European Committee for Energy Saving.

1. Introduction

The today's conventional programs are very good to calculate motors at 50Hz operation, it takes a few seconds and in a very easygoing way. But for frequency fed motors, these conventional tools could not take into account easily the phenomenon of the skin effect produced in the rotor slot. These new tools used for some years are the finite elements. Their disadvantage is that it takes longer to make a calculation and the procedure is not very easy.

2. Finite Elements Analysis

The proposed method is a combination of the classical analytical program and the additional tool of the finite elements. Figure. 1 describes an easy flow diagram procedure.

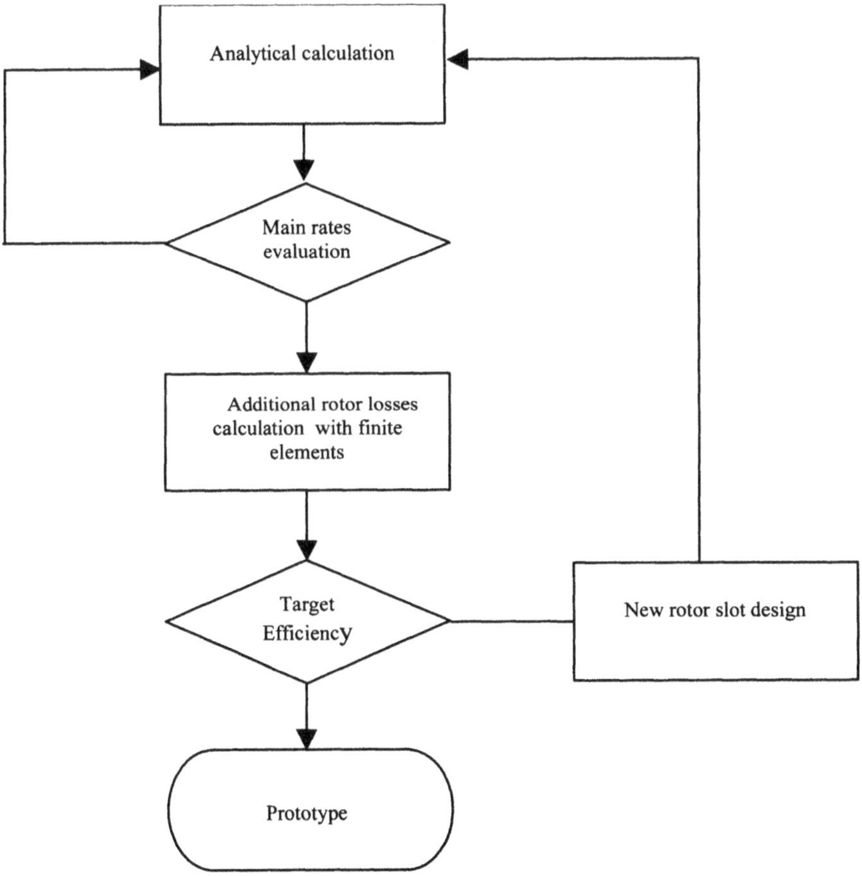

Fig. 1. Flowchart of calculation procedure

Additionally to the calculations based on the main frequency, the contribution of the harmonics due to the converter modulation has to be considered. A typical Fourier analysis spectrum of the stator current shows the amplitude and the frequency to be considered for the evaluation of losses (Fig.2). The main stator copper losses are due to the first harmonic or fundamental. At rated speed the frequency of the rotor current is very low (up to 2 Hz per second) and no skin effect is to be expected. The losses could easily be calculated by measurement of the rotor current, but with the usual die-cast aluminium rotors there is no chance to acquire its value.

To validate this theory [1], that the spectrum of the rotor current frequency is the slip multiplied by the stator frequency, a wound rotor with slip rings is used. The actual current measurement was done at the friction ring connection. A Fourier analysis of the rotor currents acquired in this way shows that the stator current harmonics induces into the rotor circuit a current called harmonics frequency (Fig. 2,3). Considering the following expression (1), the slip of this harmonics (υ) is proximate to 1. This phenomena is easy to understand, as the fictive synchronic speed of the harmonic field is much higher than the actual rotor speed and, like in a transformer the induced current has the same frequency of the primary winding. For this harmonics frequency the skin effect has to be expected, the current is concentrated in the upper part of the slot with a very high current density.

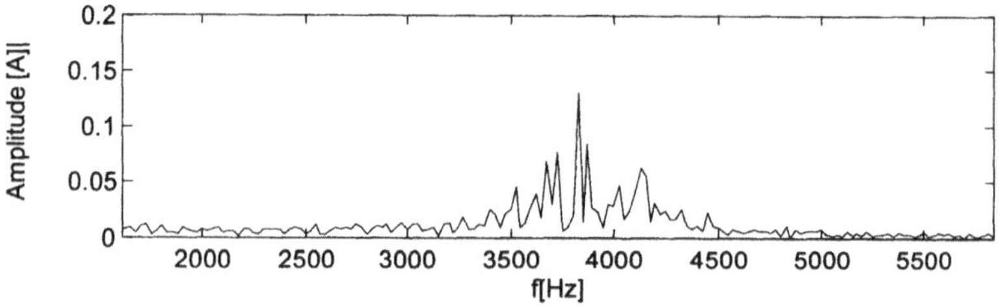

Fig. 2. First harmonic band of stator current

$$f_2 = s \cdot f_1 \tag{1}$$

$$s_\upsilon = \frac{\omega_{o\upsilon} - \omega_s}{\omega_{o\upsilon}} = 1 - \frac{\omega_s}{\omega_{o\upsilon}} = 1 - \frac{60 \cdot f_1 / p}{\upsilon \cdot 60 \cdot \frac{f_1}{p}} = 1 - \frac{1}{\upsilon} \tag{2}$$

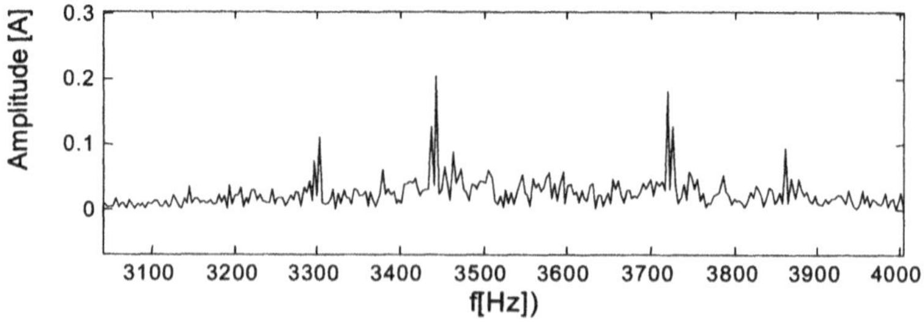

Fig. 3. Rotor induced current of the first harmonic band of the stator current

2.1 Example of Slot Optimisation

There is a not desired high frequency current harmonic generated by the PWM modulating frequency. For instance, figures 4-8 show the current distribution of a modulation of 1.9 kHz. The amplitude of this current is less than 10% of the fundamental, but it has a much higher density as this current is concentrated in the top part of slot. This results in the increase of temperature and decrease of efficiency.

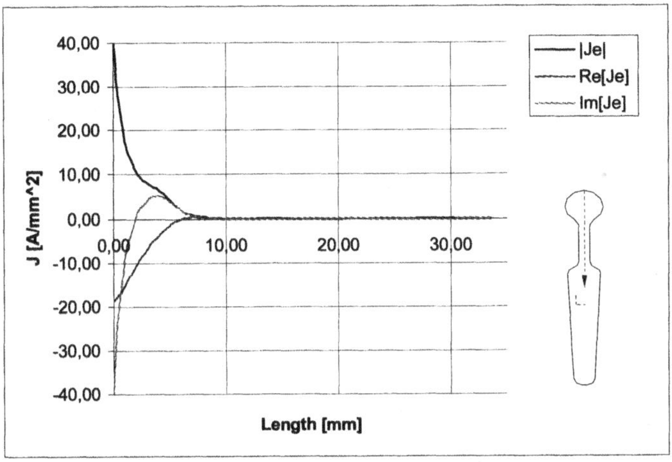

Fig. 4. Model 1 – Slot distribution of first harmonic current density 1.9kHz in double cage close die-cast aluminium rotor

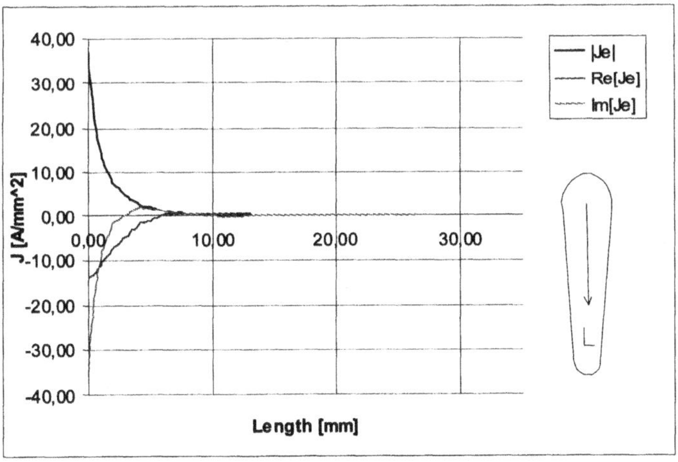

Fig. 5. Model 2 - Slot distribution of first harmonic current density 1.9kHz in single cage close die-cast aluminium rotor

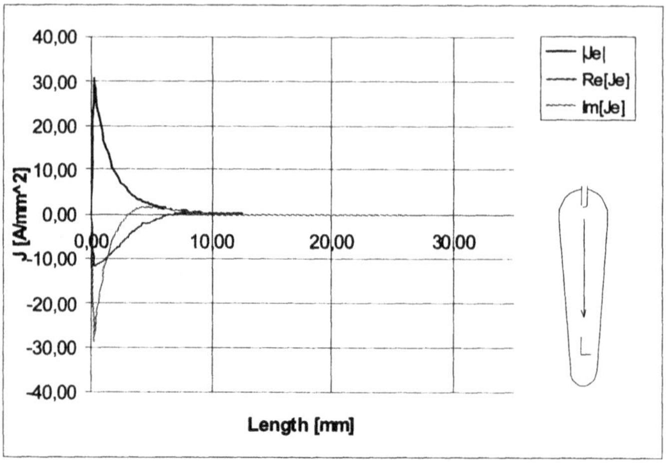

Fig. 6. Model 3 - Slot distribution of first harmonic current density 1.9kHz in single cage open die-cast aluminium rotor

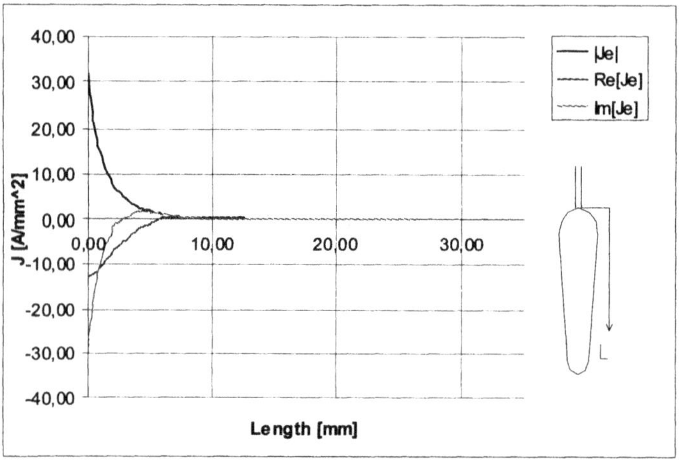

Fig. 7. Model 4 - Slot distribution of first harmonic current density 1.9kHz in single cage open die-cast aluminium rotor. Longer leakage path.

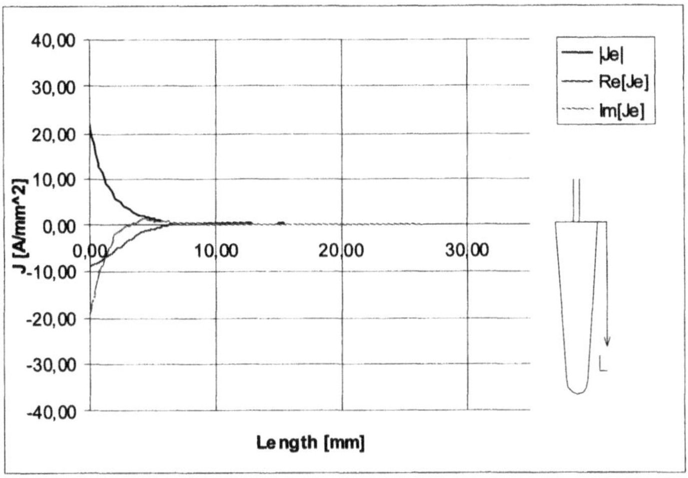

Fig. 8. Model 5 - Slot distribution of first harmonic current density 1.9kHz in single cage closed die-cast aluminium rotor. Wider top and long leakage path.

The table in figure 9 summarises the rotor slot losses considering the different models. If the motor is designed for frequency converter application only, any of the slot types from 2-5 can be chosen. For bypass applications [5] model 5 is proposed, where the starting current relationship is kept below 10 and efficiency is more advantageous. Nevertheless, production costs are higher.

Model	Unit	1	2	3	4	5
Frequency	[Hz]	50	50	50	50	50
Switching freq.	[kHz]	1,90	1,90	1,90	1,90	1,90
Rotor slot						
Surface	[mm2]	171	180	177	180	190
Harmonic losses m	[W/m]	945	719	741	684	444
Harmonic losses	[W]	321	244	252	233	151
Fundamental losses	[W]	882	711	711	711	680
%RH/RF	[%]	36,4	34,4	35,5	32,7	22,2
%RH/RHModel1	[%]	100	76	78	72	47

Fig. 9. Comparison table of the different slot models.

3. Test Method

The proposed method of efficiency determination with harmonics analysis [2] is based on the direct or indirect torque measurement and segregation of losses.

3.1 Distribution of losses

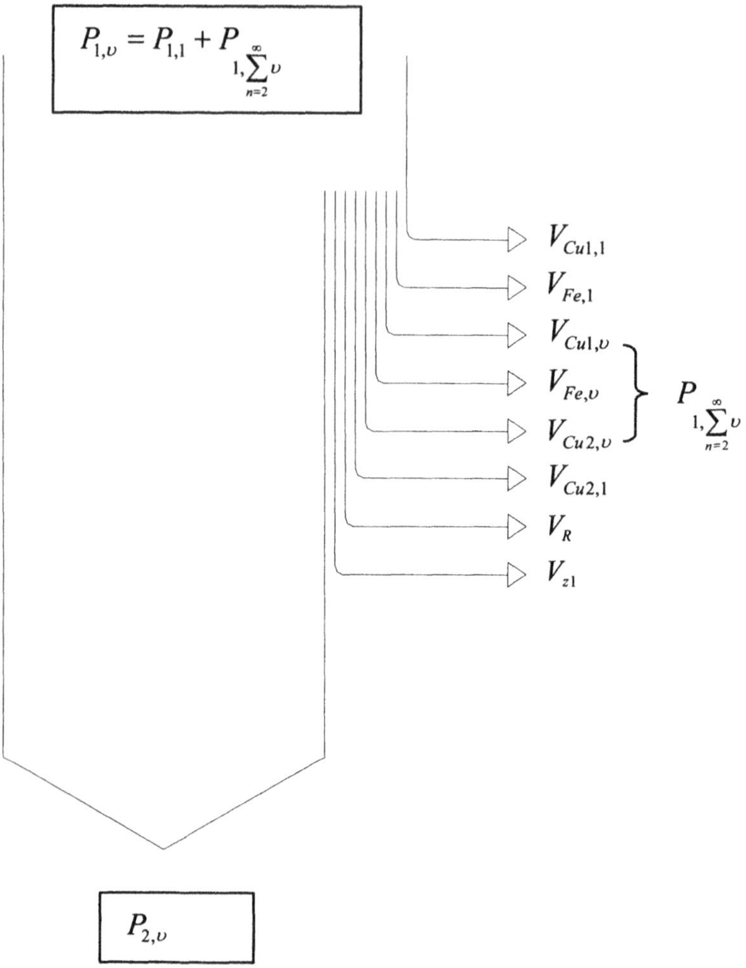

Fig. 10. Distribution of losses with the inclusion of the harmonic losses

Name	Unit	Description
$P_{1,\upsilon}$	[W]	Total input power
$P_{1,1}$	[W]	Fundamental input power
$P_{1,\sum_{n=2}^{\infty}\upsilon}$	[W]	Harmonics input power
$V_{Cu1,1}$	[W]	Fundamental stator joule losses
$V_{Fe,1}$	[W]	Fundamental iron losses

$P_{o,v}$	[W]	Total no load input power
$V_{Cuo1,1}$	[W]	No load harmonics stator joule losses
$V_{Cu1,\upsilon}$	[W]	Harmonics stator joule losses
$V_{Fe,\upsilon}$	[W]	Harmonic iron losses
$V_{Cu2,\upsilon}$	[W]	Harmonic rotor losses
$V_{o,\sum_{n=2}^{\infty}\upsilon}$	[W]	Total Harmonic losses at no load
$P_{\delta 1}$	[W]	Fictive airgap power
$V_{Cu2,1}$	[W]	Fundamental rotor losses
V_R	[W]	Mechanical losses ventilation and friction
Vz1	[W]	Additional load losses
$P_{2,\upsilon}$	[W]	Motor output

3.2 Test Procedure

Basing on the suggestions in the standards [4] and additional tips, our development group is working on defining an accurate and repetitive method to determinate the efficiency of motors fed by frequency converter. This method could be defined as electrical drive efficiency measurement aided by power analyser instruments or easily called harmonics method.

3.2.1 Definition of Main Test Drive Ratings

Voltage, frequency, output, service, type of control of the frequency converter and its switching frequency

3.2.2 Cold Resistance Measurement Winding

It is acceptable to use the resistance method to calculate the temperature of the winding. But embedded thermocouples are also recommended.

3.2.3 Preheating Period

Before carrying out any load adjustment, the test motor requires a temperature close to the operating temperature and the mechanical losses have to be stabilised. For not direct torque measurement, the rated torque has to be kept constant for about one hour. For direct torque measurement keep the output constant.

3.2.4 No Load Characteristic

To determine the mechanical and fundamental iron losses a no load characteristic is required. This test could be done with the frequency converter at about 6 points ranging from 125% of the rated voltage to 20% with constant frequency. The mechanical and fundamental iron losses are obtained by subtraction of the following expression. Segregation of the mechanical losses is carried out as usual by extending the curve to zero. The harmonic losses and the fundamental no load current is directly obtained from the instrument.

$$V_{Fe,1} + V_R = P_{01,\upsilon} - V_{o,\sum_{n=2}^{\infty}\upsilon} - V_{Cu0,1} \qquad (3)$$

3.2.5 No Load Measurement at Rated Voltage

If there is no possibility to change the voltage, the harmonic losses could be determinated by subtracting the input power at no load fed by the frequency converter to the no load sinusoidal input power at rated voltage. It has to be observed that if running time between the two tests is longer, the mechanical losses that could cause a wrong calculation of the harmonic losses are not modified.

3.2.6 Dynamometer Correction

This correction is required for not direct torque measurement.

3.2.7 Load Adjustment

For not direct torque measurement it is necessary to calculate the output by segregation of losses. This check is very good to find out any mistake in the measurements at the first stage.

3.2.8 Temperature Rise

Keep the motor with constant output for direct torque measurement. For indirect torque measurement, the calculated input power is maintained constant during this temperature rise period.

3.2.9 Warm Resistance Measurement of the Winding

The motor reaches the stabilisation temperature when the reference temperature of the frame does not increase more than 1 degree in one hour. Switch off the motor and measure the winding resistance to determinate the temperature increase.

3.2.10 Recovering Period of the Thermal Equilibrium Speed and Load Characteristic

To measure the load characteristic from 25% to 150% of the rated output, wait until the motor recovers almost the same speed as at the end of the temperature rise. If possible, measure the winding temperature a each load point.

3.2.11 Summation of Losses Calculation

With the aid of the harmonic analysis for each load point, calculate the additional losses by segregation (expression 4).

$$\mathrm{Vz1} = P_{1,\upsilon} - [P_{2,\upsilon} + P_{1,\sum_{n=2}^{\infty}\upsilon} + V_{Cu1,1} + V_{Fe,1} + V_{Cu2,1} + V_R] \qquad (4)$$

3.2.11 Measured and Calculated Output

If the precision of the measurement through direct or indirect method with dynamometric correction is accepted, an increase of the additional losses has to be taken into account, in order to match both results of the measured and calculated output. Its value could be higher than usual because any error of segregation of losses could be hidden. For negative additional losses, a mistake of the calculation of the other losses has to be considered.

3.2.12 Efficiency Determination of the Complete Drive: Motor and Frequency Converter

The efficiency test comparison has been carried out on a motor with slot model 1 and 2 (Fig. 4 and 5).

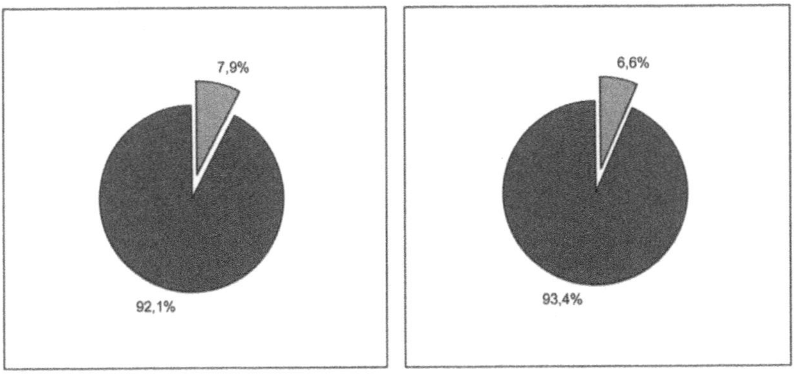

Fig. 11. Test result comparison with efficiency improvement of 1,3% for 45kW motor

4. Conclusions

This method would allow the endusers to compare the efficiency of the different motor and converter manufacturers.

For applications with converter fed motors it is not enough to derate the output of the motor. A special design is necessary taking into account the required control range and the frequency converter signal. The efficiency of the electrical drive has not to be reduced.

References

[1] NED MOHAN, TORE M. UNDELAND, WILLIAM P. ROBBINS, "Power Electronics:Converter,Applications and Design". John Wiley & Sons, 1989
[2] PETER-KLAUS BUDIG, "Stromrichtergespeiste Drehstromantriebe. Theorie und Betriebsverhalten von Asynchronantrieben". VDE Verlag, 2001
[3] DAVID MEEKER, "Finite element method magnetics". 2001
[4] IEC 60034-17 Cage induction motors when fed from converters. Application guide.
[5] A. JORNET, A. ORILLE, A. PEREZ, D. PEREZ, " Optimal design on high frequency induction motors with the aid of finite elements". ICEM 2002 Brugge.

A Critical Approach to the Iron Losses in Induction Motors

Aldo Boglietti, Andrea Cavagnino, Mario Lazzari, Michele Pastorelli

Politecnico di Torino, Dipartimento di Ingegneria Elettrica Industriale, Italy

Introduction

Since induction motors represent the more consistent electrical motors used in industrial environment, the increase of the induction motor efficiency represents a crucial aspect to obtain a consistent energy saving. In the USA, from 1997 the electric motor producers have to build high efficiency motors in according to the protocol promulgated by the Energy Policy Act (EPACT). This protocol defines the minimum value of the efficiency for general purpose, 2 or 4 poles, 400V, 50÷60 Hz, three-phase induction motors with a rated power in the range 1÷200 HP [1]. In Europe, the European Committee of Manufacturers of Electric Machines and Power Electronics (CEMEP) has realized a similar protocol to which the electric motor producers can to agree in voluntary manner [2]. An increase of the motor efficiency means a reduction of the loss contributions. The total motor losses can be divided in the stator and rotor Joule losses, in the iron losses and in the mechanical losses. A viable way to improve the efficiency of an induction motor is the reduction of the iron loss contribution, because this choice does not require a complete change of the magnetic sheet shape. A possible solution for getting a strong reduction of the iron losses could be the use of good quality magnetic materials (with specific iron losses in the range 2.7 ÷ 3.5 W/kg at 1.5 T) instead of the low or medium quality magnetic materials typically used in general purpose induction motors (with specific iron losses from 5.3 W/kg up to 8.0 W/kg at 1.5 T). In a previous paper [3], the authors have shown that a reduction of the magnetic material specific iron losses does not produce the expected iron loss reduction. In fact, better magnetic materials from the losses point of view have a worse behaviors form the permeability point of view. As a consequence, the increases of the magnetizing current and the corresponding Joule losses reduce the benefit of lower iron losses. The authors have shown that, the better compromise to reduce the no load losses, is the use of oriented sheets [4], but their correct use cannot be proposed in general purpose motors, due to the material cost and the complexity of the core assembly procedure. It is important to underline that the use of the same stator and rotor magnetic sheet shapes adopted in standard efficiency motor realizations represents the principal objective for the design of higher efficiency induction motor. As a consequence, it is possible to obtain a consistent reduction of the production cost because the realization of new expensive tools for punching new stator and rotor slots can be avoided. If the motor radial geometry not change, a loss reduction is usually researched with “no tooling cost” strategies (i.e. the used magnetic materials, axial lengthening of the motor core, special stator windings with low space harmonic contribution).

Effects of the magnetic material annealing on the iron losses

A viable solution for reducing the no-load losses is the annealing of the stator core. This technological process allows a recovery of the magnetic and energetic properties after the punching process requested for the stator and rotor sheet realization. Tests on punched toroidal cores have shown very interesting results with a reduction of the specific iron losses up to 50% after the annealing process. In the following a short description of the magnetic samples and of the performed tests is reported. Five cores have been realized using a magnetic material with a thickness of 0.5 mm and a specific losses of 8.0 W/kg at 1.5 T. The first sample is assembled overlapping 45 rings having the inner diameter equal to 120 mm and the outer diameter equal to 200 mm. The other four samples have been realized with same external dimensions of the first wound core, but using a different number of concentric rings. In particular, the second sample is assembled with two concentric rings (as sketched in Fig.1), the third one with three rings and so on for the other cores. It is evident that all the cores have the same external dimensions, but each sample presents a different length of the punched edges.
The magnetic and energetic properties of the five cores have been measured considering the same instrumentation set up usually adopted to perform the measurement on the Epstein frame. Fig. 2 shows the specific iron for the first sample with one ring and for the fifth sample with five rings. The specific iron losses are compared with the same ones obtained from a conventional Epstein frame test. It is important to remark that the magnetic sheet used in the Epstein frame is from the same coil of the toroidal core samples and that the Epstein strips are not realized with a punching process but through a sheared process. The specific iron losses curves show a significant increase for both the toroidal cores with respect the Epstein frame. On the contrary the difference between the values obtained from the one ring sample and the five rings sample is not so evident. It is important to underline that, the results obtained from the other three toroidal cores have not been reported in Fig.2 because positioned inside the restricted area between the two curves reported in the picture. An anneal process has been applied to five cores, twins with the first five ones. The anneal process has been performed in vacuum conditions at a temperature of 930 °C Celsius for four hours. The annealed cores have been tested with the same procedure previously described. The energetic properties have been compared with the same ones of the punched cores. The obtained results are reported in Fig.3 and Fig.4. From the

energetic point of view, it is well evident the consistent reduction of the specific iron losses after the anneal process up to the 60%. It is interesting to highlight that the specific iron losses after the anneal process are lower than the specific iron losses measured on the Epstein sample. The magnetic properties are better after the anneal process too, in particular for flux density lower then 1.5 T, even if in heavy saturation the differences can be neglected. These results are very interesting because an anneal process seems to have a positive effect also on a sheared material like the magnetic sheets used in the Epstein frame (Fig.3). Obviously these results cannot extended to every type of non oriented magnetic materials, but they justify the interest to next and more deep analysis on a wide set of magnetic materials.

Effects of the magnetic material annealing on the motor no-load losses

The obtained results on the annealing process (a reduction of the iron losses up to 60%) have been considered an interesting starting point to foresee an iron loss reduction in annealed stator cores. In particular, the effects of the annealing process on a stator core have been analyzed comparing two twin motors (one with a standard stator, the second one with an annealed stator). The two motors have been tested in no-load conditions at synchronous speed using a synchronous motor with the same poles number of the motors under test. The conventional iron losses have been computed using the well know relation:

$$P_{ir} = P_{no-load} - 3\, R_s\, I_0^2 \tag{1}$$

P_{ir}	iron losses;	$P_{no\text{-}load}$	measured no load losses;
R_s	stator resistance;	I_0	measured no load current.

The obtained results on the two motors are reported in Fig.5, where the iron losses reduction due to the anneal process is evident but it is not so consistent as in the toroidal samples. Several considerations can be made for justifying this result but the authors opinion is that the additional losses contribution in no load condition cannot be neglected.

Critical considerations on the conventional iron losses

The results previously discussed lead to the necessity of a critical approach to the meaning of iron losses, taking into account that the iron losses computed by Eq. 1 have to be considered as conventional iron losses. In fact, in no load condition the power balance can be written as:

$$P_{no-load} = P_{ir} + 3R_s\, I_0^2 + P_{ad-0} \tag{2}$$

P_{ad-0} additional losses in no load condition.

The additional losses are due to several contributions and in first approximation the additional losses can be considered as a function of the torque squared. This assumption is used as a reference by the standard IEEE 112b for the industrial motor efficiency measurement. The additional loss analysis is not the target of this work, but among the several contributions to the additional losses, the rotor cage joule losses due to the harmonic current induced by the winding spatial harmonics can be considered a not negligible contribution in no load conditions. In particular, this contribution can be forced to be zero using a non conductive rotor cage. In order to investigate on the weight of this type of additional losses a rotor with a plastic cage has been cast. A motor, assembled using the annealed stator core and the plastic cage rotor, has been tested in no load condition at synchronous speed. The conventional iron losses has been obtained using Eq. 1 and the results have been compared with the previous ones in Fig.5. It is evident that the additional losses due to the rotor harmonic currents are not negligible and they are comparable with the measured loss reduction in the annealed stator. Taking into account that the rotor harmonic current losses are just one of the additional losses involved in no-load condition it is evident that the conventional iron losses cannot be interpreted as the actual iron losses produced in the stator motor core. In particular, during the no load test (also at synchronous speed) the actual iron losses weight on the measured ones could be reduced by the presence of the no load additional losses. A direct consequence is that every improvement on the actual iron loss reduction could be shadowed by the additional losses contribution, leading to a not well clear interpretation of the obtained results. In addition, the presence of not negligible additional losses during the iron losses measurement can be considered as a reasonable justification of the low accuracy in the prediction of the iron losses in electrical machines [5]. In fact, also using sophisticated approach (i.e.: finite element method, hysteresis and eddy current models, etc.) the iron losses prediction is affect by consistent errors, with the predicted iron losses always lower than the measured ones. A wide discussion on this problem is reported by the authors in [6], where several references can be found. A correct answer to this problem is probably that, it is not correct to compare the predicted iron losses with measured ones because it is not possible to separate the actual iron losses by the no load additional ones.

Conclusions

In the paper the effects of the annealing process on the iron losses reduction has been analyzed on toroidal samples and on induction motors. The obtained results have shown that the consistent decrease of the iron losses in the toroidal samples (up to 60 %) is not present in the induction motors. These results have been used as starting point for a critical approach to the meaning of iron losses in electrical machine. In particular, taking into account how the international standard tests compute the conventional iron losses, the limits of the measurement procedures have been focused. Using an "ad hoc" rotor cage, the effects of the additional losses due to the harmonics currents induced in the rotor cage by the spatial field harmonics have been measured. In particular, the amount of this additional losses are not negligible in comparison with the conventional iron losses, and in the considered motors under test, their value is comparable with the iron loss reduction due the annealing process. A direct consequence of these results is that improvements on the actual iron loss reduction could be shadowed by the additional losses contribution, leading to a not well clear interpretation of the obtained results. In other words, the no-load tests cannot be considered as a resolutive test for the actual iron loss measurement, but just as a common accepted method for the conventional iron loss determination.

References

[1] A. Balducci, "EPACT Legislation - The United States experience of minimum efficiency standards for induction motor", Energy Efficiency Improvements in Electric Motors and Drives, Ed. Springer - Verlag, ISBN 3-540-67489-6.

[2] P. Bertoldi, G. Kuehneund, "The European negotiated agreement to improve motor efficiency", Energy Efficiency Improvements in Electric Motors and Drives, Ed. Springer - Verlag, ISBN 3-540-67489-6.

[3] A. Boglietti, P. Ferraris, M. Lazzari, F. Profumo, "Analysis of the magnetic material quality influence on induction motors energetic performance", ICEM 86, Munchen 8-10, September 1986, Germany.

[4] A. Boglietti, P. Ferraris, M. Lazzari, F. Profumo, "Preliminary consideration about the adoption of unconventional magnetic materials for induction motors", *Electrical Machines and Power System*, Vol. 21, N. 4 July-August 1993.

[5] G. Bertotti, A. Boglietti, M. Chiampi, D. Chiarabaglio, F. Fiorillo, M. Lazzari, "An improved estimation of iron losses in rotating electrical machines", *IEEE Transactions on Magnetics,* November 1991, Vol 27, N.6

[6] A. Boglietti, A. Cavagnino, M. Lazzari, M. Pastorelli, "Effect of the punch process on the magnetic and energetic properties of soft magnetic material", IEEE International Electric Machines and Drives Conference 2001, 17-20 June 2001, Boston, USA.

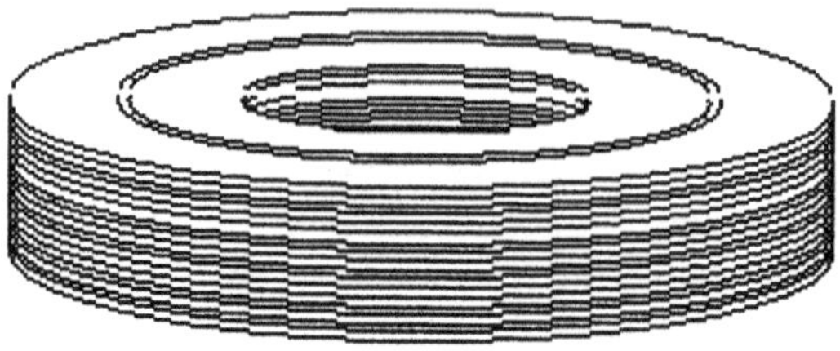

Fig. 1. Geometrical characteristic of the core assembled using two concentric rings

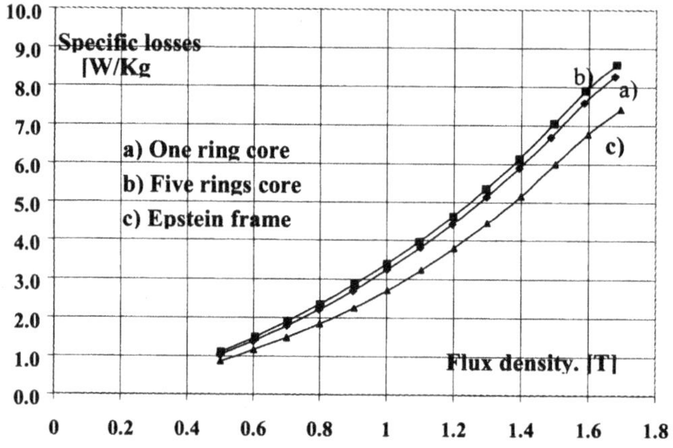

Fig. 2. Specific losses characteristics

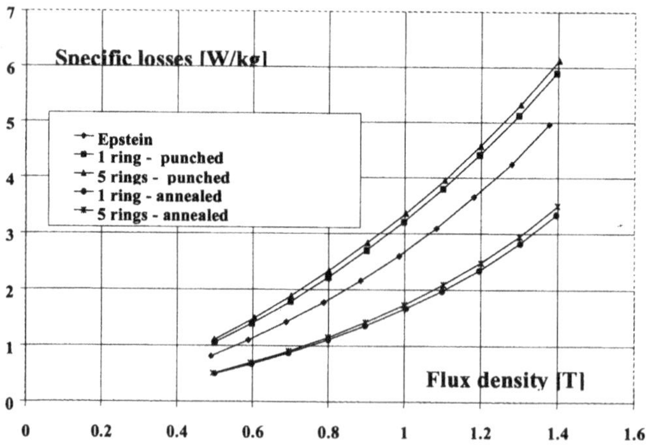

Fig. 3. Comparison between the specific iron losses in the punched and in the annealed samples

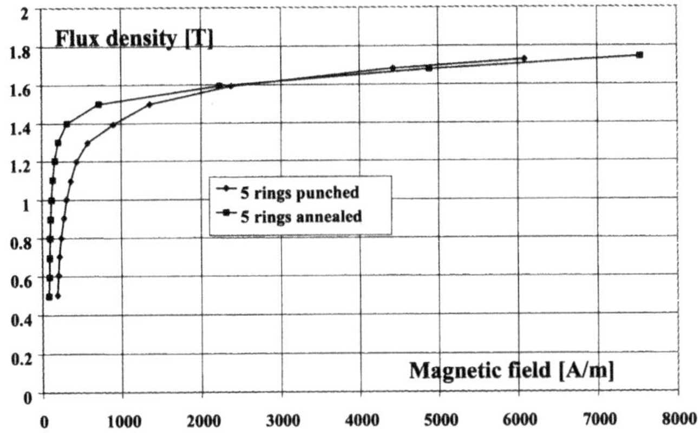

Fig. 4. Comparison between the magnetizing curves in the punched and in the annealed samples

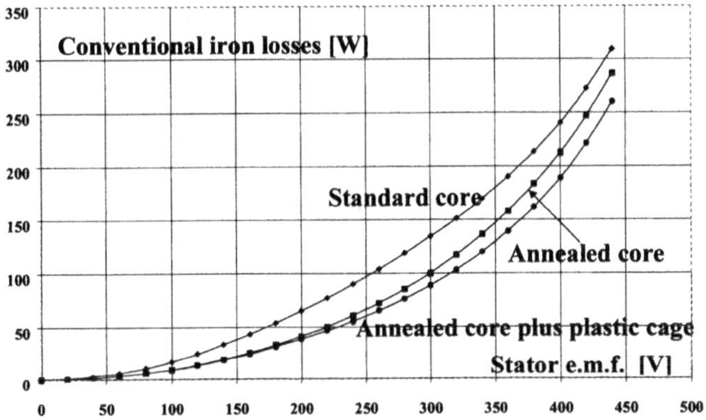

Fig. 5. Conventonal iron load losses comparison

The Efficiency of Electric Drives Supplied from Frequency Converters (PWM) and Parasitical Effects Occurring in Induction Motors

Adam Decner

Research and Development Center of Electrical Machines "KOMEL"
al. Roździeńskiego 188, 40-203 Katowice, POLAND
e-mail: info@komel.katowice.pl, http: www.komel.katowice.pl, fax: +48 32 2599948

1. Introduction

Much savings of energy can be gained by control of the rotational speed of many drives, especially of flow machines. For squirrel cage motors driving these machines it can be done by converter feed of the stator. On the other hand motors supplied with distorted voltage (from PWM converter) have worse working parameters than when grid feed. What's more, parasitical effects occur in their construction elements; some effects can be dangerous for the maintaining staff.

Supplying motor from the frequency converter causes:

- increasing of power losses in electric and magnetic circuit, what reduces efficiency of the motor, increases the temperature-rise of stator and rotor windings,
- increasing of vibration and noise emitted level,
- increasing of electromagnetic interference emitted by motor,
- rising of partial discharge between coils of stator winding,
- excitation of additional variable torques in electromagnetic torque,
- shaft voltages and bearing currents,
- voltages on the frame of the motors when isolated from the earth,
- generating to the grid higher harmonics of current.

Target of tests done at the Laboratory of Research and Development Center of Electrical Machines KOMEL, at Katowice, Poland was to determine how the motor works when supplied from 8 different converters. An asynchronous squirrel cage motor: P_N=11kW, f_N=50Hz, I_N=21A, U_N=400V, n_N=1460rpm was the object of tests. The converters were delivered by different European and Japan manufacturers.

2. Tests

Tests were performed when the motor was grid feed (sinusoidal voltage) and when supplied from eight different converters – each one from different manufacturer.

Each converter was matched for:
- optimum for mention motor,
- U/f=const. characteristic.

All converters were set as below:
- voltage of first harmonic was U_{1h}=400V when frequency was f=50Hz (at nominal load),
- range of adjustment of frequency was 1-50Hz by U/f=const. and if it was possible without deflection characteristic at low frequency,
- switching frequency was 8kHz.

Converters has got no input and output filters.

Motor was tested in two ways:
1. Torque=const and f=var.
2. Torque=var. and f=50Hz.

For all cases active power absorbed by motor, torque on the shaft and rotational speed were measured. Tests were made after reaching the thermal stability of the motor.

Measures concerned also:
- voltages on the motor's frame isolated from the ground,
- bearing currents,
- emitted noise,
- vibrations.

When motor was supplied from grid the frequency was 5, 25, 50Hz.

2.1 Idle Run and Load of Motor When F=Const

Tests were made supplying motor from grid and from 8 converters. For frequency 1, 5, 25, 50 and 120Hz motor was supplied from converters. For frequency 5, 25 and 50 of sinusoidal voltage waveform motor was supplied from synchronous alternator. A schema of measuring circuit is shown Figure 1.

The results of tests are shown as diagrams.

The Figure 2 shows results of measured form factor of voltage waveform supplying tested motor at idle run and at the 3-ed Figure during load test of the motor.

Form factor is defined as:

For voltage:

$$thd_U = \frac{\sqrt{U^2 - U_{1h}^2}}{U};$$

for current:

$$thd_I = \frac{\sqrt{I^2 - I_{1h}^2}}{I},$$

Where U, I – True RMS of measured parameter,
U_{1h}, I_{1h} – True RMS of first harmonic of measured parameter.

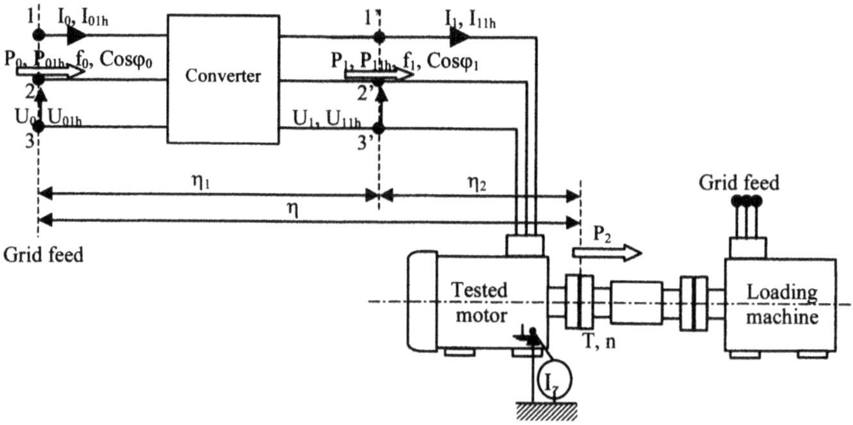

Fig. 1. Schema of measuring circuit

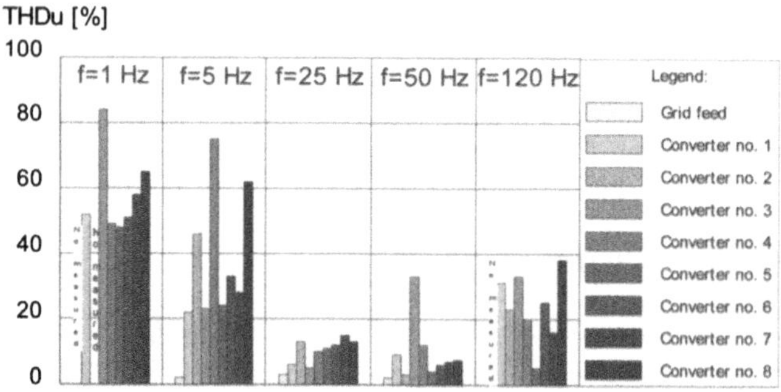

Fig. 2. Values of thd_u of voltage, when motor at idle run

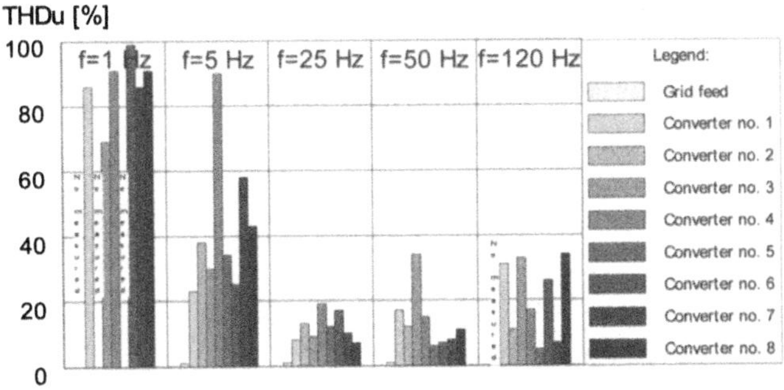

Fig. 3. Values of thd_u of voltage, when motor loaded

For frequency 1Hz to 120Hz the thd_U was from 4% up to 84% at idle run, and from 5% up to 99% when the motor was loaded.

Supplying motor with so distorted voltage has the consequences in increasing of power loses in the core and causes additional losses (in magnetic and electric circuit) and in construction elements as well. The very high voltage on the frame of the motor was observed.

The next figure shows results of measured form factor of current, when the motor was loaded.

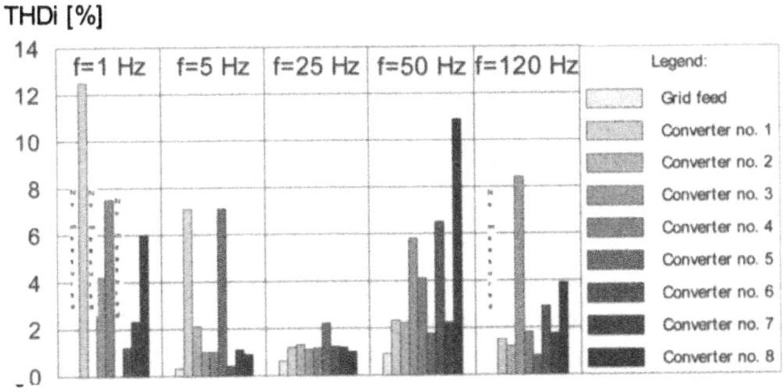

Fig. 4. Values of thd_I of current, when motor load run

For frequency 1Hz up to 120Hz thd_I was from 0.4% up to 12.5%. For all converters at 50Hz thd_I was higher then at other frequencies. Higher harmonics of current cased that true RMS value of current increased. This is the reason of increasing power losses of the stator winding.

2.2 Changing of Load when the Motor Run at 50Hz

Tests were performed by supplying the motor from grid and from 8 converters. The frequency of first harmonic of voltage was constant f=50Hz, but torque on shaft was changed from $\approx 0T_N$ up to $1,5T_N$.

Figure 5 presents results of measured thd_U of voltage supplying motor and Figure 6 results of measured thd_I of current.

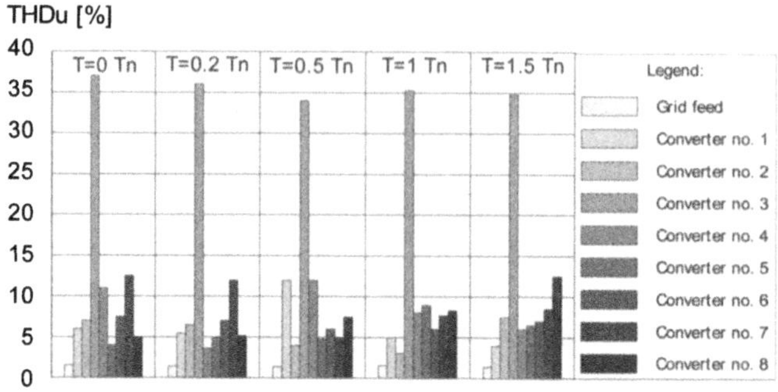

Fig. 5. Values of thd_u of voltage when torque changed from 0 to 1.5 T_N

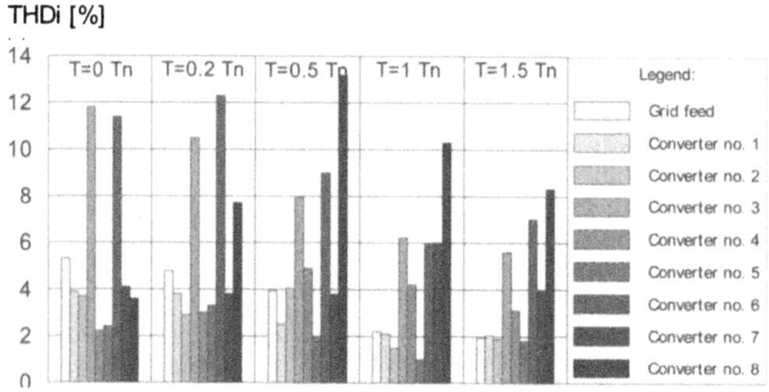

Fig. 6. Values of thd_I of current when torque changed from 0 to 1.5 T_N

When torque was controlled from $\approx 0T_N$ up to $1,5T_N$ value of thd_U changed:

- from 1.4% up to 1.5% (when motor was supplied from grid),
- from 3.0% to 12% (when motor was supplied from converters), for one converter the thd_U value was 34% up to 36%.

Value of thd_I was changed from 1% up to 13% and most frequently decreased with increasing of the load.

2.3 Determination of Efficiency

The tests were performed using 2 analyzers as in schema shown at Figure 1. The motor was loaded with nominal torque and supplied with nominal frequency and nominal voltage (first harmonic). The figure below presents efficiency of tested converters, motor supplied from converter and the complete drive.

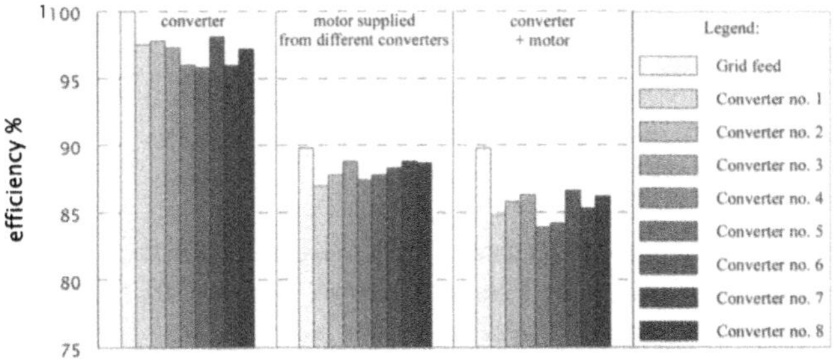

Fig. 7. Efficiency of converter, motor supplied from converter and converter with motor

Efficiency of converters was between 95,8% - 98,1% when efficiency of motor was 89,9% when the motor was grid feed. The picture indicates that the efficiency of complete drive is lower then the efficiency of the motor when it is supplied from grid. The efficiency of converter 100% means efficiency of the grid.

2.4 Voltages on the Frame of the Motor Isolated from the Earth, Bearing Currents

Supplying motor with distorted voltage causes parasitical effects, which are very undesirably. When the motor is supplied with symmetric sinusoidal voltage the parasitical effects occur only in transient state, or do not occur at all. These effects are:

- bearing currents,
- shaft voltages,
- voltages on the frame of motors isolated from the earth,
- earth currents of motors connected to earth.

There is real danger to appear a very high voltage on the frame of the motor isolated from ground, when motor is supplied from converter.

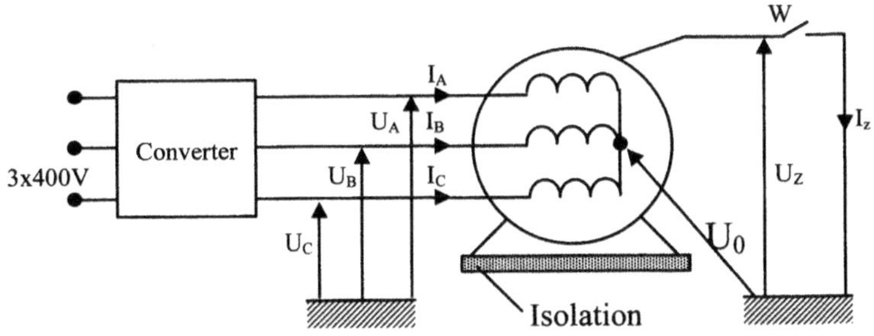

Fig. 8. Schema to measure voltage on frame and earth current

Where: I_A, I_B, I_C – currents,
U_A, U_B, U_C – phase voltages,
I_Z – earth current,
U_Z – voltage on the frame of motor isolated from the earth,
U_0 – voltage between star point and earth.

Tests were made at the idle run and the motor was isolated from the ground. Earth current was measured by shunt. The Figure 9 shows diagram of measured true RMS value of voltages on the frame when the motor was supplied from converters.

When the motor was supplied from grid with sinusoidal voltage waveform the value of this voltage was 0 (zero). The 3rd and 6th harmonic and switching frequency of waveform of supplied voltage were dominant.

The Figure 10 shows true RMS value of earth current. There is very difficult to find any regularity in those measures.

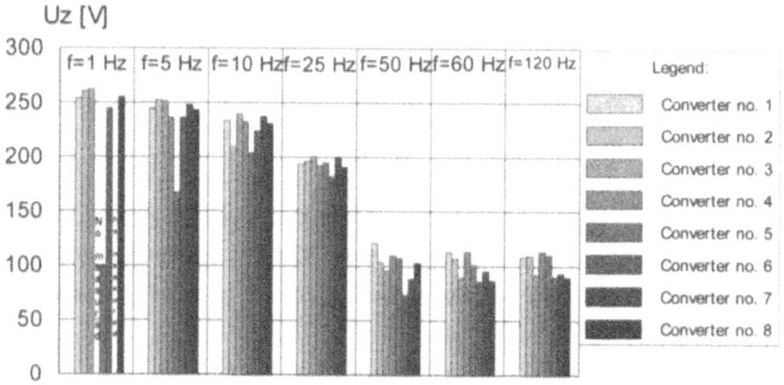

Fig. 9. True RMS value of voltages on the frame

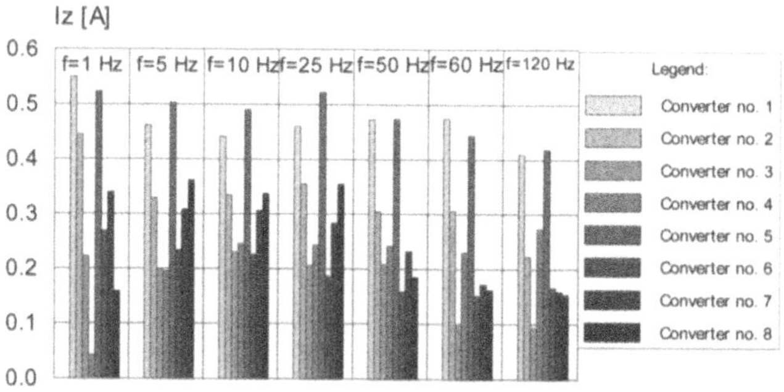

Fig. 10. True RMS value of earth currents

Asymmetrical character of voltage waveform is a source of parasitical effects. This means that instantaneous value of voltage waveform is not equal to zero.

Important is the fact, that voltage at the frame of the motor is very high (over 250V) when the motor is isolated from the ground. Analysis shows that dominant

is switching frequency. After shortcut frame to earth was done, true RMS value of the current about 0.5 A was observed. These undesirable effects may:
- be dangerous to the staff,
- destroy measuring apparatus,
- bad influence on protection apparatus,
- spark between elements of frame, what may cause explosion,
- faster destroy of isolation in motor.

3. Conclusions

- When converter and motor are compatible (good matched) supplying from converter does not decrease much efficiency of the motor.
- Converter has always bad impact to supplying grid.
- Converter feed drive has serious impact to devices in a close area.
- Voltage on the motor's frame is dangerous, that's why good connection to the ground is very important.

Although converters decrease efficiency of the motor, the total energy savings using variable speed drives can be huge. It is reached by flow control and adjusting the flow for the actual demand.

It must be noticed the motor supplied with distorted voltage waveform causes a serious danger for people and apparatuses, since high voltage on the housing may occur.

References

[1] M. Jakubiec, K. Kłapciński, A. Polak "Straty mocy, sprawność zjawiska pasożytnicze układu napędowego silnik indukcyjny-przemiennik", BOBRME Komel Nr 62/2001

[2] A. Polak "Dodatkowe zjawiska pasożytnicze występujące w silnikach indukcyjnych zasilanych napięciem odkształconym", SME 2000

Stray-Load Losses Analysis in Copper Squirrel Cage Induction Machine

Christophe Paris[1], Olivier Walti[1-2]

[1] FAVI SA , 14 rue Louis Deneux, BP 5 – 80490 HALLENCOURT, France, phone: +33 3 22 66 20 56 – fax: +33 3 22 66 20 63 – e-mail: parisc@esiee-amiens.fr

[2] ESIEE, Amiens, 14 quai de la Somme, BP 100 – 80083 AMIENS Cedex 2, France, phone: +33 3 22 66 20 00 – fax: +33 3 22 66 20 10 – e-mail: waltio@esiee-amiens.fr

Abstract

Our study aims at showing the importance and the influence of stray losses in copper squirrel cage induction machine's dynamic behaviour. Its operates an important role on the machine's efficiency but also on electromagnetic torque characteristic's look.
First, the question is to know the origin of these stray losses and after, identify the effects on the torque in terms of speed characteristic and on the efficiency in terms of effective power characteristic. Finally it could be interesting to see how to make it optimal.

Key words

Squirrel cage machine, copper die-casting, stray-load losses, electromagnetic torque, high efficient machines

1. Introduction

This present study involves a foundry, FAVI, located in the north of FRANCE, specialised in copper pressure die-casting, and which is anxious to use its expertise in squirrel cage machine rotors manufacturing and ESIEE-Amiens, an engineer school.

Generally, squirrel cages for induction machines are made of aluminium die-casting in slots of the rotor laminated frame, developing in that way, with the two short-circuit end rings, a compact entity. But FAVI Company is focused on an industrial process to manufacture copper pressure die-casting rotors (Fig.1). In fact, the single technology used to manufacture squirrel cages with copper conductors is to assemble shaped bars into the magnetic circuit slots and to weld at the extremities of both end rings.

Squirrel cage induction machine's performance can be valued by two basic characteristics : the electromagnetic torque C in terms with speed N and the efficiency η in terms with effective power P_u.

Fig. 1. Copper die-casting rotor

The improvement of these performances depends on the control of losses in the induction machine. According to IEEE standard 112-1996, those losses can be listed into four main parts : copper losses (rotor and stator), windage losses, iron losses and stray-load losses [3]. The last are the most difficult to be quantified. So this study is devoted to analyse it, in order to control it at best and make it optimal. Moreover the study of losses reduction in induction machines corresponds the general pattern of the control of energy needs, in other words in the high efficiency machines' development[1].

The study was developed into three stages : first, we tried to locate stray-load losses origin and then identify its effects on C(N) and $\eta(P_u)$ characteristics. Finally we will evoke some ways to reduce these losses.

2. Stray-Load Losses' Origin

During the production of die-casting rotors[4], the bars and the end rings are casting in the same time, the laminations package doing a mold. The isolation between bars and laminations is not perfect. Also some currents between two successive bars (fig.2a and 2b) and between bars and end rings can flow through the magnetic circuit.

These contacts are the siege of eddy currents. Induced currents whose frequency is $f_r = g.f$ and which generate an alternative flight flux from field lines closed themselves transversely into slots(Fig. 3a). Actually the variation of this flux induces in the rotor's solid bars, some eddy currents at frequency f_r which direction is set by Lenz law(Fig.3b).

3. Influence on Electromagnetic Starting Torque

These currents operate an important role on electromagnetic torque characteristic's look, mainly on transient part [2]. When admitting they grow with the frequency, we can easily think they have a significant influence on electromagnetic starting torque.

This way, current density in bar sections is not uniform and the current is concentrated in the high part of the bar (skin effect) (Fig.3 a and b). This starting current tends to change the single cage machine into a double-cage machine and consequently to increase the value of starting torque.

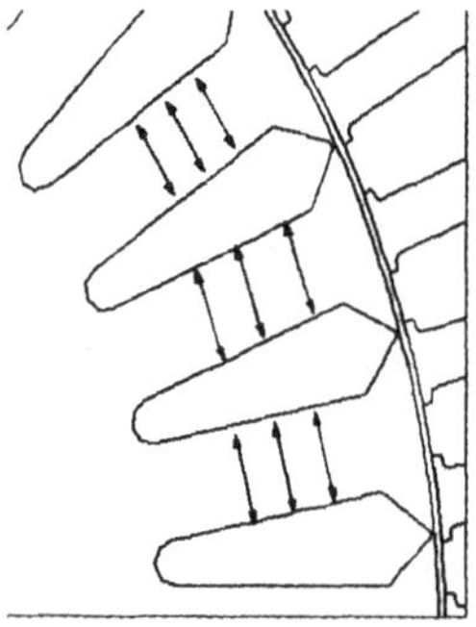

Fig. 2.a. Inter bars currents

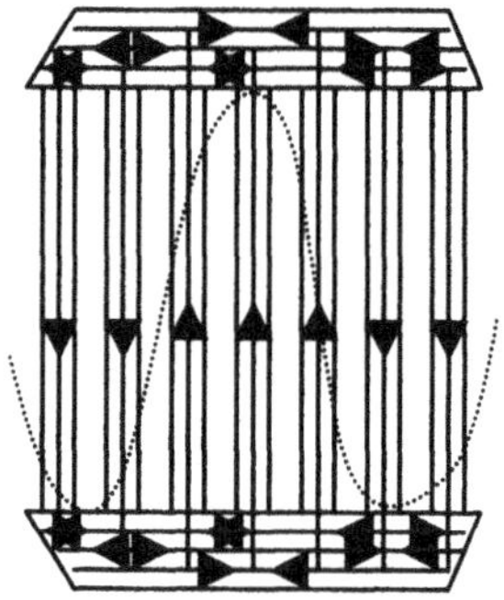

Fig. 2.b. Currents distribution into the cage

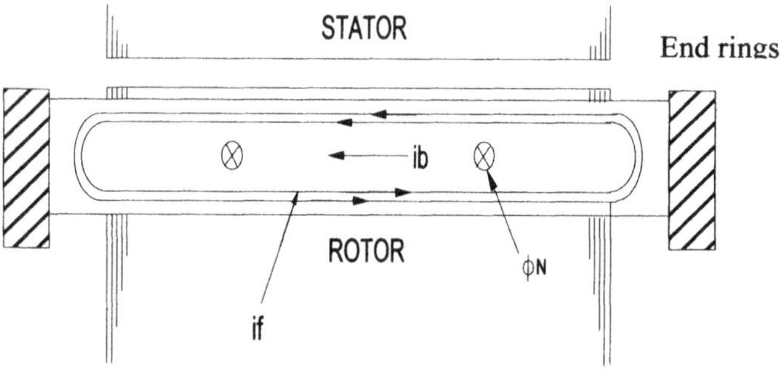

Fig. 3.a. Currents distribution into a bar at start

Electromagnetic torque is not a measurable variable; so we have to estimate it. Two solutions come at us. The first one is a calculation using Park transformation and a reduced-order model [5]. But when calculating electromagnetic torque in theory, all these currents aren't taken into account and a large gap can be noticed on values.

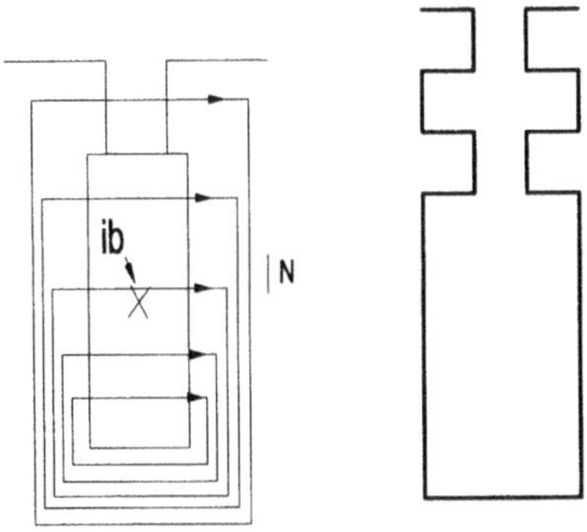

Fig. 3.b. Currents distribution into a slot at start

In order to determine the real impact of these hypothesis on starting torque values, a test bench has been set in the research labolatory of FAVI and settled in ESIEE-Amiens, allowing us to do direct starts of induction machines.

This bench is made up by a squirrel cage induction machine pulling a DC machine and a mass of inertia lengthening the starting time.

At the time, two methods open to us and allow us to determine starting torque values. First, thanks to a power analyser including a torque estimator, we are able to get the instantaneous electromagnetic torque characteristic. And then after filtering, we obtain the real electromagnetic torque representation (Fig.4 and Fig.5). The torque estimator uses the Park transformation and calculate from non reduced-order differential equations and takes supplementary losses into account. The other test permits to validate the starting torque value. It's about using a torque meter during a locked rotor test at nominal voltage steady state. The torque meter gives us the starting useful torque ,which is very closed to the real starting electromagnetic torque.

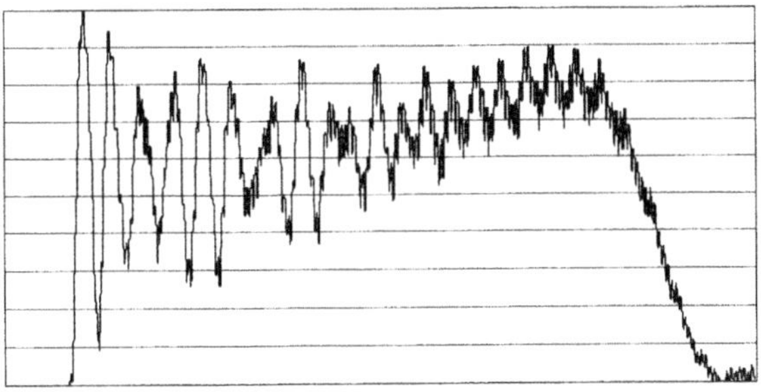

Fig. 4. Instantaneous electromagnetic torque representation

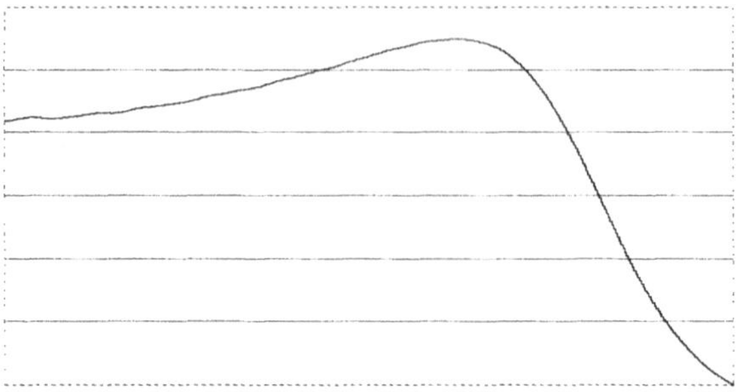

Fig. 5. Filtered electromagnetic torque representation

4. Influence on Efficiency

An important aim is to obtain high efficiency induction machine which expression is as followed :

$$\eta = \frac{Pout}{Pout+losses} \tag{1}$$

Our tests are made according to IEEE standard 112-1996 –method B- using the following losses separation (Fig. 6 And Table 1) :
Also, a stray-load losses increase is followed by an efficiency's abatement of one or two points on a 5.5 kW induction machine.

The results of our tests are listed in a table for each rotor (Table 2). This allows us to see if a rotor parameter has any influence on one of the dynamic results.

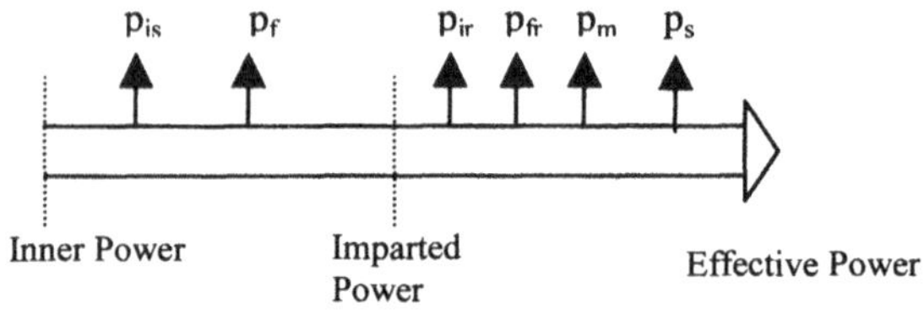

Fig. 6. Power repartition in induction machine

Table 1. Losses separation

p_{js}	stator copper losses
p_{fr}	stator iron losses
p_{js}	rotor copper losses
p_{fr}	rotor iron losses
p_m	friction and windage losses
P_{sl}	stray-load losses

Table 2. IEEE tests results

C (Nm)	U (V)	I (A)	P (W)	N (min-1)	g (%)	Pu (W)	rdt (%)	cos phi	DT (°C)
36.32	**400.0**	**12.05**	**6657**	**1469.9**	**2.01%**	**5629**	**84.56**	**79.75**	**96.01**
54.76	399.8	17.14	10317	1448.9	3.41%	8389	81.31	86.92	97.22
45.09	399.9	14.30	8346	1460.8	2.61%	6959	83.38	84.25	100.03
36.33	400.5	12.00	6631	1470.6	1.96%	5611	84.63	79.65	101.56
32.32	400.4	11.12	5925	1474.1	1.73%	5029	84.88	76.83	101.48
26.58	400.1	9.93	4894	1479.2	1.39%	4150	84.81	71.11	101.01
17.90	399.8	8.46	3385	1486.4	0.91%	2807	82.93	57.78	100.04
12.53	400.5	7.85	2488	1490.7	0.62%	1972	79.29	45.68	97.69
9.06	399.4	7.50	1912	1493.2	0.45%	1429	74.73	36.85	95.52
3.14	400.6	7.38	973	1497.4	0.17%	506	51.97	19.01	94.35

Pjs(W):	661.60	Pjr(W):	116.42	Pfer(W):	202.21	Pmec(W):	16.01	Psup(W):	31.75
	9.9%		1.7%		3.0%		0.2%		0.5%

5. Results

All the tests have been done on a tetrapolar 5.5kW-50Hz-400V squirrel copper cage induction machine(Table III). Differents rotors are tested with only one parameter changed, allowing us to know the influence of each parameter on dynamic behaviour and in particular on stray losses.

Also, on the subject of electromagnetic torque characteristic, with all of these values, we are able to calculate the electromagnetic torque representation and then we have just to compare it with the real one determined experimentally (Fig. 7).

Table 3. Nominal Specifications Of The Studied Motor

Rated Shaft Power	5.5 kW
Supply	400 V
Frequency	50Hz
Current	11.7 A
Number of poles	4
Stator slots	48
Rotor slots	28
Stator number of phases	3
Stator number of turns in series	160
Stator winding's coefficients	0.9576
Rotor number of phases	28
Rotor number of turns in series	0.5
Rotor winding's coefficients	1

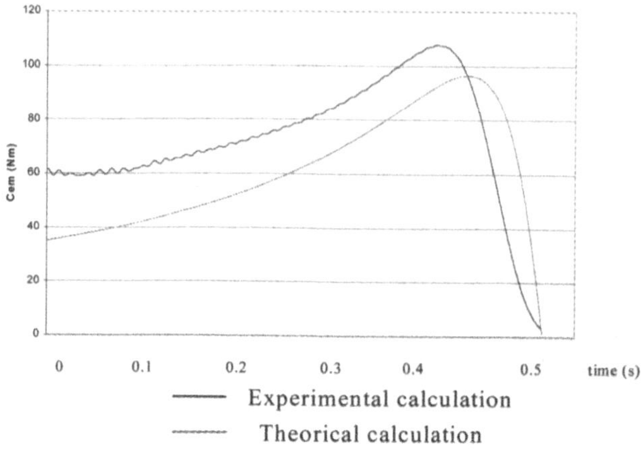

Fig. 7. Comparison between theoretical and practical methods to determine electromagnetic torque representation

We can easily notice an important gap between the two characteristics. The more we are in transient part greater is the gap. In fact the real starting torque is above than the one calculated by an approched method of nearly 50%.

During steady state, the difference is minor because there's much less eddy currents phenomenon. The two electromagnetic torque representations on Fig.7 have been plot according to time. So the slope in steady part are not perfectly comparable. If they were, the two part of curves would be practically disconcerted.

On the contrary, concerning efficiency, stray losses are not favourable. Differents parameters can have an influence on eddy currents and so, on supplementary losses. We can modify the slots'shape or the end rings' high in order to reduce bars-laminations contacts. But effectively, the quality of laminations and differents treatments, thermal as an example, are also a particular interest in the reduction of these losses and the increase of efficiency (Fig.8).

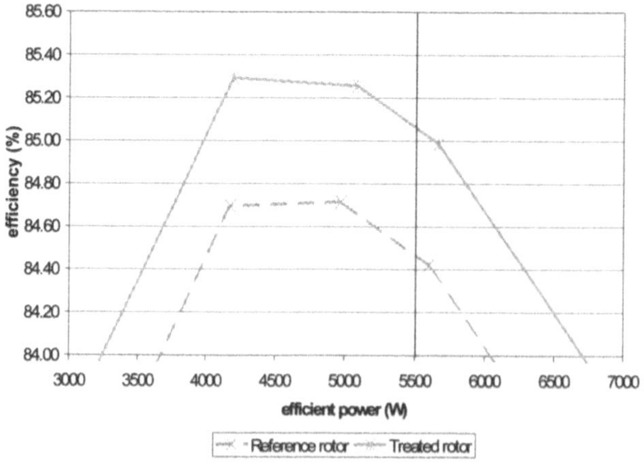

Fig. 8. Treated rotor efficiency

Finally it would seem that helix pitch plays also an important role on efficiency characteristic. Indeed it reduces the noise during accelerations and contributes to a more regular start and acceleration. But according to the tests, the stray losses of the rotors with no helix pitch are reduced by half in comparison with the reference rotor (29 Watts against 56.7 Watts). This difference feels fit up to the efficiency characteristics (Fig. 9).

6. Conclusion

Our study enters within the context of squirrel cage induction machines' efficiency. The difficulty to control stray-load losses points out that is essential to do extensive research in order to make the two main characteristics, which are electromagnetic torque and efficiency, optimal.

An accurate determination of eddy currents points also out the gaps between used models and real tests' results.

Finally, it is generally said that aluminium rotor provides a starting torque plainly higher in comparison with one made of copper. But these observations lean on simplified hypothesis. So the difference exists but is lower than we can say.

In fact, as we have seen, the main factor of the gap is stray losses. Also, these losses are superior for a copper rotor and the error comitted when calculating the torque from a reduced model is a lot higher than with an aluminium rotor.

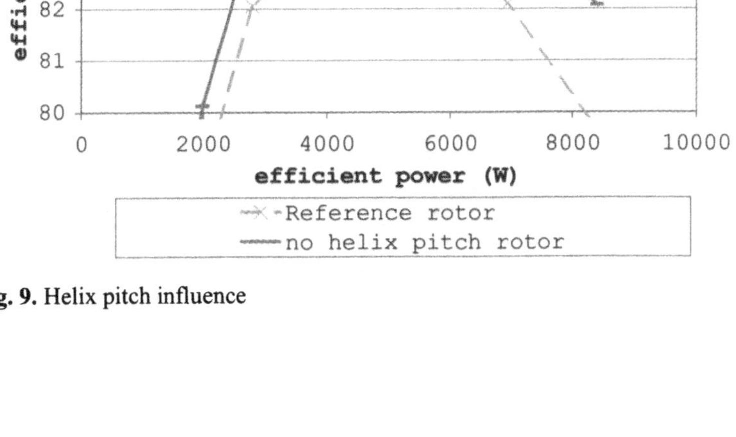

Fig. 9. Helix pitch influence

7. References

[1] J.C. ANDREAS, "Energy-Efficient Electric Motors – Selection and Application", *Ed. Marcel Dekker.*

[2] Norme IEEE Std 112-1996 , "IEEE Standard Test Procedure for Polyphase Induction Motors and Generators."

[3] T CHEVALIER, "Modélisation et mesure des pertes fer dans les machines électriques, application à la machine asynchrone", *Thèse de doctorat, décembre 1999.*

[4] A. ANSEL, O.WALTI, J.F.BRUDNY, "Influence de la coulée sous ssion d'une cage d'écureuil en cuivre sur le comportement magnétique d'une machine asynchrone triphasée ", *Thèse de doctorat, juin 2001.*

[5] T.L. SKA VERENINA, "The Theory and Effects of Neglecting Stator and Networks Transients," *Ph.D. Thesis, Purdue University, December 1979.*

Assessment of Efficiency of Low Voltage, Three Phase Motors

P. Van Roy, R. Belmans

Department of Electrical Engineering ESAT/ELECTA K.U.Leuven
Kasteelpark Arenberg 10, B-3001 Leuven-Heverlee, Belgium
Tel.(+)32.16.32.10.20; Fax(+)32.16.32.19.85
e-mail: ronnie.belmans@esat.kuleuven.ac.be

Introduction

Three-phase, low-voltage squirrel cage induction motors are the most commonly used electric motors in industry. They can be found from a few hundred watts up to several megawatts. The induction motors are characterised by data provided by the manufacturer at rated speed, power, voltage, current, power factor and efficiency. In the past, the efficiency value was of minor importance. Nowadays, with the growing emphasis on energy conservation the efficiency value has become very important and even dominant for applications in industry. Efficiency data by manufacturers are measured or calculated according to certain standards. The main differences between these standards are discussed in this paper, as well as the process of making accurate measurements. Knowing the real efficiency at rated and partial load, instead of the fictional IEC efficiency, allows choosing the motor that has the highest efficiency in the actual process where it is used. Annual energy savings of up to 50% of the purchase cost can be achieved. Some practical examples of energy saving using variable speed drives are also given.

Efficiency standards

Worldwide, there exist several standards for testing electric machinery. For induction motors, the three most important ones are IEEE Standard 112 [1], JEC 37 (Japan) and IEC 34-2 [2]. At this instant, a new IEC standard, the IEC 61972, is under development [3]. In most European countries, the standards are harmonised to IEC 34-2.

The efficiency value obtained from the different testing standards can differ by several percent, as will be shown by the measurement results. This seems in contradiction with the theoretically simple definition of the efficiency:

$$\eta = \frac{\text{power out}}{\text{power in}} = 1 - \frac{\text{overall losses}}{\text{power in}} \tag{1}$$

The second form allows the correction to a specified ambient and reference motor temperature, by correcting the individual loss components.

The first four loss components are stator and rotor copper losses (P_{stator} and P_{rotor}), iron losses (P_{Fe}), and friction and windage losses ($P_{fr,w}$).

P_{Fe} and $P_{fr,w}$ are determined by a no-load test, the copper losses are calculated based on stator resistance, slip and input power measurements under load. The values of the copper losses are corrected to the reference motor temperature.

Additional load losses have been the subject of numerous studies. In fact, these are all the losses that are not covered by the above mentioned loss components and therefore, they may be expressed as:

$$P_{addit} = \left(P_{in} - P_{out}\right) - \left(P_{Fe} + P_{stator} + P_{rotor} + P_{fr,w}\right) \tag{2}$$

The main difference between the standards emerges from the way in which the fifth loss component, the additional load losses, is treated.

Since a direct measurement of the additional load losses requires the rotor to be removed, the only correct and practical method is to measure both input and output power, and to calculate the difference between the overall loss and the four known loss components, using (2). This is the method used in IEEE standard 112 - Method B. The resulting values are linearised and corrected for zero additional losses at zero load.

Historically the torque measurement required by this method was difficult, and therefore the IEC 34.2 standard assumes a standard value for the additional load losses at rated load of 0.5% of the input power, proportional to the current squared at lower load levels. The new proposed IEC 61972 standard gives two possibilities for the assessment of the additional load losses. The first one is a determination by means of the measured output power, as in the IEEE 112-B; the second one attributes a fixed amount to every machine of the same rated power.

The Japanese JEC standard 37 completely neglects the additional load losses.

Measurements

Eighteen induction motors of 11, 55 and 75 kW were tested, in a program in collaboration with Laborelec, the Belgian national laboratory for electricity. The motor manufacturers were ABB, ACEC, Brook-Hansen, Leroy-Somer, Samco, Siemens and WEG.

Efficiency was measured according to IEEE 112-B and IEC 34.2 standards, and compared with the European catalogue value, normally based on the IEC standard. Although some manufacturers' values are reasonably accurate, others overestimate the efficiency by 3 to 4%. Measured additional load losses vary from 1.5 to 2.3 % of input power for the 11 kW motors tested (7 motors), from 0.4 to 3.0% for the

55 kW motors (6 motors) and from 0.9 to 2.7% for the 75 kW motors (5 motors). Values in the same range were found for some other motors tested in other power ratings. Similar values can be found in [4] and other references.

Given the achievable measurement accuracy, differences in efficiency of less than 0.5 % are not necessarily reliable. Therefore, it is proposed to use a standardised list of values, as used in the NEMA nameplate labelling standard MG1-12.542.

Furthermore, a difference in partial load efficiency, at 50 and 75 % load, of up to 5 % was found for motors with a difference in rated load efficiency of only 1 %. As motors are usually overdimensioned, the partial load efficiency is often even more important than the rated load efficiency, and should be mentioned as well.

Given the relatively low purchase and high energy cost, differences in efficiency of a few percent can lead to very high annual savings compared with the purchase cost, even up to 50%.

Measurement problems

The biggest problem in making accurate efficiency measurements of grid connected motors in a laboratory situation is properly setting up and aligning the motor, torque transducer and load. A very high accuracy is required in current, voltage, power, torque and speed measurement.

When measuring frequency converter drives, EMC becomes a problem. Proper cable connections and earthing schemes are needed, as well as shielding the measurement equipment. A major issue is measuring the distorted voltages and currents at input or output of the PWM frequency converter. Power analysers like the Voltech PM3000A allow high frequency measurements – incorporating the fast voltage variations – while synchronising the measurements with the much lower basic frequency.

In industrial environments, torque measurements can be performed using strain gauges. Applying these strain gauges obviously requires a drive stand still, which can be a problem for the industrial process. Contactless data transmission can be analogue or digital, but may be sensitive to EMC.

Additional load losses in the IEC 60034-2 and the new IEC 61972 standard

In the proposed new IEC standard, the additional load losses are either measured in a procedure similar to the IEEE method, or taken as a certain percentage of input power. For this second option, the only difference with the existing IEC 34.2 is that this percentage depends on the motor size, e.g. 1.9% of input power for the 55 kW motors. This may be a better average value than the 0.5 % of IEC 34.2, but this is entirely irrelevant, as indicated by the (real) example in Table 1.

Table 1. IEC 34.2, IEC 61972 and real additional load losses and efficiency

	IEC 34.2		IEC 61972		Real	
Motor	P_{addit} [%]	Eff. [%]	P_{addit} [%]	Eff. [%]	P_{addit} [%]	Eff. [%]
A	0.5	93.0	1.9	91.6	2.5	91.0
B	0.5	92.0	1.9	90.6	0.5	92.0

For example: motor A, labelled as a high-efficiency motor, has a 93 % efficiency according to the old IEC standard or 91.6 % according to the new version. The real efficiency is 91%, because the additional load losses are actually 2.5% for this motor. Motor B of the same power rating has old and new IEC efficiencies of 92% and 90.6% respectively, and a real efficiency of 92%, because the additional load losses are 0.5%. The comparison of both motors according to a method using an assumed or average value for the additional load losses is futile. This comparison would indicate motor A to be the "best". In fact, motor B is more efficient. Clearly, this method is extremely unfair to the motor manufacturers, and to the customers who want reliable information on motor efficiency. It is not important what fixed or average value of additional load losses is used: it is the difference in additional load losses among motors of the same rating that is relevant.

Determining the additional load losses by means of the measurement of the output power, as in the IEEE method and the first option in the new proposed IEC standard is the only relevant method, since additional load losses can differ significantly between motors of the same rating.

A regrettable example of ignoring the additional load losses can be found in the 1998 IEA Hi-Motors Competition. The 75 kW ABB motor designed for -and winning- the competition states reductions in stator and rotor copper losses, iron losses, and friction and windage losses, but states a "0 W reduction" in additional load losses. The method used was IEC 34.2. The ABB motor's efficiency is not disputed, but actual measurements have shown that some so-called "high-efficiency" motors – sold at premium price - achieve a high IEC 34.2 efficiency with, however, additional load losses of up to 3%.

Australia

In Australia, the Minimum Energy Performance Standards (MEPS) regulations came into effect on 1 October 2001 [5]. It defines "Minimum Efficiency" and "High Efficiency" requirements, which are the same as the proposed EU eff2 and eff1 standard respectively. All motors must meet the minimum efficiency levels. Labelling as High Efficiency is voluntary.

Australia recognises the difference between both standards, and imposes different minimum efficiencies depending on which standard is used. Figure 1 shows the required efficiency for High Efficiency motors, depending on whether the IEEE or IEC standard is used to determine efficiency. The higher requirements when using the IEC method can be very disadvantageous for manufacturers whose motors have low additional load losses. The requirements will not change for at least 4 years.

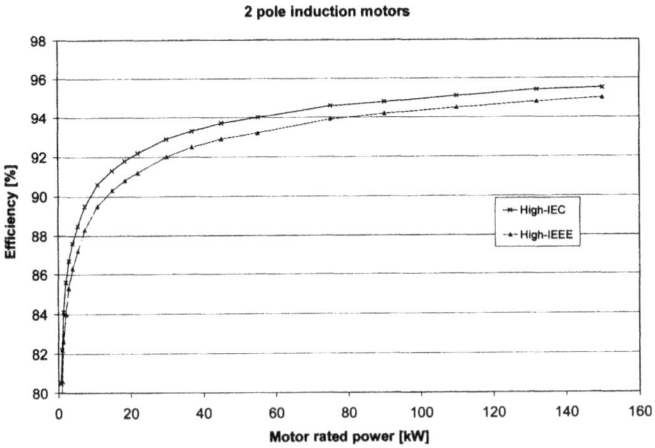

Fig. 1. Requirements for "High Efficiency" motors, Australia

Energy savings with variable speed applications

A variable speed drive, using a standard induction motor and a frequency converter, can lead to annual energy savings of up to 50%, e.g. in pump and ventilator drives, when compared with fixed speed on/off, throttle or bypass systems. At present, no standards are available to determine the efficiency of these drive systems. In this study, the efficiency of a drive is found by dividing the output by the input power. Most converters have efficiencies of 95 to 98%, even at relatively small loads. The average drive efficiency is 2% lower than the grid connected motor efficiency. However, this is less important than the energy saving potential. Some examples of industrial projects where traditional throttle valve or bypass installations were replaced by variable speed drives are given.

- 20 bar – 160 kW pressurizer: 23% savings.
- Circulating water from tower, normally operating without throttling: 24% savings due to more appropriate flow.
- Bypass at high pressure: 13% savings.
- Bypass at lower pressure: 40% savings.
- Several circulating pumps, 55 – 90 kW range: 40 to 46% savings.

Conclusions

The present IEC standard - or any method with fixed allowance for additional load losses as still present in the new IEC standard - does not provide reliable efficiency values. The additional load losses must be measured, and can in no way be replaced by any kind of fixed allowance, as the differences in additional load losses between motors of the same rating are too significant to be ignored. The difference from one motor to another can exceed 2% of input power, far exceeding the measurement error.

The partial load efficiency is just as important as the full load efficiency with respect to energy consumption, as motors are often overdimensioned. A difference in efficiency can have a large impact on the overall energy consumption. Annual energy savings of 50 % of the purchase cost are possible. In industrial processes, variable speed drives can reduce the energy consumption by more than 40 %, especially when fans or pumps are used.

Acknowledgement

The authors are grateful to the Belgian "Fonds voor Wetenschappelijk Onderzoek Vlaanderen" for its financial support of this work.

References

[1] "IEEE Standard Test Procedure for Polyphase Induction Motors and Generators," IEEE Std 112-1996, IEEE Power Eng. Society, New York, NY.

[2] "Rotating electrical Machines - Methods for determining losses and efficiency of rotating electrical machines from tests," IEC Std 34-2:1972.

[3] Method for determining losses and efficiency of three-phase, cage induction motors, IEC Std 61972:1998

[4] C.N. Glew, Efficiency Measurement Testing Standards Stray Losses, the Key to Efficiency determination, in A. De Almeida, Energy efficiency Improvements in Electric Motors and Drives. Berlin, Springer, 1997, pp. 249-265.

[5] http://www.energyrating.gov.au/manufacturers/motor1.html

[6] Bonneville Power Administration United States Department of Energy, Energy-efficient Electric Motor Selection Handbook, 1993.

[7] Gerald G. Gray, Walter J. Martiny, Efficiency testing of medium induction motors, a comment on IEEE Std 112-1991, IEEE Transactions on Energy Conversion, VOL 11, No. 3, pp 495-499, September 1996.

[8] W. Nürnberg, R. Hanitsch, "Die Prüfung elektrischer Maschinen," Springer-Verlag, 1987.

[9] H. Auinger, Considerations about the Determination and Designation of the Efficiency of Electric Machines, in A. De Almeida, Energy efficiency Improvements in Electric Motors and Drives, Berlin, Springer, 1997, pp. 284-293.

[10] A. De Almeida, Energy Efficient Motor Technologies, in Energy efficiency Improvements in Electric Motors and Drives, Berlin, Springer, 1997, pp. 1-17.

Permanent Magnet Assisted Synchronous Reluctance Motor: an Alternative Motor in Variable Speed Drives

Jorma Haataja, Juha Pyrhönen

Lappeenranta University of Technology

Abstract

The design and performance of a permanent magnet assisted synchronous reluctance motor (PMASynRM) is investigated using finite element method. The motor is called PMASynRM when the share of reluctance torque is significant compared to the PM electrical torque. The advantages of adding PMs to the synchronous reluctance motor rotor construction are the increased motor power factor and thus reduced motor stator ohmic losses. The ohmic losses represent the majority of the motor total losses. The advantage of reluctance torque is the decreased need of expensive permanent magnet (PM) material, which makes this solution thus cheaper than the respective permanent magnet motor. The advantages of PMASynRM compared to the induction motor (IM) are the synchronous rotation speed, possibility of sensorless rotation control, higher power factor and better efficiency.

1 Introduction

An interesting alternative for today's high efficiency variable speed drives is the PMASynRM drive, which belongs to the family of brushless synchronous AC motor drives. These drives consist of sinusoidal current driven machines, which use a quasi-sinusoidally distributed AC stator winding and frequency converter.

The design of brushless synchronous AC motors has not matured yet to a similar degree, on which e.g. the designing of induction machines is. During the recent years there has been a considerable increase of interest in using brushless synchronous AC machines in applications where previously asynchronous machines have been used. Traditionally brushless synchronous AC machines have been used in low-power servo drives, but with the recent development in both permanent magnets and power electronics also medium and large power drives are gaining more interest. The suitability of a brushless synchronous AC motor to a particular application is, however, dependent on the motor design. Selecting the parameter to fulfill the requirements of applications is clearly an optimization problem. The main types of brushless synchronous AC motor are surface mounted permanent magnet (SPM) or interior permanent magnet synchronous motor (IPM), synchronous reluctance motor (SynRM) and the combination of these two types namely permanent magnet assisted synchronous reluctance motor (PMASymRM). Even though the IPM motor has a reluctance torque component the share of it is remarkably lower than electric torque especially if the pole-pair number is big. The family of brushless synchronous AC machines is shown in Fig. 1. The dotted areas represent steel, the white areas represent air or some nonmagnetic material and dark grey areas represent permanent magnets.

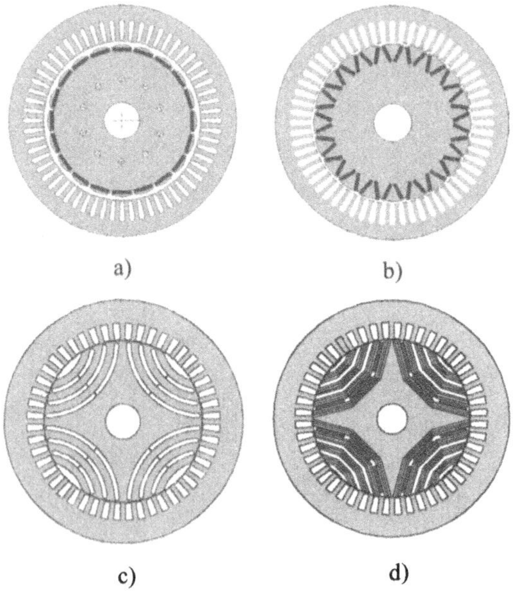

Fig. 1. Brushless synchronous AC machines a) surface mounted permanent magnet motor, b) Interior permanent magnet motor, c) Synchronous reluctance motor and d) permanent magnet assisted synchronous reluctance motor.

The SPM motor (Fig. 1a) operates purely with magnet alignment torque. The permeability of the permanent magnet material is close to air and thus the IPM motor (Fig. 1b) produces also reluctance torque because the inductances of d- and q- axis are not equivalent. Because the magnetizing inductance is inversely proportional to the pole-pair number the number of the pole pairs of SynRMs may not be too high. According to the literature the pole-pair number of three is the highest value used in SynRM type of motors and the most commonly used values are one or two. The pole-pair numbers of PM motors do not have such a limitation.

Brushless synchronous AC machines are quite robust high power-density machines capable of operating at high motor and frequency converter efficiencies over wide speed. These kinds of drives are used in applications such as traction and spindle drives. The drives for these kinds of applications require high torque-density and wide constant power speed range. High overload capacity, low torque ripple, limited temperature rise and feasibility of an axis-position control [1,2 - 4].

2 Synchronous Reluctance Motor

The performance of the SynRM is mainly dependent on the d- and q-axis magnetizing inductance values. The reluctance torque is proportional to the difference of these inductances ($L_{dm} - L_{qm}$) while the saliency ratio L_{dm}/ L_{qm} determines many of the motors' operating characteristics. It has been shown by Betz [4] that the saliency ratio determines the machines power-factor, sensitivity to parameter variation and field-weakning performance. In particular, the ideal constant-power speed range is approximately half the saliency ratio. Thus to achieve a 4:1 constant power range a saliency ratio of at least 8 is theoretically required. To maximize the d-axis' inductance, the flux guides should ideally have the same shape as the flux path shown in Fig 2a and to minimize the q-axis flux linkage, the flux barriers should be perpendicular to the flux lines as shown in Fig. 2b.

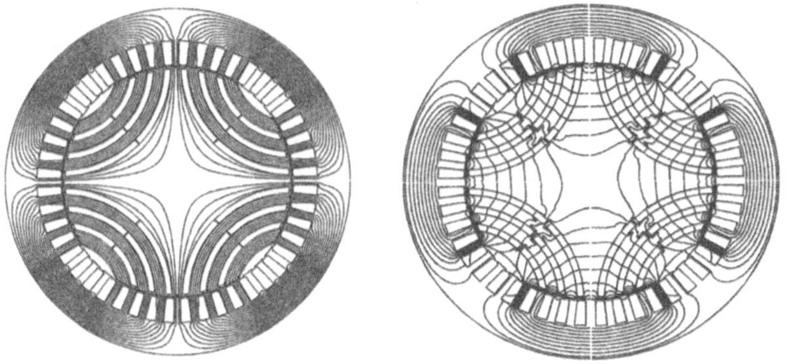

a) b)

Fig. 2. Synchronous reluctance motor flux plots a) in d-axis and b) in q-axis with $|i_d| = |i_q|$

The intrinsic magnetizing saliency ratio ξ_{mi} is the maximum possible saliency ratio for a given motor geometry. It is derived by considering the inductance of motor with no rotor and with a solid iron rotor [5]. Neglecting leakage inductance the saliency ratio is given by:

$$\xi_{mi} = \frac{L_{dm}}{L_{qm}} = \frac{a r_1}{p\delta} \tag{1}$$

where a is the rotor insulation ratio, that is, the ratio of the insulation thickness to the lamination plus insulation thickness, r_1 is the air-gap radius, p is the number of pole-pairs and δ is the radial air-gap length. Eq. (1) shows that the air-gap length and the pole-pair number exert the main influence on the saliency ratio. If the rotor radius increases the motor magnetizing inductance increases. If the average ratio of flux barrier is assumed constant the increase of rotor radius enables the increase of air-gap length without decreasing the saliency ratio. The possibility to increase air-gap length without decreasing the saliency ratio would be a benefit for higher power range motors. The small air-gap is a consequence of trying to make a compromise between the power factor and the efficiency. A small air gap is needed to produce high saliency ratio and power factor. At the same time rotor surface losses caused by the time harmonics start to weaken the efficiency.

The actual saliency ratio is considerably lower than this due to practical factors such as Carter's coefficient, the finite pole-arc, and stator leakage inductance. The measured unsaturated saliency ratio ξ_m is generally in the range $0.2\ \xi < \xi_m < 0.4\ \xi$. [5]. The practical constant power speed range of synchronous reluctance motor is generally much lower than that predicted from half the saliency ratio. By the investigations of [6] the reduction was mostly due to magnetic saturation.

2.1 Protype SynRM Compared to IM

In this chapter the efficiency and the loss distribution of a 30 kW prototype SynRM are compared to the properties of a same size IM in a variable speed drive. The model of this SynRM is used as the base of the PMASynRM structure. The SynRM's no-load values were obtained using sinusoidal voltages. The load tests were done using a DTC converter.

The SynRM's rotor structure was designed and manufactured in a standard 30 kW four-pole IM frame. Both IM and SynRM were supplied by a prototype sensorless direct torque controlled (DTC) variable speed drive. The SynRM's rotor was of a transversally laminated type. To eliminate cogging the rotor was skewed by one stator slot pitch. To evaluate the effects of the changes made, all motor configurations were tested in laboratory using IEEE 112 B test method. The variables investigated were the length of air-gap and the effect of stator semi-magnetic (μ_r=2 … 5) slot-wedges. The motors' identification numbers and the constructions are given in Table I.

Table 1. The test SynRMs

Motor no.	Features
1	Sinusoidal fed standard induction motor, air-gap 0.8 mm
2	Inverter fed standard induction motor
3	Inverter fed standard induction motor + semi magnetic slot-wedge
4	SynRM, air-gap length 0.3 mm
5	SynRM, air-gap length 0.3 mm + semi magnetic slot-wedge
6	SynRM, air-gap length 0.4 mm + semi magnetic slot-wedge

The motor number 1 is the original standard induction motor, the stator construction of which was, for simplicity, used also in the SynRM's construction. The effect of the variable speed drive to standard induction motor's efficiency was investigated in motor drive system 2. The frequency converter used to run the motor was ABB's ACS600, which applies the direct torque control (DTC) system.

The effect of semi magnetic slot wedges was investigated both in an IM (motor 3) and in SynRMs (motors 5 and 6). To achieve high enough power factors for the SynRM the air-gap length of the SynRM has to be selected smaller than the respective value of IM. The effect of the air-gap length has been investigated only for SynRMs (motors 5 and 6). The measurement results of different losses based on the load test are given in Table II.

Table 2. Estimated and IEEE temperature corrected motor data at rated power 30 kW

	Motor 1	Motor 2	Motor 3	Motor 4	Motor 5	Motor 6
Efficiency [%]	92.4	91.6	91.9	90.2	91.4	91.8
Power factor	0.80	0.78	0.79	0.77	0.82	0.72
Iron loss [W]	422	422	461	870	605	690
Mechanical loss [W]	270	270	182	300	202	189
No-load losses	692	692	643	1170	807	879
Stator ohmic loss [W]	1051	1192	1210	1418	1286	1390
Rotor ohmic loss [W]	650	650	650	-	-	-
Additional loss [W]	74	213	145	690	650	425
Load depend losses	1775	2055	2005	2108	1936	1815
Total loss [W]	2467	2747	2648	3278	2743	2694

According to Table II the total losses of the best SynRM construction are quite equal with the losses of the respective induction motor construction in the converter supply. The largest losses are found with the SynRM construction without semi-magnetic slot wedges and with the lowest air-gap length. Using the slot wedges the frequency converter supplied IM's total losses decrease to a bit lower level than the losses of the sinusoidally fed motor without slot wedges. Thus even though the iron losses are higher and power factor lower using slot wedges, the mechanical and slip losses decrease the total loses so that they get a lower value than the losses of the respective sinusoidally fed IM. This energy saving is because of the semi magnetic slot wedges.

As it can be seen in Table II, the best SynRM's (motor 6) total losses due to the absence of the rotor copper losses are quite equal with the losses of the standard induction motor in converter supply. The best efficiency is achieved with the air-gap length 0.4 mm using semi-magnetic slot wedges. According to Table II, as no changes to motor air-gap length are done, the semi-magnetic slot wedges decrease both induction and SynRMs' load losses because of the lower additional losses. In variable speed drive the standard 30 kW induction motor's efficiency is quite equal to the efficiency of the SynRMs with slot wedges.

3 Permanent Magnet Assisted Synrm

One possibility to improve SynRM's performance – especially its power factor and efficiency – is to add permanent magnets to the SynRM's rotor [3,4-6]. Using PM material may improve the motor power factor and thus reduce the motors' current. The reduced currents also reduce motor stator ohmic losses. On the other hand the reluctance torque reduces the quantity of expensive PM material needed and makes this solution cheaper than the respective permanent magnet motor.

In the PMASynRMs, investigated here, two different solutions to insert the PM material to the rotor are shown. The difference between these two lay-outs arises from the positioning of the PM material with respect to the SynRM's direct and quadrature axes.

In the first method PM material was used to support the direct axis flux and in the second method PM material was used to oppose the quadrature axis armature reaction. The layouts of these two different types of PMASynRMs are shown in Fig. 3.

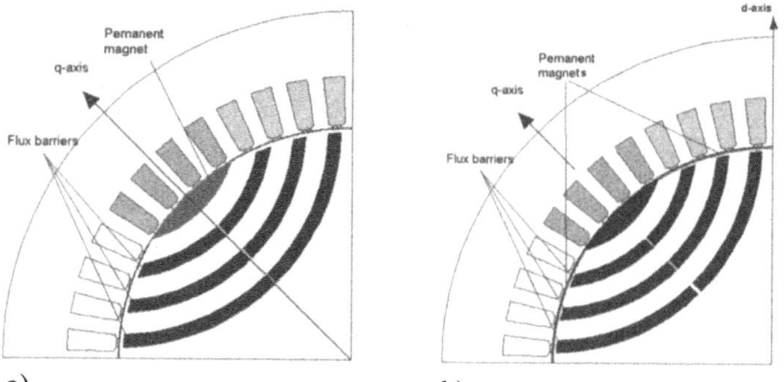

a) b)

Fig. 3. The lay-out of a PMASynRM, where a) PM opposes the q-axis armature reaction, b) PM increases the d-axis flux.

The PMASynRMs' vector diagrams have been presented in Fig. 4.

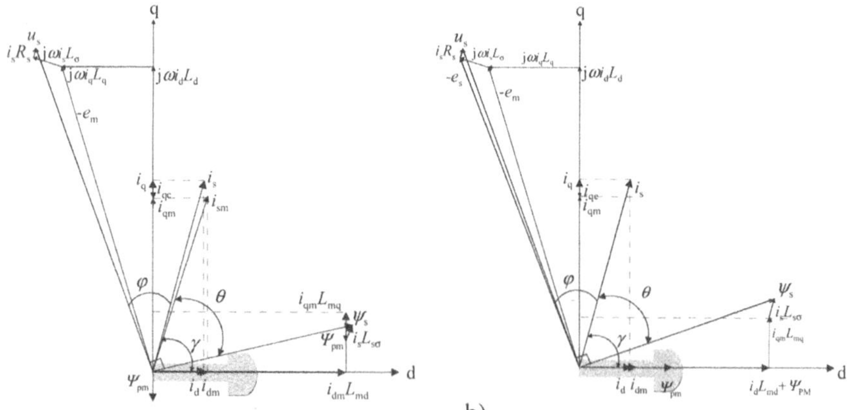

a) b)

Fig. 4. The vector diagram of PMASynRM, where the PM a) decreases the q-axis flux and b) amplifies the d-axis flux.

The d-axis PMASynRM torque components as a function of current angle are presented in Fig. 5. The current angle is between the current space vector and the d-axis. Also the SynRM torque is presented.

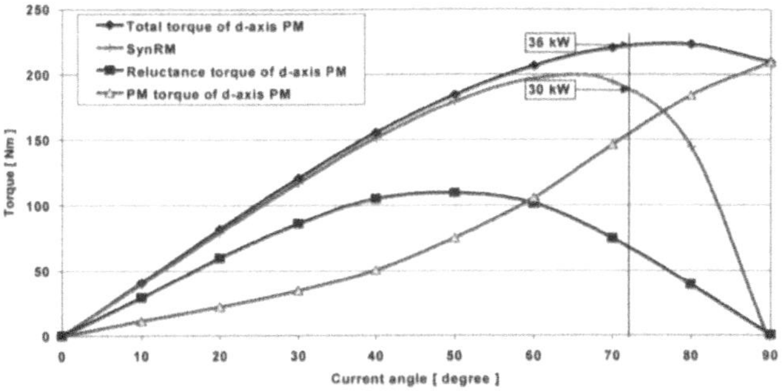

Fig. 5. The PMASynRM,s torques with motor nominal current in the case, where the usage of the PM material increases d-axis flux.

Using PM material on the rotor surface increases the effective air-gap length and reduces thus the reluctance torque remarkable. Because of the bigger air-gap, the maximum reluctance torques is achieved close to 45° current angle, which is the theoretical value for non-saturated machines. The PM increases also the motor's maximum torque value. The motor nominal power is increased from 30 kW to 36 kW using the same load current value by using PM material on the d-axis. The amount of PM material used is 2.2 kg. The PM effect on the q-axis has been studied by moving PMs into different flux barrier locations as shown in Fig. 6.

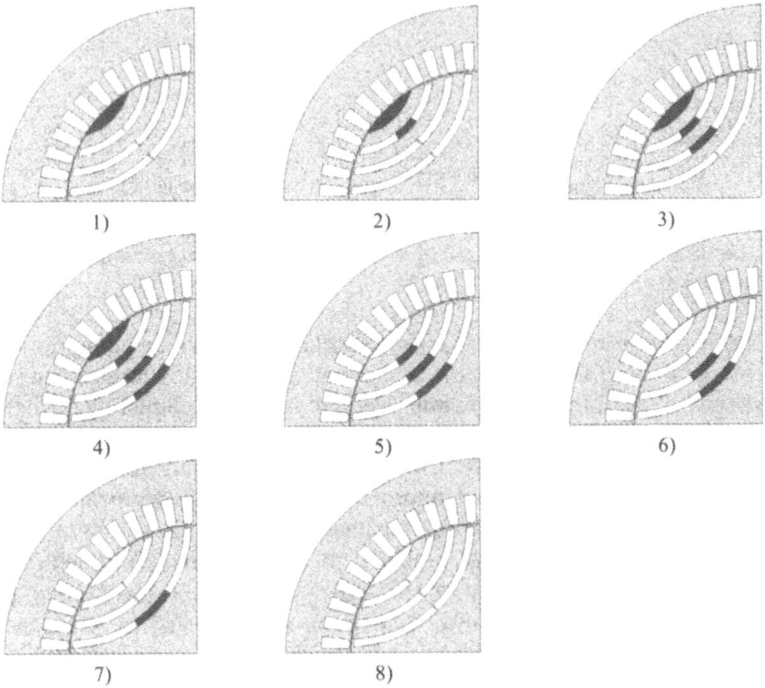

Fig. 6. Arrangement of PM in the SynRM rotor construction 1) PMASynRM-1, 2) PMASynRM-2, 3) PMASynRM-3, 4) PMASynRM-4, 5) PMASynRM-5, 6) PMASynRM-6, 7) PMASynRM-7 and 8) SynRM

The nominal current torque values of the motor types given in Fig. 6 as a function of current angle are presented in Figs. 7a and 7b.

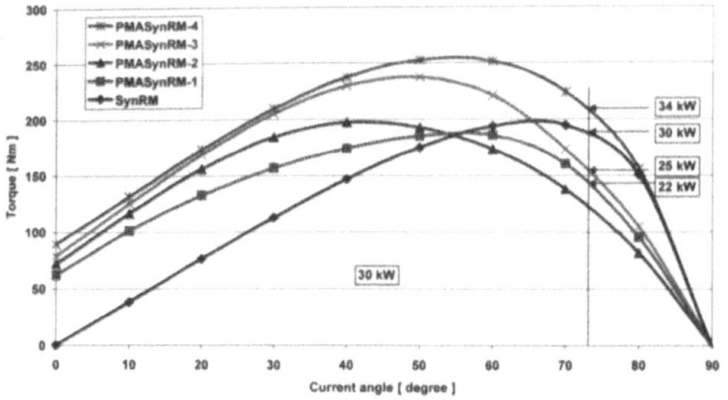

a)

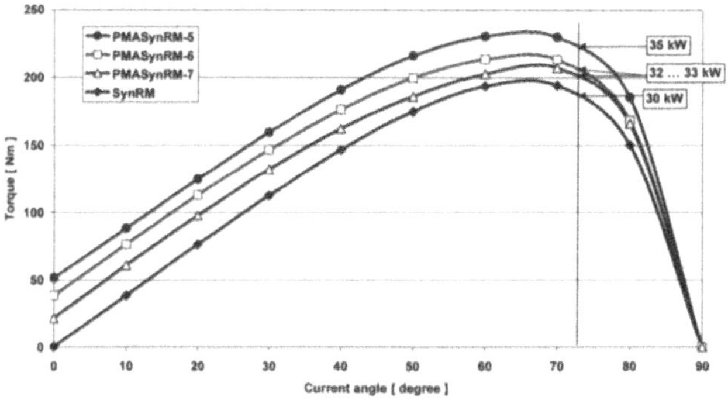

b)

Fig. 7. The FEM calculated torques as a function of current angle representing the constructions of Fig. 6.

By adding PM material in the q-axis according to Fig. 6.6 the motor power may be increased from 30 kW to 35 kW at nominal current and with the same losses. The amount of PM in Fig, 6.1 is about 1 kg and in Fig 6.5 about 3 kg. When comparing Figs. 7a and 7b it can be noticed, that if the magnets are placed close to the rotor surface the torque achieved at low current angle is higher but on the other hand the maximum torque is not improved and may even be lower than in the SynRM. The improvements in the motor may be seen in Fig. 7b by noticing the power values. The initial SynRM produces 30 at its nominal point with 73° current angle with the efficiency 0.918 and power factor $\cos\varphi$ = 0.72. The PMASynRM produce higher power with same losses. Thus the efficiency and power factor are improved. The evaluated efficiency and power factor values of the PMASynRM presented here are compared to the measured values of IM and SynRM in Table III.

Table 3. The measured and caluculated efficiency and power factor values in converter supply

	IM	SynRM	PMASynRM-d	PMASynRM-q
Efficiency [%]	91.9	91.8	93.1	92.9
Power Factor	0.80	0.72	0.86	0.84
Amount of rotor electric material [kg]	5.21 (Al)	-	2.2 (PM12)	3.0 (PM12)

4 Conclusion

The advantages of the SynRM compared to induction motor build up mainly from the synchronous rotation. In addition to this the PMASynRMs have better power factor and efficiency than the same size IM. Om the other hand PMASynRM call always for converter supply and a smaller air-gap. The maximum number of PMASynRM pole-pair is limited, because d- and q-axis inductances decrease as a function of pole number. By using PM material on the d-axis the motor power factor may be increased from initial SynRM value 0.72 to 0.86 and by using PM material on the q-axis to 0.84.

The heat sources in PMASynRM rotor are the iron losses caused by the air-gap flux harmonics. Thus if the air-gap harmonics can be decreased to a tolerable level using for example semi-magnetic slot wedges it can be assumed that the heat rise of a PMASynRM rotor is lower than in a respective IM's rotor. This is an advantage compared to IM because in closed IMs the rotor is always more difficult to cool than the stator. If a standard IM is selected for a converter drive the continuous output power must be derated.

The flux barrier apertures make the PMASynRM rotor lighter compared to a respective IM's rotor. The lighter rotor of the PMASynRM makes it suitable for servo motor drives where high power per weight ratio is required. Simultaneously, the PMASynRM rotor construction is weaker than that of the IM. The strength of the PMASynRM rotor may be increased by casting "bars" to the flux barrier holes. If these "bars" were made of conductive material like aluminium, they may be also used as damper windings. The squirrel cage can be made by short circuit these bars with conductive end rings.

The PMSynRM synchronous speed and rotor high saliency ratio L_{dm}/L_{qm} enables more accurate rotor speed control than in IM with slip. Thus high quality speed control without rotor position encoder for very low speeds and even for zero speed using SynRM or PMASynRM is possible. With the heavy fluctuations of load torque the absence of damper winding sets high challenges for the motor control system.

References

[1] Fratta A., Vagati A., Villata F., On the evolution of AC machines for spindle drive applications, Conference Record of the 1989 IEEE Industry Applications Society Annual Meeting, Vol. 1, pp. 699–704.

[2] Pillay P., Krishnan R., Application chareacteristics of permanent magnet synchronous and brushless DC motors for servo drives, IEEE Transactions of Industry applications, col. 27, Issue 5, Sept.-Oct. 1991, pp. 986-996.

[3] Leonardi F., McCleer P.J., Elantably A., Rotors for synchronous reluctance traction motors: a comparative study, Conference Record of the 1999 IEEE Industry Applications, Thirty-Fourth IAS Annual Meeting, vol. 2, pp. 835 –839.

[4] Betz R.E., Jovanovic, M., Lagerquist, R., Miller, T.J.E., Aspects of the control of synchronous reluctance machines including saturation and iron losses, Conference Record of the 1992 IEEE Industry Applications Society Annual Meeting, vol. 1, pp. 456–463.

[5] Soong W.L., Staton D.A., Miller T.J.E., Validation of lumped-circuit and finite-element modelling of axially-laminated brushless motors, Sixth International Conference on Electrical Machines and Drives, 1993, pp. 85 –90.

[6] Soong W.L., Staton D.A., Miller T.J.E., 1993. Design of a new axially laminated interior permanent magnet motor. Conference Record of the 1993 IEEE Industry Applications Society Annual Meeting, Vol.1, pp. 185–193.

Permanent Magnet Motors Find its Way to Circulator Pumps

Keld Folsach Rasmussen, Nils Thorup

K.F. Rasmussen, Senior Engineer, Ph.D., and N. Thorup, Senior Engineer
GRUNDFOS Management A/S, Poul Due Jensens Vej 7, DK-8850 Bjerringbro.

Abstract

Permanent Magnet Synchronous Motors, PMSM, have been known for a long time, but have primarily been used in high performance applications like servomotors. The development in power electronics and permanent magnet materials has now made the PMSM an attractive alternative for circulator pumps. The combination of high efficiency of this motor and ability through electronic control to adjust the pump to the actual heating requirements of the customer shows a dramatic reduction in the energy consumption. This article shows that with the new Grundfos Magna UPE 40-120 the energy consumption can be reduced to 45% of that of a traditionally uncontrolled circulator pump.

Permanent Magnet Synchronous Motors, PMSM

Throughout the past ten years the demand for more energy-efficient electric motors has been increasing. At the same time great progress has been made in the development of permanent magnets as well as the power and microelectronics, which are required for making a motor with permanent magnets rotate. This is the reason why motors with permanent magnet rotors (PM motors) are becoming more and more attractive. The PM motor combines the traditional electromagnets used in electric machines with the permanent magnets, see Fig. 1.

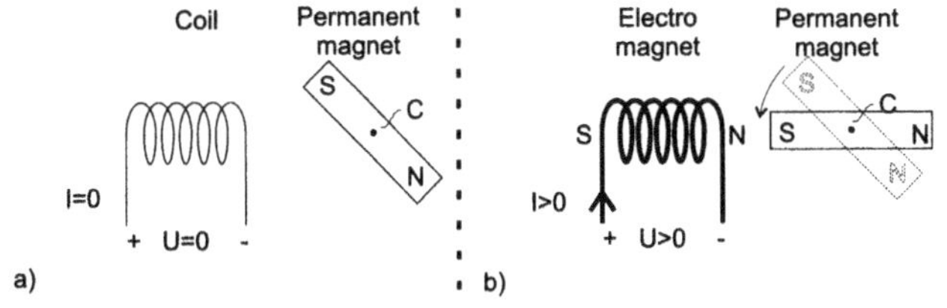

Fig. 1. An electro magnet and a permanent magnet, and how to create a motion

In a) a permanent magnet with a north, N, and south, S, pole is located next to a coil. In B) an electric current is fed into the coil, converting it to an electromagnet with a south and a north pole. As the permanent magnet can rotate around the point C, the attraction between a north and a south pole will rotate the magnet, so it aligns with the electromagnet.

With a suitable combination of electromagnets and electronic control of those, a continuous rotational movement can be made, see Fig. 2.

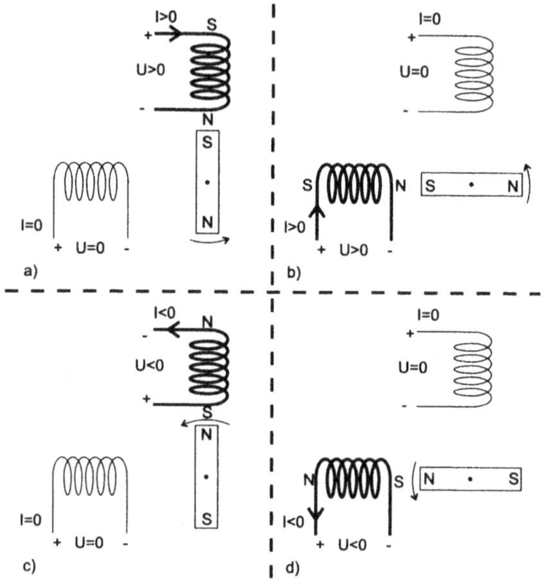

Fig. 2. Continuous rotational movement of a permanent magnet rotor. The arrow on the coil shows the directions of the current

In Fig. 2 a continuous rotational movement of the permanent magnet can be achieved by following the sequence a)-b)-c)-d)-a) etc.

The current is shifted between the two coils, so that the permanent magnet is rotated in the same direction at each step. Further, at c) and d) the direction of the current in each coil is opposite compared to a) and b). The direction of the current determines the polarity of the electromagnet, so that the end of the coil close to the magnet can be either North Pole or South Pole.

In Magna this principle is further refined, so it complies with customer requests on comfort and noise.

The Advantage of Electronically Controlled Pumps

In Fig. 3 the typical pump characteristic of an uncontrolled circulator pump is shown (Bidstrup 2002). It can be seen that at high flow, the head is low. When the flow decreases, the head increased. In applications for a circulator pump, like a central heating system, this is not what is required. In central heating systems, the flow determines the actual heating. The more heating needed the higher flow required.

The head is used to compensate for the friction losses from water flowing in the system. The friction occurs in pipes, valves, thermostats etc. As the flow (speed of water) decreases, the friction will decrease and the head needed for overcoming the friction will also decrease. This is opposite to the pump characteristic, where the head increases with smaller flow. This means in practice, that a lot of the energy fed into the pump is converted into losses in the system.

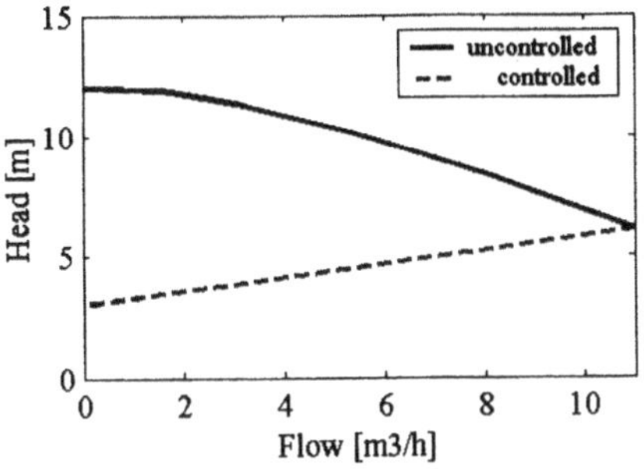

Fig. 3. Pump characteristics for an uncontrolled circulator pump and for a proportional pressure controlled circulator pump

In Fig. 3 a pump characteristic for a controlled circulator pump is shown (Bidstrup 2002). Here it can be seen, that the pump characteristic matches the system need. High flow gives high head and small flow gives lower head. In practice, only the amount of energy needed in the system is fed into the pump.

An analogy towards automobiles would be: An uncontrolled pump equals the case, where the driver runs at full throttle, and try to control the speed of the car by using the brake. For the case of the controlled pump, the driver adjusts the speed of the car by adjusting the throttle to the engine.

Designing Toward Pump for Central Heating Application

To have the highest energy efficiency for a circulator pump, i.e. having the minimum energy input for a given energy output, it is important to focus on the load profile for the pump. A circulator pump for a central heating system sees a load profile, which varies during the year. In a few days during the winter, when it is very cold, full heating is required. As mentioned earlier, full heating is achieved with maximum flow. As the outdoor temperature rises, less heating and hence less flow is required. The number of days where less than maximum flow is needed is very high. In table 1 a typical working profile is shown.

It can be seen from table 1, that app. 6% of the time, the pump is running with full flow. 6% of one year is app. 2-3 weeks. In close to 80% of the time, it is running with less than half of full or rated flow. It is therefore important, that the efficiency is high not only at rated flow but also at reduced flow.

Table 1. Yearly load profile for a pump in a central heating system acc. Blauer Engel

Flow Q [%]	Duty time [%]	Graphic presentation of Blauer Engel
100	6	
75	15	
50	35	
25	44	

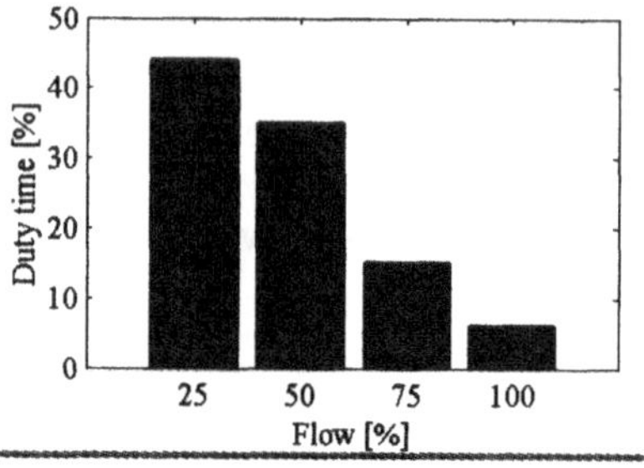

In Fig. 4 the efficiency-curves for 2 different motors are shown, one designed for high efficiency at rated flow at 11[m3/h], "motor a", and one for high efficiency in the entire working area, "motor b".

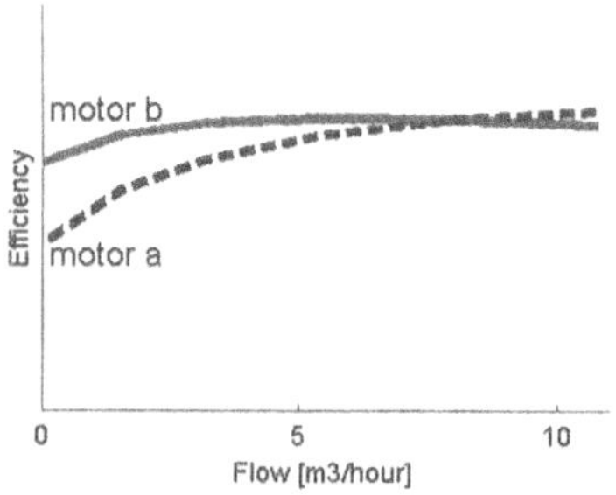

Fig. 4. Efficiency of two different motors as function of the flow

Even though motor a has the highest maximum efficiency, motor b is the best choice with respect to yearly energy consumption, since motor b has the highest efficiency at reduced flow, where it operates more than 80% of the year. The Magna is designed for high efficiency in its full working area similar to motor b.

Comparison of Power Consumption

In the following a comparison between controlled an uncontrolled circulator pumps in terms of the energy consumption is made. The pumps are the uncontrolled UPS 40-120, the controlled UPE 40-120 and the new Magna UPE 40-120.

The comparison is based on the working profile from table 1. Further, the comparison uses a German average heating year of 6840 hours. The remaining hours during the summer the pump is switched of. All data in the comparison are measured data on pumps using the factory settings for each pump.

In table 2 the results for the uncontrolled pump are shown. Table 3 shows the results for the controlled pump and table 4 for the Magna.

Table 2. Yearly energy consumption of an uncontrolled circulator pump. The pump is a Grundfos UPS 40-120 in standard speed 3.

Flow [%]	Flow [m3/h]	Head [m]	Power [W]	Time [hour]	Energy [kWh]
100	11,0	7,1	475	410	195
75	8,3	8,0	440	1026	451
50	5,5	8,7	391	2394	935
25	2,3	8,9	328	3010	988
				Total	2569

Table 3. Yearly energy consumption of a speed controlled circulator pump with an induction motor. The pump is a Grundfos UPE 40-120 in proportional pressure mode.

Flow [%]	Flow [m3/h]	Head [m]	Power [W]	Time [hour]	Energy [kWh]
100	11,0	5,0	460	410	189
75	8,3	4,9	363	1026	372
50	5,5	4,5	266	2394	637
25	2,3	4,0	195	3010	586
				Total	1783

Table 4. Yearly energy consumption of a speed controlled circulator pump with a PMSM. The pump is a Grundfos Magna 40-120 in proportional pressure mode .

Flow [%]	Flow [m3/h]	Head [m]	Power [W]	Time [hour]	Energy [kWh]
100	11,0	5,7	436	410	179
75	8,3	4,6	239	1026	245
50	5,5	4,1	164	2394	393
25	2,3	3,5	110	3010	332
				Total	1148

In Fig. 5 the energy consumption for the different pumps are summarized.

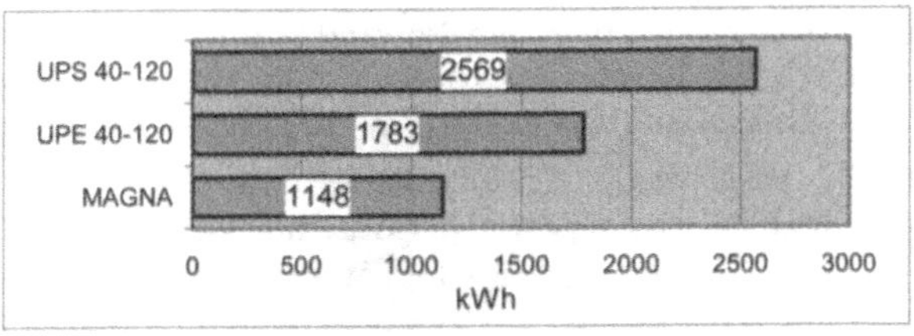

Fig. 5. Comparison of energy consumption for 3 different pumps

Replacing the uncontrolled UPS 40-120 pump with a controlled UPE pump reduces the energy consumption from 2569kWh to 1783kWh or to 70%. The Magna has reduced this even more to 1148kWh, an additional reduction of 35% by replacing the existing motor with a PM Motor. Compared to the UPS, the integration of a PMSM has reduced the energy consumption to 45%. This reduction can be directly seen as a 55% saving on the end user energy bill.

Life Cycle Cost (LCC)

Comparing the total cost of the entire life cycle for the 3 different solutions for the same problems is showed in Fig. 6.

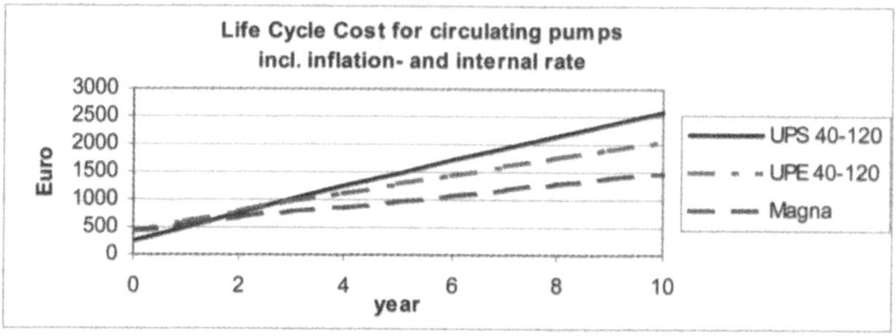

Fig. 6. Total cost of the entire life cycle for the 3 different solutions

The Magna pump have a reduction on the total LCC at 39% according to the uncontrolled traditional, even that the traditional pump only have an cost prize at 59% compared with the Magna. The payback time is app. 1.9 years. According to traditional bank rate, the internal rate of the investment changing an UPS to a Magna will be more than 50%. Grundfos estimates that the saving potential for using PM-technology in larger Circulator pumps is 6,6 TWh/year, or more than 1,5 million households electricity consumption.

Conclusion

This article has discussed some important issues related to circulator pumps for central heating systems. Topics like electronic control of and working profile for circulator pumps has been evaluated, an the importance proper control of the pump has been shown. The importance of a high efficiency over the entire working area has been demonstrated. The permanent magnet motor is without comparison the type of electric motor that has the superior performance with respect to high efficiency, especially for use in circulator pumps for central heating systems.

The Magna with a PMSM has been designed to account for all of this, and practical results show a reduction in energy consumption to 45% of that of an uncontrolled circulator pump, i.e. the end user saves 55% on the energy bill. In LCC the save is 39% or a payback time of 1.9 years.

References

Bidstrup, N., A new generation of Intelligent Electronically Controlled Circulator Pumps, IEECB 2002, Nice, France, May 2002.

High Efficiency PM Motor with Outer Rotor Construction: a Valid Alternative to a Conventional Motor Design ?

Alberto Colotti, Sandro Ceol

Servax Drives, Landert Motoren AG, Unterweg 14, CH-8180 Bülach,
Ph. +41 1 863 51 11

Introduction

The interest in high efficiency drives leads to basic question in the design phase of an electrical drive. Permanent magnet motors are generally more efficient than induction motors because of their inherent excitation. Taking a pure technical view only the difficulties to reduce the flux in the idling state or in the flux weakening region are drawbacks which could be decisive towards the choice of an induction motor when efficiency is the major requirement.

Obviously it can not be forgotten, that other aspects usually limit the usage of permanent magnet motors. There are first of all some bad attributes of the permanent magnet material like price, irreversible thermal losses or the complicated handling. Furthermore also the system integration will require more efforts than a simple induction motor used directly on a constant frequency supply.

Once the decision towards pm motors has been taken, there are still a lot of design choices to check. One of them consists in choosing between the outer rotor and the inner rotor motor topology. There are some special applications, like wheel or fan drives, where the outer rotor fits ideally because he can be connected directly to the load. With a slightly more complicated construction it's possible to use outer rotor motors in classical arrangements with flange by connecting the axis to the bell type rotor and leading it through the stator. The impact on the construction will not be discussed in this paper.

The construction type influences the air gap diameter, the winding design, the layout of the magnetic path and the amount of permanent magnet material. In respect of energy efficiency the design may have some advantages because of higher air gap diameter. To answer this question models have to be elaborated which allow a fair and reliable comparison [1].

Motor Topologies

The increased complexity of the outer rotor design (Figure 2) is evident. The bell type rotor and the housing are more expensive in production than the corresponding parts of the inner rotor design (Figure 1). The length of the motor is limited by the stiffness of the components. The inner rotor design is more compact and robust. On the other hand it's worth noting, that the winding process of an inner stator is favourable because of the good accessibility.

The main difference consists in the position of the air gap. In the inner rotor motor the gap diameter is limited by the needs of joke and winding area. Only the rotor joke and the height of the permanent magnets are outside the air gap of the outer rotor motor.

A major aspect of all motor design is the thermal behaviour. The loss path from the windings to the ambient is quite different. The outer rotor forms a thermal barrier. The losses have to flow through the whole length of the stator before reaching flange and housing. The thermal resistance is about 20 to 50% higher. The only remedy would be to use a direct liquid cooling on the inside of the stator.

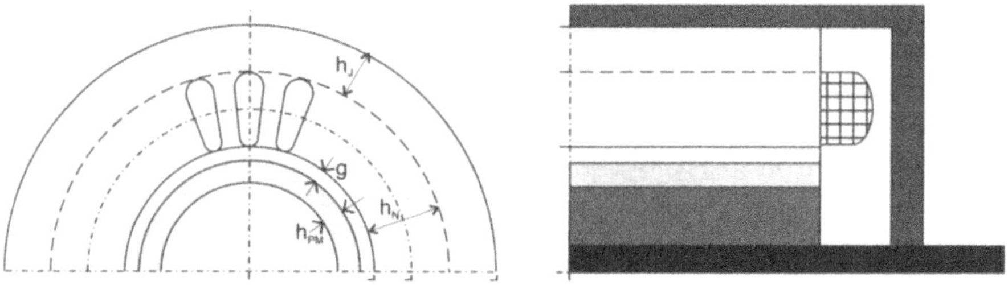

Fig. 1. Radial and axial section of the inner rotor design

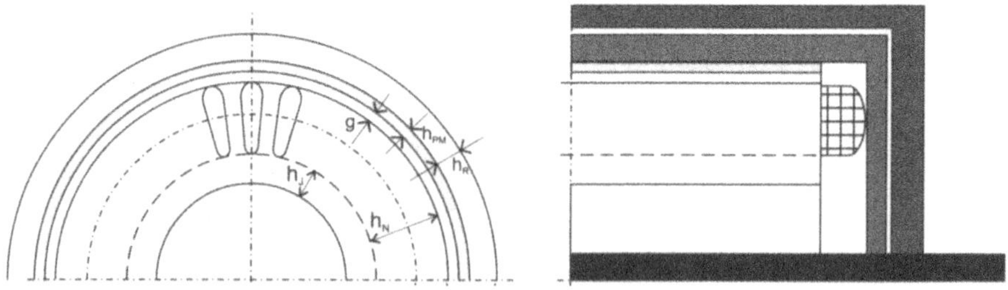

Fig. 2. Radial and axial section of the outer rotor design

Analyse

Glossary and definitions

The used terms are explained in table 1. Different meanings for outer rotor or outer stator are stated. Furthermore the design choices and their values for the following numerical analyse are mentioned.

Table 1. Definitions and design choices

Term	Symbols	Units	Remarks, definitions and calculations	
			Inner Rotor	Outer Rotor
Outer diameter	D	m	Stator lamination	Rotor bell
			D=140mm	
Stack length	l	m	l=100mm	
Air gap diameter	d_g	m	Stator bore	Stator lamination
Total height of slots	h_S	m	Design choice within limits	
Filling factor of slots	f_S	-	Limited by possibilities of manufacturing	
Air gap	g	m	Material choice, affect teeth and joke meas-	
Height of magnets	h_{PM}	m	ures;	
Remanence	B_r	T	Design choices: g=1mm; h_{PM}=2mm; B_r=1.2T	
Air gap induction and assumption for fundamental component	B_g B_{g1}	T	$B_g = \frac{B_r}{1+g/h_{PM}} = 0.8\text{T}$; $B_{g1} = 1.125 B_g = 0.9\text{T}$	
Widths of teeth (parallel sided) and slots (on air gap side)	w_T $w_{S,g}$	m	Given by flux density in the teeth: Design choice: $w_T = w_{S,g}$	
Flux density in tooth	B_T	T	$B_T = 1.4\text{T}$	
Flux density in stator an rotor jokes	B_J	T	Design choice, $B_J = 1.2\text{T}$	
No. of pole pairs	p	-	Design choice	
No. of slots	N_S	-	Design choice	
Pole pitch	τ_P	m	$\tau_P = \frac{\pi d_g}{2p}$	
Heights of joke and rotor	h_J, h_R	m	$h_J = h_R = \frac{\tau_P}{2}\frac{B_g}{B_J}$	

With the mentioned design choices the air gap diameter of the inner rotor motor is:

$$d_{g,I} = D - 2h_J - 2h_S = \frac{D - 2h_S}{1 + \frac{\pi}{2p}\frac{B_g}{B_J}} \tag{1}$$

For the outer rotor motor:

$$d_{g,A} = D - 2h_R - 2h_{PM} - 2g = \frac{D - 2h_{PM} - 2g}{1 + \frac{\pi}{2p}\frac{B_g}{B_J}} \tag{2}$$

The total slot area of the inner rotor motor can be approximated by:

$$A_{S,I} = N_S \frac{1}{2}\left(w_{S,g} + w_{S,O}\right)h_S \tag{3}$$

where $w_{S,O}$ is the outer slot width. It follows:

$$A_{S,I} = \frac{1}{2}\pi\left(\frac{d_g}{2} + d_g + 2h_S - \frac{d_g}{2}\right)h_S = \pi\left(\frac{d_g}{2} + h_S\right)h_S \tag{4}$$

The same calculation for the outer rotor motor gives:

$$A_{S,O} = \pi\left(\frac{d_g}{2} - h_S\right)h_S \tag{5}$$

Torque model

Using the slot area A_S, the number of phases m, the current loading A, as described in [2] or [3] can now be expressed as:

$$A = \frac{m}{2}\frac{4N_{Ph}}{\pi\, d_g} I = \frac{1}{\pi\, d_g} J\, f_S\, A_S \tag{6}$$

where m is the number of phases, N_{Ph} the number of winding turns per phase and J the current density. The torque is calculated by means of the air gap shear stress σ and respecting the fundamental winding factor ξ_1 by:

$$T = 2\pi\left(\frac{d_g}{2}\right)^2 l\,\sigma = 2\pi\left(\frac{d_g}{2}\right)^2 l\,\xi_1 \frac{1}{\sqrt{2}} B_{g1} A = \frac{\sqrt{2}}{4} d_g l\, B_{g1} \xi_1 J\, f_S\, A_S \tag{7}$$

Loss models

Using the average winding diameter $d_g \pm h_S$ with the plus sign for the inner and the minus sign for the outer rotor motor the copper losses can be approximated by:

$$P_{Cu} = \rho_{Cu}(\vartheta) J^2 f_S\, A_S \left(l + \frac{\pi}{2}(d_g \pm h_S)\right) \tag{8}$$

where the temperature depending specific resistance of copper $\rho_{Cu}(\vartheta)$ has to be considered. The heating of the cupper can be calculated by using the thermal resistance between copper area and the ambient $R_{Th,Cu\text{-}Amb}$ [4]. With the thermal coefficient α_{Cu} the expression for the copper losses becomes:

$$P_{Cu} = \frac{1}{\frac{1}{P_{Cu}(20°C)} - \alpha R_{Th,Cu-Amb}} \tag{9}$$

The iron losses are approximated by:

$$P_{FE} = \delta_{Fe} v_{Fe} \left(\frac{f}{f_0} \right)^{1.6} \left[\left(\frac{B_J}{B_0} \right)^2 V_J + \left(\frac{B_T}{B_0} \right)^2 V_T \right] \tag{10}$$

where the joke (V_J) and teeth (V_T) volumes of the two topologies have to be considered.

Comparison

The numerical comparison is made with both of the motors being driven at their maximal power loss rate. The assumed thermal resistances are 0.7 for the inner rotor motor and 1K/W for the outer rotor motor. These values yield to admissible losses of 150 and 105W respectively for a temperature difference of 150K.

In Fig. 3 the torque and the efficiency of the inner rotor motor are shown. The maximal point is 10.8Nm and 93.8% and is reached for a slot height of about 28mm. In Fig. 4 the same curves are drawn for the outer rotor motor. Here the maximal values of 8.7Nm and 94.6% results obviously for the maximal slot height. The third curve in this figure shows the efficiency values of the inner rotor motor for torque values which correspond to the values of the outer rotor motor. The outer rotor motor has a major efficiency of about 0.7%.

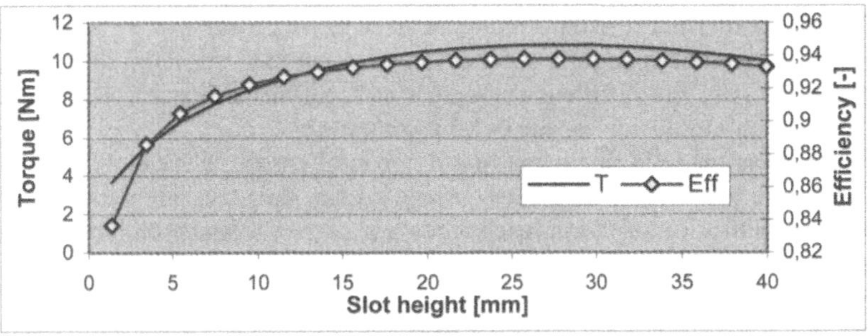

Fig. 3. Torque and efficiency for inner rotor motor for 6 Poles, 2000rpm and P_v=150W.

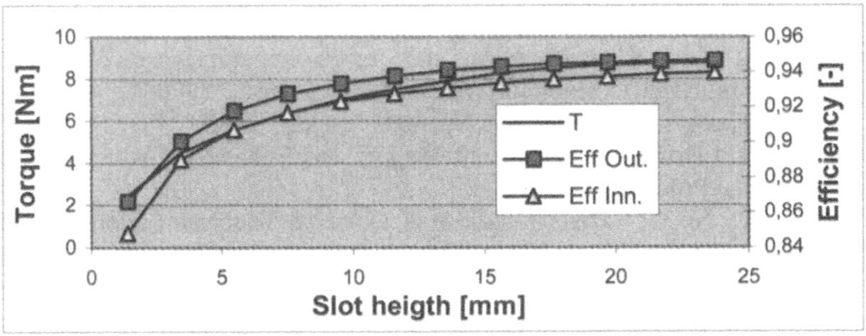

Fig. 4. Torque and efficiency for outer rotor motor for 6 Poles, 2000rpm and P_v=105W; efficiency of the inner rotor motor for same torque and slot height.

The influence of the rotational speed is shown in Fig. 5. Due to the smaller amount of iron the outer rotor motor gains with increasing speed.

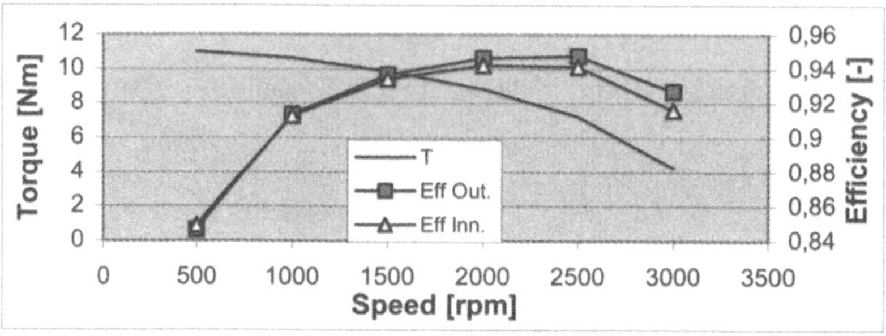

Fig. 5. Maximal torque of the outer rotor motor for 6 Poles and variable speed; efficiencies of both motor types for equal torque production and optimal slot height.

Conclusion

Outer rotor motors are not as good as they ought to be because of two reasons:

- The smaller winding diameter reduces the slot area.
- The higher thermal resistance contributes to more power loss.

The outer rotor motor is slightly more efficient when the utilisation allows both motor topologies. But the difference is very small. So the efficiency is probably never reason enough to decide for the outer rotor motor.

No mention was made in this paper about maximal torque. The outer rotor motor produces more torque with a smaller current loading than the inner rotor motor. Thus the demagnetisation of the magnets occurs at higher torques. In combination with large load inertias an outer rotor motor will offer better dynamics.

Reference

[1] Slemon, G. R., Liu, Xian: Modeling and design optimization of Permanent Magnet Motors, Electric Machines and Power Systems, 20: 71-92, 1992.
[2] Richter, R.: Elektrische Maschinen, Springer Verlag, 1930.
[3] Miller, T. J. E.: Brushless Permanent-Magnet and Reluctance Motor Drives, Oxford Science Publication, 1989.
[4] Staton, D. A., So, E.: Determination of Optimal Thermal Parameters for Brushless Permanent Magnet Motor Design, IAS, IEEE 1998.

Gearless PM Motor Drives Save Energy

Sven Sjöberg

ABB Automation Technology Products AB, Motors & Machines, LV Motors, Örjansgränd 10, 72170 Västerås, Sweden

Permanent Magnet Motors Eliminate Gearboxes

A permanent magnet (PM) motor for low speed applications has been developed by ABB. Used for decades in fast running applications due to its favorable weight-to-performance ratio, the PM motor has now been adapted to provide high accuracy and reliability at low speed, eliminating the need for speed reduction gearboxes across a range of industries. Initially, however, the paper industry will be targeted, as paper machines require large numbers of high accuracy, low speed drives.

The PM motor is the heart of a system known as Drive ^{IT} Direct Drive Solution. This consists of a Drive ^{IT} PM motor, controlled by Drive ^{IT} low voltage AC drive, ACS 600 frequency converter, connected directly to the paper machine, without gear boxes or pulse encoder. ABB introduced this new Direct Drive at the PulPaper 2001 trade show in Helsinki last year.

There have been two pilot installations running so far. The first machine has been running successfully since mid 1999 in a Finnish paper mill. January 2001 saw the second start-up using this technology for two paper machines in another Finnish mill. The start-up was trouble-free and the drives are running very well.

The first complete paper machine with this new Direct Drive will be started up in middle of September 2002 at M-Real mill in Äänekoski, Finland, in their line of manufacturing packaging materials for the medical and cosmetics industries.

ABB's PM motor is a synchronous motor, which, with no rotor slip, provides better accuracy than standard asynchronous motors. In an asynchronous motor, the slip varies according to speed and load. With a synchronous motor, it is easy to optimize the speed, while the elimination of slip compensation improves the dynamic motor control performance.

However, the traditional synchronous motor is a more complicated construction, requiring more maintenance. The permanent magnets simplify the construction by eliminating the need for the rotor windings and the brushes normally used for excitation in synchronous motors, by creating a constant flux in the air gap.

ABB has put a lot of efforts in investigating different types of PM motors, like axial, radial and transversal flux motors. This has resulted in a radial flux motor design that combines the high quality performance of the synchronous motor with

the robust design of the asynchronous induction motor. The motor is energized directly on the stator by the variable speed drive.
The PM motor can also deliver more power from a smaller unit. To drive the in-drives of a paper machine directly at 300 to 600 rpm with a conventional asynchronous motor would require a motor frame substantially larger than that of a 1500 rpm motor. The new motor type is, in most cases, the same size or even smaller than the existing induction motor.

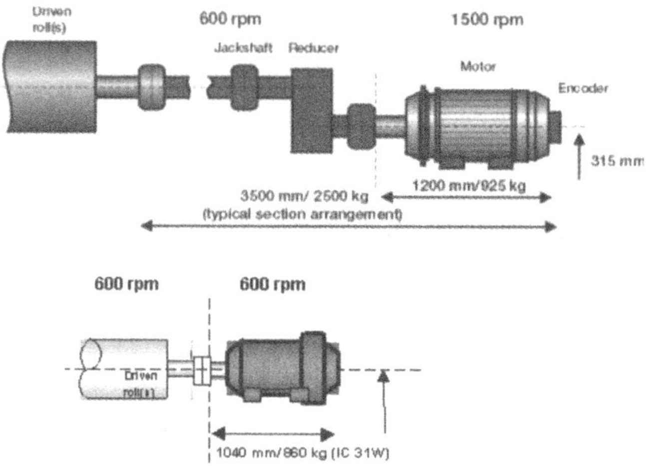

Fig. 1. Conventional drive and gearless drive

The permanent magnets are made from neodymium boron iron (NdFeB), part of a type of materials known as the "rare earth" group of elements. NdFeB is the most recent magnetic material on the market. It is the most powerful magnetic material at room temperature, with high values of flux density at very high values of magnetizing force. It is also extremely resistant to demagnetization. Compared to samarium cobalt, another rare earth material, which was popular in the 1980s NdFeB is less costly and less brittle.

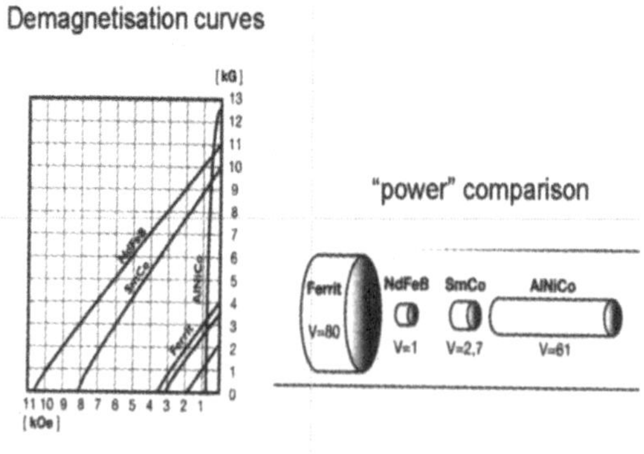

Fig. 2. Permanent Magnet Materials

Higher Efficiency and Less Maintenance

Standard asynchronous motors, normally designed to run at 1000-3000 rpm, are not particularly well suited for low speed operation as their efficiency and power factor drops with the reduction in speed. They may also be unable to deliver sufficiently smooth torque across the speed range. The low efficiency and power factor will also result in a larger and more expensive frequency converter.
This is normally overcome by using a gearbox. However, the gearbox is a complicated piece of machinery that takes up space and needs maintenance as well as large quantities of oil.
Direct Drive provides a high torque drive directly coupled to the in-drive of the paper machine section. By eliminating the gearbox, the user saves space and installation costs, as he only needs to prepare the foundations for one piece of driving machinery. This also gives more freedom for the design of the mill layout. Eliminating the wear of motor brushes and gearboxes reduces maintenance. No gearbox losses save energy.
High performance at low speed is sometimes achieved by using a DC drive. Compared to this solution, Direct Drive will save on motor maintenance, as the permanent magnet motor is very robust and, in maintenance terms, very similar to standard AC induction motors.
Fewer components and simpler configuration reduces plant-engineering hours, facilitates installation, cuts down supplier contacts, enables more efficient use of floor space and reduces spare part inventories.
Simpler configuration also improves the availability of the production machinery. Less maintenance means fewer production interruptions and start-ups, decreasing raw material waste, increasing end product quality and reducing wear of the production machinery. Maintenance and repair work can also be carried out faster.
The Direct Drive technology improves drive controllability, enabling the paper machine drive to run without a pulse encoder, as synchronized motors give very exact control without feedback. The accuracy is as good as that of an induction motor in variable speed operation with a feedback device. This means the pulse encoder can be eliminated, reducing the need for maintenance. This is particularly beneficial in the paper industry, where poor reliability of feedback devices in the past have contributed to production stoppages. It can also reduce design complexity, as the feedback devices sometimes can be difficult to integrate in the system or have to be positioned in places that are difficult to reach.

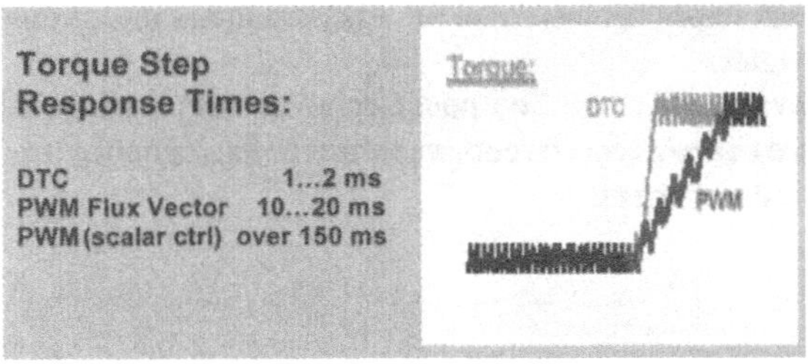

Fig. 3. Torque Response of DTC controlled PM Motor

The better electrical efficiency of Direct Drive has a direct impact on power consumption. The savings increase considerably with further reduced speed. The low speed PM motor has about the same efficiency as the conventional 1500 rpm asynchronous motor but the savings are mainly coming from the elimination of the gear box.

Familiar Motor Type in a New Role

The permanent magnet motor has been around for decades but it is only now that the technology is being applied to large motors. Thanks to their small size and high accuracy, permanent magnet motors are for instance used in wristwatches and computer hard drives. Modern cars have about 30-50 PM motors installed. By contrast, the largest permanent magnet motor weighs in at several tonnes.

The actual motor design is a radial flux construction, air or water-cooled motor, with permanent magnet rotor and conventional stator. Power ranges from 27 to 1800 kW and the motor voltage is typically 400/690 VAC.

The temperature of the PM rotor remains naturally low and, when using the standard water-cooled frame, the power density and protection class can be designed to be high. As a rule of thumb the torque can be at least doubled in a water-cooled PM motor compared to a conventional air-cooled asynchronous squirrel cage motor. The permanent magnets are located in reshaped rotor slots resulting in a simple and robust design. Special design has been used to minimize the torque harmonics to obtain smooth torque characteristics.

The main challenge in developing the new Direct Drive is related to the design of cost efficient PM motors. A completely new hardware has to be developed. The number of PM motors for potential industrial applications are small compared to the big volume of asynchronous motors produced. To minimize the investment needed and bring down the cost of the PM motor to an acceptable level, ABB has, as mentioned earlier, chosen the radial flux design which means that existing production facilities can be used with small additional investment.

As with all synchronous motors, the motor speed can only be controlled with a variable speed drive. Furthermore, the synchronous motor control must be specifically developed for permanent magnet flux control. ABB's Direct Torque Control method has now been further developed to achieve this.

The Direct Drive system is based on ABB's ACS 600 frequency converters. The motor control can use the same inverter hardware as the asynchronous motor control and can be cooled by air or water.

Water-cooling gives higher power density and compact drive cabinets, while the higher protection class enclosures allow more freedom for drive placement by reducing the exposure of the drive components.

Used in Azipod Technology

The Direct Drive is initially being targeted at the paper industry, as this sector is particularly dependent on high accuracy and high reliability in low speed applications. However, PM motor solutions have existed for some time, finding early applications as servomotors and traction motors and since some years also in driving elevators. One well known use of PM motors is the Azipod, marine propulsion system, while the first "standard" solution for low speed industrial applications is now the Direct Drive. With the PM motor, the Azipod solution becomes more compact, making it viable for smaller ships than was previously possible. The smaller outer diameter of the motor has a premium effect on the hydrodynamic properties of the pod and on the total propulsion efficiency.

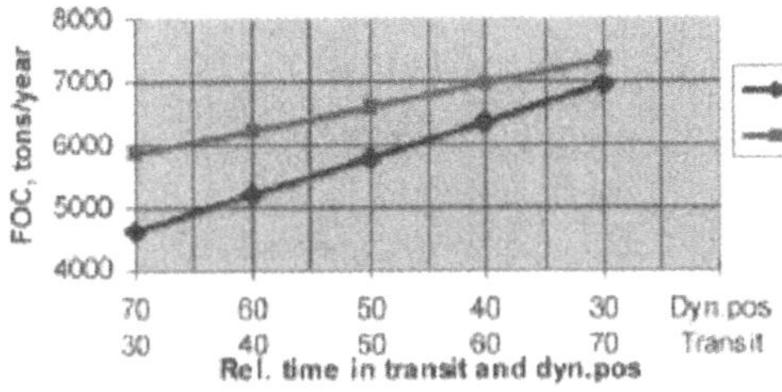

Fig. 4. Comparison of fuel oil consumption (FOC) with varying operational profiles

The experience gained of PM motors with Azipod has been used when developing Direct Drive for the pulp and paper industry.

Another growing application area for the PM machine is as generator in wind turbines, which, as a slow running application, could benefit from the removal of gearboxes. At the moment, the investment cost for a paper machine Direct Drive, due to the high cost of materials, is comparable with the investment cost of a traditional drive installation with gearbox. As outlined, the benefits lie in reduced maintenance, less space requirements, improved energy efficiency and better reliability. This will initially restrict the usage of the technology to specialist applications such as paper machines and Azipod. But in years to come, there is no reason why the permanent magnet motor should not become a more common sight in variable speed drives on factory floors.

References

[1] B. Welin and Carl-Johan Friman, "New Direct Drive system opens a new area for paper machines," Paperi ja Puu - Paper and Timber Vol. 83, No. 5, 2001, pp. 385-387

[2] J. Ylitalo, "New Thruster Concept for Station Keeping and Electric Propulsion", Dynamic Positioning Conference 2001, Houston, September 2001

[3] F. Owman, "Windformer – An integrated system for large scale wind power production and transmission", Windpower 2001 Conference, Washington, June

[4] I. Waltzer, "Technology Trends in Large Permanent Magnet Motor Applications," ICEM 2002

Die-Cast Copper Motor Rotors: Die Materials and Process Considerations for Economical Copper Rotor Production

Dale T. Peters[1], John G. Cowie[2], Edwin F. Brush, Jr.[3], Stephen P. Midson[4]

[1] Senior Advisor, Copper Development Association Inc., 27 Raintree Lane, Hilton Head Island, SC 29926 USA

[2] Vice President, CDA Inc., 260 Madison Ave. New York, NY 10016 USA

[3] BBF Associates, 68 Gun Club Lane, Weston, MA 02193 USA

[4] Formcast Development, Inc., 100 S. Pecos St., Denver, CO 80223 USA

Introduction

For a high melting temperature metal such as copper (T_m = 1083 °C) to be economically pressure die cast, one or more suitable high temperature die materials must be identified and adapted to the process. This is a meaningful objective in terms of the rotor of the induction motor where aluminum has been the material of choice for the squirrel cage structure because this lower melting metal is readily die cast. A fabrication process other than pressure die casting is difficult to imagine. Because of copper's higher electrical conductivity, substitution of copper for the aluminum would increase the electrical energy efficiency of the induction motor and potentially provide benefits in manufacturing economics and weight savings, depending on design objective. Copper rotor casting experience and performance of motors equipped with copper rotors are detailed in another paper in this conference. At the 1999 EEMODS conference, a progress report on the die material investigation was presented [1]. This paper updates the results presented in 1999 and elaborates on the die failure mechanism and how die life can be extended. More complete accounts of this work can be found elsewhere [2,3].

The principle failure mechanism in die casting dies is referred to as "heat checking" by die casters, but decarburization and softening of common die steels at the high surface temperature are also contributors [4]. Heat checking is a thermal fatigue phenomenon resulting from the rapid cyclic expansion of the die surface layer on contact with molten metal and the constraint of the surface by the much cooler inner portions of the die. In the cooling portion of each cycle, the outer layer falls to a temperature below that of the bulk of the die putting the surface under a large tensile stress that can exceed the yield point of the die material.

This surface-to-interior differential expansion and contraction is greatly exacerbated in die casting high-melting metals such as pure copper. Consideration of the thermal fatigue failure mechanism suggests that the surface-to-interior ΔT and resulting large strains can be minimized by raising the temperature of the bulk of the die insert. Herman et al [5] in 1975 and Doehler [6] in 1951 made this same suggestion. A practical system for heating and insulating the die inserts to maintain the high temperature critical to improving die life has been lacking and has been developed in this work. A thermal modeling exercise done early in this research allowed determination of the minimum temperature necessary to avoid stresses exceeding the material yield stress for various die materials.

Experimental Method

To avoid the expense and handling difficulties of using thousands of lamination stacks, a test die was devised for the die material tests. This test die consisting of six machined inserts shown in Fig. 1 allowed use of readily obtainable pieces of alloys and provided for evaluation of two or even three materials simultaneously. The gate area was designed to simulate a single gate of a multiple gate rotor die of a 15 Hp-motor to feed 0.9 kg of copper to the flat semi-circle at the top of the casting. A dry powdered release agent was applied immediately before each shot. The test die design is believed to be an aggressive test of die materials exceeding conditions the rotor die set experiences.

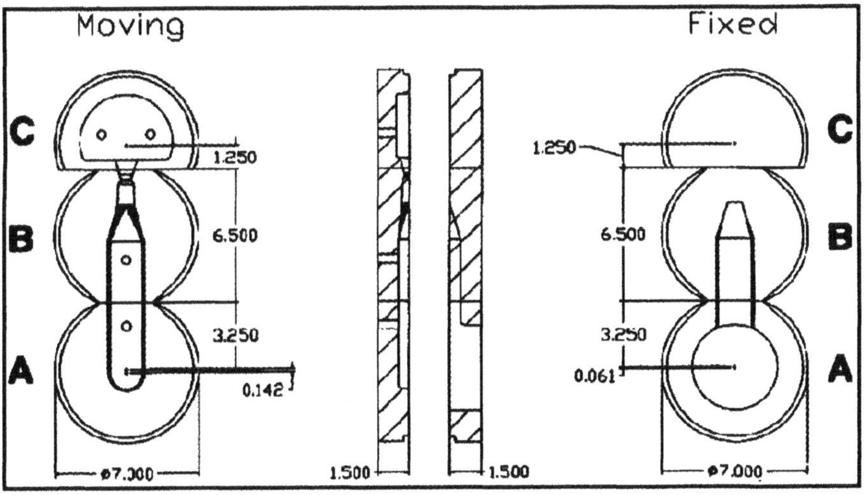

Fig. 1. Die material test die made up of six machined inserts.

Ladling molten copper from a holding furnace would make control of oxygen and hydrogen problematic. For this work, 3.6-kg charges of chopped copper wire rod (C11000 copper) was induction melted one shot at a time on a two minute cycle. This required a 60 kW power supply. Maximum utilization of the power supply was achieved by using two push-up furnaces alternatively switched to the supply. A 660 metric ton real-time shot controlled Buhler die casting machine at Formcast Development, Inc. was used for this study. For the die material trials, the ram speed during die fill was 1 m/s and the final pressure was 49 MPa.

3-D Computer Analysis of Die Thermal Profiles

A 3-D computer analysis of heat transfer in the die material test inserts vividly showed the thermal gradients generated when the die is brought into contact with molten copper (K. D. Williams, Flow Simulation Services, Albuquerque, NM). This analysis was valuable in understanding the thermal fatigue failure mechanism and how to minimize or largely avoid it. The time to gate freezing and the number of shots to reach the equilibrium temperature profile were also obtained.

Temperature profiles in H-13 die inserts were generated for this material in the test die geometry of Fig. 1. Die surface temperature distributions at the instant of filling with 1200°C molten copper and at points in time immediately thereafter were calculated. Since the die surfaces are generally coated with a mold release compound, a value for the heat resistance, R, of this coating had to be selected. This was taken as 1°Ccm²/watt, a value in the middle of the range found in the die casting literature. To avoid representational problems, the die surface temperatures calculated from the model are shown as though they were "painted" onto the surface of the test casting. An example for the case of the instant of die cavity fill with 1200°C copper is shown in Fig. 2. Representations of this type for numerous scenarios are presented elsewhere [2]. In this case the casting surface will actually be hotter than the die surface because of the surface heat conducting resistance. In fact, in this example, the investigators assumed that coating the narrow gate region would be difficult and assigned a very low heat flow resistance to this region. Thus the die surface temperature in the gate region was essentially at the melting point of copper. This means that the surface of the gate region of the die insert had risen from the initial temperature of 200°C by 880°C. Because of the low thermal conductivity of H-13 tool steel, the body of the mold was still at 200°C. This implies that the surface had a temperature-induced strain of a least 1.19%, an enormous strain to sustain on a cyclic basis.

Immediately after filling, the coated areas of the die surface were only in the range of 550-600°C, or 350°C above the initial temperature and the bulk of the insert. The surface at the biscuit area at the end of the ram in the shot sleeve was at about 800-850°C.

At 0.5 seconds after casting, results showed that the temperature in the uncoated gate area had started to drop, but the rest of the die surface was getting hotter. The metal volume in the gate is small and with R taken as a very small value in this region, heat diffusion to the die steel is rapid. The longer coated surface areas had risen to the 700 to 800°C range. After 6.5 seconds, it was found that the gate area was relatively cold but coated areas of the larger volume sections of the casting had risen to 750 to 900°C.

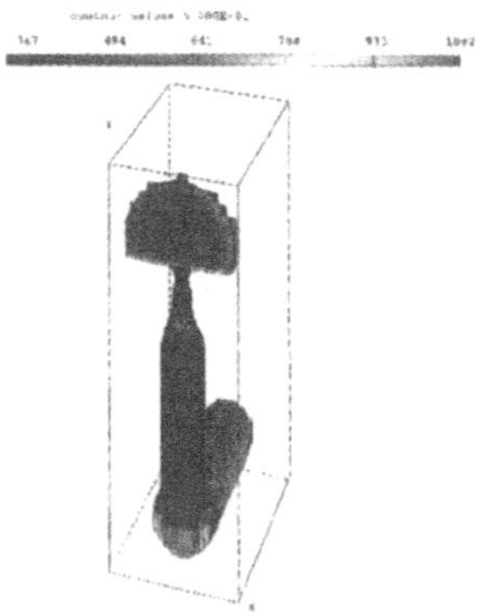

Fig. 2. Output of thermal modeling showing H-13 die surface "painted" onto the test casting at instant of fill with 1200°C copper.

Thus in the H-13 tool steel, we expect from these calculations that the die surface temperature will rise to values ranging from 825°C to over 1000°C everywhere outside the gate region (assumed to be uncoated in this example). These high temperatures occur even with a surface coating with a resistance of 1°Ccm^2/watt over these surfaces. These high surface temperatures imply that substantial surface strain occurs everywhere in the H-13 dies.

It was clear that to achieve the higher average mold temperatures required to minimize the ΔT between the die surface and interior associated with each cycle, and the resulting cyclic strain, it would be necessary to both insulate the die inserts from the backing steel and provide a source of heat directly to the inserts. Temperature distribution data of the type shown in Fig. 2 was obtained for a tungsten die set with an initial wall temperature taken as 650°C (assuming direct die insert heating and insulation). Although the gate area surface temperature was seen to be near the melting point of copper because of the low surface resistance assumed for this region, the remaining surface temperature was found to be only in the range of 750 to 800°C, only a 100 to 150°C increase. The smaller increase compared to the H-13 example is in part due to the higher thermal conductivity of tungsten.

A model prediction of temperature-time profiles in a tungsten insert preheated to 380°C was generated using a die/copper contact resistance of 0.3°Ccm^2/watt. This prediction was compared to the measured temperatures during the shot with thermocouples located near the front, center and rear of the insert [2]. The agreement was excellent. A ΔT of about 400°C between the front and rear of the insert was generated immediately after filling the die cavity. Calculations showed that this would lead to a plastic strain on each cycle. For tungsten, the minimum die temperature to assure strain and resulting stress below the yield point is 550°C. It is important to note that with the ductile/brittle transition temperature being about

200°C for tungsten, the machine operator cannot use the first few shots to achieve the operating temperature without cracking the die.

In nickel-base alloy molds, the temperature gradient and resulting surface stress will be higher due to the lower thermal conductivity of these alloys. The minimum die temperature to assure that cyclic surface stresses remain below the yield point was estimated to be about 625°C.

Die Material Evaluation Tests

H-13 Die Steel

An H-13 die set was run to failure to establish a base line and to gain experience in die casting copper. To minimize thermal shock with the first few shots of molten copper, the dies and shot sleeve were preheated to about 350°C with an oxy-acetylene torch. As expected, substantial physical damage was quickly evident after only about 20 shots. The run was continued for a total of 800 shots during which steady deterioration by heat checking, cracking and erosion at insert joints and ejector holes was taking place. The run was discontinued when ejection of the casting became difficult as copper solidified in the deep fissures. Electrical conductivity of the die-cast copper averaged 98% IACS.

TZM and Anviloy

TZM is a molybdenum-base alloy containing nominally 0.5% Ti, 0.09% Zr and 0.025% C. Anviloy 1200 is tungsten-base containing 4% Ni, 2% Fe and 4% Mo. The alloys were tested simultaneously in the test die configuration of Fig. 1. At this point in the die material investigation, the first heated die configuration had been designed and installed on the machine. This allowed preheating and maintaining the dies at 450°C. This was the maximum temperature attainable with this initial heater array design and was about 100°C below the minimum required to avoid exceeding the yield strength at the surface suggested by the thermal modeling. Failure of one or more heaters during the first run of 500 shots resulted in operation at an even lower temperature for a portion of the run. Despite these problems, no heat checking of either alloy was evident but minor cracking of the Anviloy inserts at sharp radii was noted.

A second run was carried out some weeks later with all heaters operating extending the total number of shots to 940. At this point, both the TZM and Anviloy inserts showed additional minor cracking at the ejector pin holes due to inadequate allowance for the higher thermal expansion of the steel ejector pins. Otherwise these die sets appeared to be capable of extended runs in this severe copper die casting exercise with no heat checking in the gate areas, contours, and flat surfaces. The TZM inserts did suffer serious surface degradation by oxidation because the surface reached temperatures above 700°C where the oxide melts and volatilizes. This problem makes uncoated TZM a poor choice for die casting pure copper.

This work indicates that with sufficient preheat and maintenance of the operating temperature at 550°C, Anviloy is a suitable die material for die casting of pure copper. High base material and machining costs are deterrents to its use, but Anviloy may offer a viable alternative in a part or parts of the die, such as the runner or gate, where the incoming metal temperature or flow rate are extreme.

Nickel-base Alloys

Three very different types of nickel-base alloys were evaluated with two inserts of each alloy in the test die. INCONEL alloy 617 is a 22% Cr, 12.5% Co alloy solid solution strengthened with 9% Mo. INCONEL alloy 718 is a gamma prime strengthened alloy containing 15.5% Cr, 0.7% Al, 2.5% Ti and 0.95% Nb. INCONEL alloy 754 is a mechanically alloyed 20% Cr alloy with small additions of Al and Ti. A dispersion of Y_2O_3 is the principle strengthener giving resistance to recrystallization and excellent retention of high temperature strength. In a run of 250 shots, the inserts were preheated to 350°C using the electrical resistance heaters and not permitted to fall below this temperature in the cooling portion of the cycle. Even though the 754 alloy has the highest strength at the copper melting temperature, these inserts began to show cracking in less that 50 shots. INCONEL alloy 718 began cracking in about 100 shots. Being a precipitation-hardening alloy, alloy 718 would be expected to have very low strength near the surface which would reach the melting point of copper on each cycle but maintain its high tensile and yield values in the interior and back of the insert where ductility (17-19%) is only fair. INCONEL alloy 617 showed only minor craze cracking after 250 shots at this low operating temperature (275°C below the minimum required). This test served to reveal alloy 617 as having the best combination of strength and ductility over the range of temperatures experienced by the insert.

A second extended run was done to evaluate the solid solution nickel-base alloys, INCONEL alloys 601, 617, and 625. Alloy 601 is a lower strength Ni-23% Cr alloy with 1.35% Al. It has only 14% elongation at 1177°C and yield strength of only 15 MPa. Alloy 625 has 21.5% Cr, 9% Mo and 3.65% Nb and has somewhat higher tensile and yield strengths at room and intermediate temperatures, but is not quite as strong at 1100°C as alloy 617. Ductilities of both alloys 617 and 625 are quite high (45% minimum over the range of temperature) but slightly higher in alloy 617. At the point in time of these runs, the array of heaters and insert insulation shown in Fig. 3 had been developed to the point that the preheat and operating temperatures could be maintained at 540°C and with further tweaking for the next run, to the 625-640°C range.

In the course of an extended run with this die heating equipment, it became apparent that the amount of heat checking was markedly reduced as the operating temperature was increased. Finally in the last 330 shots at the highest operating temperature, there appeared to be no further deterioration of the die set. A total of 950 shots at the several progressively increasing operating temperatures had been made in this rather severe test. Clear distinctions among the three INCONEL alloys were difficult to discern. Alloy 601 may have somewhat inadequate tensile and rupture strengths for very long campaigns at or near 650°C.

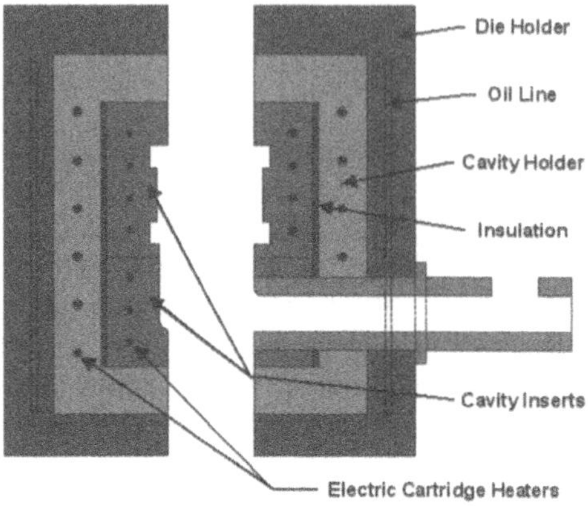

Fig. 3. Schematic illustration of placement of electric resistance and insulation in the die material test set-up developed in the course of this study.

Chemical analysis of several copper test castings showed average iron pick-up of 65 ppm, 5 ppm Ni and 0.074 wt% oxygen. The microstructures were quite sound. The electrical conductivity was higher than that of the castings from the H-13 dies averaging 99.9 % IACS. Elimination of the steel shot sleeve in favor of a nickel-base alloy sleeve would presumably further reduce the iron and increase conductivity slightly.

Conclusions from Die Material Tests

Extended production runs will be required to prove the point, but these tests show promise that the INCONEL alloys 617 and 625 operated in the 600-650°C temperature range are very promising die materials for long die life in large volume production of die-cast copper motor rotors. Although not tested in this study, Haynes alloy 230 is also a strong candidate die material. This alloy has slightly higher yield strength and ductility than alloy 617 and is weld repairable. Alloy 230 will be used in the first copper rotor production die set being built at this writing. An important conclusion from this work is that it is absolutely essential to operate at elevated temperature to extend die life. The higher die temperature reduces the surface-to-interior ΔT on each shot which in turn greatly minimizes the cyclic expansion and contraction and thus the thermal fatigue mechanism causing heat checking and more severe cracking. Temperatures above 650°C are not required and in fact would reduce productivity by increasing cooling time. A practical die heating and insulation design has been developed.

Acknowledgements

This project was sponsored by the world copper industry through the International Copper Association, Ltd. and managed by the Copper Development Association Inc. Additional funding was provided by the U. S. Department of Energy Office of Industrial Technologies and the Air Conditioning and Refrigeration Technical Institute. Formcast Development, Inc., Denver, Colorado, provided the die casting equipment and expertise. Mr. Ruedi Beck of DieTec, GmbH, Gossau, Switzerland, designed the innovative approaches to heating die inserts.

References

[1] Peters DT, Cowie JG, Brush EF Jr. (1999) Die casting copper motor rotors: mold materials and processing for cost effective manufacturing. Proceedings of EEMODS Second International Conference, London

[2] Peters DT, Cowie JG, Brush EF Jr., Midson SP (2002) Use of high temperature die materials and hot dies for high pressure die casting pure copper and copper alloys. North American Die Casting Association Die Casting Congress, Rosemont, IL

[3] Peters DT, Cowie JG, Brush EF Jr., Midson SP (2002) Advances in pressure die casting of electrical grade copper. American Foundry Society Congress Paper No. 02-002, Kansas City, MO

[4] Benedyk JC, Moracz DJ, Wallace JF (1970) Thermal fatigue behavior of die material for aluminum die castings. Trans. of SDCE, Paper No. 111

[5] Herman EA, Wallace JF, Machonis AA (1975) Copper Alloy Pressure Die Casting, p52, International Copper Research Association, New York, NY

[6] Doehler HH (1951) Die Casting, pp 163-165, McGraw-Hill Book Co., Inc., New York,NY (1951).

Die-Cast Copper Motor Rotors: Motor Test Results, Copper Compared to Aluminum

Edwin F. Brush, Jr.[1], John G. Cowie[2], Dale T. Peters[3], Darryl J. Van Son[4]

[1] BBF Associates, 68 Gun Club Lane, Weston, MA 02193 USA

[2] Vice President, Copper Development Association Inc., 260 Madison Ave., NewYork, NY 10016 USA

[3] Senior Advisor, CDA Inc. 27 Raintree Lane, Hilton Head Island, SC 29926 USA

[4] Van Son Consultants, 4500 Mount Harmony Road, Greenwood, AR 72936 USA

Introduction

Motor manufacturers have long realized that because the electrical conductivity of copper is nearly 60% higher than that of aluminum, substituting copper for aluminum in the squirrel cage of the induction motor would markedly increase the electrical energy efficiency of the machine. Most motors larger than about 200 kW and a few special purpose smaller motors are built with copper squirrel cage structures manufactured by a time consuming and costly fabrication process. The intricate squirrel cage of smaller motors is produced by pressure die casting aluminum. Alternative cost-effective manufacturing methods have not been devised. A major barrier to adoption of copper for the rotor has been the high cost resulting from the short die life of the ordinary die steels experienced in die casting copper with its high melting temperature (1083°C compared to 660°C for aluminum).

The incentive to solve the problem of short die life and resulting high manufacturing costs is the 15 to 20% reduction in overall motor energy losses that motor manufacturer models have shown to be possible if copper were utilized in the rotor. The U.S. Department of Energy has reported that motors larger than 1/6 Hp (1/8 kW) use about 60% of all electricity generated in the United States and that medium power motors (1 to 125 Hp, 0.75 to 100 kW) use about 60% of electricity supplied to all motors [1]. In another paper at this conference, we have presented the results of a major effort to identify suitable high-temperature die materials and to adapt them to cost effective copper die casting. This work showed that use of nickel-base alloy dies operated at 625 to 650°C is the path to much extended die life.

This paper summarizes the results of copper rotor die casting trials for four motor manufacturers and the results of manufacturer's tests of performance of motors equipped with copper rotors compared to the counterpart aluminum rotors. Available data from the literature on motors with copper rotors is also summarized. A more complete account of details of the rotor die casting and motor performance tests has been presented and published [2].

Experimental Method

Copper rotors were cast for four motor companies for evaluation in their own facilities. These rotors were pressure die cast in a 660-Tonne real-time shot controlled Buhler horizontal machine using H-13 die inserts. Ordinary tool steel dies were used because only a few rotors were required for testing. These die inserts were mounted in a three-platen master mold assembly of the type conventionally used in rotor die casting.

Chopped copper wire rod was inductively melted on a shot-by-shot just-in-time basis to avoid a large holding furnace and the attendant problems of control of oxygen and hydrogen in the molten copper over an extended time. The copper was heated to 1230 °C providing about 150 °C of superheat.

To maintain superheat, a heated shot sleeve surrounded with a thermal wrap was used. The shot sleeves were specifically sized for each rotor size to minimize air entrapment and porosity in the casting.

The real-time shot control capability of the die casting machine provided opportunity to study a number of die casting variables that might affect the quality of the cast copper and the performance of the rotors in motor tests. On the machine used, ram speed can be set at a number of positions and final compacting pressure and duration are adjustable. A wide range of these variables was used to assess the sensitivity of the copper die casting process to machine operating parameters [2].

Because copper is so much hotter than aluminum entering the conductor bar channels, there was some concern that the conductor bar might weld to iron laminations or that the properties of the iron would be compromised by heat treatment. Welding of laminations to the copper would increase the magnetic loss component of the total motor losses. On ejection from the machine, half the rotors were water quenched on the theory that rapid cooling would shrink the copper from the iron and would minimize high temperature annealing of the iron. The other half was allowed to air cool.

Motor Performance Tests

A total of about 140 rotors were cast for four motor manufacturers to evaluate in their own laboratories. Three companies used dynamometer efficiency tests as per IEEE Specification 112, test method B, as required in the U. S. by the National Electrical Manufacturers Association (NEMA) and the Energy Policy Act of 1992 (EPAct). The fourth company used the IEC 34-2 test method. The IEC method assumes a fixed percentage as stray load losses. The IEEE test method is a true watts in vs. watts out efficiency test that segregates the energy losses into five categories of Iron Core Losses, Stator Resistance, Rotor Resistance, Windage and Friction and Stray Load Losses.

The first four are measured directly and the remainder is in the "stray load" category. For reasons explained below, stray load losses are reduced by the copper rotor and it is therefore important to determine this loss rather than assume a value for it.

To ensure an accurate comparison with the corresponding aluminum rotor, a single wound stator was used to test all rotors in each test program.

Participating motor manufacturers were assured confidentiality. Each agreed to disclose test data, but at their request, are not identified.

15 Hp (11.2 kW) Motor

The first copper rotors cast were for a 15 Hp (11.2 kW) motor and were 5.7 inches (144.8 mm) in diameter with a 6-inch (152.4 mm) stack height containing 14 lbs (6.4 kg) of copper in the conductor bars and end rings (13.2 kg charge). It is important to note that the laminations used here were designed for aluminum; i.e. the slot design had not been optimized for copper. A number of rotors were cast covering three different injection pressures and one-half were water quenched.

Seven rotors covering a range of process variables were tested and compared to a large database of similar aluminum rotor motors averaged as a "typical" motor. The same "standard" stator was tested seven times, yielding a spread of stator resistance losses ranging from 502 watts to 522 watts. This represents an approximate plus or minus 2% testing error which was assumed to be applicable across all test data. Applying this logic, the data for stator resistance and core iron magnetic loss have been averaged and considered a constant in both copper and aluminum rotors since they are not affected by rotor material.

Test Results

The test results were remarkably consistent across all process variables. The key measure of efficiency yielded virtually no difference with 90.7% as average and variation of only plus or minus 0.1 percentage points. Rotor watts loss averaged 157 watts with a maximum variation from 153 to 167 watts. With only seven tests, no pattern could be discerned relative to any of the process variables.

The conclusion is that the process is very robust and process variations within the range tested have no predictable effect on final performance results. Although the post-casting cooling method seemed to have no effect on the results, water quenching reduced handling time to one minute versus a 20-minute air-cooling time. This would allow much faster production in a manufacturing plant.

From the remarkable consistency of the test results, we conclude that the casting process is most viable. Results variations were all within test measurement accuracy and no pattern emerged reflecting die casting variables. When compared to historical variation in aluminum rotor motors, these copper rotors were so consistent as to deem the data variation insignificant.

Table 1 shows the IEEE test results as averages for seven rotors tested. Rotor resistance losses are the key item in rotor material substitution and yielded a 40% reduction in measured losses. This represents 80% of the theoretical maximum value possible in the conductivity difference between rotor materials. This is a very good start for a first attempt at real motors and may be improved further with detail lamination slot design.

Table 1. IEEE loss segregation test results for 15 Hp (11.2 kW) motor

	Al (W)	Cu (W)	ΔW	%
Stator Resistance	507	507	0	0
Iron Core Loss	286	286	0	0
Rotor Resistance	261	157	-104	-40
Windage & Friction	115	72	-43	-37
Stray Load Losses	137	105	-32	-23
Totals	1306	1127	-179	-14

Windage and friction losses are mechanical losses retarding rotation. Although these seem to have no relevance to rotor material, they do in this case. The copper rotors cast had smooth end rings except for projections for balancing weights. They did not include cooling fins on the end rings. With a lower resistance rotor, less heat is generated to be dissipated. These rotors, lacking fins, were adjoined on the shaft with an internal circulating fan for stator cooling. These fans are more efficient as they can be sized for their circulating job without having to dissipate rotor heat. As a result, when compared to aluminum rotors with fins, total windage losses were down 37% from 115 watts to 72 watts. Friction in the bearings is assumed to be the same. The cooler running copper rotors allow reduced windage losses via a more efficient internal fan and reduce the amount of copper required by eliminating the rotor end ring fins.

Stray load losses are the cumulative effect of magnetic transfer efficiency between the stationary stator and the rotating rotor as experienced in the air gap between the two. Consistent air gap and rotor balance also affect stray load losses and there is an electrical component to the magnetic transfer efficiency. Consistency in conductivity of rotor conduction bars is critical to proper induction magnetic transfer. Porosity or nonmetallic inclusions in cast rotor bars can result in variation in effective rotor bar cross sectional area, and therefore resistance, resulting in variation in the magnetic field in the air gap. This increases stray load losses via inconsistent magnetic flux density between stator and rotor reducing overall efficiency. The seven copper rotors exhibited such rotor bar consistency so as to reduce stray load losses by 23%, from 137 watts to 105 watts. A more accurate and consistent casting process might possibly produce similar stay load improvements in aluminum rotors. It is clear that the die-cast copper rotors contributed to the overall motor efficiency via a consistency not normally achieved in typical motor production.

The substitution of copper as rotor material directly achieved 58% of the total savings and was materially involved in saving the other 24% in windage losses and 18% in casting accuracy stray load losses. The combination resulted in 179 watts of savings or a total of 14% reduction in total losses. These results support the efficacy of both the material and the process. The rotors did not require balancing weights usually used to compensate for rotor bar inconsistencies.

Other Performance Measures

In addition to the loss measurements, the test method itemizes performance issues such as temperature rise above ambient, full load speed and power factor (Table 2). These data reveal a motor having different characteristics than a typical aluminum rotor motor. Overall efficiency resulted in a solid addition of 1.2 percentage points added directly to the motor nameplate efficiency. This is significant in that 20 years of motor efficiency improvements have already utilized all of the easy things that reduce losses. Copper rotors represent one of the largest possible reductions in losses without using amorphous steels or superconducting, still exotic and very expensive alternatives.

Table 2. Performance characteristics of 15 Hp (11.2 kW) motor

	Al	Cu	Difference	% Change
Efficiency	89.50	90.70	+1.20	+1.40
Temperature Rise, °C	64.00	59.50	-4.50	-7.00
Full Load RPM	1760.00	1775.00	+15.00	+0.85
Slip, %	2.22	1.37	-0.85	-38.00
Power Factor, %	81.50	79.00	-2.50	-3.00

Temperature rise above ambient is significant in the life expectancy of the motor. The general rule of thumb in the motor industry is that for every 10 degrees Centigrade hotter a motor runs, life expectancy can be cut in half. With nearly 5°C reduction in the copper motor temperature rise, we can expect a possible 50% increase in motor life when the motor is operated near design capacity. Only real field tests and time would be able to prove this hypothesis, but similar results have appeared in premium efficiency motors. Power factor is down slightly (3%) but is very near measurement accuracy levels. Power factor is only an issue if the electric power utility measures a low power factor for the entire factory facility.

Slip is the difference between the synchronous RPM of the field rotation at 60 Hz (or 50 Hz elsewhere in the world) and the full load RPM of the rotor and shaft assembly. This difference is what creates the torque to rotate the load. The copper rotors achieve this torque point with less slip or a higher measured RPM. The implications of a "stiff" motor or one that does not slow down much under load and the higher full load RPM are discussed in Ref. [2]. Starting, breakdown and locked rotor torque values are somewhat reduced in the copper rotor motor and again are discussed in Ref. [2]. Since we have simply substituted copper for aluminum with no design change to accommodate the copper, these torque factors could be corrected with changes in the cross sectional shape of the rotor bars not necessarily requiring an increase in total copper cross sectional area and cost.

Other Motors Tested

In the larger 25 Hp (18.5 kW) motor, the end rings were 6.5 inches (165 mm) in diameter with a stack height of 9.5 inches (241 mm). The squirrel cage contained

11.4 kg of copper and required melting 17.7 kg of copper per shot. The motor manufacturer provided sufficient laminations for 14 rotors. Motor tests of this second set of larger rotors showed even more dramatic results. This in part is due to the use of a rotor lamination slot design specifically designed for copper.

Again there was remarkable consistency in the results for the four rotors tested and compared to the same motor with an aluminum rotor. The rotor losses were 40% lower in the copper rotors and the overall losses were reduced by 17.6%. When the stator was optimized for the copper rotor, overall losses were reduced by 23%. Lower losses led to reduced rotor and stator temperatures. On completion of tests, the temperature of the stator winding of the motor with the copper rotor was 32°C cooler than that of the aluminum design; the copper rotor was 29°C cooler than the aluminum rotor. Lower running temperatures mean that smaller internal cooling fans can be used and this had a significant effect in reducing the parasitic component of the friction and windage losses on this motor designed for the copper rotor. Motor temperature translates directly to motor life and maintenance costs. Motors with cast copper rotors, with proper maintenance, would be expected to last longer and be more reliable.

A set of rotors cast for another motor company were for a 4 Hp (3 kW) motor. The end ring was 3.54 inches (90 mm) in diameter, stack height 6.1 inches (155 mm) and contained 3.2 kg of copper. Overall motor losses were reduced by 21% with the copper rotor compared to the conventional aluminum.

Copper rotors for a 5 Hp (3.7 kW) motor were cast for a fourth manufacturer. Rotor I^2R losses were reduced by 38% compared to the aluminum counterpart motor, but surprisingly, the iron core loss component was much higher for the motor with the copper rotor. This was apparently due to insufficient consideration of the rotor and stator lamination designs from the aluminum rotor motor used. It appears the iron was almost totally saturated in the aluminum design. The higher current in the copper rotor could not further magnetize the iron and appeared as a large apparent loss.

As shown in Table 3, I^2R losses for all motors fitted with copper rotors from this test program showed rotor reductions of about 40% with one smaller motor showing an even greater reduction.

Table 3. Rotor I^2R losses – copper vs. aluminum

HP	KW	Poles	Al	Cu	Difference	%
4	3	4	221	92	129	-58
5	3.7	4	--*	--*	--*	-38
15	11	4	262	157	104	-40
25	19	4	410	292	118	-40

*Actual loss values not reported

Metallurgical examination of cast copper rotors confirmed that there was no interaction between the copper conductor bars and iron laminations. Chemical analysis revealed that small amounts of iron (10 to 11 ppm) and oxygen (0.084 to 0.163 wt. %) were picked up during casting. The combined effects of the presence of microstructural defects and chemical contamination reduced the electrical conductivity of the cast copper conductor bars only slightly to 96.8 and 98.7% IACS in the two measurements performed on the first set of rotors cast.

Porosity in the far end ring of the first set of copper rotors appeared to be 2 to 3 percent but did not extend into the conductor bars. The uniformity of conduction paths in these copper rotors shows up as a reduction in stray load losses and had not been expected. No balancing to compensate for uneven weight distribution was required. The larger rotors of the second group cast were more of a problem in this regard showing as much as 25% voids in the first shots and 8 to 10% in the rotors tested for electrical performance. This is apparently due to inadequate venting at the far end ring and excessive oxygen pick-up during the very long melting time (about 13 min.) resulting from the small power supply available. This porosity had little apparent effect on the performance of these copper rotors. Die cast aluminum rotors very often have considerable porosity requiring use of extra aluminum to compensate for porosity and always require balancing.

Copper Rotor Literature Data

Table 4 summarizes the overall motor efficiencies and loss reductions observed in motors fitted with copper rotors where comparisons with aluminum are reported in the literature including the data of this study. A broad range of motor power from 4 Hp (3 kW) to 270 Hp (200 kW) is covered. A clear pattern of increased efficiencies with higher values for the larger motors and loss reductions averaging 14.7% is evident.

Table 4. Overall motor efficiencies and loss reductions via copper rotors – data from this study and the literature

Hp	kW	Poles	Eff. Al	Eff. Cu	Diff.	Loss Reduction, %	Reference
4	3	4	83.2	86.4	3.2	19.0	This study
7.5	5.5	4	74.0	79.0	5.0	19.2	3
10	7.5	4	85.0	86.5	1.5	10.0	4
15	11.2	4	89.5	90.7	1.2	11.4	This study
25	18.8	4	90.9	92.5	1.6	17.6	This study
40	30	4	88.8	90.1	1.3	11.6	5
120	90	2	91.4	92.8	1.4	16.3	5
270	200	4	92.0	93.0	1.0	12.5	3

Conclusions

The motor performance tests reported here have verified years of calculations on the part of motor manufacturers about the prospective benefits of incorporating copper in the squirrel cage structure. The results show conclusively that overall motor energy losses are reduced by an average 14% and the nameplate efficiency is increased by at least a full percentage point.

Acknowledgements

This project was sponsored by the world copper industry through the International Copper Association, Ltd. and managed by the Copper Development Association Inc. Additional funding was provided by the U. S. Department of Energy Office of Industrial Technologies and the Air Conditioning and Refrigeration Technical Institute. Several major motor manufacturers underwrote the costs of rotor lamination material, die inserts and in-house dynamometer testing of motors equipped with die-cast rotors. Formcast, Inc., Denver, Colorado, under Dr. Stephen P. Midson's direction, provided the die casting equipment and the casting expertise. Mr. Ruedi Beck of DieTec, GmbH, Gossau, Switzerland, served to provide the die casting tooling design and innovative approaches to the heated nickel alloy die insert technology that is to be used in commercial production of copper rotors.

References

[1] DOE/CS-0147 – U. S. Department of Energy (1980) Classification and evaluation of electric motors and pumps.

[2] Peters DT, Van Son DJ, Cowie JG, Brush EF Jr.(2002) Improved energy efficiency and performance through the die-cast copper rotor. International Conference on Electric Machines, Brugge, Belgium

[3] Lie S, Di Pietro C (1995) Copper die-cast rotor efficiency improvement and economic consideration. IEEE Trans. Energy Convers. 10 No. 3: 419-424

[4] Poloujadoff M, Mipo JC, Nurdin M (1995) Some economical comparisons between aluminum and copper squirrel cages. IEEE Trans Energy Convers. 10 No. 3: 415-418

[5] Private communication with manufacturer.

Design of High Efficiency Induction Motors with Die-Casting Copper Rotors

Francesco Parasiliti, Marco Villani

Department of Electric Engineering, University of L'Aquila, 67040 L'Aquila, Italy

Abstract

The paper deals with the use of copper cage in three-phase low voltage induction motors and gives a design guideline to optimize their efficiency, according to the new European classification scheme. An accurate motor design allows to "move" the motor from lower to upper efficiency classes without affecting the starting performance.

Introduction

The European Committee of Manufacturers of Electrical Machines and Power Electronics (CEMEP) and the European Commission have agreed to a joint classification scheme that enables all OEMs (Original Equipment Manufacturers) and the other customers and users of induction motors to have a simple appreciation of the efficiency of this component [1]. This agreement should stimulate the Manufacturers in the development of new ranges of high efficiency motors that requires an accurate motor design [2], the adoption of new materials (e.g. premium steel [3]) and innovative technologies.

It is well known that incorporation of copper for the rotor bars and end rings in place of aluminum would result in attractive improvements in motor energy efficiency.

Tool steel moulds as used for the aluminum die casting process have proved to be entirely inadequate when casting higher melting point metals including copper. Lack of a durable and cost effective mould material has been the technical barrier preventing manufacture of the copper cast rotor.

In order to win acceptance by motor manufactures, they need to be produced with the same equipment that is currently in place and the copper industry has to find a way to deal with the thermal difference between molten copper and conventional technology and its effect on die life.

Few years ago, a consortium of USA partners has been assembled, having the aim to design, fabricate and demonstrate moulds suitable to withstand the copper motor rotor die casting environment for an economically acceptable life. Advances are being made and availability of durable and cost effective mould materials is expected in the next future [4], [5], [6].

The adoption of die-casting copper rotor, requires obviously to review the motor design criteria. Moreover, because the starting torque is proportional to its rotor resistance, the starting performance specifications tend to limit the amount by which the rotor ohmic loss can be reduced in favor of efficiency.

For this reason, particular attention has to paid to the new motor design in order to ensure "feasible replacement". Manufacturers are careful to follow designs that retain the critical elements of motor performance necessary to function within starting torque, starting current, breakdown torque, temperature considerations and other pertinent motor performance requirements. By maintaining these standards, replacement motors do not risk incompatibility with the current applications.

The authors have deeply investigated the use of copper rotor cage and have developed a design guideline to optimize the efficiency in three-phase low voltage induction motors, by choosing several design strategies: the simple substitution of copper for aluminum has been tested, then the motor with copper cage has been optimized by changing accurately some motor dimensions.

This activity falls in a research program supported by the MIUR (Italian Ministry for Education, University and Research) and concerning the analysis of motors and drives energy efficiency increase in industrial and civil applications.

The Substitution of Copper for Aluminum

A first analysis has concerned the evaluation of motor performance by substituting copper for aluminum, without chancing neither the electrical steel or overall design of motor: this test has allowed to verify the achievable efficiency according to the European classification scheme (Fig. 1).

Three low power sizes have been chosen, and particularly: 1.5, 3 and 7.5 kW, 4 pole, 50 Hz, 400 V, TEFC, single-cage rotor.

These motors, represent commercial motors with aluminum cage rotor and are typical motors produced by European manufacturer: their rated efficiencies belong to the Class Eff 3 (Fig. 1) and they can be labeled "Low efficiency motors".

The motor performance have been evaluated by an analytical procedure where the physical description of the motor is reduced to equivalent parameters such as resistance and inductances: the adopted model takes into account the influence of saturation on stator and rotor reactances and the influence of the skin effect on rotor parameters. The effects of the temperature on motor are computed on the basis of a detailed lumped-parameters thermal model. The validity of the mathematical model has been verified by means of experimental tests on several three-phase induction motors.

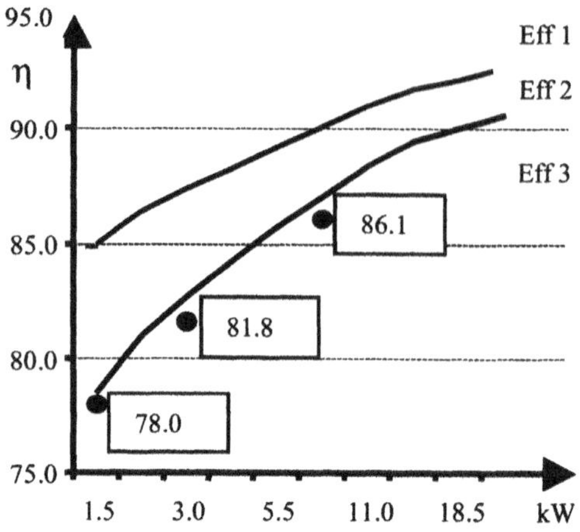

Fig. 1. The European classification scheme and the efficiency of investigated motors with aluminum rotors

The results concerning the substitution of copper for aluminum are shown in the Tables 1, 2 and 3 (see the column "initial"): for the evaluation of active material cost, the following average costs have been assumed: 0.62 €/kg for the electrical steel, 3.13 €/kg for the stator winding, and 1.90 and 2.60 €/kg respectively for the aluminum and copper die-casting (in this analysis, the additional cost for the copper die-casting and moulds have not been taken into account).

It is evident an efficiency improvement that allows to move all sizes from the class Eff 3 to the class Eff 2 (see the triangular symbols in Fig. 2): this is due mainly to a significant reduction of losses in rotor bars of 50%. Moreover, the motor with cooper die-cast rotor has slightly lower line current and stator winding losses, light higher breakdown torque and good power factor: as drawbacks, it has poor starting performance. In fact, it is evident a slight increase of starting currents (about 7% for the small size and 4% for the others) and a drastic reduction of starting torque (13% for the 1.5 kW, 21% for the 3.0 kW and 25% for the 7.5 kW).

The Optimized Design of Induction Motors with Copper Squirrel-Cage

Further improvement on motor performance can be achieved if the substitution of copper for aluminum is associated with an accurate motor design that allows to exploit the advantages of a copper cage, without affecting the starting performance. Moreover, it is interesting to evaluate the achievable improvement on the efficiency according to the European classification scheme.

The additional costs associated with making higher efficiency motors, compared to standard efficiency motors, will vary between companies and design strategies: the biggest differences will be between those companies that base High Efficiency Motors on existing standards designs and those which use completely new designs and tool replacements. For this reason three strategies have been introduced respect to the amount of the additional cost, each of one affects the number of design variables for the optimization procedure. They are labeled as follows:

S1) copper cage + new stack length and stator winding;

S2) S1 + change of lamination;

S3) S2 + change of all motor dimensions, and stator diameters.

In the first case (S1) the stator and rotor slots dimensions and the inner and outside stator diameter are unchanged. The challenge is to physically fit more active material into the motor to reduce iron and copper losses. The cost of tooling for the new designs is effectively the same of the traditional design since the need for costly new lamination punch tools or stator housing tools are avoided (except the additional cost for copper die-casting).

The second case (S2) foresees also the change of lamination (stator and rotor slots dimensions) and this requires a renewal of the lamination tooling even if the inner and outside stator diameters are unchanged.

The third and more expansive level (S3) requires to change all motor dimensions, inner and outside stator diameters included. In this case a final option is the use of a larger frame size with modification of the existing housing.

It is important to underline that in all strategies, the change of electrical steel has not been foreseen and it is the same of the original design: moreover, the variables "Stack length", "Outside Stator Diameter" and "Inner Stator Diameter" have been varied with reference to "normalized values" only, according to the Manufacturer suggestions.

Several constraints have been introduced that concern the typical motor performance but above all the starting performance. For each optimization, a low cost motor design was involved by means of an appropriate algorithm developed by the authors [7]; the optimization was formulated as constrained maximization of the objective function "rated efficiency" expressed in terms of the motor design variables.

Results

The final results for the optimized designs with copper cages are summarized in Tables 1, 2 and 3: they show the main geometric dimensions, the motor performance, the weight and the cost of active materials.

The first comment concerns the efficiency values whose differences, respect the original design with aluminum rotor, are shown in Tab.4: it is evident a gradual increase that allow to move the motors toward the upper efficiency class Eff 1. The efficiency values of new motors with reference to the European classification scheme have been reported in Fig. 2.

For the 1.5 kW motor (Tab.1), the strategies S1 and S2 give rise to new designs that are within the Eff 2 class and only with S3 has been possible to achieve the lower limit of the Eff 1 class: this is due to the available dimensions on stator diameters, stack length and housing, that have not allowed "further movements". Obviously, the use of a "premium steel" combined with an accurate motor design allows, for this size, to reach easily the Eff 1 class.

For the other sizes (Tab. 2 and 3), it is evident how the solution S1 is sufficient to move the motors in the Eff 1 class, and even with S2 and S3 highest levels can be reached. It is important to underline that all these movements have been achieved without chancing the electrical steel (that is without using any "premium steel").

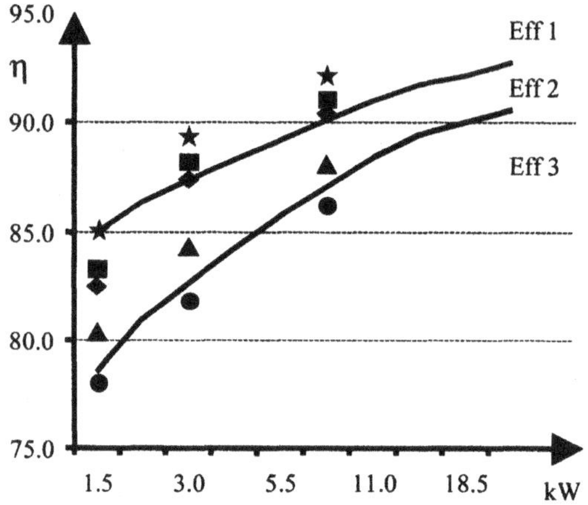

Fig. 2. The European classification scheme and the efficiency of new designs with copper rotor

Table 1. 1.5 kW: Comparison between the initial and optimized designs

1.5 kW		Initial		Optimized (Copper cage)		
		Al ●	Cu ▲	S1 ◆	S2 ■	S3 ★
Stack length	[mm]	70		80	80	100
Inner stator diameter	[mm]	90		-	-	90
Outer stator diameter	[mm]	145		-	-	160
New stator winding		-		x	x	x
New lamination		-		-	x	x
Efficiency	%	78.0	80.3	82.5	83.5	85.1
Power factor		0.83	0.82	0.81	0.82	0.82
Rated current	[A]	3.30	3.29	3.23	3.17	3.09
Starting current	[A]	16.8	18.0	18.6	18.5	18.7
Starting torque	[Nm]	18.7	16.2	17.4	18.9	18.3
Breakdown torque	[Nm]	23.2	24.3	25.7	26.4	26.5
Losses [W]: Stator winding		275	258	210	186	165
Rotor bars		73	34	30	31	25
Iron		43	53	55	56	47
Weight [kg]: Gross iron		11.2	11.2	12.9	13.0	19.6
Stator winding		1.46	1.46	1.71	1.77	1.88
Rotor bars		0.65	2.14	2.28	2.14	2.93
Cost of active materials	[€]	12.8	17.1	19.3	19.2	25.7

Table 2. 3 kW: Comparison between the initial and optimized designs

3 kW		Initial		Optimized (Copper cage)		
		Al ●	Cu ▲	S1 ◆	S2 ■	S3 ★
Stack length	[mm]	130		150	150	150
Inner stator diameter	[mm]	90		-	-	90
Outer stator diameter	[mm]	152		-	-	178
New stator winding		-		x	x	x
New lamination		-		-	x	x
Efficiency	%	81.8	84.3	87.4	88.2	89.2
Power factor		0.81	0.80	0.79	0.79	0.79
Rated current	[A]	6.52	6.37	6.33	6.18	6.50
Starting current	[A]	30.1	31.6	33.4	31.1	35.0
Starting torque	[Nm]	35.6	28.0	33.8	34.1	35.7
Breakdown torque	[Nm]	49.0	49.6	54.2	53.9	59.0
Losses [W]: Stator winding		356	330	209	179	189
Rotor bars		163	77	64	61	60
Iron		115	115	128	121	96
Weight [kg]: Gross iron		22.7	22.7	26.2	26.1	36.0
Stator winding		1.94	1.94	2.8	3.50	3.04
Rotor bars		0.74	2.43	2.67	3.08	3.39
Cost of active materials	[€]	21.5	26.5	29.4	35.1	40.6

Table 3. 7.5 kW: Comparison between the initial and optimized designs

7.5 kW		Initial		Optimized (Copper cage)		
		Al ●	Cu ▲	S1 ◆	S2 ■	S3 ★
Stack length	[mm]	160		178	180	180
Inner stator diameter	[mm]	127		-	-	150
Outer stator diameter	[mm]	200		-	-	240
New stator winding		-		x	x	x
New lamination		-		-	x	x
Efficiency	%	86.1	88.0	90.2	90.7	92.1
Power factor		0.82	0.81	0.82	0.82	0.84
Rated current	[A]	8.83	8.69	8.41	8.33	8.04
Starting current	[A]	53.8	56.1	53.4	51.5	47.0
Starting torque	[Nm]	104	78.0	98.7	98.0	100
Breakdown torque	[Nm]	152	153	162	158	147
Losses [W]: Stator winding		603	566	349	315	230
Rotor bars		267	128	116	109	62
Iron		242	242	251	256	233
Weight [kg]: Gross iron		48.4	48.4	53.9	54.5	77.4
Stator winding		4.08	4.08	5.57	5.97	7.55
Rotor bars		1.45	4.78	5.17	5.45	8.35
Cost of active materials	[€]	45.5	55.2	64.3	66.6	93.3

Table 4. Efficiency variations and percentage loss reduction (respect to the aluminum rotor)

	Initial Al ➔ Cu	Optimized (Copper cage) S1 ◆	S2 ■	S3 ★
		ΔEfficiency		
1.5 kW	2.3	4.5	5.5	7.1
3 kW	2.5	5.6	6.4	7.4
7.5 kW	1.9	4.1	4.6	6.0
		ΔLosses (%)		
1.5 kW	11.2	22.7	28.0	36.3
3 kW	16.6	34.5	39.6	42.0
7.5 kW	14.6	32.6	35.6	46.4

The percent difference on total losses are shown in Tab.4: the significant losses reduction led to lower temperature on rotor and stator windings and it means that smaller internal cooling fans can be employed with effect in reducing the friction and windage losses. Moreover, motor temperatures translate directly to motor life and maintenance costs.

The new motors present a reasonable breakdown torque, starting torque and starting current: particularly, the optimization algorithm has found new designs with a starting torque comparable respect the initial design with aluminum rotor ones.

The increase on active material cost is due mainly to the use of copper in the rotor bars and the increase of amount of iron in the stator and rotor core. The comparison of the optimized designs with the initial one with copper cage, points out how the designs S1 and S2 present a reasonable cost even if, for the 1.5 kW, these solutions do not guarantee the achievement of the upper class Eff 1.

It is important to underline that, in the proposed examples, the active material costs represent 30÷40% of manufacturing costs (active material, labor and structure costs): consequently, the increases on the active material costs reflect on an increase on manufacturing costs whose percentages are in the range 10÷50% (the higher values correspond to the S3 designs).

Conclusions

Copper rotor motors could be the next step in a steady line of motor efficiency improvements and may be able to achieve highest efficiency values according to the European classification scheme.

The proposed examples point out how the design strategies reflects significantly on efficiency and allow to "move" the motor from lower to upper efficiency classes.

Starting from the Eff 3 class, the use of copper in place of aluminium, allows to reach the Eff 2 class, but the starting performance are very poor, with a significant reduction on starting torque (up to 25%). The highest efficiency level Eff 1 can be achieved by a design optimization of copper cage motor, and this result can be achieved with low additional costs.

Obviously, these improvements are affected by the cost for the copper die-casting and the cost of mould material and these aspects represent now technical barriers preventing manufacture of the copper cast rotor. If advances will be done on availability of durable and cost effective mould materials, the motors with die-casting copper rotor will gain certainly more and more interest in the future European market that shall require only energy efficiency induction motors.

References

[1] P. Bertoldi, "EU/CEMEP Classification Scheme for Motors and Negotiated Agreement", Energy Efficiency Improvements in Electric Motors and Drive, Springer, June 2000, ISBN: 3-540-67489-6, pp. 369-375

[2] F.Parasiliti, M.Villani "Evaluation of the Design Options and Cost Impact of Improving Motor Efficiency", Energy Efficiency Improvements in Electric Motors and Drive, Springer, June 2000, ISBN: 3-540-67489-6, pp. 514-528.

[3] F. Parasiliti, M. Villani "Technical and economical evaluation of electrical steels for high efficiency motors", Transworld Research Network, Recent Res. Devel. Magnetics, n. 2 (2001), pp. 47-54.

[4] Dale T. Peter "Die-Casting Background", Die-Casting Copper Motor Rotor: Workshop and Technology Demonstration, CDA-Copper Development Association, Denver (USA), January 2002.

[5] E. Brush "Rotor Die-Casting in Copper To Date", Die-Casting Copper Motor Rotor: Workshop and Technology Demonstration, CDA-Copper Development Association, Denver (USA), January 2002.

[6] A.Ansel, O.Walti, J.F.Brudny, "Influence of copper pressure die-casting on induction machine magnetic behaviour", International Conference on Electrical Machines, (ICEM), Helsinki, September 2000.

[7] A. Daidone , F. Parasiliti , M. Villani , S. Lucidi, "A New Method for the Design Optimization of Three-Phase Induction Motors", IEEE Transaction on Magnetics, 34 (1998), pp. 2932-2935.

This work was developed in the ambit of the COFIN 2001 Program and supported by MIUR (Italian Ministry for Education, University and Research).

A New Technology to Make Rotors with Copper as Magnetic Conductor

Christophe Paris[1], Olivier Walti[1-2]

1 FAVI SA, 14 rue Louis Deneux, BP 5, 80490 HALLENCOURT, France
Phone: +33 3 22 66 20 56, Fax: +33 3 22 66 20 63, E-mail: parisc@esiee-amiens.fr

2 ESIEE-Amiens, 14 quai de la Somme, BP 100, 80083 AMIENS Cedex 2, France
Phone: +33 3 22 66 20 00, Fax: +33 3 22 66 20 10, E-mail: waltio@esiee-amiens.fr

Abstract

Our study aims at showing a new technology to make rotors with copper as magnetic conductor. It deals with copper's pressure die-casting rotors production. It will be very interesting also to compare these new technology to the usual to make rotors with copper as magnetic conductor; but also to study the electromagnic's behaviour of the copper die-casting rotor in comparison with the aluminium technology.

1. Introduction

Generally, squirrel cages for induction machines are made by die-casting aluminium in slots of the rotor laminated frame, developing in that way, with the two short-circuit end rings, a compact entity.

FAVI Company, located in the north of France and specialised in copper alloys pressure die casting, focused on an industrial process to manufacture copper pressure die-cast rotors. Actually, the single technology used to manufacture squirrel cages with copper conductors is to assemble shaped bars into the magnetic circuit slots and to weld at the extremities of both end rings. We will make in our paper a comparison between these two ways of production with pointing out the obvious gain fathered by the die-casting process as for its simplicity as well as for the very large possibilities of filling of the magnetic circuit slots.

We would like to present in this article, the contribution of our technology by comparing the functional characteristics obtained from two induction machines with the same geometry and whose the only difference is the nature of the rotor conductor material, that is to say the change from aluminium to copper.

Moreover, we will develop the impact of our technology on the machine magnetic behaviour, more particularly on the change of the characteristics due to the thermal shock during the die-casting operation.

Finally, we will conclude with a technical and economical study from datas given by the EFF1 motors range construction standard, which is regulating the manufacturing of high efficiency electrical motors. We could therefore point out times of return of investment, simply fathered by the induced energy saving.

2. Two Methods for the Production of Copper Rotors

To the best of our knowledge, the most commonly used method to manufacture rotors with copper conductors is referred to as « welded assembly construction ». This part of the article will concentrate on the comparison of the two different manufacturing processes of the rotor using either pressure die casting or « welded assembly construction ».

In order to synthesize these two processes, we have presented the principle operations of the process.

The welded assembly method requires the following operations (Fig. 1)

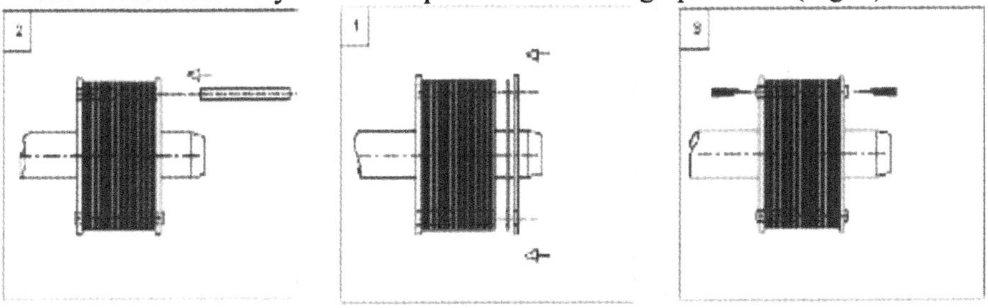

a) Assembly of the magnetic laminations + two copper flanges on an axle (definitive or otherwise).

b) Positioning of the bars streamlined to the geometrical shape of the rotor slot.

c) Mechanical coupling of the bars onto the rings.

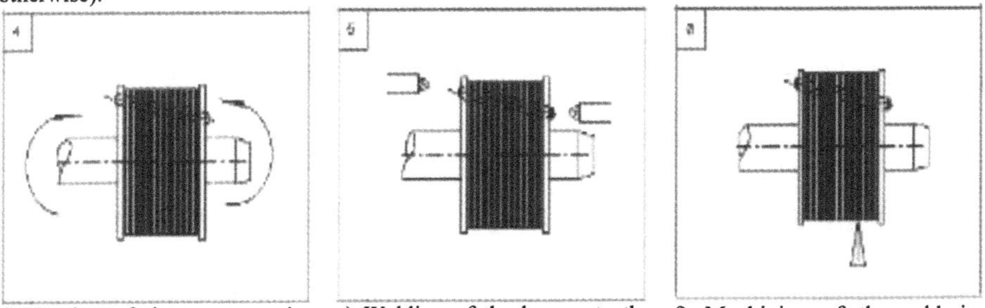

d) Torsion of the rotor to give the helix angle.

e) Welding of the bars onto the rings.

f) Machining of the soldering bosses onto the rings.

Fig. 1. Operations required for welded assembly method.

Moreover, in order to facilitate the placing of the bars within the steel plates it is essential that the copper streamlined bars used meet with the strict geometric demands. Added to this, these bars are machined at each end in order to facilitate the welding process between the ring and the bar.

Finally, in order to increase the height of the rings large flanges or for certain applications, rings made up of several basic layers, are used. The welding process in this case is delicate. In fact when the ring is solid it is vital to heat everything in order to obtain a perfect contact between the bars and the ring, which imposes heavy restraints on the magnetic sheet plates.

In the case of a laminated ring, there will always be a slight gap between each layer, naturally inducing a thermic insulator which creates unfavourable conditions for the diffusion of the heat.

The die casting method can be summed up in three main points (Fig. 2).

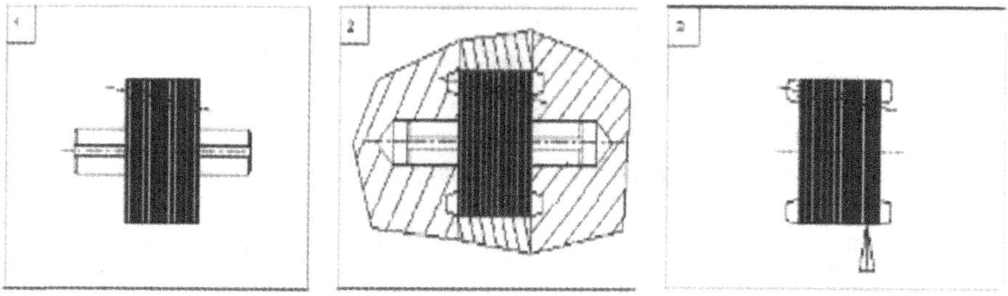

a) Assembly of the magnetic laminations on an axle creating the helix angle.

b) Injection of the metal in order to produce the bars and the two rings.

c) Machining of the rotor in order to remove the ingates resulting from the injection.

Fig. 2. Die casting method main points.

The injection of the metal requires specialized equipment and know how in this field. It also required five years of perfecting the technique in order to maintain the quality of the metal and ensure that the end products are faultless.

In order to underline the advantages gained from the injection method in comparison with that of the welded assembly method, we will now, quite simply describe some of the differences between the two methods (TABLE I).

Before analysing these differences, it must be noted that an injected rotor is made more quickly, due to the dramatic reduction in the number of operations to be carried out.

3. Copper's Contributing in Comparison with Aluminium

3.1 Improvement of the electromagnetic behaviour [1]

The purpose of this paragraph is to compare the characteristics torque / speed and efficiency / effective power of two machines perfectly identical as regards geometry, the only difference standing in the nature of the material conductive to the rotor. Testing has been carried out on a F category machine with a rated power of 5.5 kW, power defined in the case of a rotor composed of aluminium conductors.

Table 1. Comparison of the two methods

General observations	Die-casting Method	Welded Assembly
Filling of rotor slots	**Optimum**	Depends on assembly
Shape of conductors	**Adapted to needs**	Rectangular or circular
Shape of rotor slots	**Adapted to needs and allowing optimisation of the motor**	Depends on the section of the bars
Heating of sheet metal	**Homogenous**	Concentrated at the two ends
Speed of process	**Gain in productivity**	Numerous operations
Contact between	**Excellent and full**	Depends on quality
bars/rings	**proof**	of the welding

Materials chosen for this comparison are aluminium, material classically used in the manufacturing of rotor cages and our Cu97 alloy (99.7% of copper), specially finalised for this application.

The torque, in an asynchronous cage machine, is initially inversely proportional to the resistance of the rotor conductors concerning speeds which stand close to the nominal working point. When geometry from slots to the rotor is identical, the only influent parameter is the material resistivity.

Furthermore, the diversity of copper alloys allows to adapt the functional characteristics of the motor to its future application by gambling on the material resistivity (modification of the nominal torque or starting torque), on the alloy additives in order to improve the mechanical stability. These different data will be commented in a next newsletter.

Due to the decrease in the resistivity between the both selected materials, a Cu97 asynchronous rotor cage machine generates a torque 50% higher than the torque obtained for an aluminium rotor in the stable use zone of the both machines (Fig.3).

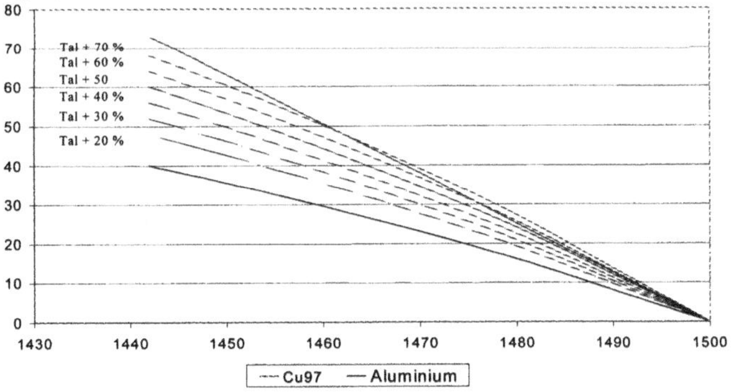

Fig. 3. Torque characteristics in the stable use zone

We can see that at the nominal speed of this machine defined for the technology of the Aluminium rotor, the torque obtained in the case of a Cu97 rotor is 80% higher than the torque obtained for the Aluminium rotor.

Furthermore, as the global efficiency of the machine is slightly higher in the case of a copper rotor (Fig.4), it seems interesting to specify that a FAVI asynchronous cage machine allows a wider application range.

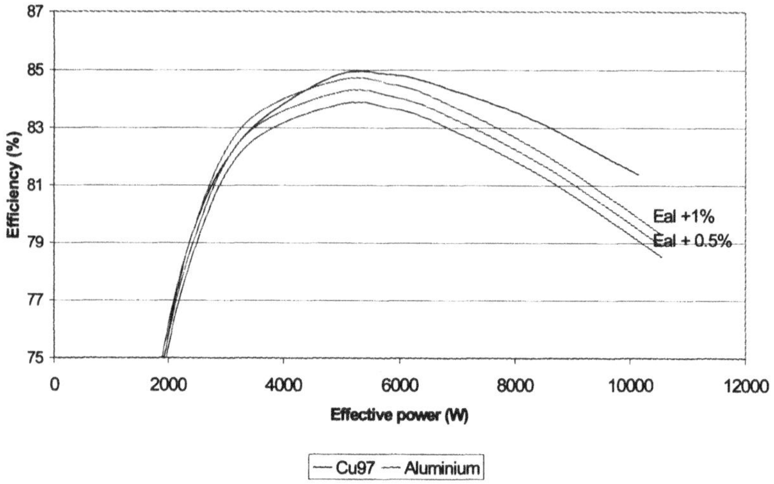

Fig. 4. Efficiency characteristics

3.2 Reduction of increases in temperature

The loss of Joules in a conductor is proportional to its resistance strengh and is thus directly related to its ability to resist and its dimensions.

We compare the temperature readings of two asynchronous machines, which are completely identical from an geometric stand point while they are in operation. The only difference between them will be the make up of the conductor material, in our case, aluminium or Cu97, which the rotor's bars and rings are made up of.

Otherwise, everything else in the make up of the two motors is equal. Tests carried out in the laboratory have confirmed **that the amount of Joules that the rotor loses, decreases by 40%** when we replace the aluminium rotor by the Cu97 rotor.

A Cu97 rotor consumes less energy and perfectly respects the energy control policy which is perfectly relayed by international agreements in this domain in addition to the 14001 norm which concerns the environment.

This decrease in the loss of Joules affects the distribution of temperatures within the machine which in turn augurs ***a lesser drop in temperature which allows the fabrication of watertight motors.*** In fact, the temperature readings from a 7kW machine which is running prove that this is the case.

A reduction in this temperature is very beneficial in harsh environnments in which thermal interaction between the elements is a crucial parameter. We believe that the ***Alterno-Starter*** perfectly overcomes this constraint.

On reading these results and on looking back to the definition of the nominal power of one of the machines it seems clear to us that an asynchronous motor with a Cu97 rotor has a much higher nominal strength. This is even without having modified the geometry of the machine. The volumic strength has been increased. This is an interesting point for heavily loaded applications watertight motors or to increase the range of ways that the machines can be used, for example energy reducers.

Otherwise, if we suppose that the average lifespan of an electrical machine depends on the temperature that it runs at and, that an empirical law stipulates that if the temperature is brought down by 10°C it will multiply the life of the machine by two, we then can certifiy that a Cu97 rotor increases the life of asynchronous machines.

3.3 Influence of thermal shock generated by copper injection on the magnetic properties of laminations [2], [3].

Before and after injection, we noticed that the colour of the laminations changed (Fig.5). Therefore, we needed to research the influence of thermal shock on the magnetic performance of the « asynchronous machine ».

Moreover, the copper melting point is far higher to the curie point of the electric steel generally used.

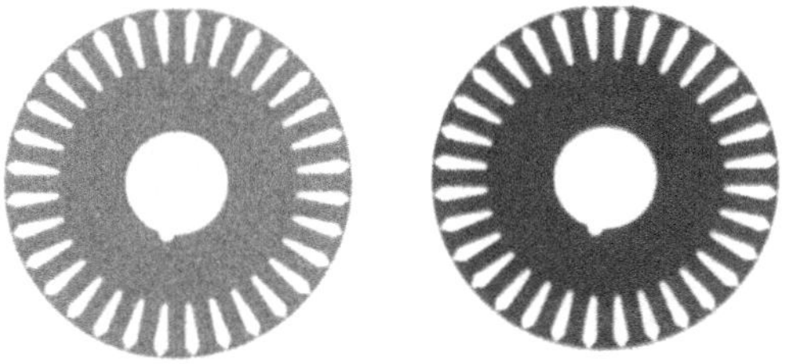

Fig. 5. Lamination before and after die-casting

In order to characterise this thermal effect on the magnetic performance, we wanted to use a normalised protocol. Then we set up a bench for characterisation from a standard Epstein square (Fig.6). This device corresponds to a vacuum transformer whose the magnetic circult created by samples easily interchangeable.

Furthermore, our system is entirely guided by a software developed under LABVIEW in order to answer to repeatability criterions and to allow an easier research.

This protocol comes up to the CEI 60404 – 2 standard of 1996.

We controlled the secondary stress of the Epstein square in the prospect of carrying out a materials analysis under controlled sinusoidal induction. Therefore, this parameter will be our reference.

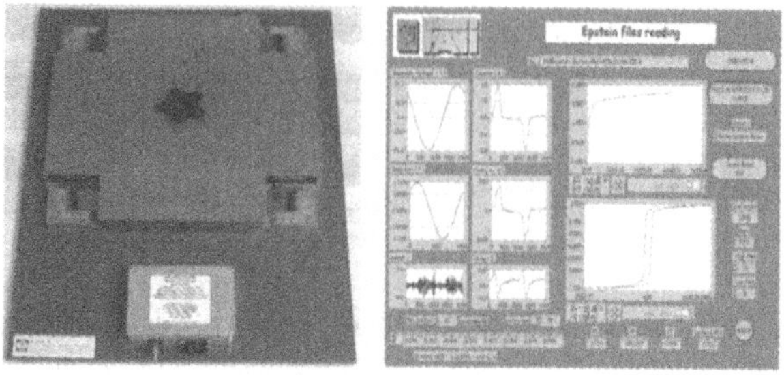

Fig. 6. Epstein square and supervision software

We defined the following step in order to quantify the influence of a thermal shock.

First, thanks to an adapted software, we modelized the thermal performance of the lamination during a copper injection phase (Fig.7).

This modelling enables us to know the "isotherms" through the heat diffusion in the laminations.

The following figures show the results depending on the lapse of time that the lamination was subjected to the heat source (the melting copper).

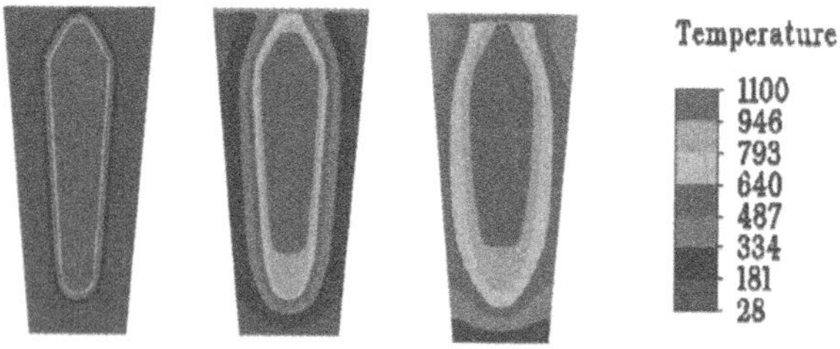

Fig. 7. Thermal shock modelisation

Thanks to these data, we elaborated a device which enables us to introduce a set of Epstein samples in a temperature controlled chamber. Temperatures have been chosen according to the results during the simulation.

Of course, the thermal stresses are rather far from these imposed by the injection.

The temperature level remains the same, only time and exposure mode can be discussed.

We selectionned the 550°C, 750°C, 800°C, 920°C and 1100°C temperatures.

In fact ,the Curie temperature of the studied magnetic material was about 760°C.

Furthermore, we always compared our results to a material which has not been heated up.

It does not seem that the magnetic characteristics were affected by this thermal shock (Fig.8). We can notice that the iron losses are slightly reduced. This could be explained by the release of the mechanical stresses generated by the lamination cutting.

Moreover, we show a very big increase of the relative permeability in stationary field.

We know that it is very difficult to transpose these results in stationary field in comparison with the rotating field.

Nevertheless, we begun this study in a doctoral thesis. This initial results are encouraging. We will present these results in a next document «Info Rotors » as soon as we obtain their validation thanks to very specific tests (for example on the machine synchronism). It is sure that the magnetic model of the machine must fitted out for the rotor Cu97 use. The first results we obtained during the analysis of the starting torque go in a right direction.

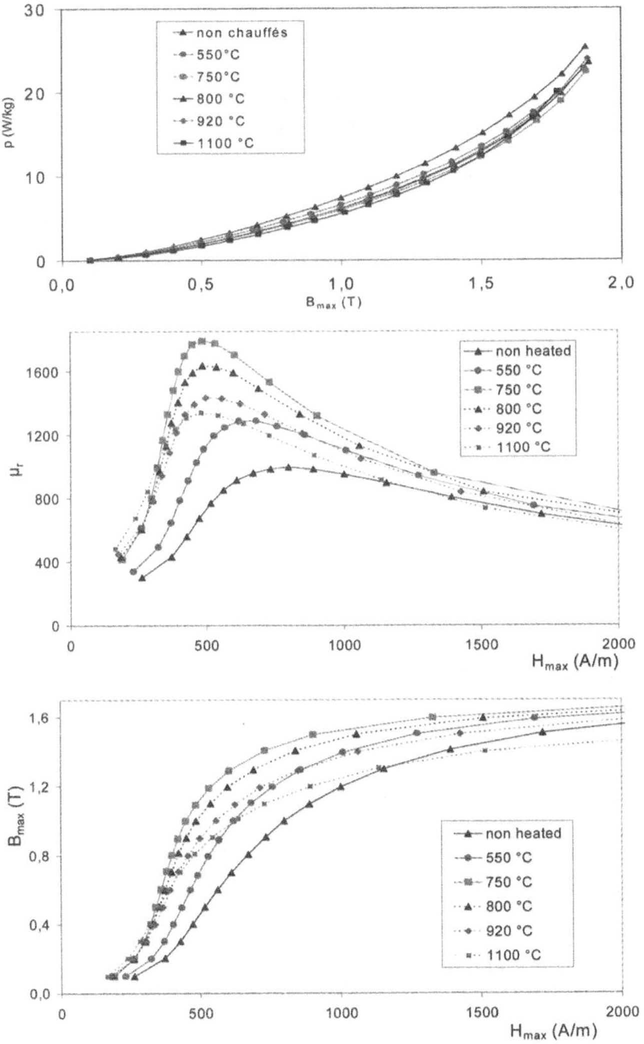

Fig. 8. Magnetic characteristics

3.4 Economically viable

This paragraph sums up our different observations obtained during energy trials. We measured the energy consumption over a 24 hour period and estimated the economy made based on the cost of energy at 7.62 euro cents per kW/h. These trials were carried out using a 5.5kW asynchronous machine, the nominal power defined for aluminium rotor conductors. We have therefore chosen a type A5 aluminium rotor and a copper type CU97 rotor for comparison. Nevertheless, due to restraints imposed by the process it would be limiting to base the difference solely on the one single parameter that the resistance represents.

We kept the operating points at a constant torque and at a constant wattage rating.

For each of the trials , we took a reading of the speed of the shaft, the wattage rating of the MAS, the cos φ of the stator during supply without a selector (directly on the 400V three-phased voltage network), the real power consumption of the machine over a 24 hour period, in addition to its average temperature when operating (a reading was taken at the inductor heads) (Table 2).

Table 2.a Constant Couple Trial. C=35.7 N.m

	Speed (Tr/mn)	Power (W)	Cos φ	Consumption (kW/h).	Efficiency	T (°C)
Alu.	1450	5415	0.78	156.483	82.98%	92.5
Cu97	1470	5490	0.77	156.342	84.28%	83.6

You can note that for the same consumption, the efficiency of a machine with a Cu97 rotor will be improved, thus allowing a higher nominal power to be obtained. In addition to this trial, if we imposed a constant heat of 100°C in the machine we obtained a wattage rating which was 4% greater to that of the aluminium Cu97 technology.

Table 2.b Constant Wattage Rating Trial. Pu=5.5 kW

	Speed (Tr/mn)	Torque (Nm)	Cos φ	Consumption (kW/h).	Efficiency	T (°C)
Alu.	1445	36.3	0.79	158.446	83.38%	89
Cu97	1470	35.65	0.77	156.342	84.28%	83.6

We can see that there is a difference in consumption between the two technologies. The Cu97 rotor diminishes the running costs of the machine by 16 eurocents every day it is in use. Nevertheless, this result can be improved if the machine is constructed following a schedule of conditions including the Cu97 rotor technology.

Furthermore, during these trials we noted down the areas of improvement concerning the additional losses in the machine.

In order to complete this trial, we wanted to characterize the N=1445 rev/mn constant speed machine at the same time as keeping the wattage rating constant.

We took readings of the power consumption of the 2 main parts of the machine (the motor and speed selector), detailed in the table below. Note that in this particular case, the supply to the machines is carried out with the help of a speed selector allowing a quench frequency of 12kHz..

Table 2.c Trial with selector N = 1445 Rev/mn – Power = 5.5 kW

	Consumption (kWh) Motor	Frequency (Hz)	Consumption. (kWh) Motor+Selector	T (°C) Motor
Alu	162.03	50.81	165.987	108
Cu97	158.59	49.45	163.168	96

A much more noticeable difference can be seen between the two technologies. The Cu97 rotor provides an economy for the user of 26 eurocents per day of usage. Moreover, the Cu97's operating temperature is much cooler, creating a definite advantage in the case of vacuum-tight motors and distinctly increases the life span of the machine.

4. Conclusion

We could therefore highlight the following points :

- A rise of 50% for the torque while working at constant speed.
- A reduction in machine overheating.
- A rise of about 1 to 2% for the efficiency at nominal point.

Due to a technical and economical study , We can show that copper die casting rotors provide more respects in regards to the data given by the EFF1 motors range construction standard, which is regulating the manufacturing of high efficiency electrical motors.

Finally, all these values allow the time needed for return on investment to be easily identified and prove that it is possible to compare the two technologies, Aluminium and Cu97.

5. References

[1] *J.C. ANDREAS*, "Energy-Efficient Electric Motors – Selection and Application", *Ed. Marcel Dekker.*

[2] A. ANSEL, O.WALTI, J.F.BRUDNY, "Influence de la coulée sous pression d'une cage d'écureuil en cuivre sur le comportement magnétique d'une machine asynchrone triphasée ", *Thèse de doctorat, juin 2001.*

[3] *J. ANCEL*, M. POLOUJADOFF "Nature de la résistance de contact entre les barreaux et la tôlerie d'une cage en aluminium coulé.", *Revu générale de l'électricité – Avril 1968 – pp.368 – 376.*

High Speed Induction Motor Drives with Active Magnetic Bearings for Special Submerged Gas Processing

Landson M. C. Mhango

Prof. Landson M. C. Mhango, Airscrew Limited,
Windmill Road, Sunbury on Thames, Middl., TW16 7EF, UK ,
E-mail: landson.mhango@airscrew.co.uk

Abstract

The combination of advances in active magnetic bearings, recent developments in power electronics technology, new design techniques of electrical machines and advances in iron steel materials is contributing to new designs of submerged high speed electric drives which are suitable for gas processing applications well into this century. Typical applications include submerged industrial high pressure and high temperature gas processes, calibration of gas meters, gas circulation in nuclear power reactors etc.

The rapid increase in gas pressure and temperature in gas processing has brought about the requirements for optimisation of impeller, motor and pressure vessel designs. This means optimising the aerodynamic performance of the impeller. This is achievable if on one hand the impeller speed is variable over a reasonable wide range and on the other if the choice of speed is unlimited. However, increase in pressure, temperature and speed bring considerable dynamic problems such as noise, vibration, high frequency losses in the motor, increased mechanical forces and increase in physical size of the pressure vessel.

This paper will look at the results of case studies of high-speed drives and will discuss their benefits and advantages. The paper will also describe the basic design concepts of innovative high-speed induction motors, their development and experimental results. The results are derived from the research and development programmes that were carried out in the form of case studies and span a period of four years. Based on the individual operational requirement of the drive, each case study adopted a slightly different constructional design for the motor as well as for the impeller whilst using the same design approach. The case studies were confined to speed range of between 6000 r/min and 24000 r/min and they examined the evidence of the advantages as well as identifying the new dynamic problems associated with high-speed operation. The case studies were based on typical industrial applications; examples of these will be presented.

1. Introduction

In recent years, the sudden increase in both the operational temperature and the pressure in gas processing applications has initiated the re-examination of design considerations for all the major component parts of the gas circulators. The drive motor, the impeller and the pressure vessel have all been re-examined. The design techniques and the method of operation have also undergone significant changes. The optimisation of aerodynamic performance of the impeller is considered as the main factor that influence the new changes. Variable speed operation is also viewed as the basis for achieving good aerodynamic performance. Operational considerations, reliability and cost are equally important. In the submergible gas processing systems, the physical size of the components has great influence on cost, efficiency and reliability. These factors have led to the development of high speed drives and to the use of active magnetic bearings.

2. Conventional Drives

A typical example of a system in which gas is processed at high pressure and high temperature is in a nuclear reactor where gas is used as a primary coolant. In the United Kingdom, a number of MAGNOX and Advanced Gas-cooled Reactors (AGR's) use carbon dioxide as primary coolant at high pressure and high temperature to produce electricity. A practical example of such a reactor system is shown in Fig 1. and it circulates carbon dioxide around the main primary loop of the reactor. The typical rating of these machines is 6.0MW with a rotational speed of 3000 r/min.

The conventional bearing options have largely been limited to either gas bearings or oil lubricated bearings. Gas bearings have been restricted to small drives due to their technological limitations to handle large rotating mass. As a result, large gas circulators use and will continue to use oil-lubricated bearings for some time to come.

The gas circulators that are used in the AGR's, in UK, are not only complex technically but they were first designed at the time when there were many technical limitations. Since the oil lubricated bearings require complicated re-lubrication and sealing systems, the control of oil ingress into the circulator compartment that avoids oil ingress into the reactor system is also complicated and expensive. In practice, the circulator compartment is separated from the reactor primary loop by use of a labyrinth and a buffer gas system. These additional mechanical features result in cost penalties and create maintenance problems.

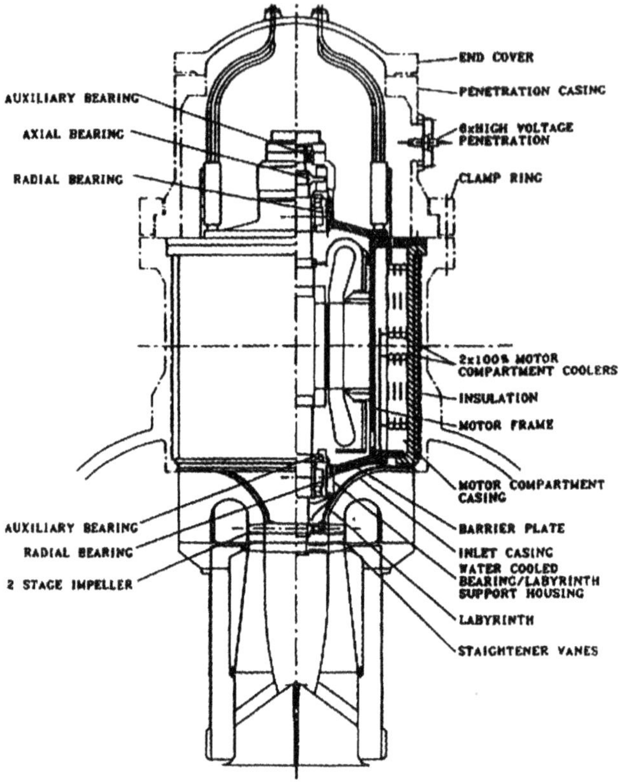

Fig. 1. Layout of the Reactor Main Gas Circulator

3. New Developments

Significant development initiatives of the main gas circulator for gas cooled reactors have been undertaken in the support of the modular helium reactor (MHR) steam cycle system. A typical example of this is the pebble bed modular reactor (PBMR). These new generation of gas cooled reactors been developed to use lubricant free bearings. A proposed design of MHR is a reactor rated at 350MW thermal with 135MW electrical output [2].

The new forms of gas reactors prompted a shift from using oil bearings to the introduction of active magnetic bearings for both small and large rotating machinery. In parallel with the developments in magnetic bearings, static PWM inverter drives have also been developed which provide variable voltage and variable frequency supply for variable speed operation of AC machines. During the same period, active magnetic bearings have been developed for vertical and horizontal orientations of the rotating machines.

Since the 1980's, active magnetic bearings have been available for use in a wide range of special designs of rotating machinery. Many of the active magnetic bearings use the principle of controlled dc electromagnetic suspension. These types of bearings have found wider practical use in high speed drives. A particular speed range of 10:1 is easily achievable. In practical terms, electrical rotating machines of ratings ranging from a few killowatts to several megawatts and covering speed range of up to 60,000r/min have been manufactured and are operational [3]. Another machine, a 23MW motor/compressor drive with the full shaft train supported on magnetic bearings has been reported by R Jayawant [4]. This drive is comprised of two radial magnetic bearings and a double acting thrust bearing in the compressor together with three radial magnetic bearings in the motor. The system operates at a variable speed to a maximum of 6300r/min and develops a maximum power of 23MW.

Active magnetic bearings generate their own losses. These losses can be classified into two groups: stator losses and rotor losses. Stator losses consist of eddy current loss, copper loss, switching loss of power devices, conduction loss of power devices and conduction loss of power cables. Rotor losses consist of windage loss, eddy current loss and hysteresis loss. Most of the losses can be minimised by an appropriate selection of magnetic materials and switching devices. However, the magnetic bearing losses are a small percentage of the total losses of the motor/bearing assembly. A combination of water and forced gas cooling are often adequate to keep the drive temperature within acceptable design limits.

4. Speed Selection and Variable Speed Operation

The relationship between the speed of operation, the type and size of the impeller, and the physical size of the motor is an extremely important consideration for high speed submerged applications. As the operational speed increases, the sizes of both the impeller and the motor reduce for a given aerodynamic duty. The head generated is a function of the impeller tip speed. In general, the diameter of the impeller varies inversely with the speed. The shape of the impeller also changes with speed. Therefore, the basic factors for the selection of the impeller are the shape, size and speed of operation. Constructional shapes of impellers tend to fit into three basic groups; centrifugal impellers for low specific speeds, mixed flow impellers for medium specific speeds and axial flow impellers for high specific speeds. It follows, therefore, that in a design process, the selected operational speed which satisfies the intended duty, defines the impeller specific speed from which straight forward scaling laws can be applied to scale the selected impeller in order to optimise the size and shape of the impeller. The selected operational speed is then used to design the motor as well as to define the supply frequency.

5. Experimental Model of Motor/Bearing Assembly

A small variable speed blower was designed and developed for use in auxiliary helium systems of field reactors. The blower has a shaft weight of around 10kg, runs anywhere between 1000r/min and 24000r/min and has a shaft power of 10kW. The aerodynamic characteristics were to deliver 0.6kW at the high speed low pressure end. A summary of the duties is shown in Table1. This is a clear illustration of the fact that the duties have wide variation within the required pressure range, all to be achieved at constant flow and also justifies the need for a wide variable speed drive with active magnetic bearings.

Table 1. Duties for the New 10kW High Speed Blower with AMB

Gas	Units	He	He
Pi	Bar	70	0.8
ρ	kg/m^3	12.54	0.14
ΔP	Kpa	50	6
Q	m^3/s	0.03	0.03
$\Delta P/\rho$		3.98	42.8
Gas power	KW	1.5	0.18
N_1	r/min	13,698	81,345
N_2	r/min	5,596	33,233

The layout of the experimental motor/bearing assembly is shown in Fi 2. The motor was designed to run in both vertical and horizontal orientations. However, the magnetic bearings need to be tuned to suit the orientation.

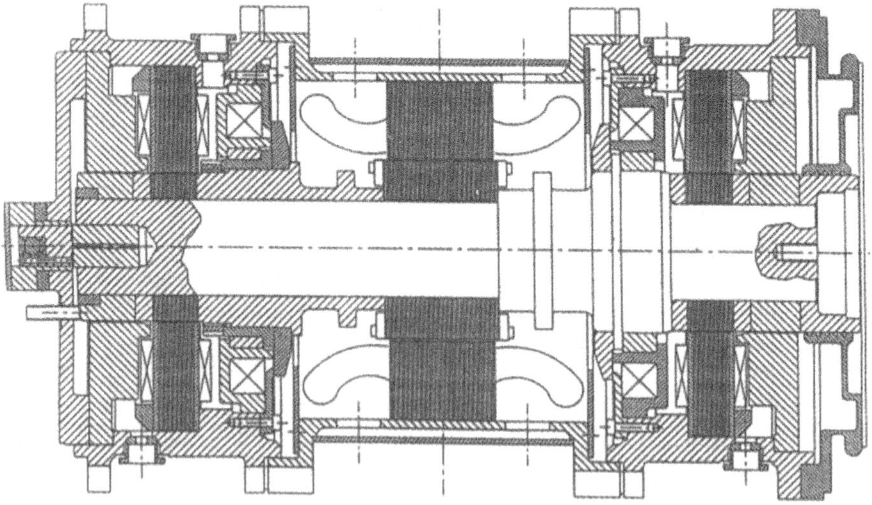

Fig. 2. Layout of experimental high speed induction motor with AMB

The layout of the motor shows the positions of both the radial and thrust bearings. These control the radial and axial movement of the rotor with respect to the stator. The experimental motor/fan assembly is shown in Fig 3.

Fig. 3. Experimental model of 10kW, 24,000 r/min motor/fan assembly

The performance of the motor was predicted at the design stage and the predicted values were compared with the experimental results. The summary of the comparison is given in Table2, showing good correlation.

Table 2. Comparison of predicted and experimental results at full load

Parameter	Units	Predicted	Experimental
Power output	KW	10.00	10.00
Motor torque	Nm	4.06	4.05
Speed	r/min	23,500	23,578
Input current	A	17.20	17.30
Efficiency	%	93	93.5
Power factor	p.u	0.95	0.94
Power input	kW	10.75	10.70

Fig 4 shows the comparison of estimated and measured full performance characteristics of the motor/impeller blower. It can be seen that there is good agreement of the results from no load to peak torque point. Between stall and peak torque there are noticeable errors due to partly the difficulty to produce a more accurate model in this region and partly due to problems to motor instability during experimental tests.

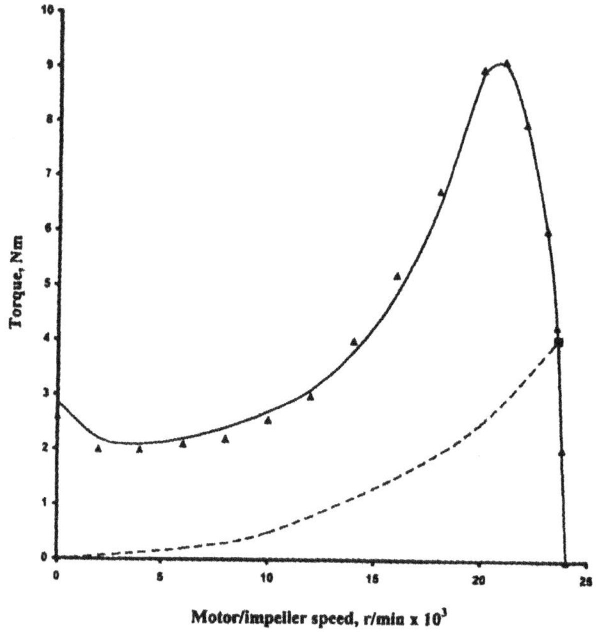

Fig. 4. Motor/Impeller performance characteristics of 10kW drive

_____	Motor torque/speed curve (experimental)
- - - -	Impeller curve
	Impeller operating point
Δ	Motor torque/speed points (calculated)

6. Benefits of Variable Speed and AMB

Typical duty ranges for fans and pumps are up to 4:1. However, the new helium cooled reactors require duty ranges of up to 40:1, arising primarily from large performance variations. This can be achieved by using a variable high-speed drive. A typical life target of 40 years, in helium cooled reactors, can also be achieved using a high-speed drive which is supported by active magnetic bearings. High -speed operation reduces the physical sizes of the motor and that of the impeller as well. The variable voltage/frequency control avoids large inrush currents to flow during starting period. The theoretical modelling of quasi-analog computer simulations were undertaken to examine the behaviour of the rotor/shaft assembly including the effects of critical speed. The success of these stability studies proved to be the key to the operational behaviour of the active magnetic bearings [3]. Operational aerodynamic instabilities such as surge and stall are minimised considerably.

The use of magnetic bearings, in high speed and high power density applications, requires the implementation of an effective catcher bearing system to contain the rotor in the event of failure of the bearings or the failure of supply to the bearing control. Hence, the development of magnetic bearings is automatically accompanied by the development of catcher bearings. However, high-speed operation of induction motors brings about many dynamic problems. Use of aluminium die-cast rotor is limited by surface speed. It is, now, time to look very seriously into the development of copper die-casting technology. It may also be helpful to investigate the possibility of removing the skew from the rotor to the stator. This work is in progress and results will be reported as they become available.

7. Conclusions

This paper has put forward a case for using new type of high-speed induction motor drives for submerged gas processing involving high pressure and high temperature. The new modular helium cooled nuclear reactor application is one of the typical practical examples. The case studies have shown that the ability to run at high speed has brought many benefits. The system functionality has increased. There is good reduction in running cost and capital outlay. The operational duty range is increased as well as system pressure and temperature. The overall physical size of the blower system has reduced.

References

[1] Mhango, L M C and Rennie, A J "Recent developments and benefits of high-speed pwm inverter-fed induction-motor drive for gas filtration in nuclear power plants" Proc. Int. Conf. on Electrical Machines and Drives, IEE Conf. Pub. No.376, Oxford University, September 1993

[2] Rennie, A J and McDonald, C F "Active magnetic bearings for rotating machinery in future gas cooled reactor plants" 4th Int. Symposium on magnetic bearings, ETH Zurich, Switzerland, August 1994

[3] Mhango, L M C and Perryman, R "Innovative high-speed machines with active magnetic bearings for special submerged gas processing" Proc. of Instn. of Mech. Engrs, Vol.216, Part C, March 2002

[4] Jayawant, R "Electrical machines fitted with Glacier Magnetic Bearings" IEE Colloquium , Digest No. 1997/164, London 1997

Comparative Study of Axial-type Bi-2223 Bulk Motor with the Use of Different Stator Windings

Itsuya Muta, Taketsune Nakamura, Hun-June Jung, Tsutomu Hoshino

Department of Electrical Engineering, Graduate School of Engineering, Kyoto University, Yoshida-Honmachi, Sakyo, Kyoto 606-8501, JAPAN

Introduction

Using high temperature superconducting (HTS) bulk materials, some types of bulk motors have been proposed and developed [1-4]. We have been studying the characteristics of electric motors with the use of Bi-2223 bulk disk rotor [5-8]. In order to do some preliminary researches, the stator of the motor has been constructed to be two-pole and three-phase concentrated armature winding wound around salient magnetic poles. The synchronous speed of the rotating magnetic field is in principle 3600 rpm. The test results, however, showed that the motor rotated with large slip and the torque decreased with increasing speed. As one of the main reasons for the results, we suppose the effect of the space harmonics of the air-gap magnetic field distribution. It is well known that magnetic space harmonics have the detrimental effect to motor performances. Therefore, the investigation of this effect is essential in order to realize the high performance HTS motor.

In this paper, a new type of stator is introduced for the Bi-2223 bulk motor. The stator has four-pole distributed winding, and produces rotating magnetic field with the speed of 1800 rpm at 60 Hz frequency, expelling space harmonics. The tests are carried out for concentrated and distributed stator windings with the same Bi-2223 bulk rotor disk. Then, the test results are compared in order to examine the effect of space harmonics upon the motor characteristics. Furthermore, the same tests are also performed with use of copper rotor disk with the same dimensions of Bi-2223 rotor disk for the investigation of the effect of superconducting Bi-2223 bulk on the motor property.

Experimental Method

$Bi_{2-x}Pb_xSr_2Ca_2Cu_3O_y$ (hereafter, stated as Bi-2223) disk is used for the rotor. The dimension of the disk is, respectively, ϕ120 mm and 7 mm thick. The hole with the bore diameter of 26 mm is also made at the center of the disk for the shaft. The specification of the Bi-2223 rotor bulk is listed in Table I. Further, the copper disk with the same dimension as the Bi-2223 disk is also used as rotor for comparative study.

(a) Concentrated winding stator

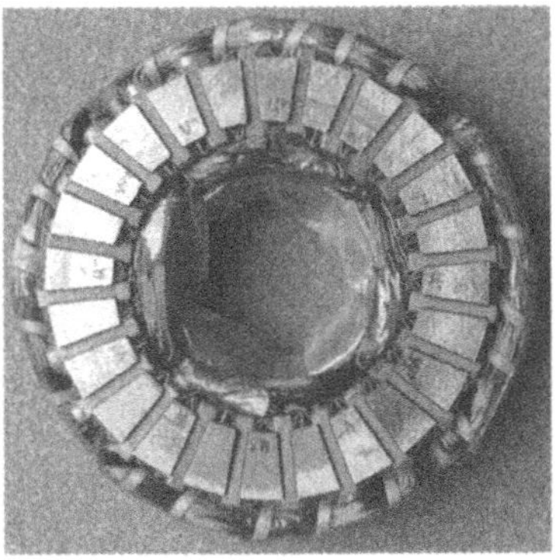

(b) Distributed winding stator

Fig. 1. Photographs of the stationary armature winding.

Two types of stator windings are used in this study. One is two-pole and three-phase concentrated winding as shown in Fig. 1 (a). The speed of the produced rotating field is 3600 rpm at 60 Hz frequency. Another one (Fig. 1 (b)) is four-pole and three-phase distributed winding, and the corresponding rotational speed of the field is 1800 rpm. Fig. 2 shows the photograph of the motor set-up that uses the distributed stator winding. As shown, the bulk rotor is sandwiched in between two stators those are connected in parallel each other. The air-gap length is set to be 1.5 mm. This system is installed in the metal cryostat, and then immersed in liquid nitrogen without the application of the magnetic field. All the tests for Bi-2223 bulk motor are carried out at the temperature of atmospheric liquid nitrogen, i.e., 77.3 K.

Results and Discussion

Air-Gap Magnetic Flux Density Distribution

Firstly, the distribution of air-gap magnetic flux density is measured and analyzed in order to examine the space harmonics. The measurements are performed by means of transverse type hall sensor that is fixed on the rotor surface at room temperature. The signal from the hall sensor is transferred to the transient recorder by applying three-phase (60 Hz) rotating magnetic field with the fixed condition of the shaft. After one measurement is finished, the shaft is rotated every one degree, and then the same measurement is repeated 180 times. Fig. 3 (a) and (b) show the typical results for concentrated and distributed stator windings, respectively. The electrical angle of the results is 0°. As can be clearly seen, the spatial distribution of the magnetic flux density due to the concentrated winding is largely distorted (Fig. 3 (a)). On the contrary, the distribution for the distributed winding is more like sinusoidal as expected (Fig. 3 (b)).

Fig. 2. Photograph of motor set-up

Table 1. Specification of the Bi-2223 rotor bulk.

Diameter / mm	120
Thickness / mm	7
Bore diameter of the hole / mm	26
Critical temperature / K	105
Critical current density (77.3 K) / Am^{-2}	3×10^7

These spatial distributions are analyzed by Fast-Fourier-Transform (FFT) as shown in Fig. 4 (a) and (b). In case of the concentrated winding, large amount of 5th and 7th harmonic component are included. As is well known, these harmonics greatly influence the characteristics of the motor. In particular, since the 7th harmonic component produces a positive sequence rotating magnetic field, its large amount causes to fix the rotational speed of around 500 rpm due to so-called crawling phenomenon. On the other hand, the 5th as well as 7th components are suppressed for distributed stator winding. Therefore, It can be deduced that the motor characteristics will be improved by using the distributed winding. This is to be confirmed in the following section.

Characteristics of Copper Disk Motor

Fig. 5 shows the lock test results for the copper disk rotor at room temperature. One can see that the phase current that induces the same value of torque is about half in case of distributed stator winding compared to concentrated one.

Load test results are also shown in Fig. 6 for some values of the phase voltage. As can be seen, the characteristics in the concentrated winding (Fig. 6 (a)) behave like the induction motor with large resistance, i.e., the larger the rotational speed is, and the smaller the torque is. Further, the rotational speed reaches only about 700 rpm even at the no load condition. That is, the motor is operated with large slip mode, tremendously caused by the 7th component. As is already discussed in the former section, the air-gap magnetic flux distribution is largely distorted in the concentrated stator winding. Therefore, it can be deduced that the results for the concentrated stator winding ascribe to these harmonic components of the air-gap magnetic flux density. Then, in order to confirm this consideration, the same tests are performed for the distributed stator winding as shown in Fig. 6 (b). One can clearly see the improvement of the motor characteristics. That is, the rotational speed at no load condition is around 1700 rpm. The speed of the rotating magnetic field in this winding is 1800 rpm, and then the corresponding slip is around 0.06.

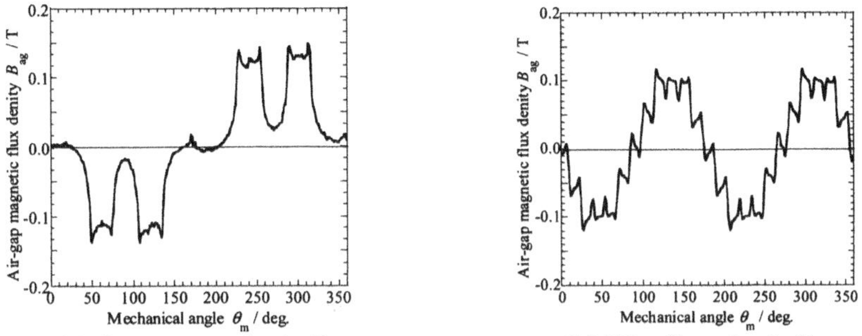

(a) Concentrated winding (b) Distributed winding

Fig. 3. Spatial distribution of the air-gap magnetic flux density for the electrical angle of 0° at room temperature.

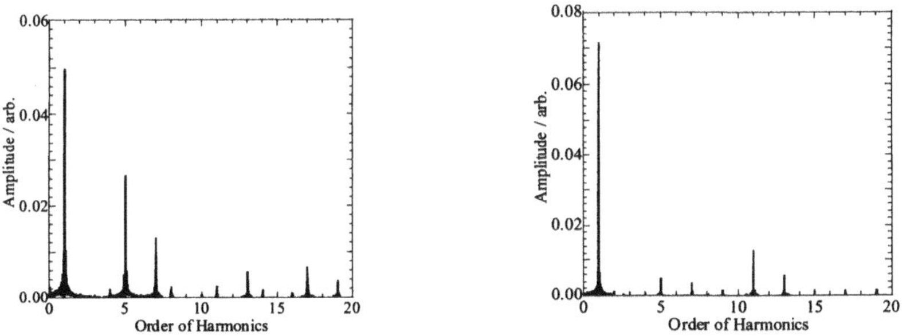

(a) Concentrated winding (b) Distributed winding

Fig. 4. Fast-Fourier-Transform (FFT) results for Fig. 3.

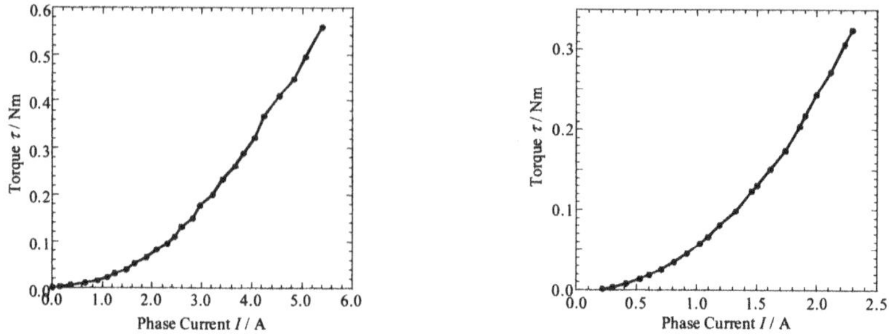

(a) Concentrated winding (b) Distributed winding

Fig. 5. Lock test results for copper disk motor at room temperature.

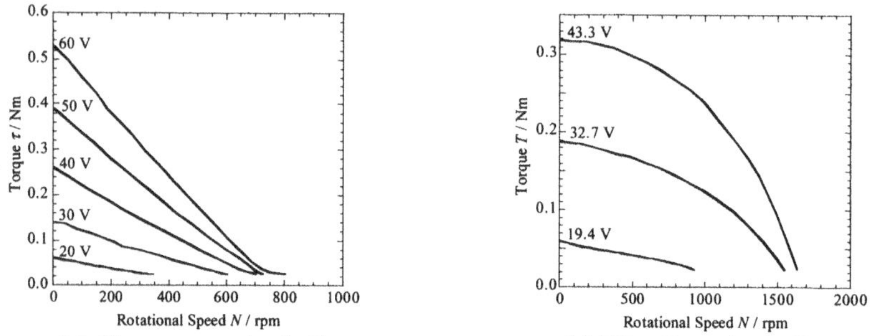

(a) Concentrated winding (b) Distributed winding

Fig. 6. Load test results for copper disk motor at room temperature. The values in the figure show the phase voltage.

Characteristics of Bi-2223 Disk Motor

In this section, the same tests are also carried out for Bi-2223 bulk motor. Fig. 7 shows the lock test results for Bi-2223 bulk motor operated at 77.3 K. As shown, the torque for concentrated winding is about 0.3 Nm even at the phase current of 12 A (Fig. 7 (a)). In this case, temperature of liquid nitrogen is largely fluctuated, i.e., large dissipation occurs. On the other hand, the torque attains 1 Nm at 8 A in case of distributed winding (Fig. 7 (b)). Further, load test results are also shown in Fig. 8. It is obvious that the characteristics for distributed winding is excellent compared to those for the concentrated winding due to the aforementioned reasons. The main mechanism of the torque generation is considered to be the hysteresis property of Bi-2223 disk. The rotation, however, does not reach the synchronous speed even at the no load condition (Fig. 8 (b)). The reason can be discussed with the pinning property of the bulk. This will be presented in another papers in the near future.

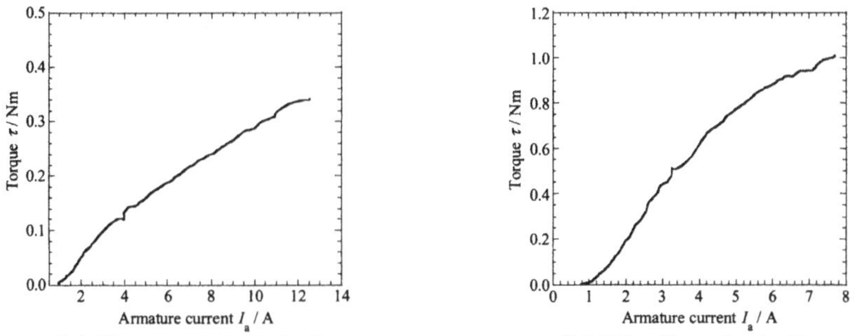

(a) Concentrated winding (b) Distributed winding

Fig. 7. Lock test results for Bi-2223 disk motor at 77.3 K.

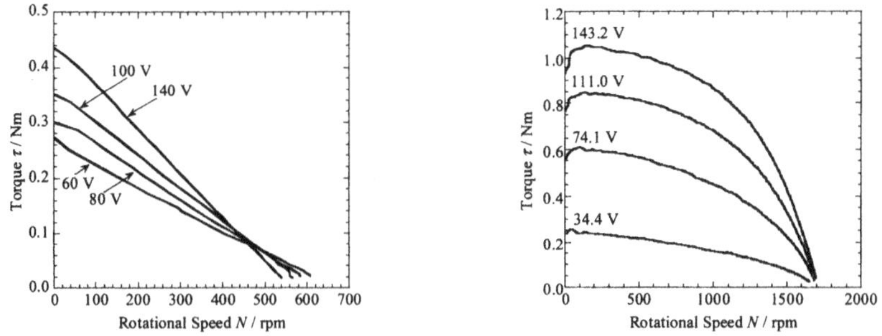

(a) Concentrated winding (b) Distributed winding

Fig. 8. Load test results for Bi-2223 disk motor at 77.3 K. The values in the figure show the phase voltage.

Conclusion

Superconducting Bi-2223 bulk motor was fabricated and tested with two different stator windings in order to examine the space harmonics of the air-gap magnetic flux distribution upon the motor characteristics. The same tests are also performed for copper disk with the same size of Bi-2223. It was shown that the characteristics for the distributed stator winding improved wonderfully compared to those for concentrated winding. These results originated from the harmonic components of the air-gap magnetic flux density in the concentrated winding, and these were confirmed by the measurements with hall sensor. Therefore, the reduction of the space harmonics is crucial in order to realize the high performance HTS motor. Further, from the experiments up to now, it should be noted that Bi-2223 bulk motor has the torque-speed curves quite similar to ones of solid (bulk) rotor type induction motors.

Acknowledgement

This work has been supported in part by Grant-in-Aid for Scientific Research (No. 13450113) from The Ministry of Education, Culture, Sports, Science and Technology, Japan.

References

[1] L. K. Kovalev et al., Physica C, vol. 354, 2001, 34.
[2] L. K. Kovalev et al., Physica C, vol. 357, 2001, 860.
[3] G. J. Barnes et al., Supercond. Sci. and Technol., vol. 13, 2001, 875.
[4] H. Ohsaki et al., Journal of Mater. Process. Technol., vol. 108, 2001, 860.
[5] I. Muta et al., IEEE Transactions on Applied Superconductivity, vol. 11, No. 1, 2001, 1964.
[5] I. Muta et al., Physica C, vol. 54, 2001, 100.
[6] I. Muta et al., Physica C, vol. 372-375, 2002, 1535.
[7] I. Muta et al., Proceedings of 15th International conference on Electrical Machines (Brugge, Belgium), 2002, 190.
[8] H. J. Jung et al., Preprint of Applied Superconductivity Conference 2002 (Houston, TX, USA).

High Efficiency Pumps for Building Applications

Thorsten Kettner, Frank-Hendrik Wurm

WILO AG, Research and Technology Center
Nortkirchenstrasse 100, D-44263 Dortmund, Germany

Introduction

87 Million heating pumps with power input less than 250 W installed in building applications in Europe induce a total energy consumption of about 41 TWh/a [1], not taking into account solar applications, floor heating or sanitary hot water circulation. 41 TWh corresponds to the yearly energy, which is produced by 4 power stations each with 1.2 GW. To produce the needed electrical power three times higher primary energy is needed in case of conventional power stations. Therefore any optimization of pump technology involves a high overall energy saving potential.

In today's heating installations mostly canned pumps with asynchronous motors are used. These pumps are robust, maintenance free and silent, but unfortunately the efficiency is limited due to the construction principle.

In this article a new pump generation based on electronic commutated DC motors (EC motors) is introduced. The use of these pumps in building applications will give the chance for a significant reduction of energy consumption.

Development Steps of Heating Circulators

In the past different development steps have been realized to reduce the energy consumption of heating pumps. Fig. 1 illustrates the development of the yearly energy consumption for pumps with a motor power of 180 W as an example. In a first approach a speed selector has been added to constant speed pumps to enable a manual adjustment of the pump power to the needs of the installation. In a next step an electronic control of the differential pressure has been introduced [2]. The controller adjusts the differential pressure of the pump to the needs of the system. A further improvement was the realization of automatic recognition methods of the flow temperature [3]. Using these algorithms the pump detects the period of reduced system temperatures, e.g. during night, and decreases the power input to a minimum.

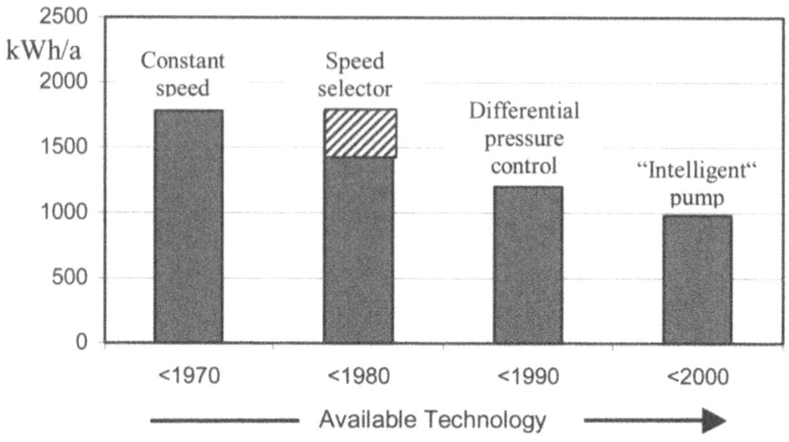

Fig. 1. Yearly energy consumption of heating pumps (example)

Though different technological solutions for energy saving are actually available the main sales and stock quantities are still pumps with constant speed or speed selector. On the other hand, using the existing technologies based on asynchronous motors, a further significant reduction of energy consumption is not to be expected.

New Pump Generation Using Ec Motors

To overcome the limits of today's pump solutions the asynchronous motor can be replaced by an electronic commutated DC motor. The benefits of EC motors are well known from different applications like fans, copiers or disk drives. In these devices dry runner motors are used. To use EC motors in canned pumps in wet runner technology some important tasks have to be solved:

- Corrosion of magnets in water
- Reliability of magnets for high and low temperatures
- Reduction of eddy current losses
- Reduction of commutation noise.

The solutions chosen for Wilo-Stratos are shown in Fig. 2. The high efficiency of the pump is mainly achieved by the permanent magnet rotor in combination with high speed rotation and a 3D impeller. The permanent magnet rotor is protected against corrosion with a stainless steel sleeve. The use of a carbon fiber hybrid can eliminates the eddy current losses. The difficulty is to design the can for a pressure of 10 bar at temperatures of 110° C. The commutation noise can be reduced by an electronic control with sine wave commutation. Fig. 3 shows the high efficiency pump Wilo-Stratos. This pump is suitable for heating and air condition applications with a medium temperature range between –10° C and 110° C.

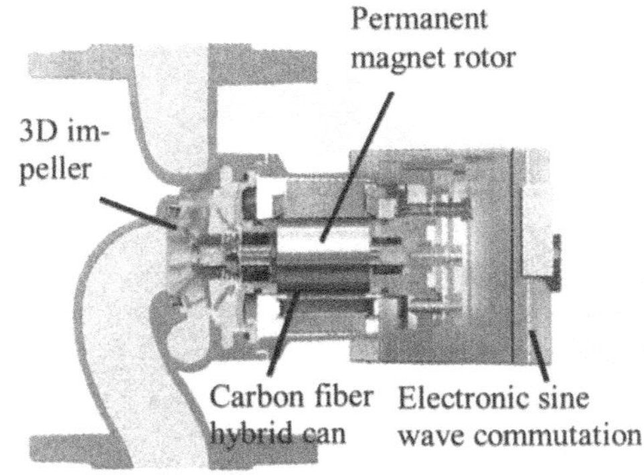

Fig. 2. Realization of a wet runner EC motor

Fig. 3. High efficiency pump Wilo-Stratos

The efficiency advantage of wet runner EC motors in comparison with AC motors is shown in Fig. 4. The two dots mark the pump size shown as an example in Fig. 2 and Fig. 3. The efficiency gain using EC motors increases for lower motor power and is almost two times higher for pumps with 15 – 40 W motor power used in one- or two-family houses today. The use of wet runner technology in pumps gives advantages up to about 1200 W motor power. Above this motor size dry runner motors are used. Table 1 summarizes some specifications for pumps of identical hydraulic performance based on an asynchronous motor (AC) and an EC motor.

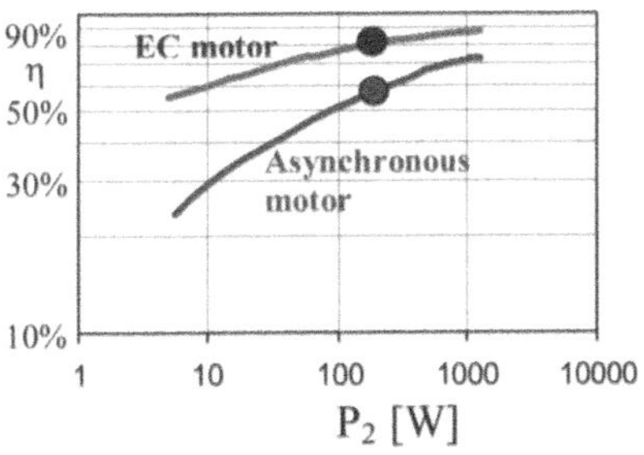

Fig. 4. Efficiency curves of AC- and EC-wet runner motors

Table 1. Comparison of AC and EC pump for identical hydraulic performance

Parameter	AC pump	EC pump
$P_{1,max}$	400 W	270 W
$P_{1,min}$	50 W	10 W
n_{max}	2.800 rpm	4.800 rpm
Weight	8 kg	6 kg

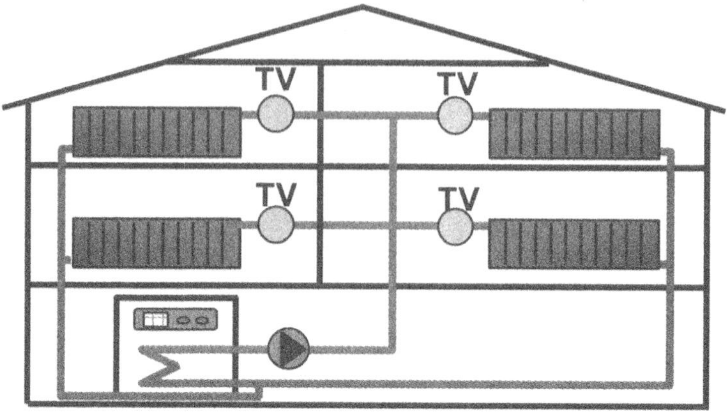

Fig. 5. Principle heating installation

Application Advantages

Beside the above mentioned advantages of EC pumps the behavior of these pumps in building applications is important. Fig. 5 shows a typical heating installation. The room temperature control can be organized either automatically by thermostatic valves (TV) or by manual adjusting of valves. Both solutions lead to a variable flow in the pipes. An electronic pump reacts on the variable flow and adapts the differential pressure to the needs of the pipe system like shown in Fig. 6.

A comparison of the needed input power along the electronic control curve for a pump with asynchronous motor and EC motor gives an interesting result, illustrated in Fig. 7. For 100 % flow, which is the nominal point of the system, the reduction of the input power is 35 %.

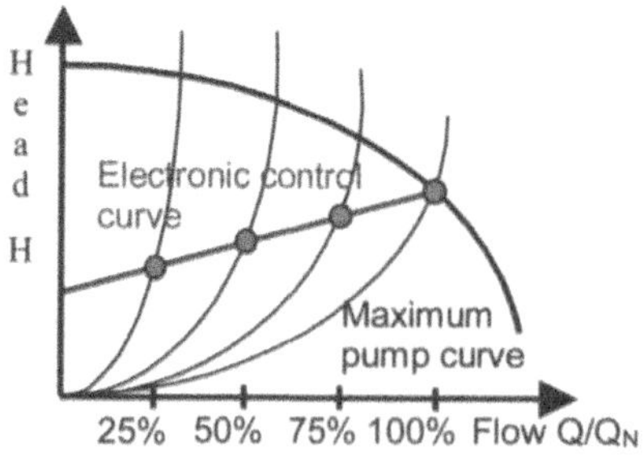

Fig. 6. Δp-variable control curve

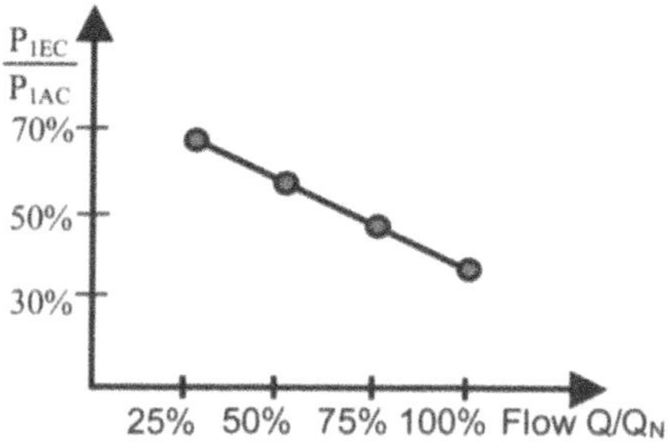

Fig. 7. Reduction of power consumption of EC pump in comparison to AC pump

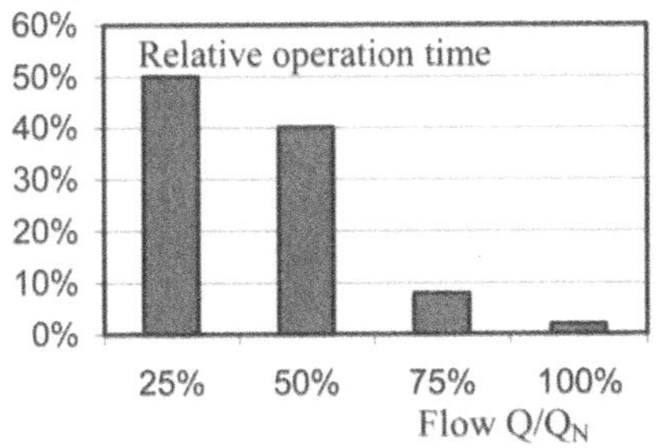

Fig. 8. Simulated load profile for a typical heating system (5.500 operation hours/year)

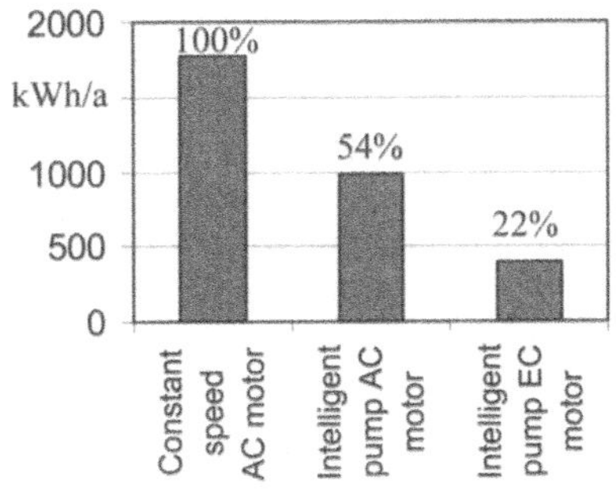

Fig. 9. Yearly energy consumption based on simulated load profile

This reduction increases with decreasing flow and reaches almost 70 % for 25 % of the nominal flow. The EC pump has an excellent part load behavior in comparison to the AC pump.

To estimate the effects of this significant reduction of power consumption in part load it is important to know, for which period of time the pump is operating at 25 %, 50 %, 75 % or 100 % of the nominal flow in one year.

Fig. 8 presents a typical load profile for a heating system. The load profile is a representative result of an analysis by simulation [4]. The total operation time of the heating pump is 5.550 hours in this example, valid for Germany. It is quite remarkable, that around 90 % of the operation time in one year the pump is working with maximum 50 % of the nominal flow. For other European countries the total operation time may be different, but the distribution of operation points will be similar.

Taking these results into account for a comparison of the yearly energy consumption of heating pumps, significant differences can be found. **Fig. 9** shows the energy consumption based on the load profile of **Fig. 8** for three different pump technologies. While the electronic control of an AC pump already leads to a reduction to 54 % of the consumption of the constant speed AC pump, the "intelligent" EC pump only needs 22 % of this energy per year.

Similar results can be observed in other building applications like air condition systems, where the room temperature control leads to variable flow and a big share of part load conditions also, comparable to the above described heating installations.

Summary

The difficult transfer of EC motors into wet runner technology has been solved with the Wilo-Stratos range. Canned pumps with EC motor enable high efficiency for the design point of the pump compared to pumps with asynchronous motor.

The difference in efficiency between these motor types increases for decreasing motor power.

Beside the improvement of the best point efficiency, EC motors are characterized by an excellent part load behavior. This feature is very important for pumps used in building applications, since the pumps are operated mainly in part load conditions. A comparison between pumps realized with asynchronous and EC motors applying simulated load profiles show significant differences in power consumption.

Transferring these results to the above mentioned quantity of pumps installed in Europe in heating applications we can estimate an energy saving potential of more than 20 TWh/a, assuming that all pumps would be realized in EC motor technology. Therefore a significant economical and ecological effect could be achieved.

References

[1] "Promotion of Energy Efficiency in Circulation Pumps, especially in Domestic Heating Systems", Final Report, EU SAVE II Project, Contract No. 4.1031/-Z/99-256.

[2] H.-G. Schmalfuß, "Heizungspumpenregelung", IKZ, Heft 21, 1981, pp. 48 - 56.

[3] S. Greitzke, T. Kettner, F. Albers, "Temperature responsive power control for an electric pump unit", European Patent EP 0 736 826, Oct. 1996.

[4] R. Hirschberg "Bestimmung der Belastunsprofile von Heizungsumwälzpumpen in der Gebäudetechnik", VDMA report, May 2001.

Rewinding Induction Motors up to 225 kW with Minimum Loss of Efficiency: a Joint AEMT/EASA Project

John Sheppard Allen, Austin Herbert Bonnett, Keith John Bradley, David Graham Walters

Introduction

It has been estimated [1] that rewind efficiency reductions cause annual energy losses of 1500 to 2000 GWh/annum in the EU, which represents a cost to EU industry of between €100M and €135M/year in wasted energy. These losses, taken over an average motor life of 15 years are much higher – approximately 30 Twh costing €1800M.

Whilst best practice repair methods to minimise rewind losses have previously been developed and published by the Association of Electrical & Mechanical Trades (AEMT) in the UK [2], the tests on which this work was based were mostly conducted using small motors up to 22 kW. Since there was no direct evidence that the same best practice repair methods would apply to larger motors, it was felt desirable to extend the original UK (AEMT) based research programme to include motors up to 225 kW.

If sufficient machines were to be included to give statistically reliable results, this was obviously going to be an expensive exercise. However, following informal contacts between AEMT and US DOE representatives at EEMODS 99, an approach was made to the US repair trade association (EASA) who agreed to join the AEMT in part-funding a project and in putting forward a joint proposal to seek outside funding from both US and UK Governments and from motor users. Sufficient funding was obtained and the final project sponsors comprised:

- The Electrical Apparatus Service Association (EASA) – USA
- The Dept of Energy – Washington – USA
- The Association of Electrical & Mechanical Trades (AEMT) – UK
- The Energy Efficiency Best Practice Programme (EEBPP) – UK
- Ministry of Defence Ships Support Agency – UK
- UK Water Industry Research Ltd.
- British Nuclear Fuels plc – UK

In summary, the project results show that there is no significant loss in efficiency of machines up to 225 kW using best practice methods with conventional repair shop equipment, even when some machines were burned out and rewound three times. This paper describes the procedures and the tests, which led to this conclusion.

The Project Organisation

To eliminate as many variables as possible, all the rewind work was carried out by one company, Dowding & Mills PLC, and all the testing (except for the "round robin" testing - see par. 4.3 below) was carried out by Nottingham University UK. Dowding & Mills' Technical Director acted as Project Director and one of his staff acted as Project Manager. They were supported by an ad hoc Technical Committee drawn from both manufacturers and repairers, and they reported to a steering committee made up of Sponsors' representatives and chaired jointly by the Technical Consultants to EASA and the AEMT. The Steering Committee conducted all its business by e-mail based on periodic reports from the Project Director. The project organisation is shown in Fig 1 below.

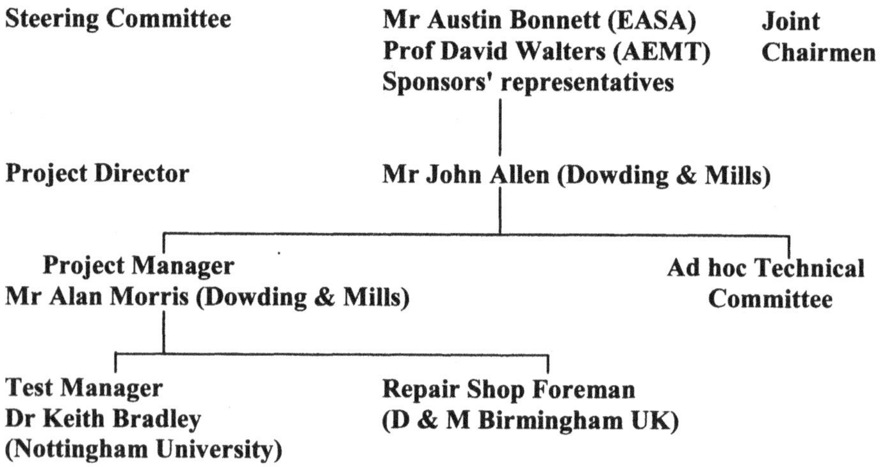

Fig. 1. Project Organisation Chart

Motors

Ten motor manufacturers provided motors, technical data and assistance for the study. They were ABB, Baldor, Brook Crompton, GEC (now Alstom), Leeson, Reliance, Siemens, Toshiba, US Motors and VEM. Twenty-three 2- and 4-pole motors rated between 5.5 and 225 kW were actually used in the trial; these included 50 Hz IEC and 60 Hz NEMA machines with IP 23 and IP 54 enclosures. One other 30 kW machine was used for "round robin" testing – see para. 4.3 below. All but one of the motors was low voltage (380V-460V); the one 3.3 kV machine included in the trials behaved in a similar manner to the LV machines.

Motor Testing

Test Facilities

The Dept of Electrical and Electronic Engineering at Nottingham University was chosen to carry out the tests because it was believed to have one of the best independent test laboratories for machines of this size. Fig.2 shows a schematic diagram of the test rigs used and illustrates some of the equipment.

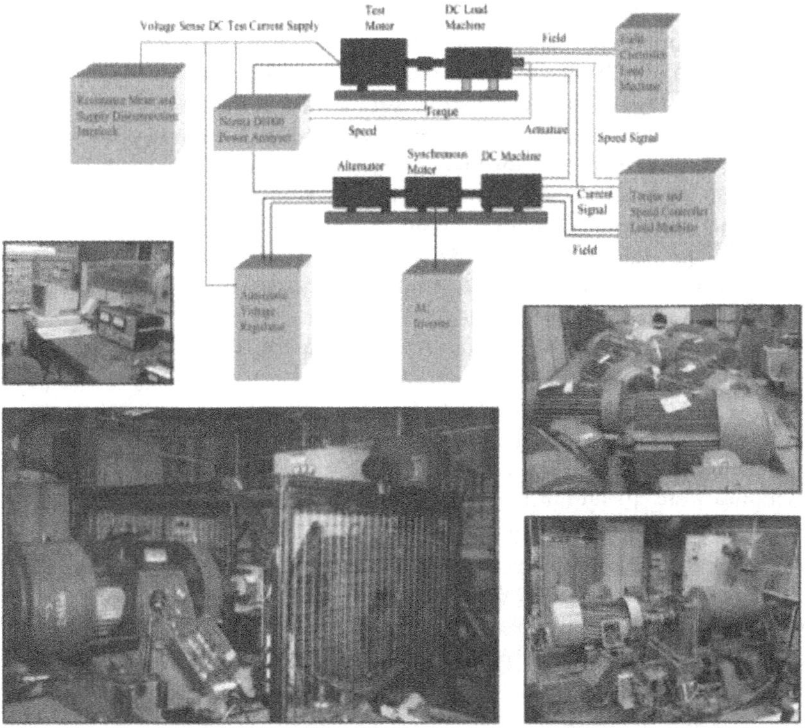

Fig. 2. University of Nottingham Test Facility

Test Methods, Accuracy and Repeatability

There are four basic contributors to the accuracy and repeatability of motor tests. These are the supply system, the mechanical loading system, the instrumentation and the test procedure.

The test rig of Fig.2 has a dedicated A.C. generator to supply the test machine. The AVR for this generator senses the voltage at the test motor terminals. The generator is driven by a synchronous motor supplied from an inverter. A constant amplitude sinusoidal voltage of almost perfect balance and waveform purity with a frequency stability of 50 parts per million is supplied. The supply is repeatable.

The loading method is by D.C. machine in the Ward Leonard configuration. This system provides smooth torque control even at light load with continuous D.C. current. Active electronic control of the field current and the armature current of the D.C. load machine effectively enable constant torque to be applied to a test machine to a resolution of better than 0.05%. The load is closely repeatable. No-load tests used a disc brake mounted directly on the motor shaft to stop the machine quickly to facilitate measurement of winding resistance. The small, known, windage-loss of this brake was used to correct the results.

The most appropriate of a range of in-line torque transducers with full-scale ratings of 140, 340, 550, 820 or 2000 Nm was mounted as a floating shaft between stainless steel diaphragm couplings linking test motor and loading system. Having no support bearings between the test machine and torque transducer ensures accuracy and repeatability. Torque transducers and electronics were individually calibrated in a static rig to an accuracy rating of 0.05%. Electrical measurements were made by Norma D6000 power analyzer with an accuracy rating for power, voltage and current of better than 0.1% and with a frequency and speed measurement accuracy rating of 0.01%. The winding resistance was measured at the motor terminals by four wire Valhalla electronic bridge of 0.02% accuracy rating.

Because it was suspected that rewinding could significantly alter stray loss, it was decided that all machines should be tested in accordance with IEEE Method B in which stray losses are calculated by loss separation at no load and at different load points with input and output power measurement.

Test procedures, time between readings during tests and thermocouple location on the winding were all optimised to provide accurate results before the test sequence started. The procedure then remained unaltered.

The techniques and equipment defined above ensured repeatability to within 0.1% in efficiency for tests on the same, unmodified, motor.

Motor efficiencies were also calculated in accordance with IEC 60034-2 and comparisons of efficiency calculated by the two methods are shown in Section 6.4.

The "Round Robin" test

To satisfy all parties of the accuracy of the test facilities and methods at Nottingham University, one 30 kW 4-pole motor was tested at 50 Hz and 60 Hz (with appropriate voltage adjustment to maintain constant flux) at five different locations in the UK and the US. Tests were in accordance with IEEE 112 Method B. The results of these tests are shown in Table 1 below.

Table 1. Round Robin Test Results of 30 kW 4 pole motor

Test Location	Test	Full load efficiency	Full load power factor	Full load amps	Temperature rise	rpm
Baldor	400v/50Hz	91.8%	86.8%	54.0	69.4°C	1469
Nottingham	400v/50Hz	92.3%	87.0%	54.2	68.0°C	1469
U.S. Electric Motors	400v/50Hz	91.9%	86.7%	53.5	59.0°C	1470
Nottingham	461v/60Hz	93.5%	85.9%	47.0	53.9°C	1776
Oregon State	460v/60Hz	92.6%	85.9%	47.0	50.0°C	1774
U.S. Electric Motors	460v/60Hz	93.1%	86.4%	46.5	42.0°C	1774

It will be noted that the maximum efficiency variation for all 50 Hz tests (the frequency for which the motor was designed) did not exceed 0.5 percentage points; the maximum variation for the 60 Hz tests was 0.9 percentage points. As a benchmark, this was compared with earlier "round robin" tests conducted by the NEMA membership, which had experienced a variation of 1.7 percentage points. Against this criteria the results were judged satisfactory. However, this level of efficiency variation, which depends entirely on the test location, equipment and procedure rather than the machine itself, could lead to a potentially serious problem concerning compliance with standards and may have affected results in some of the earlier studies which showed rewinding to adversely affect efficiency.

For example, BS EN 60034-1:1998 allows a measured efficiency tolerance of 15% of the losses for a 30 kW machine (US standards allow 20%). The 0.5 percentage point efficiency variation recorded in the 50 Hz "round robin" tests represents between 6.1% and 6.5% of the losses (depending upon which of the tested efficiencies you take as the baseline). Thus a specific motor which had an actual efficiency reduced by, say, 10% of the losses compared with the type test (and would thus be within the tolerance band) could easily be well outside it when tested at a different location from where the type test was carried out. The situation is much worse if one considers the 1.7% efficiency variation recorded in earlier NEMA "round robin" tests.

Since the motor was unchanged for all the tests, the efficiency variation can only be accounted for by outside factors such as voltage waveform, instrument accuracy, or lack of uniformity in the way the tests were actually carried out. If so, this is a weakness in existing test standards and is one which needs to be addressed when drafting future standards.

Motor Rewinding

Motor Rewind Groups

All motors were rewound by the same experienced workforce in the same workshop. However, six motors were merely sent to the workshop with no specific rewind instructions other than to work to the original AEMT Good Practice Guide; the remainder were sent with specific instructions concerning winding configuration, copper section, MLT, etc. The efficiency of the rewound motors in the second (controlled) group was significantly higher than that of those in the first (uncontrolled) group – see Section 6. These results emphasise the need for good practice training even for experienced rewind personnel.

Rewind Processes

It was decided from the outset only to use repair and rewind methods which were already commonly used for this type of machine. These include:

- **Coil cut off.** The connection end of the coils of the old windings were cut off as close as possible to the core, using a circular saw to avoid damage to the core.
- **Winding burnout** was by means of a controlled temperature burnout oven (see fig. 3). Initially, based on earlier work done by the AEMT, the thermostat was set to 350°C (662°F) but at this temperature it was found difficult to remove the windings from some machines without splaying out the stator teeth at the end of the core. A small increase in temperature to 360°C-370°C (680°F-700°F) largely overcame this problem and did not appear to have any significant effect on iron loss.

Fig. 3. Photograph of burnout oven used for tests

- **Core cleaning** (i.e. removing the residual insulation from the slots after the winding had been removed) was done using a number of commonly-used methods including:

a) careful scraping using a sharp blade;
b) high pressure washing;
c) burn-out using an oxy-propane lance;
d) brushing with a medium/soft wire brush.

Rasps and slot files were **not** used.

- **Rewinding** All motors were rewound by hand mostly using double layer short pitched lap windings because these are generally favoured by repairers. However, lap windings were not substituted for 3 tier concentric windings because this was not recommended in the original AEMT Best Practice Guide [2].
  All motors in the controlled group were rewound with additional copper section and reduced mean length of turn (MLT) compared with the original windings. Whilst this produced satisfactory results in these tests, reducing the MLTs in through ventilated (IP22/23) motors should not be overdone because the reduced endwinding length can reduce the heat transfer surface area between the winding and the cooling medium (air).

- **Multiple burnout and rewinds.** Three motors were subjected to three complete test/burnout/rewind/test cycles, and two other motors were subjected to two similar cycles. Each motor was fully assembled and re-tested in accordance with the NEMA IEEE 112B standard after each cycle. Two small motors were subjected to 5 burnouts but were rewound only once. The results of these tests are included in Section 6.

- **Core loss tests,** using two commercial core loss testers, were carried out on most of the motors before and after stripping out the old windings. Unfortunately, the results of these tests proved inconclusive.

- **Mechanical Damage tests**
  Work is in progress to investigate the effects on motor efficiency resulting from core damage (e.g. caused by core rubs and earth faults) and the processes used to repair such damage.

- **Bearings**
  Although it is normal practice in the industry to replace bearings when a motor is rewound, this was not generally done in these tests because of the short time the motors had been run. However, the bearings on two motors were overgreased, which resulted in an efficiency reduction of about 0.5%. Further testing with regard to the effects of lubricant quantity and machine efficiency are being carried out.

Test Results Summary

Table 2. Low Voltage wire wound motors

Phase 1: Motors rewound with no specific control on stripping or rewind

Motor description	Efficiency before rewind	Efficiency after rewind	Efficiency change*	Winding before	Winding after	Comments
1A 100 hp, 60 Hz, 4 pole	94.1%	93.1%	-1.0%	3TC-1L	3TC-1L	Bearings regreased
			-0.8%			Drive end bearing cleaned
			-0.6%			Both bearings cleaned
			-0.5%			Bearings changed
2B 100 hp, 60 Hz, 4 pole	92.9%	92.4%	-0.5%	L-2L	L-2L	575 volt machine
3C 100 hp, 60 Hz, 2 pole	94.5%	93.5%	-1.0%	GC-2L	L-2L	Bearings regreased
			-0.5%			Drive end bearing cleaned
			-0.5%			Both bearings cleaned
4D 100 hp, 60 Hz, 2 pole	95.0%	94.5%	-0.5%	L-2L	L-2L	
5E 150 hp, 60 Hz, 2 pole	92.3%	92.0%	-0.3%	L-2L	L-2L	
7B 150 hp, 60 Hz, 2 pole	93.7%	93.3%	-0.4%	L-2L	L-2L	

Table 3. Low voltage wire wound motors

Phase 2: Motors rewound with control on rewind process

Motor description	Efficiency before rewind	Efficiency after rewind	Efficiency change*	Winding before	Winding after	Comments
6F 150 hp, 60 Hz, 2 pole	94.4%	94.3%	-0.1%	L-2L	L-2L	
9E 60 hp, 60 Hz, 2 pole	90.1%	89.9%	-0.2%	L-2L	L-2L	No core cleaning
10D 125 hp, 60 Hz, 4 pole	95.4%	95.2%	-0.2%	L-2L	L-2L	
11F 200 hp, 60 Hz, 2 pole	96.4%	96.3%	-0.2%	L-2L	L-2L	
14H 55 kW, 50 Hz, 4 pole	89.9%	89.2%	-0.7%	2TC-1L	L-2L	Defective core (OEM)
16H 150 kW, 50 Hz, 4 pole	95.4%	95.6%	+0.2%	GC-2L	L-2L	
18G 55 kW, 50 Hz, 4 pole	94.2%	94.2%	0.0%	3TC-1L	3TC-1L	
19H 132 kW, 50 Hz, 2 pole	93.0%	93.0%	0.0%	2TC-2L	L-2L	
20H 45 kW, 50 Hz, 2 pole	93.9%	93.9%	0.0%	3TC-1L	3TC-1L	
21J 75 kW, 50 Hz, 2 pole	93.7%	93.9%	+0.2%	GC-2L	L-2L	

Key to winding configuration abbreviations

3TC-1L	Three tier concentric	1 coil side per slot
2TC-1L	Two tier concentric	1 coil side per slot
2TC-2L	Two tier concentric	2 coil sides per slot
L-2L	Lap winding	2 coil sides per slot
GC-2L	Concentric coil groups placed in slots in rotational sequence	2 coil sides per slot
GC-1L	Concentric goil groups places in slots in rotational sequence	1 coil side per slot

Table 4. Low voltage wire wound motors

Phase 3: Motors with multiple rewinds under controlled conditions

Motor description	Efficiency before rewind	Efficiency after rewind	Efficiency change*	Winding before	Winding after	Comments
4D 100 hp, 60 Hz, 2 pole	95.0%	94.5% 94.9% 94.8%	-0.5% -0.1% -0.2%	L-2L L-2L L-2L	L-2L L-2L L-2L	1st rewind – uncontrolled 2nd rewind 3rd rewind
12F 150 hp, 60 Hz, 2 pole	95.9%	95.9% 95.9% 95.8%	0.0% 0.0% -0.1%	L-2L	L-2L L-2L L-2L	1st rewind 2nd rewind 3rd rewind
15J 75 kW, 50 Hz, 4 pole	93.0%	93.6% 93.6% 93.7%	+0.6% +0.6% +0.7%	GC-2L	L-2L L-2L L-2L	1st rewind 2nd rewind 3rd rewind
8C 200 hp, 60 Hz 4 pole	95.7%	95.1% 95.6%	-0.6% -0.1%	GC-2L	GC-1L L-2L	1st rewind 2nd rewind
13G 110 kW, 50 Hz, 4 pole	94.8%	94.6% 94.6%	-0.2% -0.2%	3TC-1L	3TC-1L 3TC-1L	1st rewind 2nd rewind
17H 5.5 kW, 50 Hz, 4 pole	86.7%	86.9%	+0.2%	2TC-1L	L-2L	5 burn outs at 360°C 1 rewind only
22H 5.5 kW, 50 Hz, 4 pole	83.2%	84.0%	+0.8%	2TC-1L	L-2L	5 burn outs at 360°C 1 rewind only

*Each of the percent changes is relative to the "before rewind" efficiency

Table 5. Medium voltage motor with formed coils

Motor description	Efficiency before rewind	Efficiency after rewind	Efficiency change*	Winding before	Winding after	Comments
23K 225 kW, 50 Hz, 4 pole, 3300 volts	95.7%	95.5%	-0.2%	L-2L	L-2L	-

Commentary on Results Summary

- **Difference in efficiency loss between controlled and uncontrolled groups**
  The average efficiency loss for motors in the uncontrolled group was 0.62% (range -0.3% to -1.0%). This included two motors where the bearings had been overgreased. When the bearings were purged and regreased correctly, the average efficiency loss for the group was 0.45% (range -0.3% to -0.5%).
  The average efficiency loss for motors in the controlled group was 0.10% (range +0.2% to -0.7%). The one motor which recorded a 0.7% loss in efficiency had been supplied new by the manufacturer for these tests and was subsequently found to have faulty interlaminar insulation – if this result is discarded the average efficiency reduction was 0.03% (range +0.2% to –0.2%).

  These results are shown graphically in Fig 4 below **and clearly illustrate the benefits of best practice repair methods.**

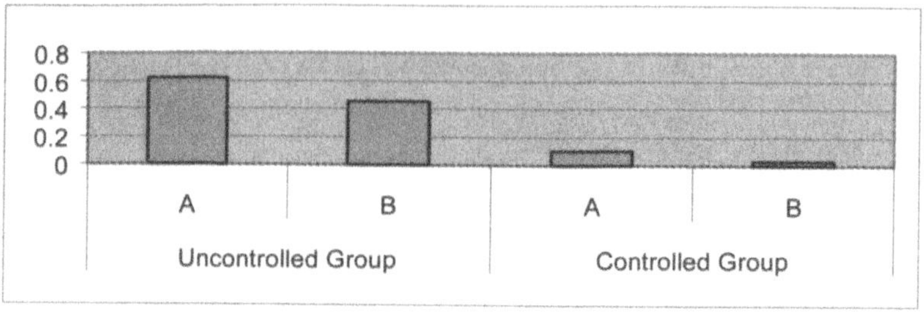

Fig. 4. Controlled & Uncontrolled groups - average loss of efficiency

A	**-**	**All results.**
B	**-**	**Results after motors with overgreased bearings had been corrected, and omitting 55 kW motor with faulty interlaminar insulation.**

- **Multiple Rewinds**
  Hitherto, motor users have often been led to believe that if a motor lost say 0.5%-1.0% efficiency when it was rewound, then this loss would be cumulative over successive rewinds and that after one or two rewinds the only sensible action would be to replace the motor if or when it failed again. This "cumulative loss" hypothesis never did seem very sound (if it had been, more cases of motors overheating after multiple rewinds would surely have come to light), but there did not seem to be much hard evidence to disprove it. Multiple rewinds (2 or 3) were therefore carried out on 5 motors in the range 75 kW-150 kW (100hp-200hp) and five burnout cycles at 360°C followed by one rewind were carried out on 2 small 5.5 kW motors. The laminations of one of these two motors had steam blued (oxide) interlaminar insulation – the other had laminations coated with an organic/inorganic varnish (modified L3 type) which had been shown in earlier AEMT work [2] to be capable of withstanding temperatures up to 500°C without damage.

The results of all these tests (see Table 4) are significant and may surprise some. They can be summed up as follows:

a) **Provided Good Practice Repair Methods are followed, up to three multiple rewinds do not adversely affect efficiency.**
b) **Even if Good Practice is not followed, as in the case of motor 8C where at the first rewind the winding was changed from a 2 layer concentric to a single layer lap, this loss can be recovered by a second rewind using correct rewind practice. (N.B. This might not be true if other elements of Good Practice – e.g. working within recommended burnout temperature limits – were not followed).**
c) **There is no evidence that further Good Practice rewinds would adversely affect efficiency – indeed the results from tests on the two 5.5 kW motors would suggest otherwise.**

These are important results for motor users and the repair industry alike and they should to be taken into account in any "Repair or Replace" decision-making by customers or in advice offered to users by regulatory and other authorities.

- **Medium voltage motor rewind**
  One 225 kW 3.3 kV motor was rewound to check whether there were any factors specific to medium voltage machines which would invalidate some or all of the repair methods used on LV machines. Whilst the tests could hardly claim to be exhaustive, nothing was found to indicate that Good Practice Repair Methods developed for LV machines were not equally suitable for MV formed coil machines also. The motor used was a highly-rated late 1990s machine with formed coils and global VPI.

Loss analysis

Whilst the summary results shown in Tables 2, 3 and 4 are sufficient for most people outside the motor repair industry, an examination of the average increase or decrease in each element of the loss of the controlled and uncontrolled motor groups yields further information of use to repairers, and demonstrates clearly the value of using Good Practice Repair methods. These results are shown in Table 6 below and are plotted graphically in Fig.5.

Table 6. Average percentage change in losses – controlled & uncontrolled groups

	Average percentage change in loss after rewind				
	Stator copper loss	**Rotor loss**	**Core loss**	**Friction & Windage loss**	**Stray loss**
Uncontrolled Group	0.5	1.65	11.5	19.05[i]	24.01
Controlled Group	(5.2)	1.87	1.06[ii]	2.01	3.2

Notes:
[i] Includes losses caused by overgreasing bearings on two motors.
[ii] Excludes results from one motor subsequently found to have faulty interlaminar insulation as supplied by manufacturer.

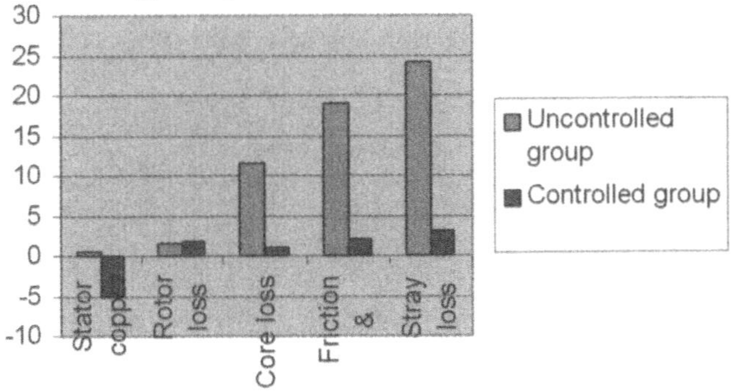

Fig. 5. Average % change in losses for controlled & uncontrolled groups

Reasons for improvements in Controlled Group

Stator I2R	Reduced MLT. Increased copper section
Core loss	Burnout temperature raised to 360°C-370°C making coils easier to remove
Friction & Windage	Care with bearing greasing, seal lubrication, bearing fits, assembly, etc.
Stray loss	Increased burnout temperature (360° – 370°C) making coil removal easier with less damage to core during cleaning.

Comparison of efficiencies calculated in accordance with IEC 60034-2 and IEEE 112B

For 13 out of the 23 motors tested, efficiencies before rewind calculated in accordance with IEC 60034-2 were higher than those calculated in accordance with IEEE 112B (for the remainder they were either the same or lower).

Post-rewind, the pattern was similar with 16 out of 23 motors showing higher efficiencies calculated in accordance with IEC 60034-2 compared with those calculated to IEEE 112B.

The average difference in efficiency for each of the three groups (uncontrolled, controlled and multiple rewind) ranged between 0.45% and 0.11% and was perhaps lower than expected, but there was a considerable spread of figures in each group. It was difficult to detect any pattern for these differences (e.g. by motor size, polarity or manufacturer), although it was noticed that motors which had a large difference in efficiency calculated by the two methods prior to rewind mostly had a similar difference after rewind. The averages and ranges of efficiency differences for motors in each group are shown in Table 7 below.

Table 7. Differences in efficiencies calculated in accordance with IEEE 112B and IEC 60034-2;( ) = IEC efficiency lower

Motor Group	Before rewind				After final rewind			
	Average efficiency		Ave. diff. % point	Range of points	Average efficiency		Diff. % point	Range % point
	IEEE 112B %	IEC 60034-2 %			IEEE 112B %	IEC 60034B %		
Uncontrolled	93.75	94.13	0.38	0-0.8	93.3	93.75	0.45	(0.2)-1.0
Controlled single rewind	92.11	92.50	0.39	(0.5)-0.9	92.20	92.50	0.30	(0.4)-0.9
Controlled multiple rewind	94.06	94.11	0.17	(0.4)-1.2	94.01	94.12	0.11	(0.3)-0.9
MV motor	95.7	95.7	0	N/A	95.9	95.8	0.1	N/A

The average difference for the uncontrolled group rose following rewinding due to the increases in stray losses in this group (see Fig.5) whilst those of the other groups fell due to better control of stray and other losses. However, it is difficult to draw any further conclusions from these comparisons.

Conclusions

The work on which this paper is based probably represents the most wide ranging and in-depth analysis yet done on the effects of motor repair processes on motor efficiency. It demonstrates quite clearly that provided good practice methods are followed by repairers:

- **motors up to 225 kW can be rewound without significant loss of efficiency.**
- **multiple rewinds do not produce cumulative reductions in efficiency.**
- **rewind equipment commonly used by most repairers is satisfactory provided that it is used correctly.**

However, the "round robin" test also showed that introducing variables such as different test beds, instruments and test personnel can significantly affect results. This may affect the validity of some earlier work by others on motor rewind efficiencies.

Certainly these results make a myth of the oft quoted statement that "each rewind costs 1% in motor efficiency" – and "motors should never be rewound more than once". The authors suggest that motor users and regulators alike should rethink some of their policies on repair versus replacing motors.

The difference in results between the controlled and uncontrolled motor groups in this project is significant and shows that the performance of even experienced repair personnel can be improved if Good Practice is followed closely. Both the AEMT and EASA intend to use the knowledge gained from this work to improve the skills of their membership. In this way, with only a little effort, induction motors can be rewound with minimum loss of efficiency.

References

[1] Walters D.G. Minimising Efficiency Loss caused by Motor Rewinds. EEMODS '99. Springer Energy Efficiency Improvements in Electric Motors & Drives, 2000, pp 168-189.

[2] AEMT (UK) The Repair of Induction Motors – Best Practices to Maintain Energy Efficiency.

The Effect of Rewinding on Induction Motor Losses and Efficiency

Ziba Kellum

Advanced Energy, 909 Capability Drive, Suite 2100, Raleigh, North Carolina, 27606 USA, E-mail: zkellum@advancedenergy.org

Abstract

Repairing electric motors when they fail can be a cost-effective option in many situations. The decision to repair a failed electric motor or replace it with a new motor can be evaluated economically if the cost and efficiency of the two options are known. Cost of the two options and the new motor efficiency are easily obtained. The efficiency of the motor following repair is not readily available nor are motor repair facilities equipped to measure it with a high degree of accuracy. Very little accredited efficiency data exists in the public domain to fill this need.

Economic evaluations of repair versus replace generally rely on an assumed value for repaired motor efficiency. These assumptions range from using the manufacturers original efficiency value stamped on the motor nameplate, sometimes subtracting an assumed amount to represent damage incurred during the repair process, to simply selecting an assumed value for efficiency from a table based on the motor rating. The aim of this paper is to provide a comparison of laboratory efficiency test results, including segregation of five categories of losses, of induction motors before and after rewinding to serve as the basis for estimates of repaired motor efficiency for use in repair versus replace economic evaluations.

This paper will provide a comparison of the efficiency and losses of 13 motors rated between 20 and 200 hp that were tested in the Advanced Energy NVLAP accredited motor test lab before and after being rewound. The motor rewinds were performed at several North American motor repair facilities. The paper will discuss how the motor rewinds affected specific motor losses and the effect of these losses on motor efficiency and reliability. Also presented and discussed will be corrective actions taken which in some cases restored the motors to their pre-repair efficiency. Three motors, which were rewound and retested multiple times, will be used to illustrate the effects of multiple rewinds to a single motor.

Project Scope

- A total of thirteen TEFC, 4-pole AC induction motors ranging from 20 to 200 hp were evaluated.
- Motors were baseline tested, the winding damaged and then sent to a repair facility for rewinding. Some motors were rewound more than once. The motors were retested after each rewind.
- The rewinds were performed at several North American motor repair facilities.
- Participating repair facilities were aware that the motors were being tested before and after rewinding.
- All testing was conducted at the Advanced Energy Motor Resource Center between December 1997 and July 2002.
- Motors were stored (not operated) when not being tested or rewound.
- All motors when new, and after each rewind, were operated a minimum of 8 hours prior to testing. This was to allow for normal bearing/grease break-in.
- From the test results, a total of 22 before and after rewind comparisons were evaluated that include: Efficiency at rated load, the five categories of losses segregated at rated load, stator winding resistance (@ 25°C) and rated load temperature rise as determined by stator winding resistance measurements.

Results

Comparison of "New Motor" to "First Rewind" Test Data

Of 13 motors tested when new and after the first rewind, 4 motors had lower efficiency after rewind, up to 0.9 of a point of efficiency (Table 1). 3 motors increased in efficiency by up to 0.7 of a point of efficiency (Table 3). An efficiency change of less than 0.2 points was not considered significant enough to evaluate. 6 motors fell into this category (Table 2). As an overall average, the 13 rewound motors changed very little relative to their "new" performance (Table 4). Tabulated values represent the change in motor efficiency resulting from the change in each loss category.

Table 1. Comparison of "new" and first rewind test results – motors that increased losses

Sample Label	HP	Change in Temperature Rise(°C)	% Change in Stator Resistance	Change in full load efficiency point(s)						
				Full Load Efficiency	Core Loss	F&W Loss	Stator Loss	Rotor Loss	Stray Load Loss	Total Loss
12AB	25	3,2	13,2%	-0,9	0,03	-0,61	-0,27	-0,03	-0,02	-0,90
7AB	20	3,4	1,5%	-0,7	0,06	-0,68	-0,07	-0,13	0,12	-0,70
6AB	20	3,1	5,5%	-0,5	0,20	-0,17	-0,26	-0,04	-0,22	-0,50
3AB	100	3,7	13,1%	-0,3	0,02	-0,54	-0,36	0,05	0,53	-0,30
Averages:		3,4	8,3%	-0,6	0,1	-0,5	-0,2	0,0	0,1	-0,6

Table 2. Comparison of "new" and first rewind test results – motors with no significant change in efficiency

Sample Label	HP	Change in Temperature Rise(°C)	% Change in Stator Resistance	Change in full load efficiency point(s) Full Load Efficiency	Core Loss	F&W Loss	Stator Loss	Rotor Loss	Stray Load Loss	Total Loss
2AB	20	-0,9	8,6%	-0,2	0,13	0,11	-0,19	-0,14	-0,11	-0,20
1AB	50	6,7	5,8%	-0,1	0,02	0,09	-0,25	-0,02	0,07	-0,10
10AB	20	-7,4	-1,4%	-0,1	-0,04	-0,30	0,09	0,02	0,14	-0,10
4AB	100	1,6	14,1%	0,1	-0,08	-0,02	-0,22	0,00	0,42	0,10
8AB	20	-5,3	-3,7%	0,1	-0,03	0,01	0,11	0,05	-0,04	0,10
9AB	20	-4,4	1,0%	0,1	-0,11	0,25	0,01	-0,05	0,01	0,10
Averages:		-1,6	4,1%	0,0	0,0	0,0	-0,1	0,0	0,1	0,0

Table 3. Comparison of "new" and first rewind test results – motors that decreased losses

Sample Label	HP	Change in Temperature Rise(°C)	% Change in Stator Resistance	Change in full load efficiency point(s) Full Load Efficiency	Core Loss	F&W Loss	Stator Loss	Rotor Loss	Stray Load Loss	Total Loss
5AB	20	3,8	-4,0%	0,4	0,00	-0,18	0,11	0,06	0,42	0,40
13AB	200	-13,5	-6,6%	0,6	-0,07	0,12	0,15	0,10	0,28	0,59
11AB	30	-10,8	-0,1%	0,7	0,26	-0,09	0,06	0,12	0,35	0,70
Averages:		-6,8	-3,6%	0,6	0,1	-0,1	0,1	0,1	0,3	0,6

Table 4. Average of changes for all motors from "new" to first rewind

Sample Label	HP	Change in Temperature Rise(°C)	% Change in Stator Resistance	Change in full load efficiency point(s) Full Load Efficiency	Core Loss	F&W Loss	Stator Loss	Rotor Loss	Stray Load Loss	Total Loss
Averages:		-1,3	3,6%	-0,1	0,0	-0,2	-0,1	0,0	0,1	-0,1

Motors That Increased Losses During Rewinding

For the 4 comparisons with a significant increase in losses, the largest area for increased losses was in Friction and Windage (F&W), averaging 0.5 point drop in efficiency for this loss alone. The next significant area of increased loss was in Stator I^2R losses, averaging a 0.17 point drop in efficiency. Change in other loss categories had an insignificant average affect on efficiency.

The change in temperature rise for this group averaged a 3.4°C increase. This increase implies a reduction in motor life, since a 10°C increase will, as a rule of thumb, reduce winding insulation life by half. A similar thumb-rule states that the life of the bearing lubricant is reduced by half for every 15°C increase in temperature.

Motors That Reduced Losses During Rewinding (Table 3)

For the group as an average the only significant pattern of change was in the area of Stray Load Losses (SL Loss), averaging a 0.3 efficiency point improvement in efficiency for this category. Other variations in losses were apparent individually. 5AB reduced stator loss while increasing F&W loss, 13AB improved in all areas except core loss and 11AB improved significantly in core loss and rotor losses.

The change in temperature rise for this group averaged a 6.8°C decrease. This decrease implies an improvement in motor winding insulation and bearing lubricant life.

A Word About Stray Load Losses

SL Loss is derived from the measured total losses obtained by subtracting Power output (Pout) from Power input (Pin). Pout and Pin may be easily determined with an uncertainty on the order of ±0.2% of the meter full-scale reading. However, when subtracting two large numbers the difference between the numbers is subject to the uncertainty of both measurements. Applying this uncertainty to the measured difference between Pin and Pout creates an uncertainty of ±2.8% of the total losses (assuming an efficiency of 90). SL Loss is then determined by subtracting the four other losses, determined by other methods, from the total losses determined from Pin – Pout. This, in effect, concentrates the uncertainty of the measured total losses (±2.8% or ±0.28 efficiency points,) into the measured value of SL Loss. The SL Loss regression method employed by test standards, IEEE 112, Method B and CSA C-390, Method 1, reduces this uncertainty by some extent, but can still result in significant uncertainty in the calculated value of SL Losses.

Over-Greasing Effect on F&W Loss

After the first rewind and retest of comparisons 12AB, 7AB and 6AB (in Table 1), we had the opportunity to inspect the bearing cavities and found them to be packed full with grease. We reduced the grease in the cavity to about ½ the cavity volume, wiped excess grease from the bearings, reassembled the motors and retested them after an 8 hour run-in. These results are shown in Table 5 with the "new" motor test data compared to test data acquired after degreasing. These comparisons are relabeled 12AC, 7AC and 6AC. This clearly shows that excess grease was the cause of the increase in F&W losses in these motors.

Table 5. Comparison of "new" and first rewind test results – motors that increased losses – retest after removing excess bearing grease and additional 8 hour run-in

Sample Label	HP	Change in Temperature Rise(°C)	%Change in Stator Resistance	Change in full load efficiency point(s)						
				Full Load Efficiency	Core Loss	F&W Loss	Stator Loss	Rotor Loss	Stray Load Loss	Total Loss
12AC	25	2,2	13,4%	-0,2	-0,02	0,23	-0,27	0,02	-0,17	-0,20
7AC	20	1,6	1,5%	0,1	-0,02	0,12	-0,03	-0,09	0,11	0,10
6AC	20	2,4	4,6%	-0,4	0,04	-0,01	-0,22	-0,02	-0,20	-0,40
Averages:		2,1	6,5%	-0,2	0,0	0,1	-0,2	0,0	-0,1	-0,2

Stator Winding Resistance and Stator Loss

A comparison of the change in stator resistance to the change in stator losses showed a strong correlation. In only one case (11AB) was there a significant increase in stator loss despite no change in stator resistance. One theory is that there may have been a winding design change that produced this result. This theory is still being investigated.

Motors That Were Rewound More Than Once

Table 6 lists efficiency comparisons for one motor that was rewound twice (7AC, CD) and two motors that were rewound three times (8AB, BC, CD and 10AC, CD, DE). Each subsequent rewind following the initial rewind is compared here to the performance of the previous rewind. Although there are some significant variations in the efficiency and losses of subsequent rewinds there is no pattern that would indicate that motors become more susceptible to efficiency loss with repeated rewinding.

Predicting Efficiency Change of Repaired Motors

It would be very convenient if motor users could use a simple test to determine if efficiency of a rewound motor had decreased from the pre-rewind efficiency. To this end a comparison of the percentage change in no-load power input versus the change in rated load efficiency was evaluated. This data revealed that motors with greater than a 10% increase in no-load watts had suffered efficiency degradation. Of note however was that one motor with a 0.4 point rated load efficiency drop did not exhibit any change in no-load power input.

Table 6. Motors that were rewound more than once

Sample Label	HP	Change in Temperature Rise(°C)	% Change in Stator Resistance	Change in full load efficiency point(s)						
				Full Load Efficiency	Core Loss	F&W Loss	Stator Loss	Rotor Loss	Stray Load Loss	Total Loss
7AC	20	1,6	1,5%	0,1	-0,02	0,12	-0,03	-0,09	0,11	0,10
7CD	20	-1,2	-2,8%	-0,1	-0,08	-0,07	0,03	0,03	-0,01	-0,10
8AB	20	-5,3	-3,7%	0,1	-0,03	0,01	0,11	0,05	-0,04	0,10
8BC	20	5,1	1,2%	-0,4	-0,03	-0,32	-0,11	-0,04	0,10	-0,40
8CD	20	-7,1	-0,1%	0,2	-0,23	0,19	0,08	0,02	0,13	0,20
10AC	20	-8,5	-1,3%	0,4	0,08	0,07	0,09	0,02	0,14	0,40
10CD	20	-0,6	-2,0%	-0,8	-0,18	-0,68	0,06	0,00	0,00	-0,80
10DE	20	-0,7	1,3%	0,8	0,12	0,68	-0,06	0,00	0,06	0,80
Averages:		-4,2	-0,5%	0,2	-0,1	0,1	0,0	0,0	0,1	0,2

Conclusions

This comparison of the measured efficiency of new and rewound motors demonstrated a full load efficiency change that varied between a loss of 0.9 and a gain of 0.7 efficiency points. From the average efficiency change of –0.1 point one could conclude that rewinding does not reduce motor efficiency, in the aggregate. This conclusion should be tempered by the fact that the repair facilities participating in this study were aware of the efficiency-testing program. What can be concluded with greater confidence is that induction motors can be rewound using proper processes and procedures, even multiple times, without any degradation in efficiency.

The results also show that core losses are not necessarily the principle category for rewinding losses when they do occur, but that increased F&W due to bearing over-greasing and increased Stator losses due to higher stator winding resistance may be more common.

A table of actual measured values of efficiency and losses along with graphs used in the presentation may be obtained by email request to zkellum@advancedenergy.org. Other motor information may be obtained at www.advancedenergy.org.

Practical Issues in the Manufacture of High Efficiency Electric Motors

Peter John Tavner[1], Paolo Caselotti[2]

[1] Technical Director, FKI Energy Technology

[2] Technical Director, Marelli Motori S.p.A.

Abstract

European motor manufacturers seeking to deliver motor ranges which comply with the current voluntary efficiency agreement have to choose between developing a new range, improving an existing range or obtaining their range from another source.

Some manufacturers have introduced new ranges and in general these motors incorporate an increase in active materials, electrical steel and copper, to increase their efficiency. These increases in materials naturally increase the base cost of the motor. Other manufacturers, including the authors, have taken an existing range and modified aspects of the design to optimise performance and achieve a higher efficiency. Further manufacturers have sought to buy higher efficiency motors from a lower labour cost source, where they can offset the increase in material cost of the higher efficiency motor with a substantial reduction in labour cost for its manufacture. In the latter case an important component in achieving the higher efficiency is related to maintaining the quality of the delivered product.

This paper will describe the issues associated with optimising the efficiency of an existing range of induction motors, utilising changes in design, new materials and optimised manufacturing processes. It will pay particular attention to practical problems which occur, their elimination and the economic impact of those improvements.

Introduction

The TEFC motor has evolved, over 80 years, with designs to International Standard dimensions and outputs in a tough, blow-over cooling enclosure, which gives a rugged, interchangeable machine that can be used in a wide variety of applications.

Manufacturers have evolved their products to achieve requirements as economically as possible and a previous paper, Ref 1, has shown that this has achieved machines with progressively improving weight-power ratios, as shown in Fig. 1, taken from Glew, Ref 2 therefore becoming intrinsically cheaper.

However, Fig. 1 also shows that the weight-power of TEFC motors has not reduced at the same rate as larger machines. In recent years there has been a tendency for the weight-power of TEFC motors to increase. This is due to the constraints of International Standards and to the poor heat transfer in the TEFC machine, compared to modern large machines, which have air-blown internal active parts. So it has been progressively harder for manufacturers to make TEFC motors more competitively, by lowering the weight-power ratio, without deteriorating the noise and efficiency of the machine. In other words the limit of present knowledge of thermal performance is being reached and economic improvements can only be achieved by:

1. Lowering performance, or
2. By using more materials, increasing cost to improve performance but then finding a cheaper place to make the motor, or
3. By increasing present knowledge, using modern methods for optimal design to achieve better performance with less material or changing to better but cheaper materials.

The challenge of the efficiency initiatives in North America and the European Union is that option i) is not possible. Manufacturers are deploying a variety of methods in ii) above, the effect on costs has been analysed by Parasitili et al in Ref 3 but this was done on the basis that manufacturers increased weight/power. We are not aware of any activity in iii). But have we reached the true limit of weight-power? This paper believes not.

The situation is analogous to that in which Japanese car manufacturers found themselves in USA in the 1980s, where their conventionally-sized, 2 litre, compact car engines produced less power than the American market demanded. The easy solution was to build bigger engines, in the American fashion. However, the Japanese answer was to use the latest design techniques to get more power out of the existing 2 litre package. This is now widely accepted as the appropriate solution and those same engines have shown the way to achieve more stringent environmental requirements. This paper follows from Ref 1 to describe what can be done practically to address the fundamental problems in design and manufacture for an existing range of TEFC motors from an established manufacturer.

Interaction Between Losses, Efficiency & Weight-Power

The losses in a motor are classified as follows:

1. Load Dependent Losses including:
   - Joule losses in the stator winding;
   - Joule losses in the rotor winding;
   - Stray losses.
2. Load Independent or Constant Losses, including:
   - Iron loss;
   - Mechanical losses including fan, bearing and shaft seal.

In small and medium sized motors at normal load the Joule losses represent the significant proportion of total losses, typically above 50%. Therefore their reduction should be one of the principal objectives in raising motor efficiencies. Stray losses also represent a significant proportion of losses, occurring in both rotor and stator, but they are difficult to predict and therefore to reduce.

Haataja in Ref 4 implied the reduction of losses is the only way to develop high efficiency motors and this must increase the volume of active parts and therefore increase the weight-power. But this is only so if currents, fluxes and materials remain constant. No self-respecting manufacturer improves his machines by keeping everything the same.

Heat Transfer & Losses in TEFC Machines

The problem with the TEFC geometry is that all losses from the active parts, the core, rotor and stator windings, must be extracted from the interior of the machine, passed to the ribbed barrel and removed by convection by the blow-over air.

There are four weaknesses in this arrangement:

1. Internal air does not remove much heat from the active parts, so there is little assistance to heat transfer from internal fluid circulation.
2. The heat from the rotor to the ribbed barrel crosses 3 thermal resistance interfaces, the air gap, the stator insulation and the core/barrel interface.
3. The losses for smaller motors are dominated by Joule losses developed in the windings which is mostly extracted across 2 of these interfaces.
4. A large proportion of the second major source of loss, the strays, are concentrated in the rotor and must be extracted across 3 of these interfaces.
5. Production factors can worsen the thermal resistance across each of the interfaces.
6. The convective heat transfer across the final interface, the ribbed barrel, isnot optimal.

In consequence the bulk heat transfer coefficient for a TEFC motor, relating total losses to stator winding temperature rise and motor surface area, is typically 15 W/m^4K, much lower than that for larger electrical machines, typically 300 W/m^4K.

This paper is arguing that one can lower the weight-power, by raising the bulk heat transfer coefficient in the machine from 15 W/m^4K. This could increase loss density, with the threat of reducing efficiency. But lowering temperature will decrease loss. If the materials & design are also improved then efficiency can be increased. The following sections describes the changes that can be made.

What to Do ?

To reach a higher performance competitively one must optimise:

1. Performance parameters such as:
   - Losses & efficiency.
   - Noise.
   - Starting current & torque.
2. Production factors
   - Improved processes to reduce loss.
   - Improve the choice of core frame diameters.
   - Lower the cost of production by removing unnecessary processes.

To improve the variability of design & manufacturing processes and materials the following areas for optimisation should be addressed:

1. Design calculations;
   - Identification and prediction of stray losses;
   - Thermal & ventilation modelling to improve performance.
2. Materials;
   - Use improved steels;
   - Reduce material variance;
3. Processes;
   - Better impregnation to improve heat transfer;
   - Limit tool wear to reduce loss & wear variance;
   - Improved stator winding to reduce loss;
   - Improved rotor cage construction;
   - Improved core/barrel fit.

Reducing Losses

It has been shown that Joule losses are an important part of machine loss. These can be reduced by using high-permeability, rather than low-loss, steels, punched with sharp tools. Additional benefit can be obtained by using semi-processed steel, annealed after punching, to eliminate the punching losses.

It is also clear that one of the crucial problems for reducing stray losses lies in the rotor from which heat has to be removed across 3 thermal interfaces.

A significant contributor to these losses is high frequency current circulation in the smooth surface of the rotor, due to stator slot harmonics. Usually rotors are machined after diecasting the cage in the closed slots. Such machining is frequently done at high speed in one cut, leading to distortion and burring of the rotor lamination edges, enhancing the losses due to these currents. Experiments have shown that by reducing the speed and cut size it is possible to reduce these surface losses.

There are also papers that address the problem of stray losses due to rotor slot skew. For example Williamson et al, Ref 5, has shown the substantial improvement in efficiency that can be achieved by eliminating interbar currents, Fig 2. Experiments with motors without skew, where the rotor and stator slots numbers have been chosen to avoid noise and cogging, show a significant reduction in stray losses, which reduces motor temperature and raises efficiency.

Optimising Heat Transfer

There are some references that address the heat transfer weaknesses of TEFC machines, for example Pickering et al Ref 6. This dealt with the major area for improvement, the heat transfer from the ribbed body, which depends upon the dimensions of the rib in relation to the frame and the airflow over the ribs. This airflow should be high enough for scrubbing action but not so high as to reduce heat transfer. The fan dimensions need to be optimised in relation to the fan-cowl, which must have a good exit shape over the ribs.

The work described by Hayward in Ref 7, on the motor range considered here, showed how the fan, fan-cowl and rib dimensions can be optimised by modelling and test. Fig. 4 shows the experimental Air Flow Test Rig based on the current range of motors on which Computational Fluid Dynamics predictions were also made. Fig. 5 shows good comparison between the full size Air Flow Test Rig and the CFD predictions, confirming the modelling. The CFD was then used to optimise rib dimensions and air flow so that the heat transfer coefficient on that surface could be improved.

Production Issues

An important issue in production process is to limit the variability of efficiency values attributable in a population of motors to periodic variations in tools, processes and materials. We consider this to be a vital issue in improving performance.

The above section has referred to the important production issues of reducing core losses by maintaining sharp punching tools and if possible eliminating such losses by annealing the laminations after punching.

We have also seen that we can reduce stray losses by improving rotor machining, using two light cuts in opposite directions rather than one severe cut. A similar effect can be achieved by minimising the machining at the stator outer diameter but the stray loss intensity there, away from the air gap, is small. However the resultant surface finish on the core back is important to ensure a high heat transfer coefficient from the core to the cast stator frame, as is the machining of the frame itself. Roughness figures of Ra 1.5 are recommended.

An important issue for heat transfer is the quality and repeatability of impregnation of the stator winding. The usual criterion for impregnation is the dielectric withstand but of course impregnation with a high and repeatable coefficient of heat transfer is essential.

There are some papers that address the effect of manufacturing processes on these matters, for example by Walters.

Conclusions

Improving the efficiency of TEFC motors does not necessarily mean increasing the volume of material and therefore cost, but does necessitate optimising design and production factors. The areas to attack are:

1. Reduce core & Joule losses by optimising magnetic steel material to give high permeability and eliminate punching losses by annealing after punching.
2. Reduce stray losses particularly on the rotor by reviewing machining and the need to skew the slots, making sure that rotor and stator slot selections are optimised to avoid noise and cogging problems.
3. Improve manufacturing processes to give good heat transfer particularly in respect of:
    - Optimising rotor machining to minimise rotor surface losses and guarantee a repeatable and accurate airgap.
    - Improve stator core and stator frame machining to ensure good stator to frame fit.
    - Improve stator winding impregnation to ensure good heat transfer.
    - Reducing variability in manufacturing process.
4. Optimise stator frame cooling by appropriate fan, fan-cowl and rib design to raise the convective heat transfer at the stator.

The result is that an existing range of motors with a low weight-power ratio, can now also achieve a higher efficiency, Fig. 6.

These motors have also been incorporated into integrated products, which can obtain high driven equipment efficiencies from inverter supply, see Shakweh et al, Ref 8.

Acknowledgements

This paper is published with the permission of the Directors of MarelliMotori SpA and FKI Energy Technology.

References

[1] Caselotti P, Conchetto A, Tavner P J, High efficiency motors – performance, economy & reliability, by optimisation, 2nd EEMODS Conference, London, 20-22 September 1999, Springer Verlag, pp95-101.
[2] Glew N, Design and manufacture of energy efficient and environmentally friendly large machines, IEE Colloquium 1999, pp1/1-1/5.
[3] Parasitili F, Villani M, Evaluation of the design options and cost impact of improving induction motor efficiency, 2nd EEMODS Conference, London, 20-22 September 1999, Springer Verlag, pp514-528.
[4] Haataja J, Pyrhonen J, Improving three phase induction motor efficiency in Europe, the challenge for manufacturers, IEE EMD Conf, Cambridge, 1997, Publn 444, pp190-194.
[5] Williamson S & Smith AC, Influence of interbar currents on harmonic losses and skew in cage induction motors, IEE PEMD Conf, Bath, 2002, Publn 487, pp 369-374.
[6] Pickering S, Lampard D, Hay N, Roylance T F, Heat transfer from the stator end windings of a low voltage concentric wound induction motor, IEE EMD Conf, Durham, 1995, Publn 412, pp 477-481.
[7] Hayward J, The Ventilation Performance of an Electric Motor, Nottingham University, MSc External Project, F34SP6, 2000.
[8] Shakweh Y, Owen G H, Hall D J, Miller H, Plug and play integrated motor drives. IEE PEMD Conf, Bath, 2002, Publn 487, pp 655-661.

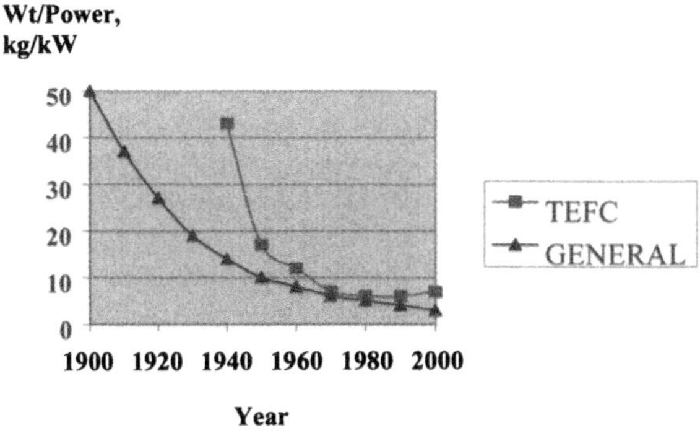

Fig. 1. Weight per unit output as a function of year, from Glew

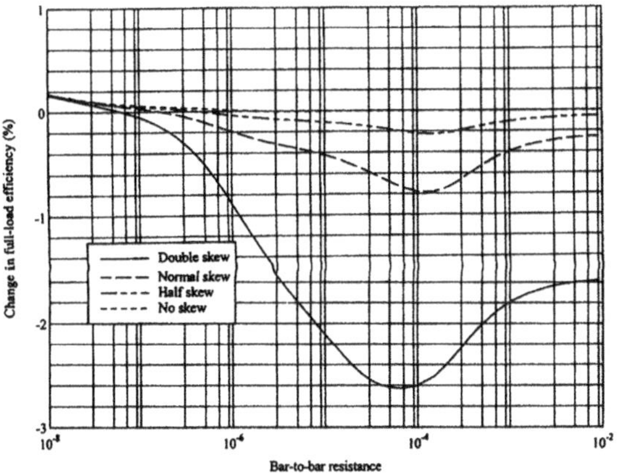

Fig. 2. Effect on motor efficiency of reducing rotor skew, taken from Williamson

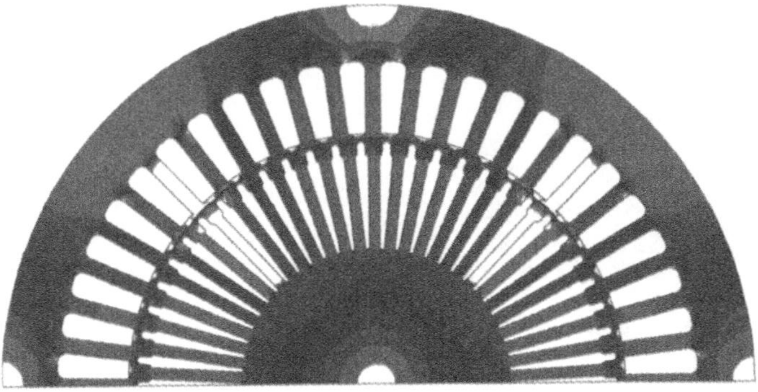

Fig. 3. Example of Finite Element Analysis of the Flux Distribution in an existing Design of 4 pole Motor, Showing the Opportunity to Optimise the Flux Pattern

Fig. 4. Photograph of an Air Flow Test Rig used to Optimise Ribbed Frame Heat Transfer

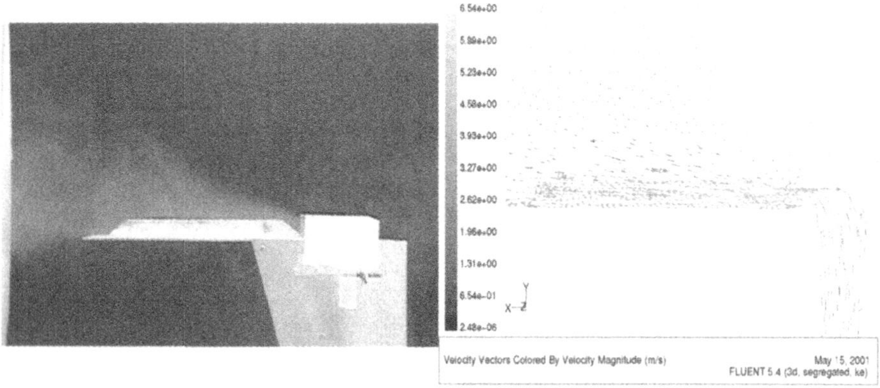

Fig. 5. Top-Smoke Test in the Original Geometry; Bottom-Corresponding CFD data from FLUENT, showing close agreement with Air Flow Test Rig.

Fig. 6. Example of a new high efficiency motor optimised from anexisting range

Fan and Fan Systems – LCC in Swedish and Norwegian Industries

Agneta Persson[1], Thor Henning Gulbrandsen[2]

[1] ÅF-Energikonsult AB, Sweden

[2] IFE Energy Systems Department, Norway

Abstract

This paper discusses life cycle cost calculations and LCC simulations on fans and fan systems. The paper also shows the results from an effect-chain survey recently carried out among Swedish industries regarding knowledge of energy-efficient motors and attitudes and behaviour concerning LCC calculations in conjunction with motors and fans.

Introduction to Life Cycle Assessment in Ventilation Systems

An industrial ventilation system used to be a rather simple system, often merely consisting of a fan extracting air from one place and deliver it somewhere else. This is not the case anymore. The modern industrial ventilation systems are complicated facilities with many purposes. They might supply the area with clean and temperate air and keep an air balance in different sectors or consist of a single system for spot extract. To achieve this a tailor made ventilation system must be designed.

Based on the facts that indoor environment is a crucial question concerning health we are now beyond discussions that compromises on this matter. What is interesting to discuss is the cost of the ventilation system, the cost for operation and maintenance, the coefficient of performance (COP) of the ventilation system, the energy cost, the working hours of the ventilation system, and find the optimal cost efficient bench mark for the life time period.

System boundaries

To achieve the best results in ventilation all parts has to be designed and adapted to each other, the electric motors, fans, dampers, ducts, elbows, diffusers, heat recovery systems and controlling equipment. In many heavy industries it might be an advantage to set boundaries on ventilation systems. The most important part to focus on within a LCC view of the ventilation system must be the energy intensive parts and the most maintenance sensitive accessories.

Life cycle economy

Life Cycle Cost, LCC, is the total cost connected with purchase, operation and maintenance of a product or production plant. LCC is a powerful tool that calculates the costs of a product over its entire lifetime. You may include many different types of costs such as design, production, warranty, repair and disposal. The boundaries for the LCC calculation must be thoroughly defined and described in order to draw the correct conclusions and make correct comparisons between different products or plants.

An LCC audit is thus a method to analyse the most cost efficient solution for building or building or purchasing a product, based on the total lifetime or part of it. For those companies which are aware of the potential you can find in a systematic LCP approach has to undertake accurate planning in the design and purchase within the project framework.

Special attention has to be paid to motor and drive system connected to the fans, to the heat recovery system and circulation pumps. Secondary focus has to be set on the fans itself and the controlling equipment. Taking into consideration the importance of the surrounding environment, which might course great damage to the equipment, the planning for design and purchase can take place.

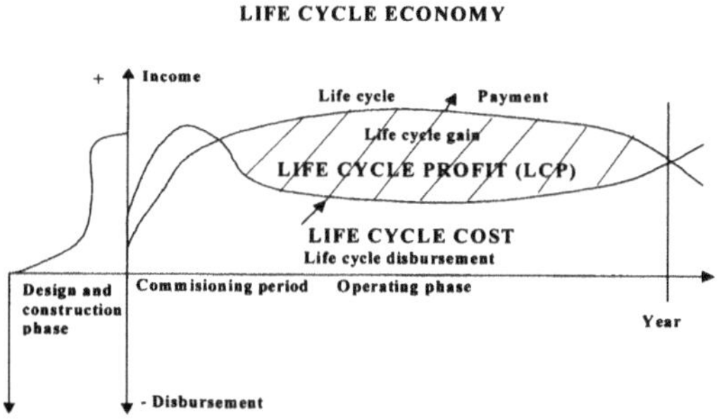

Fig. 1. An illustration of life cycle economy

Survey on industrial motors and motor driven systems

The study includes three industrial branches, pulp and paper industry, chemical base industry, respectively iron and steel industry. The analysis is based on interviews with representatives from 200 plants with 50 or more employees, which corresponds to 80 % of the total number of plants in Sweden within these three branches.

Motors and motor driven systems stand for approximately two thirds of the industrial electricity end use. The electricity end use in the three industrial branches included in this study is 28 TWh per year, which means that motors and motor driven systems in these industries use 19 TWh electricity per year. The potential for energy-efficiency measures is likely very high.

There are between 20 and 25 actors on the Swedish and Norwegian motor market of today. Two of these actors have a very dominating position. According to the motor manufacturers approximately 30 % of the motors are sold separately and 70 % goes to original equipment manufacturers (OEM) applications.

A survey among Swedish industries regarding knowledge of the energy-classification system of electrical motors has recently been carried out. This survey included an effect-chain analysis on attitudes and behaviour concerning LCC calculations in conjunction with electrical motors and motor driven systems, in deep interviews with motor manufacturers and motor dealers and an assessment of Swedish information activities regarding efficient motors.

Industrial knowledge of the energy-classification system

The study shows that the knowledge on the motor energy classification system is very low in general within the three studied industrial branches. Furthermore the study shows that the knowledge of the European motor database EuroDEEM is virtually non-existing.

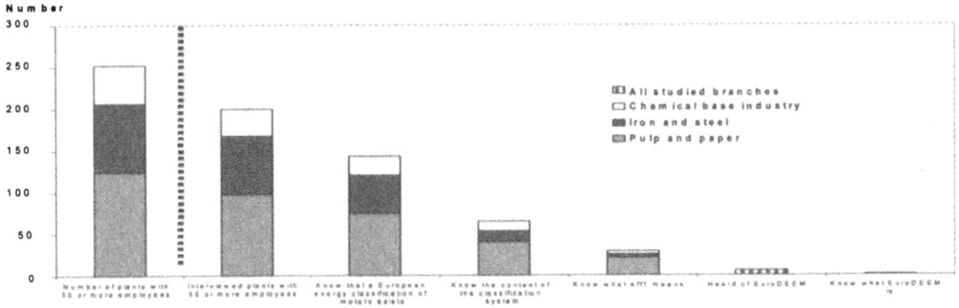

Fig. 2. Knowledge of the European motor classifications system and Euro-DEEM among 200 Swedish industrial plants (50 or more employees)

Almost 75 % of the interviewees claim they have heard of a European motor classification system. However, only one third of the interviewees can give an answer to what the classification system means. And when asked to give a definition of effl a mere 15 % of the 200 interviewees could answer the question. Finally when asked about EuroDEEM it was just three persons who ever had heard of it.

Effect-chain analysis

The effect-chain analysis is a methodology that seeks to simplify a complex reality in order to analyse and show different correlations between carried out activities and impact through various effects on the surroundings. In general an effect chain can be said to describe the impact an increased level of knowledge have on the attitude, and how this change of attitude affects the behaviour. The effect chain methodology is well suited for identifying possibilities for the use of LCC calculations and analysing policy instruments for reinforcing LCC implementation in industry.

The National Energy Agency's work on energy-efficient motors can be described as an effect chain. The first level of the effect chain could e.g. be meetings or seminars with representatives from the industry. The second level would then be that the knowledge and insights these representatives have achieved lead to a change in attitude towards energy-efficient motors. The final level in the effect chain is that changed attitudes are manifested in a different behaviour towards an increased use of energy-efficient motors.

Generally the study reveals a low level of knowledge regarding the classification system and energy-efficient motors, which makes it hard to ascertain confirmed effects of the effect chain analysis. Knowledge is also low among the interviewees regarding which energy class the last purchased motor has. This is the case for both separate motors and motors for OEM applications.

However, one confirmed effect from the study is that the interviewees having received directly addressed information has a higher level of knowledge of energy-efficient moors and the classification system.

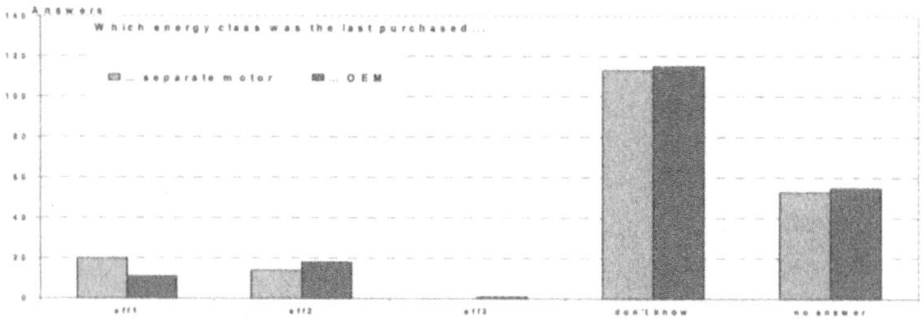

Fig. 3. Energy class of the last purchased motor. The awareness of energy class is very low for both separate motors and motors for OEM applications

The results from the study show that information seems to have an effect on the attitude. The analysis shows that energy-efficient motors are chosen more frequently when having received specific information. This is definitely valid for separate motors and it seems to be the case for motors for OEM applications as well. However, the OEM conclusion is based on very few observations.

Another conclusion from the study is that representatives from larger industrial plants (more than 200 employees) often have a higher degree of knowledge of energy-efficient motors than representatives from the smaller plants.

In the case of separate motors the information seems to have had an impact on the purchasing behaviour towards a larger share of effl motors. For motors for OEM applications we see a trace that information might have had an impact.

The overall observation from this study is that the unawareness of energy-efficient motors and the energy classification system is huge. More than three of four interviewees were e.g. not aware of the energy class of the last purchased motor. Merely 14 % of the interviewees could with certainty say that the last purchased motor was an effl motor. Pulp and paper industry were more aware of energy class of the last purchased motor than the other two studied branches.

Factors of importance for the industry when purchasing motors

The 200 industrial representatives were asked which factors they consider being of importance when purchasing motors. They were given eight fix alternatives to grade between "very important" and "of no importance". They were also asked to add other factors of importance. The eight fixed alternatives were:

- Low operation cost compared to a standard motor
- Short pay off time
- LCC (Life Cycle Cost)
- Technical life time of the motor
- Low maintenance cost
- Replacing same model
- Recommendation from motor manufacturer
- Purchasing cost

Lifetime is the factor ranked highest by the industrial representatives, almost 90 % agree with lifetime being very important or important. Low maintenance cost is the second most important factor when purchasing motors according to this study (84 %) closely followed by low purchasing cost (83 %) and following the manufacturers recommendations and replacing the motor with the same model (80 %).

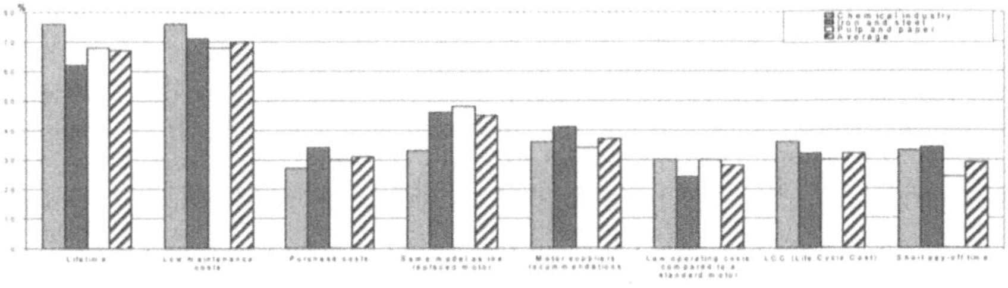

Fig. 4. Interview results of factors of importance when purchasing motors, answers "very important"

But then there is a gap in importance of the factors to the following three. Only two thirds say that low operating cost and LCC are important factors when purchasing motors. The factor ranked least important by the industrial representatives in this study was short pay off time with only 60 % considering pay off time being important or very important.

If we isolate the answer "very import" there are only two factors of major significance to the industrial representatives in the study. These are lifetime and low maintenance cost, with 70 % or just below that fully agreeing. Low operating cost is then ranked the least important with 27 % of the interviewees agreeing.

One dominating reason for low operating cost and LCC being factors of least importance is probably the split between investment and operation budget industries live with. Another important reason is lack of information, awareness and knowledge. Also lack of experience from efficient motors is mentioned in the survey. Purchasing an energy-efficient motor for OEM applications seems to be a matter of no interest to most of the industrial representatives in the study. Merely 5 % of the interviewees could with certainty say that the last purchased OEM motor was eff1 classified, while 80 % had no idea of the energy classification of the last purchased OEM motor. The awareness of energy efficiency classified motors is thus even lower when talking about OEM applications than for separate motors. This is also confirmed by the in deep interviews with the motor manufacturers and dealers. The few cases traced in the study where eff1 motors were purchased for OEM applications were motivated by policy decisions and low operating cost.

Conclusions

The Swedish motor study reveals a large degree of lacking knowledge of energy-efficient motors and the motor classification system, and virtually no knowledge at all of EuroDEEM. One confirmed conclusion from the effect-chain analysis is that directly addressed information leads to a more positive attitude towards energy-efficient motors.

The Swedish and Norwegian configurations on ventilation in industry has many similarities due to practice and regulations. LCC as a tool to achieve long term savings in both countries has very low preferential according to the Swedish motor study and to discussions with Norwegian consultants and contractors. However, the awareness of savings is obvious within all environments dealing with energy intensive issues, and industry may take life cycle costs into account though not consciously thought of in these words. A systematic view on purchase, design, operation and disposal for an optimal cost frame of a ventilation system, has still a big potential in an increased using of the LCC approach in Sweden and Norway.

References

Assessment of knowledge of energy classification of electrical motors (in Swedish), Persson A., et al., ÅF-Energikonsult AB, Stockholm, Sweden, 2002.

Industrial energy end use (in Swedish) the Swedish national Energy Agency and the Swedish Environmental Protection Agency, Eskilstuna, Sweden, 2000.

Energy Audit Procedures for Electric Motor Systems, Final report to the European Commission (DG Joint Research Centre), 15521-1999-12 F1EDA ISP GB.

EuroDEEM, internet address: http://energyefficiency.jrc.cec.eu.int.

Measuring Market Transformation, Suvilehto, H-M, Alopaeus Sandberg, T, Nilsson, H, Persson A, ECEEE 1997 Summer Study, Proceedings, Sp. Mlyn, Czech Republic, 1997.

Energy-efficient operation of electrical motors – a foundation for competitiveness (in Norwegian), Thor Henning Gulbrandsen IFE, Oslo, Norway.

Fan Systems in the European Union: Energy, Emissions and Policy Actions

Peter Radgen

Fraunhofer ISI, Breslauerstr. 48, 76139 Karlsruhe, Germany,
Email: peter.radgen@isi.fhg.de

The European Union, which has become the largest promoter of the Kyoto protocol during recent years, has shown that emission reduction is not just fiction, but instead, can become reality if all actors work co-operatively together. To achieve the reduction targets, substantial efforts are necessary to achieve the market transformation towards an energy efficient society.

Two main strategies have been identified, the use of renewable energies, and the use of energy saving technologies, which allow significant emission reductions and at the same time will allow for further economic growth. Energy efficiency should therefore be a key element in the energy policy of the European Union since it reduces the emissions related to energy consumption and, at the same time, saves energy costs and contributes to extending the remaining lifetime of our natural resources.

Among the cross cutting energy savings technologies, electric motor systems are by far the most important type of electric load. Motors are used in all sectors in a wide range of applications, such as fans, compressors, pumps, or conveyors. Electricity consumption in electric motor systems accounts for more than two thirds of all electricity consumed in industry. With energy costs making up more than 70 % of the life cycle costs of a motor system, even small improvements in the energy efficiency of motor systems will produce large energy savings across the EU. This article will pinpoint the existing opportunities in fan systems across Europe.

Fan and Fan Systems

From the fan in the kitchen hood at home, to the fans used in industrial applications such as material transport, smoke and dust extraction or dryers, fans play an essential part in our daily life. But because they are often not seen, the huge energy consumption for which they are responsible is largely invisible.

This article will answers the question of how large the consumption is, which efficiency improvement potentials exists, what the cost the savings might be and what can be done to overcome existing barriers. The results presented here are based on the fan study for the European Union [1]. However as can be seen from other motor studies, such as the compressed air study [2], the fan should not be dealt with as a single piece of equipment, as the efficiency of the system is not determined by the fan alone. Therefore the interaction between the system and the fan should be taken into account

Fans Electricity Consumption

Fans are rotary bladed machines that are used to maintain a continuous flow of a gas, typically air. The most common types are the axial and the centrifugal fan. Their efficiency is mainly influenced by blade geometry and casing shape.

Fans energy consumption can be estimated using both a **bottom-up** and a **top-down** approach. While the bottom-up approach suffers often on incomplete information, the top-down approach can be applied successfully. The main sectors analyzed are industry, transport and the tertiary sector. The industry sector is further divided into the energy intensive sub-sectors of industry, to identify the key sectors for possible actions.

One main factor influencing the result of the top-down approach is the share of fans on electricity consumption. These values can be determined on the basis of the analysis of individual energy audits in companies and larger case studies from the EU and other countries, such as the US motor challenge [3]. Table 1 summarizes the figures obtained for the share of electricity consumption for fans on the total electricity consumption of each sector.

Table 1. Share of Fan Electricity Consumption by Sector

	Share fans		Share fans
Industry		Tertiary	
Iron and Steel	12.0 %	Service	12.0 %
Non Ferrous Metals	9.5 %	Households	5.0 %
Chemical Industry	10.5 %	Agriculture	8.0 %
Glass, Pottery and building materials	9.5 %		
Mining	10.0 %	Transport	6.0 %
Food, Beverages, and Tobacco	11.0 %		
Textile, wearing apparel and leather	10.5 %		
Paper, pulp and printing	17.5 %		
Engineering and other metal industry	9.5 %		
Others	11.0 %		

The sectors with the highest share of fan electricity consumption are the pulp and paper industry, the iron and steel industry and the food industry. When comparing industrial sectors on the basis of electricity consumed for fans, the consumption is highest in the pulp and paper industry (20.1 TWh pa), the chemical industry (18.1 TWh pa) and the iron and steel industry (12.5 TWh pa). The electricity consumption for fans in all sectors of industry sums up to 100 TWh.

In the tertiary sector the fan electricity consumption is mainly due to ventilation requirements in buildings such as offices and shopping malls. The fan electricity consumption in the tertiary sector has been calculated to be around 93.7 TWh. Therefore both sectors have the same importance, but the technology used will be more homogeneous in the tertiary sector.

The consumption for fans is smallest in the transport sector, where about 3.4 TWh are consumed in addition. The fans in the transport sector have only a share of 2 % on the total fan consumption.

When dividing the fan electricity consumption by country, the consumption in Germany, France, Italy and the UK make up over 50 % of the European consumption, cf. Fig.1.

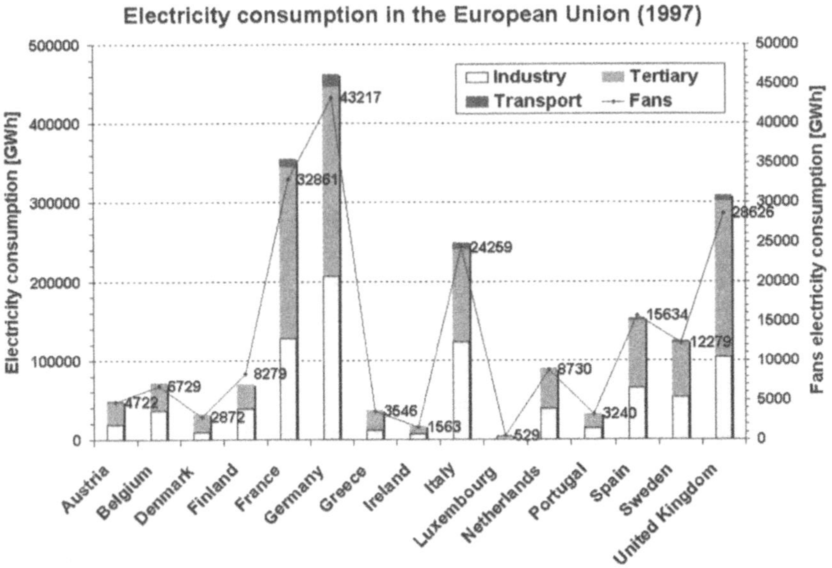

Fig. 1. Electricity Consumption and Fan Electricity Consumption in the European Union by Country

The further development of the fan electricity consumption until 2020 is depending on different factors such as the future total electricity consumption which have been taken from [4]. The electricity consumption in the EU will increase from 2054 TWh in 1997 to about 3094 TWh until 2020, an increase of about 50 %. Therefore in a "Business As Usual (BAU)" scenario with no measures and improvements implemented the fan electricity use will also increase by about 50 % compared to the base year. In the "Improved Awareness (IA)" scenario it will be assumed, that technical improvements at the fan itself will be implemented. In a more optimistic scenario, the "Global Thinking (GT) scenario" it will be assumed that technical measures and system measures will be fully implemented by all actors.

The "GT" scenario is clearly an unrealistic scenario, nevertheless it shows the upper boundary for the achievable savings. The calculate energy consumption for fans for all scenarios is shown in Fig.3. In all scenarios the energy consumption for fan systems will increase between 13 and 50 %, based on the actual consumption. This is mainly due to the fact that the energy savings achieved will be outweighed by the increased number of fans in use. However, in the "Global Thinking" scenario the total electricity consumption for fan systems will be 62 TWh lower than in the "Business as usual" scenario. If this difference in electricity consumption is translated into a CO_2 emission equivalent it will be equal to an emission reduction of about 27 million tons of CO_2 annually.

If the forecast is broken down further by sector, it becomes clear, that the electricity consumption for fans in industry will stay about the same, but we can see significant increases in the transport and the tertiary sector.

In addition it should be noted that based on stricter heat consumption regulations for buildings in different countries, the trend will be towards forced convection ventilation systems with heat recovery. Therefore new buildings will be equipped with fans at a much higher share than nowadays which has been not taken into account in the forecast.

Compared to other motor driven systems such as compressed air systems [2] and the pump systems [5], actions should focus more on the Tertiary sector, while for compressed air and pump systems attention should be directed towards the Industrial sector.

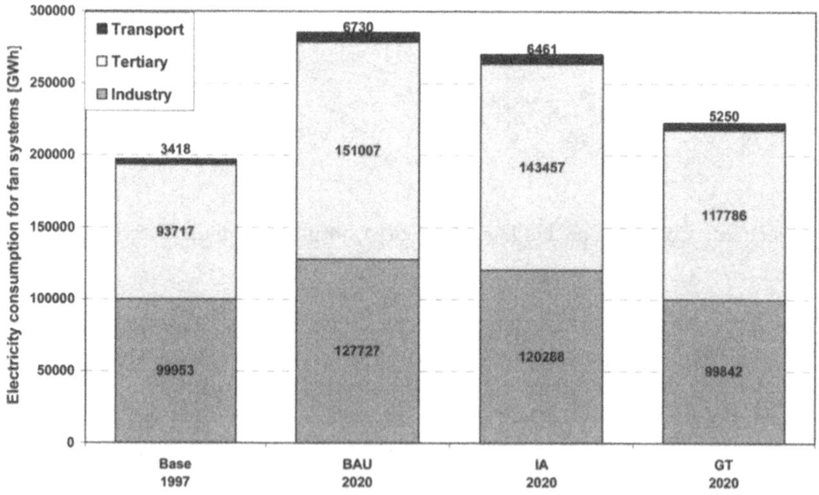

Fig. 2. Electricity consumption for fan systems in the European Union by sector up to 2020

Efficiency Improvement Potential

The efficiency potential in fan systems, which have already been used for the forecast of future energy consumption can be divided in two distinct areas for improvements. One is the better fan technology and the other the system optimization.

Improvement Potential for Fans

The improvement potential for fans is dependent on the type of fan under consideration, as the efficiency and the improvement vary by fan type. E.g. backward curved fans are much more efficient than forward curved ones. The propeller type is a simple axial rotor, while ducted fans are inserted in a duct. Forward curved bladed centrifugal fans differ from radial blade centrifugal fans according to the blade exit angle. For each fan type additional differences are based the choice of the blade profile, work design law and various geometry parameters such as gaps, casing shape etc. In addition, for each fan design the fan may rotate at different speeds. The improvement potentials given are based on the current state of the art. Table 2 summarizes the efficiency gain in percentage for each fan type. The third column gives an estimate of the cost increase as percentage of the actual cost, if these improvements would be realized. For each point of efficiency gain the cost of the fan will increase by about 1 %. If life cycle cost (LCC) are taken into account, this means that for each Euro spent about 4 Euro will be saved.

Table 2. Technical efficiency improvement potential by fan type

Fan type	Efficiency gain	Additional cost
Propeller fans	15 – 20 %	2 – 12 %
Tube-axial	2 – 10 %	4 – 10 %
Jet fans	4 – 8 %	5 – 7 %
Forward curved	5 – 15 %	7 – 10 %
Radial tip	10 %	6 – 10 %
Backward curved	2 – 5 %	8 – 12 %
Radial fans	5 %	4 – 7 %
Other fans	5 – 10 %	8 – 10 %

System Improvement Potential

Besides improvements on the fan itself, the improvements potential from the system has to be taken into account. The efficiency gain factor (EGF) describes the saving potential for one given system. However the measure might not be applicable to all systems or is already applied in some installations, therefore a market penetration factor (MPF) should be defined. This factor describes the share of systems in which the measures can be realised and therefore the savings given can be achieved. If all possible measures are applied to a system, the total saving for this system is derived by multiplying the savings of all measures. If all savings are added, an overestimation of the saving potential will be made, as in this case all savings will be calculated on the initial consumption. The fan system saving potential is summarized in Table 3. The improvement potential of fan systems based on the given measures is 17.5 %. Together with the saving potential for the fan itself, an overall saving potential between 20 and 25 % can be obtained. About two thirds of this improvement potential comes from system improvements and only one third is associated with the fan itself. Similar results were obtained in an earlier study on compressed air systems in the European Union [1] but with an overall saving potential of about 33 %.

Table 3. Fan system improvement potential (Management Opportunities and system optimisation

Management Action	Efficiency Gain Factor (EGF) System	Market Penetration Factor (MPF)	1-EGF*MPF
Operation Schedule	30 %	20 %	94.0 %
HEM	5 %	20 %	99.0 %
Correct Sizing of Motor	15 %	5 %	99.3 %
Optimising Transmission	8 %	60 %	95.2 %
Optimising Ducting	15 %	30 %	95.5 %
Additional Savings	12 %	15 %	98.2 %
	Product of (1-EGF*MPF) factors		82.5 %
		Energy Savings	17.5 %

CO_2 Emission Reduction Potential and Cost Savings

An important question to be addressed is the possible effect on the CO_2 emissions in the European Union. The European Union has made a commitment in the Kyoto protocol to reduce its CO_2 emissions in the period from 2008 to 2012 by 8 % based on the emissions in 1990. So it is important to analyze which portion of the emission reduction may be achieved by improving the efficiency of fan systems in the European Union. Based on an electricity consumption for fans of 197 TWh and a saving potential of 43.2 TWh, the emissions in the European Union could be reduced by 18.8 millions tonnes CO_2 per year, if the existing improvement potential in fan systems would be realised (spec. emission factor for power generation in Europe is 435 g CO_2 /kWh (1999).

Besides emission reduction, an important factor is the cost savings associated with energy savings. Implemented measures will reduce electricity consumption and therefore the cost associated with the purchase of electricity. As the energy price varies widely among the European countries and in some cases also depends on the time of the day or the season in which energy is required three values, 0.04, 0.06 and 0.08 €/kWh, have been assumed for the calculations of possible cost savings. The overall cost saving for fan electricity consumption will therefore be in the range between 1.7 and 3.5 Billion Euro per year.

Conclusions

The results presented have show a large economic energy saving potential by improving fans and fan systems. While there is some awareness that more efficient fans are available, there are many reasons why customers continue to purchase cheaper, "standard efficiency" fans and fan systems. While technical measures and management opportunities are considered to be more profitable than many other investments, there are many reasons for these not being implemented, as shown in Table 4.

Table 4. Barriers to energy efficient fan systems

Technicl	Managerial	
No definition of efficient fans	Presenting projects	
Differing efficiency test methods	Split budgets	
Hard to assess system efficiency	Ownership of electricity bill	
Integrated units	Senior management commitment	
Duty not properly known	Procurement, detail design or functional orientated contracts	
Special duty requirements	Procurement, end users buys seldom and few units	
	Lack of follow-up	Reduced budgets
	Bad experiences	Efficiency of low importance
	Limited equipment life	Maintenance

Market transformation to energy efficient fan systems should therefore be catalyzed by policy measures. Suggested measures for minimum standards, good / best practice and advancing the development of more efficient fans are summarized in Fig 3.

MINIMUM STANDARDS	GOOD PRACTICE	BEST PRACTICE	NEW PRODUCTS
– Regulation – Voluntary agreements – Procurement standards	– General information – Case histories – Benchmarking – On-site assistance	– Procurement specs – Endorsement – Financial incentives – Case studies	– Research & Development

PRODUCT LISTING
Product definition, labeling, databases

Fig. 3. Measures to improve energy efficiency of fans

A mixture of educational programs in combination with well designed legislation and/or financial incentives are needed to transform the market for fans. The six recommended policies that should be further considered by the EU are:

- Encourage the adoption of a single EU test standard for fans
- Develop system efficiency standards
- Develop "procurement line" information to help in the better selection of fans
- Work with manufacturers and users to develop a fan labelling scheme
- General Educational programmes aimed at all parts of the market
- R&D to develop cost effective fan efficiency improvements.

A main issue in achieving market transformation is the creation of awareness by the motor system user. Therefore initiative on the European level such as the Motor Challenge Program (MCP) or on a national level such as the Compressed Air Program “Druckluft effizient” in Germany, which are promoting the uptake of energy efficient motor systems will be very important and should be strongly supported.

References

Radgen, P. (Ed.): Market Study for Improving Energy Efficiency for Fans. Fraunhofer IRB, Stuttgart, 2002. (Online: www.isi.fhg.de/e/publikation/fans/fans.htm).

Radgen, P.; Blaustein, E.: Compressed Air Systems in the European Union Energy, Emissions, Savings Potential and Policy Actions. LOG_X Publisher, Stuttgart, 2001. (Online : www.isi.fhg.de/publikation/c-air/compressed-air.htm).

OIT: United States Industrial Electric Motor Systems Market Opportunities Assessment. Office of Industrial Technologies, US DOE, 1998.

DG Energy: European Union Energy Outlook to 2020. The Shared Analysis Project. European Commission, DG Energy, Brussels, 1999. (Online: www.shared-analysis.fhg.de).

Falkner, H. et al.: Study on improving the energy efficiency of pumps. Final report to the European Commission, June 2000.

Experimental Study on the Operation of an Industrial FC Fan

Chiara Palomba, Pierpaolo Puddu, Franco Nurzia

Dipartimento di Ingegneria Meccanica di Cagliari, Piazza D'Armi 09123 Cagliari

Abstract

Centrifugal forward curved (FC) blade fans are widely used in air ventilation and conditioning and are by far the largest fan category of all. Their success is mainly due to low cost, compactness, and low noise operation. Their peak efficiency, however, falls typically in the range 50%-70%, somewhat lower than that of other centrifugal and axial fans. The vast numbers produced and sold worldwide and the cumulative effect of so many low efficiency installations raises the question of whether it is possible to enhance their efficiency. The present paper is concerned with the investigation of the three dimensional flow field in a double inlet centrifugal industrial FC fan commonly used for handling air. An open loop facility has been designed and constructed at the Department of Mechanical Engineering at Cagliari University to house, drive and test the fan. Its operation will be studied with the aid of hot-wire probes.

Introduction

Over the past few years the Department of Mechanical Engineering at Cagliari University has been involved in the SAVEII project regarding fan usage efficiency in Europe. The increasing attention being focussed on energy efficient applications to reduce the environmental impact of human activities was the prime motivation for this work. During the course of the project typical applications of all types of fans were investigated also considering their efficiencies. This paper is concerned with the operation of one specific fan type: the multivane centrifugal forward curved fan (FC) also known as "Squirrel cage fan" or "Sirocco fan". Of all the cases examined there is a specific interest to experimentally investigate their operation in order to develop more efficient machines.

These fans are used in a variety of applications involving the handling of air such as in furnaces, for heating, ventilating and conditioning both in residential and automotive applications and for cooling electronic equipment where low running speeds are desired to prevent vibrations. They represent the largest of all fan categories built, sold and installed. Their unrivalled success is due to the fact that for equal dimensions and rotational speed they are able to handle almost twice the flow rate or, conversely they may run at half the speed to yield the same performance as other fans. Reduced speed also means reduced noise and this is another strong point of these fans. Therefore, compactness, lower running speeds and lower first cost make these fans preferable for the small/medium size range. They are less attractive in the larger sizes due to increasing running costs and to the overloading brake power characteristic that raises motor size and cost.

Despite their popularity their peak efficiency ranges from 50% to 70 %, somewhat lower than that of other fans. This has been considered of secondary importance due to the small-to-medium power range and due to the extremely low cost and compactness. Nowadays, millions of these fans are used worldwide and the cumulative effect of their poor performance becomes important for justifying efforts to design more efficient FC fans and exploit the potential energy saving.

These fans impart energy to the fluid by strongly increasing the kinetic energy of the absolute flow. Due to the small radial extension of the fan rotor and of blade design the static pressure increase across the rotor is negligible. Therefore these fans need to be housed in a cage that collects the fluid and partially converts the kinetic energy into static pressure energy. Moreover, the rotor design differs substantially from other turbo-machines which makes any attempt to apply already developed design concepts impractical. Another important aspect is rotor cage interaction as for economic reasons the same cage is often used for fans of different sizes, neglecting any optimisation.

Thus, in order to design more efficient fans, a systematic experimental investigation of the flow field both behind and in front of the rotor and in the cage itself is necessary. In fact, very little has been published on this topic and more information needs to be provided. Earlier works have highlighted the strongly three dimensional flow field and suggest that there may be regions of the rotor where back-flow occurs. The interaction with the cage is also important. A systematic test campaign has started at the Department of Mechanical Engineering at Cagliari University for the purpose of investigating these aspects of the flow on a double inlet industrial FC fan and this paper presents the preliminary results.

Test Facility and Experimental Campaign

The experimental campaign was conducted on the industrial double inlet FC fan depicted in Fig. 1 where the main geometric characteristics and axis directions are also shown. The rotors consist of 42, 144 mm long FC blades. This fan has been tested on an open loop facility designed and constructed at the Department of Mechanical Engineering at Cagliari [7]. The fan is driven by a 7.5 [kW] motor, while rotational speed may be varied and controlled through an inverter. A single slanted hot wire probe has been used to measure the three velocity components by applying the multirotation technique [6]. The measurements were taken in three different sectors (1, 3 and 4 in Fig.1); along the whole blade span, 5 mm downstream of the rotor and near the best efficiency point at 980 rpm for a flow rate of 2 m^3/s. The ensemble average, synchronised with rotor revolution, was performed over 250 rotor passages for 3 blade passages acquiring 210 points obtaining a good resolution of the flow field.

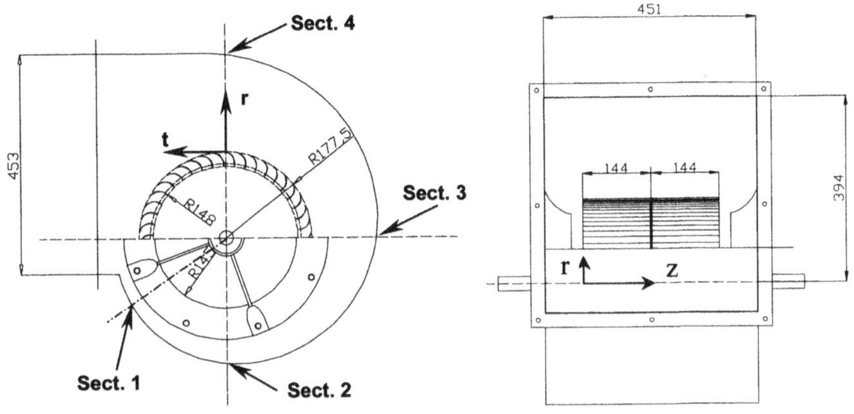

Fig. 1. Sketch of fan A) Side view B) Front view

Presentation of Results

The three relative velocity components in front of the rotor are represented in the axial tangential plane for all investigated sectors in figs. 2, 3 and 4. Blade positions are indicated by dashed lines and the letters P and S show the pressure and suction side of the blades respectively. Therefore the rotor moves from right to left. The space between the lines is the blade passage.

First of all the radial flow component distributions for the three sectors, fig 2 will be discussed and compared. In the three plots one can observe in the disk region high values for the radial flow component inside the blade passages, decreasing close to the blades. This trend is typical: trailing behind each blade is a viscous wake in which the velocity modulus is reduced producing the so called jet-wake velocity profile. As one moves axially toward the shroud, this structure gradually changes and disappears along about 40% of the blade length close to the shroud. In this region, the radial flow component responsible for flow rate is strongly reduced. This flow structure was observed in all three sectors though some differences do exist. For example, in sector 1, the radial component is smaller over the whole blade length and the jet-wake structure is limited to the first 40% of the blade close to the disk. In the remaining part, broad regions of low radial flow component are observed close to the suction side indicating the possible presence of separation zones. This sector is the one with the smallest radial clearance between the rotor and the cage.

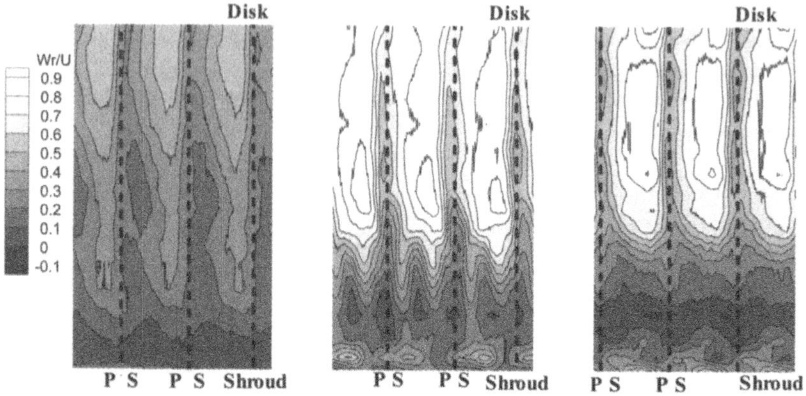

Fig. 2. Radial flow component (from left sector 1, 3 and 4)

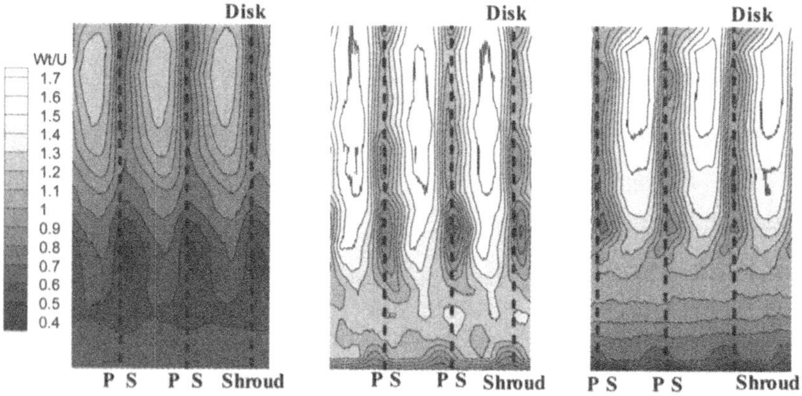

Fig. 3. Tangential relative flow component (from left sector 1, 3 and 4)

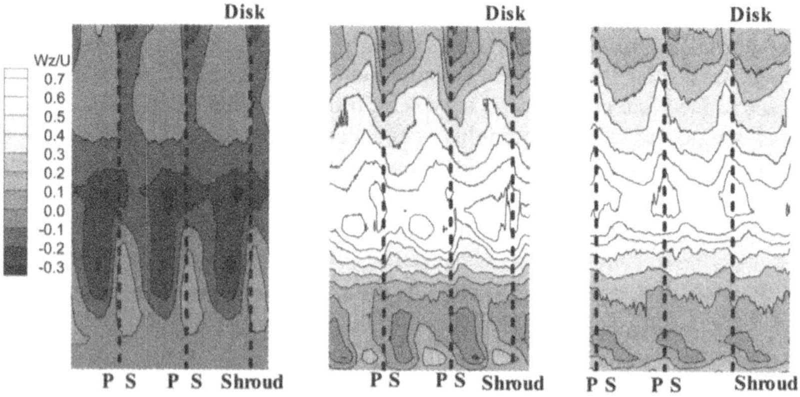

Fig. 4. Axial flow component (from left sector 1, 3 and 4)

Further important information about flow structure can be drawn by observing the relative tangential velocity, fig. 3. Note that the relative tangential component is fairly large because of the manner in which this rotor imparts energy to the fluid. If one were to consider absolute flow then the tangential component would always be greater than blade speed attaining values almost three times as high.

The jet-wake behaviour can also be observed for the relative tangential component in the disk region for all sectors with high values in the blade passage decreasing close to the blades, while, as one moves toward the shroud, the relative tangential velocity component becomes smaller and more uniform in the tangential direction. Interestingly the structure of the low flow region close to the shroud is less well defined in sector 3 away from the tongue (region where the narrower and larger parts of the cage meet) becoming better defined nearer to it. The tangential flow component in sector 4 is uniform, meaning that the whole passage has the same flow. Nevertheless, in spite of the small relative tangential flow component, the absolute flow component is still greater than blade speed even in this low flow region.

Ideally the flow reaching the rotor through the axial inlet should change from axial to radial and, therefore one would expect the axial flow component to be zero over the whole blade span in the rotor exit plane where measurements have been made. In reality, this change is hard to achieve completely and especially in machines with such small radial and large axial dimensions. So it is not completely surprising that the axial flow component, shown in fig. 4, is far from uniform. Examination of sectors 3 and 4 shows that very low values of the axial flow component are found at the shroud where the other components are also very small. The same is true in the disk region where the geometry constraints the flow to become radial. In the mid-span regions, however, the axial flow component is quite large indicating that the flow change is not completed as the flow passes through the machine. The situation is completely different in sector 1 where the axial flow component is very small and negative over practically the whole blade span. Negative axial flow component means that the flow at the exit of the rotor moves toward the inlet region from the disk to the shroud.

These results already show that there are strong variations in the working conditions along the blade span and along the circumference. The blades are subjected to constantly changing flow conditions and it may not be possible to achieve highly efficient operation for all possible situations. One could consider varying the blade geometry along the span to match the different flow conditions in the axial direction but when the blade changes circumferential position again a different situation is created and optimum operation may not be reached for all rotor positions.

To complete the analysis the incoming flow conditions will be evaluated. Due to the fan inlet geometry, a special probe carrier has been designed and installed at the inlet for inserting a custom 5 hole pressure probe to measure the mean flow direction in the different sectors. Measurements made in one sector provided some general indications about the flow conditions at the inlet: qualitatively it could be seen that the absolute flow remains axial for most of the blade span turning to radial nearer the disk. The absolute flow at the inlet seems to have no tangential component.

Conclusions

- The flow field is strongly three dimensional. Significant variations are found along the blade span and along the circumference.
- In the first portion of the blades close to the shroud the flow rate is much lower and flow is almost tangential without clear distinction between through flow and wake flow.
- The circumferential variations observed imply that the blade passages continuously change their working condition depending on their position relative to the cage and the flow is therefore always unsteady.
- The cage has a strong influence on flow as demonstrated by the behaviour observed in sector 1 where all the flow components are strongly reduced.
- Future work will certainly have to investigate the inlet region in order to evaluate incoming flow conditions and establish the local rotor-fluid work exchange. The cage flow should also be examined in two steps: first the cage flow needs to be measured to evaluate the rotor-cage interaction effects and then the influence of different cage geometry on the main flow should be studied.

Bibliography

[1] Bleier, F. B.; " Fan handbook selection, application and design", McGraw-Hill Editor,1997 pag.3.12

[2] Daly, B. B.; " Woods practical guide to fan engineering", Woods of Colchester Limited 1992 pag 123-124

[3] Eck, B.; "Fans", Pergamon Press 1973, pag 139,151

[4] Kind, R.J. and Tobin M.G.: "Flow in a Centrifugal Fan of the Squirrel-Cage Type", Transaction of the ASME, Jan. 1990 vol 112, pag. 84-90

[5] Raj, D. and Swim, W.B.; "Measurements of the Mean Flow Velocity and Velocity Fluctuations at the Exit of an FC Centrifugal Fan Rotor", Journal of Engineering for Power, Apr. 1981, vol. 103, pag. 393-399

[6] Puddu, P. "Misure di turbolenza in elementi fissi e rotanti di turbomacchine con l'impiego di sonde a filo caldo", MIS-MAC II Firenze Italia 1993

[7] Mandas, M., Puddu, P. Palomba, C. and Cucciari, S.M.; " Un set-up di misura per lo studio del flusso in un ventilatore centrifugo a pale in avanti", MIS-MAC VII Cagliari 2001

Ship Unloader Trolley's Cycle Optimization

Ernesto Brambilla

Siemens S.p.A. Italy Consulting.
Corso Matteotti 33/B 23900 Lecco Italy
E-Mail: ernebra@tin.it

Abstract

The application described hereunder will show how to save time, energy and money using a smart strategy control in order to drive the trolley's electrical motor in a crane environment. Very often in the ports it is necessary to unload ships standing along the docks. The load can be containers full of wares but often the ship is full of row material such as pellets (small balls made of iron and coal) that must be dropped over a moving rubber belt feeding a steel reduction device. In both cases the problem is still the same: how to reduce the swinging of containers or of the hoist and grab. It is very important to point out that by using a smart driving strategy in order to reduce the swinging time, it can be possible to minimize drastically the unloading time of a ship saving not only the dock space, but even the working time of the crane. The first aspect is important because the ship can reduce the unloading time, the second point because the duty time of the crane motors will be smaller. Just to have an idea let us consider an unloading cycle:

1. Grab up to the hold full of material
2. Trolley from ship to hopper
3. Waiting time to swinging stop
4. Grab down and unloading
5. Empty grab up
6. Trolley back to hold of the ship
7. Waiting time of swinging stop
8. Grab down to material
9. Grab loading material

Realistically it will be necessary a thousand cycles in order to unload a ship, and each cycle takes 120 seconds when a good operator drives the crane. If you reduce the time of points 3 and 7, using a proper control strategy of the trolley's motor, it will be possible to save up to 60 sec. each cycle for a total of approximately 70 hours. Since the drives of the grab, the hoist and the trolley have a very big rated power (i. e. 1000 kW) and consequently the motor blowers have a big power too, it is possible to switch off the blowers as well as all the lighting and conditioning systems for 70 hours every single ship unloading operation; it is easy to get how big the energy saving is.

Introduction

In many industrial fields such as steel factories, paper factories, converting paper as well as in dock operations it is necessary to shift loads from a place to another. Many kinds of different machines are used: overhead traveling cranes with trolley and hoist with a linear movement, gantry cranes, rotating boom cranes and so on. Each of those machines has its own particular motion low according to the industrial operation needs, but a common behavior is present: sways of load are not required. A typical overhead traveling crane is used to pick up steel wired reel just out of the reel-maker and to drop into the warehouse; this operation is done manually and the sways of the reel makes the operation slow. The same event happens with paper reel, synthetic fiber beams and so on. The experience of the crane's operators can minimize the sway's time but still the losses of time accumulated into the working day is relevant. In order to solve this problem or make the impact of the swinging less heavy, many control strategies are applicable, such as space loop control superimposed to velocity loop, model of crane and pre-filtering techniques, fuzzy and neural-fuzzy strategies. Here down a real case of motion control having the purpose to reduce swinging in a gantry crane is described.

System Description

The crane taken into consideration is a gantry crane and the purpose of this machine is to unload a ship full of "pellets" (small bolls made of iron and cook). The grab picks up the row material from the hold and after a travel of about 70 m drops the pellets into a hopper. The material will be carried out of the hopper to the steel reduction process with a moving rubber belt. It is clear that according to the speed of the rubber belt and hopper capacity the grab's operations will be fast enough to assure the minimum stay in term of time of the ship along the dock.

Crane Description

The crane consists of a steel tower 70 m tall that can be moved along the dock in order to explore all the ship standing along the same dock. and in a horizontal steel framework with a trolley that can move right or left (from hold to hopper and vice versa). The grab is hanging from a hoist that is fixed to the trolley center with steel cables und the grab and hoist can move up and down according to the operation's needs. The operator's cabinet is placed beside the trolley way and can move or stay in a fixed position according to the operator command. Inside the operator's cabinet, the operator's chair with all the levers as well as the diagnostic monitor and alarms panel are fixed. All the operations of the different parts of the crane are done by operator but the trolley, hoist and grab working cycle can be done in semi-automatic mode according to a software able to realize a velocity trolley profile that can minimize the unload operation. All the electrical drives, PLC, supervisor system as well as the medium to low voltage cabinet transformer, the conditioning and the lighting system are placed in a cabinet on the top of the crane.

Model Description

From the crane description it is possible to assume that the trolley and the grab are a kind of pendulum system consisting of a moving fulcrum point, a steel cable connecting the pendulum center with a mass. In particular:

M_0 =Trolley weight (ton)

M_1 = Cables weight (ton)

M_2 = Grab weight(ton)

M_3 = Load weight (ton)

L = Length of steel cable (m) changing in the last part trolley's travel

L_0 =Length between the fulcrum and the center of the weights of grab, load, hoist and cables.

φ = Deviation angle (degrees) between cables and the perpendicular dawn the trolley's center

x = Trolley absolute position (m) assuming as zero the end of trolley travel at ground site.

$v = \partial x/\partial t$ = trolley's velocity (m/s)

$a = \partial v/\partial t$ = trolley's acceleration (m/s^2)

g = Gravity acceleration = 9.8 (m/s^2)

T = Pendulum period (s)

In a frictionless environment the period of the pendulum doesn't depend on the weight but only on the length between the fulcrum and the center of the weights. It is important to notice that the distance is not the same of L but is changing with the grab condition (full and close or empty and open, the geometry changes and the position of the center of the weights changes too), and more: the grab, the hoist and the cables are moving in the air and some friction is present, but approximately we can consider the system as a pure pendulum. [1], [2].

Using the simplified model the pendulum period is:

$$T = 2 * \pi \sqrt{L_0 / g} \qquad (s)$$

In order to take into consideration the changing of L_0 (grab full of material down into the hopper or grab up and open) practically it is possible to have: L_0 = 41 m in the first situation and L_0 = 40 m in the second situation. Applying the above formula we can find:

$$T = 2 * \pi \sqrt{41/9.8} = 12.84\,(s)$$

$$T = 2 * \pi \sqrt{40/9.8} = 12.69\,(s)$$

that means a variation of about 1.17%.

When the trolley is at the beginning of the trip, down into the ship's hold, the distance between the pendulum's fulcrum and the weights center becomes longer (i.e. L_0 = 45 m). In this case the period is:

$$T = 2 * \pi \sqrt{45/9.8} = 13.48\,(s)$$

that means a variation of about 5%.

In order to do a very accurate operation it is possible to take into consideration the period change doing a periodical evaluation according to the crane configuration.

Operation description (manual mode)

- Grab open hanging up to the hold (start position) $t_0 = 0$ s
- Grab down on the row material $t_1 = 2$ s
- Grab close (row material inside) $t_2 = 3$ s
- Grab up $t_3 = 2$ s
- Trolley acceleration at max. speed $t_4 = 10$ s
- Trolley travel at max. speed $t_5 = 12$ s
- Trolley deceleration till stop up to the hopper $t_6 = 10$ s
- Waiting time due to grab swinging $t_7 = 20$ s
- Grab down to the hopper and grab open $t_8 = 2$ s
- Open grab up $t_8 = 2$ s
- Trolley acceleration at max. speed (reverse direction) $t_{10} = 10$ s
- Trolley travel at full speed $t_{11} = 12$ s
- Trolley deceleration till stop to the hold $t_{12} = 10$ s
- Waiting time due to the swinging of the grab $t_{13} = 20$ s

The total operation time for each cycle = 115 s.

It is very important to note that the time t_7 as well as time t_{13} are much bigger than the other times and the value of 20 s is not fixed but is an average value only taking into consideration an operator with a very good experience in driving cranes. In fact the operator moving smartly the trolley in right and left directions around to the point up to the hopper and up to the hold can reduce the load swinging time at the written values; the free oscillation without any correction takes minutes before stopping!

Operation Description (Semi Automatic Mode)

The purpose of driving the trolley in automatic is to eliminate the swinging time of the grab and to save about 40 s each cycle and at the same time to optimize all the other parts of trolley operations using the max. values of acceleration and speed according to the mechanical friction between the trolley's wheels and track. How is it possible to do that? One of the most used strategies is the "bang-bang" strategy [3].This strategy basically consists of the following phases regarding the trolley's trip and the hoist/grab operations:

- Trolley acceleration till half of max. velocity set value. (in our example 2 sec.)
- One period T/2 long (6sec) during witch the trolley is running at constant speed ($v_{max}/2$)
- A second acceleration step always of 2 sec. till to reach the max velocity's value.
- A constant velocity running time during witch the deviation angle of grab is zero.
- A deceleration time in three step (2+6+2) sec.

The fig.1 shows the velocity vs. time diagram of the trolley (theoretic) The fig.2 shows the practical velocity profile of the trolley. The fig.3 shows the velocity profile of the grab and the hoist. The fig.4 shows the deviation angle between the cables and the perpendicular line down the fulcrum. It is important to notice that the grab, hoist and load never over pass the hopper and that means more safety for the operators. After discharging, the grab goes back to he initial position and the control strategy is the same. During commissioning it is possible to adjust the software in order to take into consideration the T changing along the grab trip.. Practically it is obvious that the velocity diagram of the trolley and the grab as well as the deviation angle doesn't have any knees due to inertia and jerk control, the curves are smooth. [4].

Drive and PLC Architecture

The crane has many drives with a variable speed; some of them are AC drives and some are DC drives. As usual very big motors (the hoist and the grab can reach 1000 kW each) are DC motors with DC converters and four quadrants operations capability. All the other motors are AC inductance motors and are driven by vector control inverters. Each motor has an incremental encoder as a feedback sensor. An absolute encoder provides to inform the PLC about the position of the trolley along the trip. Usually all the AC drives have a common DC bus provided through a reversible power supply but for safety reasons (i. e. during a temporary lack of electric power during the braking time) a braking unit is fixed across the

DC bus itself. For safety reasons the hoist and the grab motors have a very high efficiency mechanical brake. All the drives as well as PLC, absolute encoder, grab's weight sensor and other ancillary devices are linked together with a field bus. Using this system it is possible to save a lot of connecting wires and all the equipments can exchange information according to PLC software and can provide an efficient diagnostic system. A supervisor device provides to show all the parameter values as well as all the alarms to the crane's operators.

Control Strategy

In order to perform the above described trolley's velocity profile it is necessary to have first of all a model of the system. All the values of the physical quantities described above are written in the model: max. speed, acceleration, acceleration's time. The trolley actual position, the instantaneous speed, the length of the cables between the fulcrum pendulum point and the center of the weights are monitored in real time by proper sensors and those values are acquired by the PLC and the motion controller via field bus. Since the total length can change according to the position of the grab over the hold, the operator, before starting each cycle, stores the initial values using a push bottom and as the final point is fixed (the center of the hopper), the controller knows the total length of trolley's trip. The controller can also acquire the total length of the cable and the grab's geometry and the load's weight and consequently the free oscillation period T. The max value of the speed must be set according to the mechanical environment.

Using the above mentioned values, the motion controller generates an initial set-vector done by position, velocity and torque (practically motor current) and provides to inject this vector as the first set point of the position loop, the velocity loop and the current loop respectively.[5]. The drive starts to work and every fixed time (i.e. 2 msec.) the model provides to make a refreshment of the set-vector; in this way the drive can perform the proper motion profile.(fig.5) The grab follows the trolley with an initial delay. After that it catches the trolley and, when the trolley starts to brake, it later overpasses the trolley itself till it reaches the hopper but in this position the trolley reaches the grab and no swinging is produced. It is very important to note that in the described case the real position of the grab is not detected because it is very difficult to find out a sensor which is smart enough to perform this measure. However some cranes have this capability using a scanner camera that can give the grab position every msec and the controller is able to evaluate the instantaneous deviation angle between the grab and the trolley. Other detection methods are based on magnetic sensors located around the grab's cables. The position of the grab can be used for a fine correction of the set-vector.

The real experience shows that it is also possible to perform high accuracy and efficient motion profile even without that position detector. The trolley's motion profile and the grab's motion profile can not be appreciated using a normal drawing, the effect is much more evident with an animation program , using this visualization method it is possible to see that the trolley reaches the grab over the hopper without swinging; it is possible to say that the trolley and the grab reach the hopper (position control) but they both get the hopper at the same time (position control with moving target). It is important to note that in case of the wind blowing in a direction different from the trolley's travel, the grab can start to swing and it is impossible to reduce such a kind of sway because the crane has only one controlled travel unloading direction.

Energy Saving

In order to evaluate the time saving by applying the above mentioned control strategy it is possible to consider the following example:

Total ship capacity	50000	m^3
Row material specific weight (average)	5.5	ton/m^3
Total weight to be unloaded	275000	ton
Weight of row material for each travel	40	ton
Travel number	6875	
Travel time	77	sec
Total time	529375	sec.
Total time in hours	147	h
Cycle time with operator (average)	117	sec
Total time with operator (average)	804375	sec
Total time in hours	223	h
Saving time	76	h
Power of ancillary equipment	75	kW
Energy saving per ship	5700	kWh/ship
Possible total saving (assuming 40 ship/year)	228000	kWh

Other Types of Crane

The above described crane is a "gantry crane" in which the trolley has one direction movement's capability and the load can only go up or down with the grab and the hoist; other types of cranes especially those on board of a ship can have two or three movement capabilities at the same time: rotation, up-down, right and left. In those cases the dynamic is more complex and it is necessary to perform two or three motion profiles with some kinds of interpolation in between. Anyway, regardless of the motion control software the problem of saving time and energy is still the same.

Conclusion

The paper points out that the use of different type of cranes such as gantry cranes, boom cranes, over head traveling cranes etc. always brings up a matter: the free sways of load. When the load is swinging up to the target it is necessary to wait until the oscillations are small enough to do the operations and the waiting time due to the natural oscillations of the load can persist for minutes. Above all when a heavy load is swinging the situation around is dangerous for the operators. In order to reduce the swinging many control strategies are applied: space control, fuzzy logic and neural-fuzzy network. position, velocity and torque loop with a crane model generating an input motion vector. The paper describes one specific case where a Siemens AC vector control drive working with an appropriate software drastically reduces the swinging of the load at zero value in order to unload a ship in the minimum time, saving the standing time of the ship along the dock. The example described is a real case coming from a commissioning done by the author.

References

[1] Meccanica Razionale Vol. II pag. 193. Bruno Finzi Ed. Zanichelli, Bologna, 1959.

[2] Anti-Sway Crane, M. Valesek, Paper of Mechanical Engineering - Prague University

[3] info@smartcrane.com

[4] Siemens: Der direkte Weg Hipac bringt's auf den Punkt. Intelligentes Pendel-Regelsystem mit automatischer Lastpositionierung.

[5] Contraves Intelligent Motion Controller CMP 2000 description.

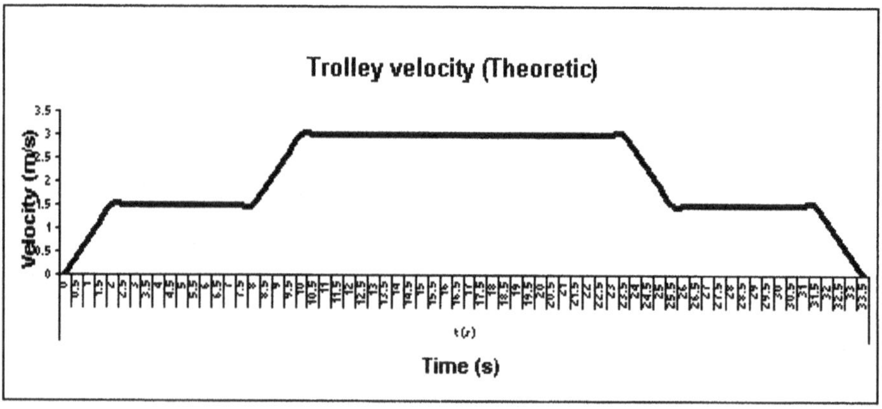

Fig. 1.

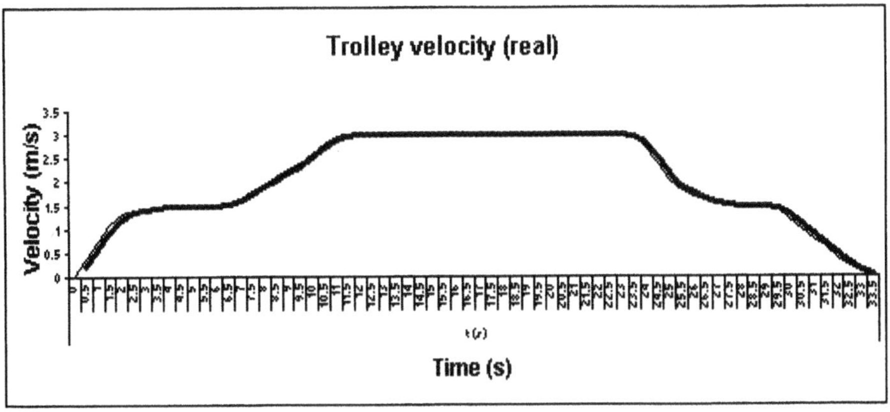

Fig. 2.

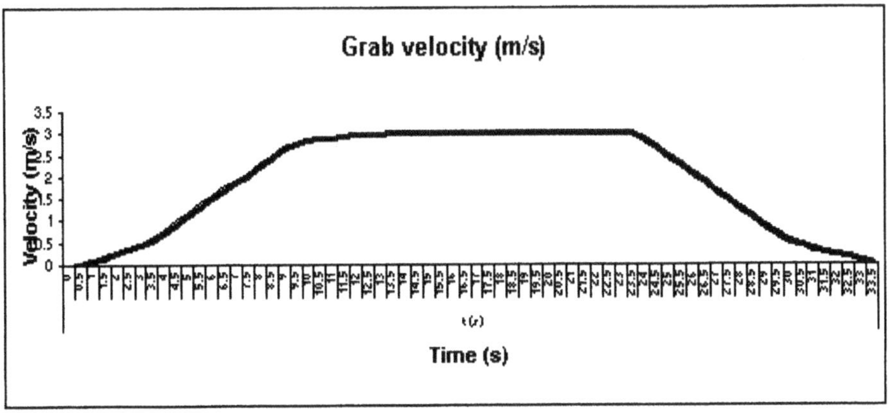

Fig. 3.

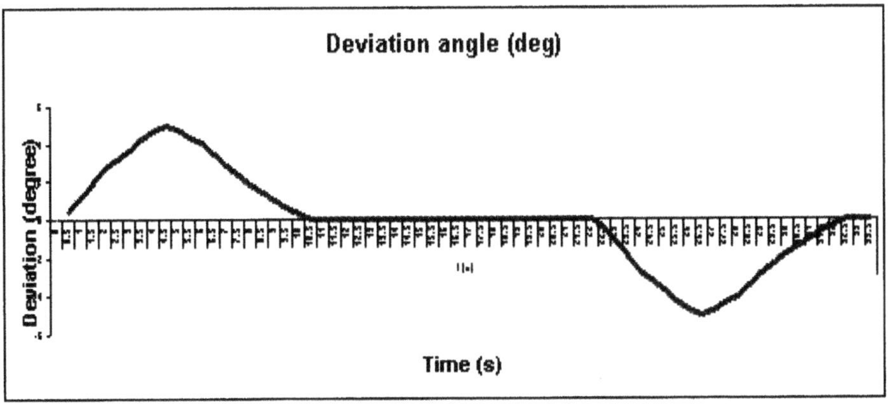

Fig. 4.

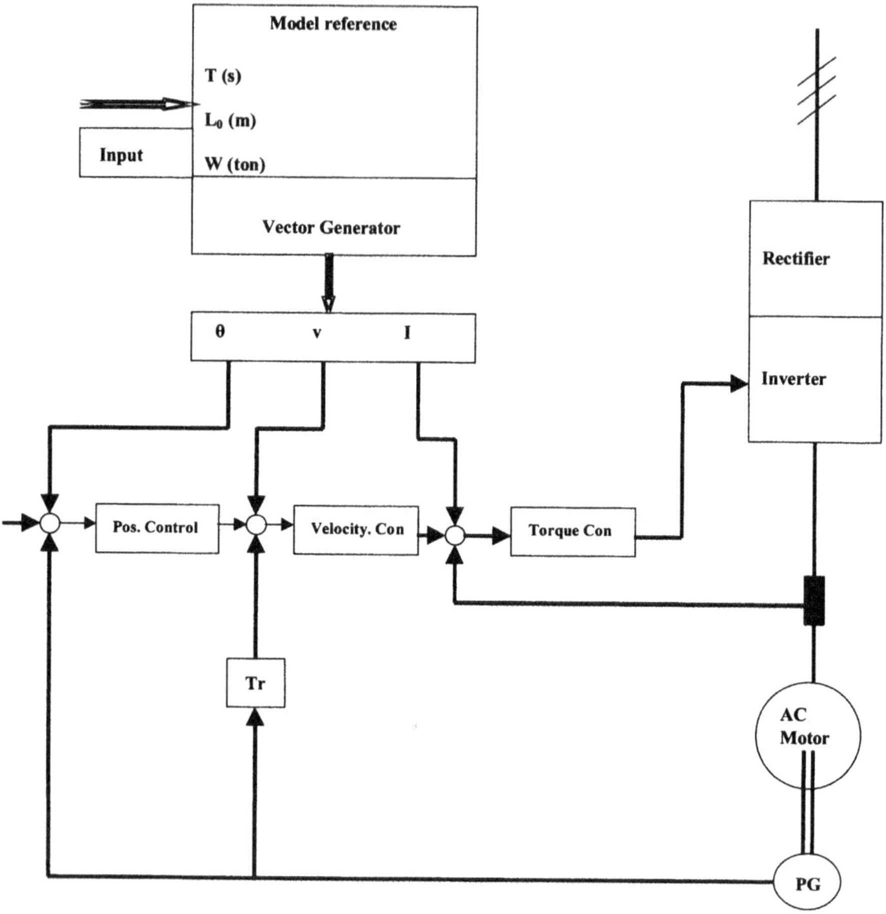

Fig. 5. Block Diagram

Motor Decisions Matter – A U.S. Campaign to Encourage Better Motor Management

Theodore W. Jones, Marc G. Hoffman

Consortium for Energy Efficiency, One State Street, Suite 1400, Boston, MA 02109

Why Motor Decisions Matter

When a motor fails it is often the first of two failures. The second is the choice of a replacement for the failed motor, usually undertaken in panic mode – with little or no forethought. Research from the U.S. Department of Energy (DOE) shows that only 11 percent of the nation's corporations plan for motor contingencies. And even in this small group, energy efficiency is rarely a consideration during the decision-making process. This is quite surprising when one considers that the financial and environmental ramifications of these decisions are usually rather significant; many motors remain in service for 20 or 30 years.

Industrial managers have the opportunity to trim motor operating costs 1-4 percent through the purchase of "premium-efficiency" motors and by 5-6 percent through proactive motor management practices, including motor evaluation, procurement, inventory and repair. These seemingly small percentages can translate into substantial savings for motors operating a large number of hours. In fact, the annual energy cost for running a motor – operating continuously at or near full load – can be 10-25 times its purchase price. This makes the incremental price increase of a premium-efficiency motor insignificant.

This paper describes how the Consortium for Energy Efficiency (CEE) and a variety of motor industry stakeholders are advancing these opportunities in the U.S. market.

Promoting Premium-Efficiency Motors

In 2000, CEE convened a working group of motor industry stakeholder to develop recommendations for moving the market for premium-efficiency motors forward. The working group included representatives from electric utilities, state and regional motor programs, DOE, Environmental Protection Agency (EPA), the motor repair industry, and the National Electrical Manufacturers Association (NEMA). The working group concluded there was a strong need for:

- A common, voluntary definition of "premium-efficiency" products across the motor and utility industries
- A common brand to help customers easily identify qualifying motors, eliminating the need for confusing tables
- Education of upper-level decision-makers about the benefits of motor management

In response to the first two issues, CEE (representing state and utility energy-efficiency programs) and NEMA (representing motor manufacturers) agreed to align their definitions of "premium efficiency" by adopting a common specification, called NEMA Premium™. In addition, both CEE and NEMA agreed to promote qualifying motors under the NEMA Premium brand.

As an energy-efficiency definition and brand, NEMA Premium is a significant milestone because, for the first time, the energy-efficiency community (state and utility programs), motor distributor/service centers and the manufacturing community have a common and standardized definition for premium efficiency. With NEMA Premium as a focal point, the working group could start to educate senior-level business leaders about the benefits of motor management.

Reaching the Right Audience

Helping industrial customers understand the decision-making process in motor purchasing is critical. Facility managers and engineers in most companies do not have the time or resources to think about motor management. When motors fail, the highest priority is returning the equipment to service quickly, not performance optimization.

CEE has found that rushed decision-making after motor failure can be costly, resulting in higher operating costs, poor equipment performance and unreliable service. Motor management techniques can improve the effectiveness of motor-related decisions by helping managers examine important cost and operational issues *before* motors fail.

Equally valuable is the recommendation of appropriate policies, plans and procedures that improve the likelihood that the most cost-effective decisions will be implemented.

The pitfalls of rushed decision-making can be avoided if senior-level managers are aware of the costs associated with this type of "hit-or-miss" strategy and institute motor management policies for the entire company.

Motor Decisions Matter^{SM} Campaign

Motor Decisions Matter (MDM) is a nationwide campaign designed to promote motor management as a cost-cutting tool and a means to improve performance. Launched in 2001, the campaign has developed a high-level "business" message that translates the benefits of motor management into terms that senior plant and corporate management will understand – and find compelling.

Campaign Sponsors

Currently the campaign has 22 sponsors, spanning a wide array of stakeholders:

Motor manufacturers
ABB, Baldor, NEMA's Motor and Generator Section (17 manufacturers)

Electric utilities
Conectiv, Long Island Power Authority, National Grid USA, Northeast Utilities, NSTAR, SMUD, Southern California Edison

Energy conservation organizations
Advanced Energy, CEE, Midwest Energy Efficiency Alliance, Northeast Energy Efficiency Partnerships, Northwest Energy Efficiency Alliance

Energy-efficiency program administrators
NYSERDA, Wisconsin Focus on Energy

Trade associations
Copper Development Association, EASA, NEMA

State and federal government agencies
U.S. Dept. of Energy, U.S. Environmental Protection Agency

Sponsors support the campaign in several important ways. First, they have made financial commitments over a minimum three-year period. Second, they have the opportunity to sit on an advisory committee that develops the campaign's strategy. Third, they provide credibility to the campaign's message because of their diverse perspectives and interest in the market.

This diversity provides an important means for delivering the campaign into the field. As shown below, the members of CEE, the Electrical Apparatus Service Association (EASA) and NEMA represent uniquely different sectors.

Organization	Primary Membership
CEE	State and utility energy-efficiency programs
EASA	Motor distributors and service centers
NEMA	Motor manufacturers

The likelihood of the campaign's message getting through to customers improves dramatically if he or she hears the same MDM message from multiple industry sources, as well as through the national media.

The campaign's message provides value to each of the major sponsor groups. It is good news for motor distributors and service centers because awareness of MDM's bottom-line benefits will likely expand the market for premium-efficiency motors and motor-related services. Electric utilities are also interested in this message because it will reduce peak demand while providing value to their customers. And motor manufacturers are interested in helping to change the way motors are bought and sold in the market.

Campaign Message

Through the campaign, Motor Decisions Matter sponsors are working to promote the benefits of motor planning and provide tools that enable commercial and industrial customers to develop a motor plan. A motor plan addresses common motor decisions, such as:

- Whether to repair or replace a motor
- How to specify and procure premium-efficiency motors
- Motor sizing
- Proper application
- Setting up inventory

Experience shows that these issues need to be addressed *before equipment failure* to optimize motor availability, reduce downtime and lower energy costs. The principal message of the MDM campaign is to anticipate these decisions before they occur and to plan for them ahead of time. Senior-level managers, who want to reduce costs and help the environment, can capture motor savings by supporting the development of a motor plan and making sure that it is implemented when motor-driven equipment fails.

Motor Planning

To be effective, motor management must be supported at both the corporate and plant levels. Most motor-related decisions are made at the plant level, where there are typically few resources available for motor management. For this reason, the availability of established corporate motor policies can really make a difference. These polices often provide needed energy-efficiency specifications, decision-points, recommended best practices and other technical guidance.

Corporate motor policies are a powerful way to tap into motor management savings because they:

1) Efficiently communicate the importance of motor efficiency to all plant personnel

2) Provide the justification needed to develop and implement motor plans
3) Send a uniform message across all departments. For example, the purchasing department may want to know why someone in engineering or maintenance is recommending equipment that may have a higher first cost but a lower lifetime operating cost.

Encouraging and supporting motor planning is the central theme behind the MDM campaign. A motor management plan can be fairly simple to develop and can take a variety of forms, such as:

- A set of written specifications for all new motors (including NEMA Premium energy-efficiency levels)
- Documentation of repair/replace decisions on all critical application motors
- Collection of nameplate information on all motors (or critical motors only) and entering them into a database for planning purposes.

Motor plans may also reference third-party resources, such as Motor Decisions Matter^{SM}, NEMA Premium^{TM} and the Department of Energy's Best Practices program to use in discussions with motor distributors and service centers.

The analysis needed to develop a motor management plan can be done internally or with third-party assistance. A customer's repair center/distributor is in an excellent position to assist customers with the development of customized motor plans (including repair/replace policies). By sharing this information, motor repair centers and distributors can take steps to ensure that the right motors and motor repair services are available for customers when needed.

Campaign Resources

The MDM campaign seeks to increase general awareness of motor management and its many benefits and to assist in the motor planning process. To this end, the campaign offers the following resources:

- **The MDM Web Site** (www.motorsmatter.org) includes information about the campaign, links to campaign sponsors and a press room and an online distributor marketing kit. The Web site receives approximately 100 visitors a day.
- **A Motor Management and Planning Kit.** The kit contains information, statistics, helpful links, and tools specially designed to assist companies in developing motor management plans that reduce energy costs. It also includes comprehensive information about developing a motor management plan, quantifying decisions about motor repair/replacement, making motor replacement choices, specifying repairs, developing plans for critical motors and building a motor inventory. More than 2,000 people have downloaded this planning kit from the MDM Web site.

- **Motor Management Case Studies.** Seven case studies, posted on the MDM Web site, illustrate the benefits of motor management and planning.
- **MotorSlide™ Calculator.** Through easy step-by-step instructions, the slide calculator computes a motor's annual energy cost, the annual energy savings with a premium-efficiency motor and a motor's efficiency. Information obtained by using the calculator is an important beginning to developing a motor plan. More than 13,000 motor slide calculators have been disseminated through the campaign and its sponsors.
- **E-mail Promotion.** A special campaign promotion provided information about the campaign to motor distributors, highlighting the benefits of motor management and planning, and providing a direct link to the on-line motor planning resources. This e-mail effort reached more than 760 motor distributors.
- **Public Relations.** News releases on new program components and customer success stories are disseminated to trade and national publications on a regular basis. Campaign spokespeople are available to provide quotes and additional information for feature stories.
- **Information Clearinghouse.** The U.S. Department of Energy's Office of Industrial Technologies Clearinghouse provides the campaign with a toll-free number where customers can call with questions and receive more information on motor-efficiency, motor repair and motor management. More than 500 respondents have contacted the clearinghouse and have received a Motor Decisions Matter Planning Kit.

Conclusion

Motor Decisions Matter represents a unique approach in promoting energy-efficiency and its co-benefits in the marketplace. It provides a single, coordinated voice, explaining the benefits of motor management to a variety of stakeholders. The campaign aims to convince senior-level managers that motor management is a strategic opportunity for their company and is worth their support. The process of developing the motor management plan has several benefits: it saves time, it evaluates the value of higher-efficiency motors (and repair services) and it notifies a local motor professional about future needs. All of these factors can help boost the market for high-efficiency motors and motor services while helping end-users become more profitable.

An Energy Management System to Saving Money and CO_2 Emissions

Soren Rise

Danish Standards Association, Kollegievej 6, 2950 Charlottenlund, Denmark

Introduction to Energy Management

The resource conscious organization, public as well as private, is concerned with controlling its energy consumption. It has an energy policy and sets energy targets in order to use energy more efficiently. Energy targets are followed up with vigorous plans to achieve greater energy efficiency.

Energy management is a systematic and structured approach to improve energy efficiency. Energy management strengthens the overall environmental profile of the organization and can considerably reduce costs for e.g. electricity, oil or natural gas.

The result of energy management is continual improvement in the form of a reduction of the emissions of CO_2 by using energy more efficiently. The organization undertakes to continue to reduce the relative consumption of energy, e.g. the energy consumption per unit produced. There is no requirement to reduce the absolute consumption of energy.

Financial, technical and organizational aspects within the organization will determine the speed and the extent to which the organization is able to improve energy efficiency of the aspects identified by energy management, but the result will always be an improvement of the energy aspects of the organization.

Stage 1 Energy Policy

The policy forms the basis for the setting of energy target and it shall be sufficiently clear to be capable of being understood by internal and external parties, i.e. employees, customers, authorities investors etc. The energy policy can be regarded as the "constitution" of the energy management system.

The organization shall insure that there is a consistent thread running from the energy policy to the significant energy consumption identified through the targets set and extending to include plans of action and key figures to allow the efforts to be centred in the areas where optimum effect is achieved.

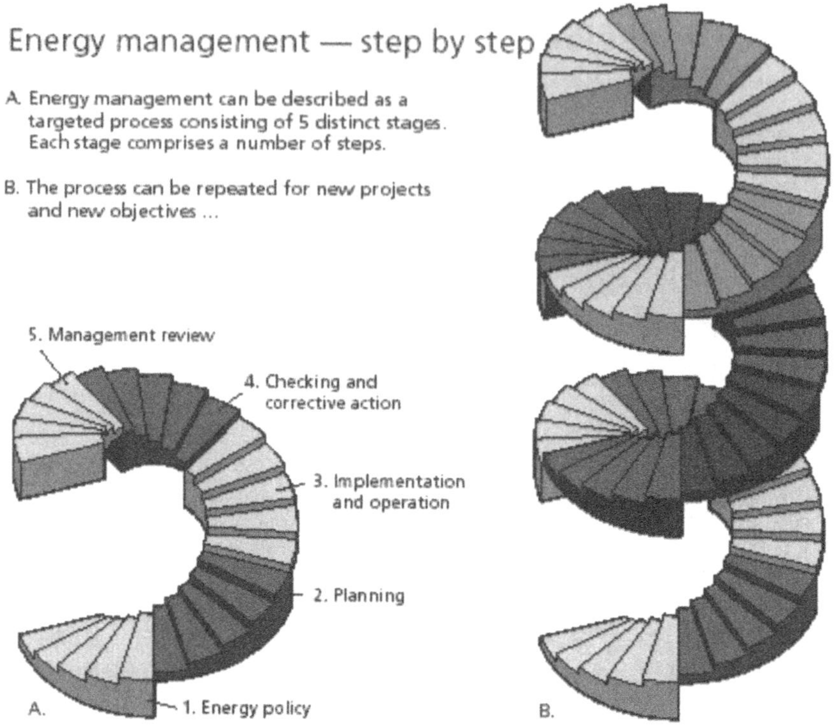

Fig. 1. Model of the energy management system

Stage 2 Planning

Review of Energy Aspects

The organisation shall make an initial review of the energy aspects. The review shall form the basis of an assessment of the energy aspects to identify the significant energy consumption. Relevant action areas – for improvement or more through analysis – shall be identified and prioritised.

The purpose of reviewing the energy aspects of the organization is to form an overview of the significant energy consumption i.e. the machinery, equipment and activities which account for the highest energy consumption or which offer the most considerable potential for energy savings. The review forms the basis for determining the order of priority of the efforts that are to result in actual savings.

Legal and other requirements

The organization shall keep up to date on and register laws and regulations etc., which affect the energy consumption of the organization. The purpose is to ensure that all relevant laws and regulations are observed.

Performance and targets

The organization shall set energy targets for the functions, which are important in relation to the significant energy consumption.

The purpose of setting energy targets is to prepare the way from policy to action.

By means of the energy targets the organization formulates how it intends to follow the energy policy over time. The energy targets shall challenge, commit and motivate the organization to implement improvements regularly and thereby maintain the progress of the efforts to improve energy efficiency.

The energy targets ensure that the organization has formulated concrete success criteria for the efforts to improve energy efficiency.

Energy Management Programmes

The organization shall draw up and maintain energy management programmes for achieving its targets.

The purpose of the energy action plans is to ensure that a concrete schedule is drawn up for achieving all energy targets. It is expedient for the organization to include the energy action plans in the overall planning. To the extent possible, the plans shall be incorporated into the organization's existing managements system, e.g. by incorporating the energy action plans into investment plans, budgets etc.

The energy action plans help organizations to improve their energy performance. The plans shall be dynamic and shall be revised regularly to secure a goal-oriented approach and effective achievements of targets. The plans shall contain a description of the initiatives and resources to be allocated to achieving the targets.

Stage 3 Implementation and Operation

Structure and Responsibility

Roles, responsibility and authorities shall be defined and communicated in order to facilitate effective energy management.

The purpose of defining roles, responsibilities and authority is to ensure that all employees involved are aware of their responsibility and to ensure optimum operation of the energy management system.

Training, Awareness and Competence

Personnel whose work affects the significant energy consumption shall be qualified to take part in the energy management activities. The organization shall define the required qualifications and identify employees or groups of employees that require training, instruction or education. The possibilities of the employees of influencing the organization's energy consumptions shall form the basis of the identification.

Communication

The organization shall inform and have a dialogue with the personnel regarding the energy aspects with the purpose of ensuring that all employees know the energy policy, targets and action plans of the organization. Relevant information to employees on the energy management system contributes to motivating and committing employees to comply with the organization's energy targets.

The organization shall ensure that all employees at all levels within the organization have the possibility to make proposals for improvements and ask questions and that all communications are responded to.

Energy Management System Documentation

The organization shall establish and maintain an overall description of the energy management system with the purpose of providing employees with a general view of the system. Other stakeholders such as customers, authorities, neighbours, investment companies etc., may also have an interest in knowing the structure and operation of the energy management system.

The description of the energy management system can be presented in form of an energy management manual so that all relevant information about the energy management system can be found in one place and is available to all employees.

Document Control

The organization shall establish and maintain procedures for controlling record and other documents required to run the management system. The purpose of the document control is to ensure that all documentation conforms to the practice of the organization. The organization shall therefore continually revise procedures, records etc. and ensure that current documents are available at appropriate locations in the organization.

Operational Control

Operation and Maintenance of Equipment with Significant Energy Consumption

To ensure optimum operation the organization shall establish and maintain procedures for the operation and maintenance of machinery, equipment and facilities having significant energy consumption.

Energy Conscious Purchasing

The organization shall establish and maintain procedures to incorporate and assess the energy consumption when purchasing energy consuming equipment and raw materials.

The purpose of energy conscious purchasing is to ensure that energy consumption is considered when decisions are made for the purchase of machinery, equipment, raw material and services. In connection with purchasing, the organization shall investigate whether more energy conserving alternatives exist, taking into account the technical requirements for the equipment as well as financial considerations.

By informing suppliers that the energy consumption of purchased products is included in the basis for the decision the organization will encourage dialogue with the supplier regarding the possibility of improving energy efficiency.

Energy Conscious Design Activities

The organization shall establish and maintain procedures for assessing the energy aspects when designing or modifying and renovating energy consuming facilities, including buildings.

The purpose of energy conscious design activities is to assess the energy consumption in connection with the design of future production lines, productions plants, extensions, reconstructions or equivalent. This ensures that future facilities and processes become as energy efficient as possible from a financial and practical point of view.

Stage 4 Checking and Corrective Action

Monitoring and Measurements

The organization shall establish and maintain documented procedures to monitor and measure on a regular basis, the significant energy consumption and all activities that can have a significant impact on the energy aspects.

The purpose of the monitoring and measurements is to ensure that the significant energy consumption of the organization is continually recorded, monitored and controlled. Monitoring and measurements can furthermore be used to document the energy savings that have been achieved.

Non-Conformance and Corrective and Preventive Action

In connection with the energy management system the organization shall identify any non-conformances and ensure that they are investigated and remedied. Non-conformances exist when e.g. the organization's energy policy, targets, action plans or the documented procedures are not followed. Examples include energy targets which have not been achieved within the specified time limit or failure to document servicing of a facility according to applicable procedures.

Recording

The organization shall establish and maintain procedures for documenting that the requirements of the energy management system have been met.

The purpose of the recording is to ensure that the necessary documentation is provided to substantiate the achievement of targets, action plans and other requirements of the energy management system.

Internal Audits of the Energy Management System

An audit programme shall be planned, taking into consideration that status and importance of the processes and areas to be audited.

An internal audit is a "health check" to find out how the energy management system works. The purpose of the internal audits is therefore to carry out a systematic review of the energy management system and assess whether the system operates in accordance with the guidelines that have been drawn up. This means to find out whether all parties live up to the responsibilities that rest with them and whether the organization meets the targets it has set with respect to energy.

Stage 5 Management Review

On the basis of the audit results, the revision of the energy review and other information, the top management of the organization shall periodically evaluate the energy management system to ensure its continuing suitability.

The purpose of the management review is to ensure continual improvement and adaptation of the system so that the system operates as intended. The review implies that the individual elements and overall operation of the energy management system are evaluated in a critical manner in relation to the ability of the system to comply with the energy policy and the energy targets.

The review shall be performed with a view to assessing the suitability of the energy management system and identifying need for changes and improvements.

References

Danish Standards Association, "DS 2403 E Energy Management – Specifications"

Danish Standards Association, "DS/INF 136 E Energy Management – Guidance on energy management"

Application of High Efficiency Motors in China's Petrochemical Industry Enterprises and Prediction on Future Application

Wumin Jiang

China Shengli Oil Management Bureau,
C/O Mr. Victor Zhou, Rm 1016 Canway Building, 66 Nan Li Shi Lu, Beijing 100045, China – Tel: +86 10 68042450 ext. 15, Fax: +86 10 68020990
Email: vzhou@public.bta.net.cn

1. Application of Motors in Petrochemical Industry Enterprises (Taking Shengli Oil Field as an Example)

China's petrochemical industry enterprises mainly consist of two parts, one is China Petroleum Gas Group Co. and the other is China Petrochemical Industry Group Co. The target of Shengli Oil Field is to get crude oil from underground by mechanical oil extraction and send it to petroleum processing and refinery sections.

Shengli Oil Field is the second largest oil field in China and is also a big power user. The total power consumption in the oil field is about 7 billion kWh, about 85% of which is for asynchronous motors of various types, out of which about 60% is for medium and small motors under 200kW. At present, most motors are of Y series with high power consumption and low efficiency (equivalent to eff_2 in European standard). The total quantity of motors of this sort is about 14,000. The application of asynchronous motors in other petrochemical industry enterprises is more or less the same as that in Shengli Oil Field and a large number of Y series motors are also in use at present.

2. Analysis by Comparison Between Y_{x2} High Efficiency Motor And Normal Y Series Motor

(1) Normal Y Series Motor

Y series motor comes out in early 80s of the last century. It has 11 frame numbers and 65 types. Housing protection = IP44 and insulation = Class B, its power level and mounting dimension conform to IEC standard. It is the leading product among small asynchronous motors in China at present, taking 80% for market share.

(2) Y_{x2} High Efficiency Motor

There are quite a few motor producers in China who are able to produce eff_1 motors as European producers do. Y_{x2} high efficiency motors are a product from one of the producers in China. This product conforms to IEC standard and its index for efficiency is equal to or better than that of EPACT in USA. Cold rolling laminated silicon steel sheets with good quality are adopted for Y_{x2} high efficiency motors so as to reduce iron loss and improve efficiency. The structure of air route is optimized and the fan diameter is properly reduced with the consideration that the temperature rise is ensured within the specific limit. The noise from fan and ventilation is reduced. The accuracy for component processing and quality for assembling are improved so as to minimize mechanical wear. Low harmonic windings and air characteristics for motor are properly selected. New and advanced methods for rotor processing and surface treatment are adopted so as to effectively control the stray loss for motors. The motor is characterized by high efficiency, low noise, low vibration, big margin for temperature rise, long service life and reliability, etc.

Besides, the specification of Yx2 high efficiency motors is the same as that of Y series motors, easy for erection and maintenance. The most advantage for Yx2 high efficiency motors is high efficiency and power saving, and only the price for Yx2 high efficiency motors is a bit higher than that of Y series motors with the same capacity.

(3) Comparison for Performance Between Y_{x2} High Efficiency Motor and Y Series Motors

A. Comparison for efficiency and cos Φ performance at the rated power (Fig. 1-1 for Efficiency comparison and Fig. 1-2 for cos Φ)

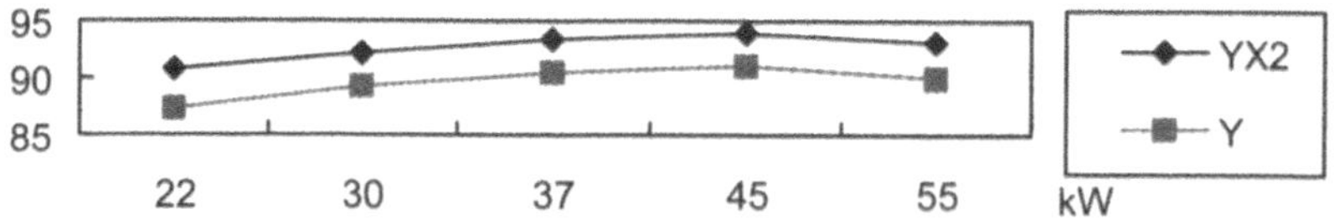

Fig. 1-1.

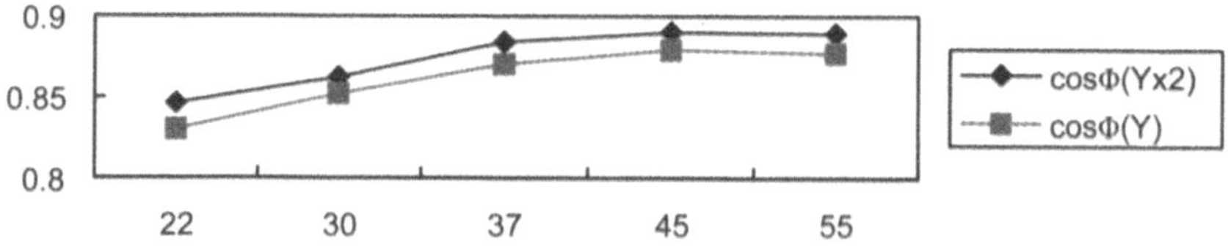

Fig.1-2.

The above show that the average efficiency of Y_{x2} high efficiency motors is 3% higher than that of Y series motors at the rated power point. For power factor cos Φ of Y_{x2} high efficiency motor is 1.2% higher than Y series motor.

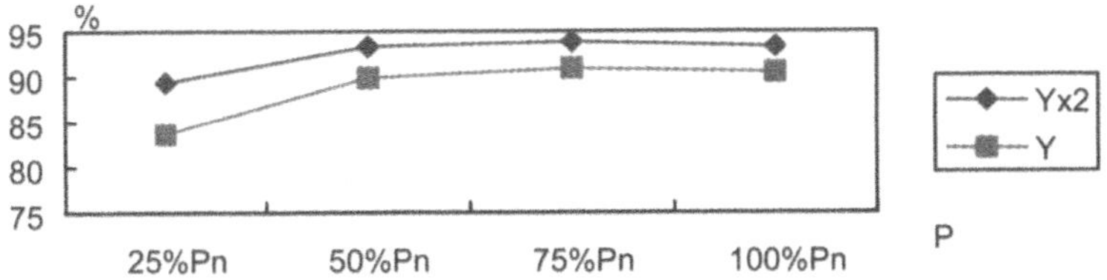

Fig. 2-1.

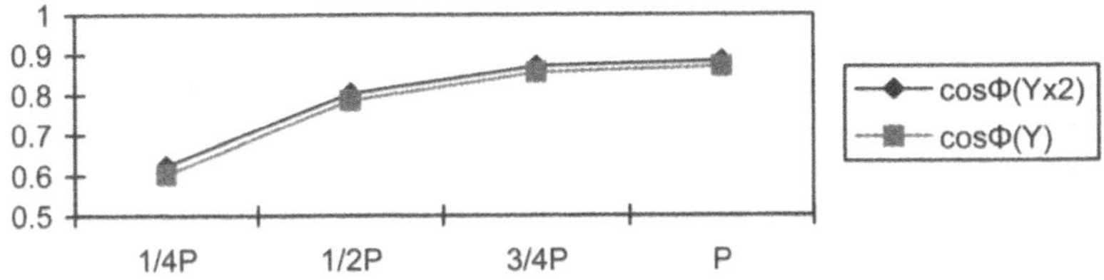

Fig. 2-2.

B. Comparison for efficiency and performance at different load (Fig. 2-1 for Efficiency comparison and Fig. 2-2 for cos Φ)

It is not difficult to see that the average efficiency of Y_{x2} high efficiency motor is 4% higher than that of Y series motor at non-rated power section. For power factor cos Φ, Y_{x2} high efficiency motor is 1.8% higher than Y series motor. The above shows that Y_{x2} high efficiency motor is a lot better for performance than Y series motor. It is necessary to note that as the driving motor for machinery in petrochemical industry enterprises runs at an average load of about 20% only, i.e. running at the low load zone, Y_{x2} high efficiency motor has more potential for power saving than Y series motor.

3. Practical Application of Y_{x2} High Efficiency Motors

In order to demonstrate the fact of power saving for Y_{x2} high efficiency motors, Gudong Mine replaced Y series motors at oil wells in its No. 2 mine with Y_{x2} high efficiency motors in the same capacity (2 for 22KW each, 1 for 30KW, 2 for 37KW each, 2 for 45KW each and 1 for 55KW) in May of 2001. Measurements were taken before and after the replacement, showing a remarkable result for power saving.

(1) Data Collected from Practical Measurement

Table 1. Power consumption before and after the replacement

Before the replacement			After the replacement		
Total	Output (m^3)	Power (kWh)	Total	Output (m^3)	Power (kWh)
	12296.8	44779		12400	38869
Unit consumption by calculation	3.64kWh/m^3		Unit consumption by calculation	3.13kWh/m^3	

The load (output) from the oil well before and after the replacement is basically the same, but the power consumption is obviously different.

Percentage variation for output before and after the replacement is 0.83% which is very small. The output after the replacement is higher than that before the replacement, making it more convincing for power saving.

(2) Saved Power Volume and Power Saving Rate

The average power saving rate:13.2%
Monthly saved power: 44779*13.2%=5910.8 kWh
Annually saved power: 5910.8*12=70929.6 kWh

(3) Economic Result

Recovery for investment: 1.61 years
NPV = 210,000 yuan (discount rate = 6%)
IRR=30%

(4) Economic Influence by Using Y_{x2} High Efficiency Motor Upon Production Cost of Crude Oil

The reduction of annual payment for power is 175,480,000 yuan.
The production cost is decreased by 0.5% .

(5) Social Benefits

Reduced emission of 70.2 tonnes of CO_2 and 1.06 tonnes of SO_2.

4. Market Potential for HE Motors in Petrochemical Industry

There is a great demand for asynchronous motors, specially for medium and small motors under 100 kW in all oil fields of petrochemical industry enterprises. The application of high efficiency motors is still at the initial and demonstrating stage. The total installed motor for the petroleum sector may reach 5 GW on the basis of rough calculation and the cost for the power consumption is over 20% of production cost. By calculation, for the whole industry if all Y series motors are replaced with Y_{x2} high efficiency motors of the same capacity, over 2 billion kWh/a can be saved.

5. Suggestions on Promotion of HE Motors

1. More communication with high level decision making leaders from petrochemical industry enterprises shall be made in China on benefits of HE motors.
2. More demonstrations for high efficiency motors shall be provided in petrochemical industry enterprises.
3. Promotion and sales from high efficiency motor producers shall be highlighted.
4. Support for the application of high efficiency motors from governmental decision making departments is needed.
5. Execution of DSM and energy service program shall be encouraged.
6. Motors in use shall be evaluated for efficiency and be timely replaced.

The article is based on the practice of "demonstration project for high efficiency motors" in Shengli Oil Field, which is completed under the cooperation between Shengli Oil Field, International Copper Association (ICA) and International Institute for Energy Conservation.

Energy Saving in Chemical Applications: Case Study: Lonza AG, Visp

Stefan Troger[1], Gilbert Schnyder[2], Ronald Tanner[3]

[1] Lonza AG, CH-3930 Visp

[2] Schnyder Engineers Ltd., CH-3940 Steg

[3] Semafor Informatik & Energie AG, CH-4056 Basel

Electricity Consumption at Lonza

The Lonza AG chemicals factory in Visp/Lalden is one of the largest consumers of electricity in Switzerland, with annual electricity bills amounting to around 40 million Swiss francs.

At Lonza, electricity is primarily used for driving electric motors (94% of its total consumption), which mainly power electric pumps (44% of consumption) and compressors (38%).

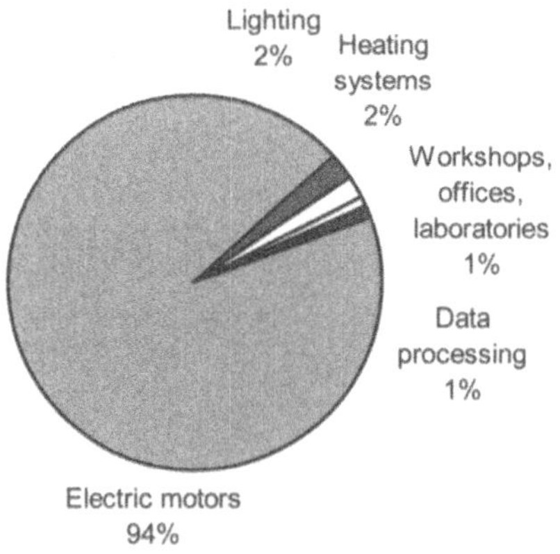

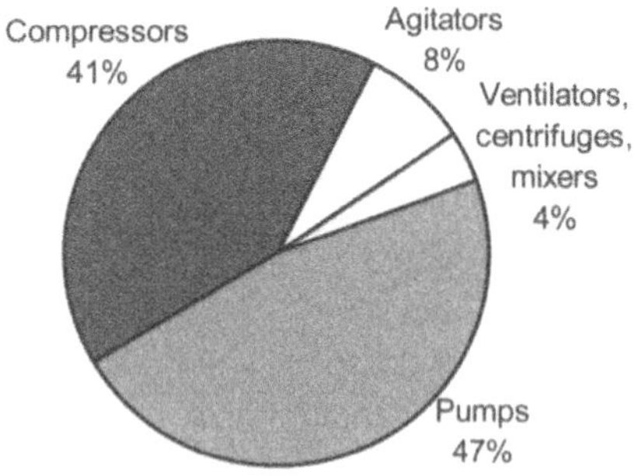

Types of Electric Motors in Use

Lonza uses electric motors with an operating voltage of 400V (low voltage),

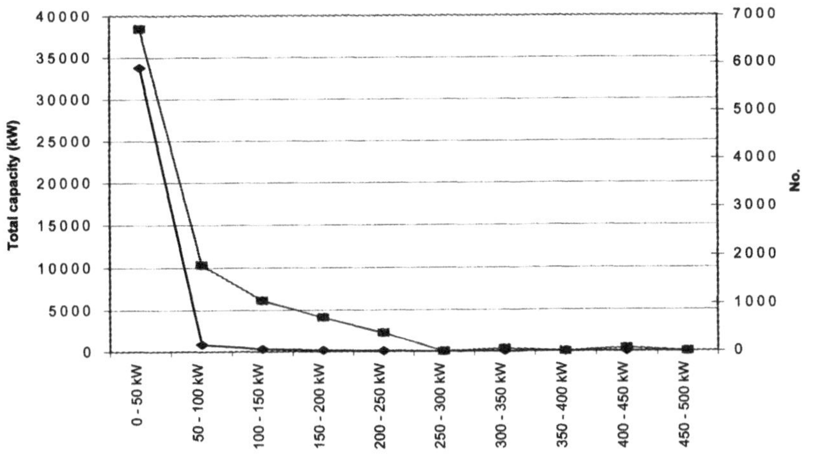

plus others with a medium high voltage of 6,200V for driving large-scale devices.

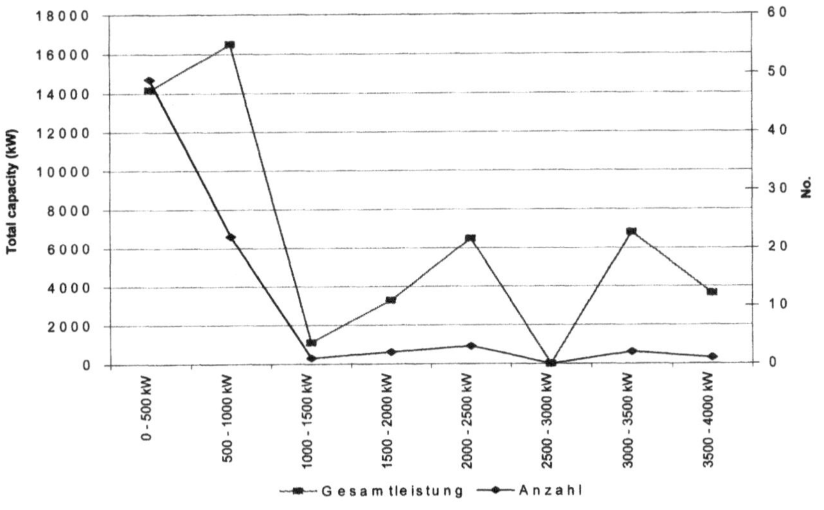

Savings Potential in the Areas of Production and Distribution

The existing savings potential in the areas of production and distribution of electricity is low, since these systems have already been optimised and are operated at a high level of efficiency η (generators: $\eta > 95\%$; transformers: $\eta > 98\%$; transmission lines: $\eta > 99\%$).

Savings Potential in the Area of Conversion

Generally speaking, there is little savings potential in the area of conversion of electrical energy into mechanical energy (electric motors), light (lamps) and heat (heating systems).

The only significant savings potential exists in the area of electric motors, especially when these have been significantly over-dimensioned (load < 30%).

There would also be a general savings potential of around 2% through the use of energy-optimised motors (with the aid of the OPAL software package). However, there is little benefit in utilising this additional potential in view of the use of standardised electric motors.

Savings Potential in the Area of Energy Use

Considerable savings potential exists in the area of energy use (end consumption).

Within the scope of three dissertations (electrical technology, process technology and economics), during 2001 and 2002 a typical modern small-scale tank depot and a typical older large-scale system (ammonia production) were studied from the point of view of energy efficiency.

Tank Depot

The tank depot concerns a modern and typical small-scale system with 11 standard pumps with thermal cables and lighting. It has been carefully optimised in terms of standardisation and flexibility, but too little attention was paid to energy efficiency during the planning stage. By reducing the operating times (control modifications), within the scope of this project it was possible to cut the overall energy consumption of the tank system by around 25% at very little cost.

The existing system possesses an economically utilisable potential of up to 60% by adjusting the pumps (trimming the impellers) on some of the drive systems to optimal operational point. However, when a pump has to be replaced, the trimmed impellers would have to be exchanged or disposed of due to wear and tear, since there is little purpose in storing modified impellers (standardisation).

In new systems, the use of variable speed drives can be beneficial with pumps with different operational points and lengthy operating times.

However, little attention is paid to energy efficiency due to the relatively low annual energy costs of the tank depot (approx. 6,000 Swiss francs p.a.), but drives of this type consume electricity to the tune of 5 to 10 million Swiss francs p.a. throughout the entire factory.

Ammonia Production

This system is one of the biggest energy consumers at Lonza. Its electricity consumption for the compression of synthesis gas and the auxiliary equipment required for this purpose amounts to approximately 2 million Swiss francs p.a. It is in operation all year round, without interruption.

The ammonia plant has been optimised a number of times over the past few years in terms of energy consumption, but very little material is available that documents the measures that have been taken.

The findings referred to in the tank depot case study with respect to the significant savings potential with small and medium-sized pumps were fully confirmed in the ammonia production study. Here, considerable energy savings potential exists if the operating time is reduced and the pumps are optimally adjusted to their effective operational point. Example 1: reduction of operating time of booster pumps (/3x9 kW) by 90%. Example 2: reduction of energy consumption of cooling water pumps (/2x25 kW) by 30% through the use of variable speed drives.

Over the past few years, effective additional technical measures (valve adjustment) have been implemented in large-scale systems (ammonia compressors) in order to improve the level of efficiency of the compressors. Further-reaching measures aimed at enhancing efficiency (variable speed drives) are not feasible for economic reasons.

Optimising the energy efficiency of the ammonia production process was difficult since not all the necessary measurement data (criteria/measuring points) are available, and comparisons of yield and energy requirements are also lacking.

In view of the imprecise measurement of the gas mixture, exact data concerning the current and optimum operational points of the production plant were not available. It was possible to achieve operation with considerably lower fluctuations by slightly improving the measurement/control procedures in the ammonia production process. On top of this, it was possible to improve the prerequisites for determining the optimum operational point more precisely and thus to increase the efficiency of the process.

Estimated Energy Savings Potential at the Lonza Factory in Visp

The results of the studies carried out in Visp showed that considerable potential for saving electricity exists in all installations and can be beneficially utilised.

The estimated utilisable potential in systems that are permanently in operation and use small-scale consumers is 10% to 30%, especially if operating times are reduced and pumps are optimally designed.

The estimated utilisable potential in systems that use large-scale consumers is 5% to 20%, especially if the system is operated at the optimum efficiency level or if it is configured for flexible use (partial load operation).

Potential can best be utilised during the planning stage for new systems (systems engineering). However, it should be borne in mind that designing a system entails a careful balance between detailed knowledge of the future installation, flexibility, standardisation and reserves, as well as planning costs and energy efficiency.

It would be possible to bring about more energy-conscious planning of systems by sensitising project engineers.

In existing systems, utilisation of available potential is considerably more complex and, in view of the already existing components, does not offer the same economic benefits since it would be associated with much greater outlay on the part of systems and operations engineers. The various measures that have been implemented show that (alongside any necessary adjustments to the system), for each Swiss franc of annual electricity costs that is saved, 1 to 2 additional francs have to be spent on engineering measures (i.e. hard-earned savings!). The main reasons why energy savings potentials are not being utilised are lack of time, inadequate methodology and a lack of specialised know-how.

Next Steps

Savings in energy consumption recur each year and remain effective until the end of the service life of the system concerned (sustainability).

Over the next few years, they will grow considerably more important due to regulations restricting CO_2 emissions (CO_2 Act), quantities of cooling water, etc.

Creation of "Energy Challenging" Position

In view of the results of these studies, a new position called "Energy Challenging" was created in February 2002. Job description:

- To enhance the energy efficiency of selected appliances
- To support the planning of new investments
- To support systems operators with the optimisation of existing installations

- To increase sensitisation on the part of planners and operators through training and further education

Role of Systems Planning (Process Engineers)

The greatest savings potential is to be found in energy-conscious planning and design of new systems.

It is possible to significantly reduce energy bills by using lower-cost energy carriers (cooling with river water instead of brine; heating with steam instead of electricity), as well as by precisely dimensioning equipment to the optimum operational point and keeping operating times as short as possible. However, dimensioning a system always means finding a suitable balance between outlay, uncertainty and flexibility.

Alongside energy savings, more energy-conscious dimensioning often offers the advantage that certain components are consequently smaller and therefore result in lower investment costs (smaller pipelines, pumps, valves, etc.).

Some of the above-mentioned disadvantages (uncertainty, reserves, engineering outlay) can be eliminated through the use of variable speed drives, especially with lengthy operating times and large capacities.

Systems operators are also able to more effectively monitor energy efficiency through the definition of the relevant parameters and dimensions.

Role of Systems Operators (Operating Engineers)

It is possible to identify energy savings potential as well as ongoing deterioration through systematic monitoring of the energy efficiency of the various electrical appliances (pumps, compressors, etc.).

Optimisation of systems (energy efficiency) is often neglected once they are in daily use.

With existing systems, reducing operating times is often the most effective method of saving energy. This may only entail minor adjustments to control settings.

Role of Energy Suppliers (Electricity, Steam, Water, Refrigeration, etc.)

Energy suppliers periodically record their customers' consumption data.

By systematically preparing all energy consumption figures and passing these on to their customers, the latter would already have access to much of the data they require for making energy efficiency calculations. These figures can also be supplemented with stationary or temporary measurements.

Cost Effective Conditions for Replacing Existing Electrical Motors with High Efficiency Motors

Pedro Maldonado, Alfredo Muñoz

Energy Research Programme, University of Chile, Echaurren 750, Santiago, Chile
Ph: +56 2 6897528, Fax: +56 2 6897528,
E-mail: pmaldona@prien.tie.cl, amunoz@prien.tie.cl

Abstract

Standard electrical motor replacement by efficient motors is not always profitable for industries and mining companies because the energy savings involved could not be enough to compensate the investment required by the new motor. The main objective of this paper is to identify under which conditions efficient motors can replace the existing standard ones.

Usually when a motor faces a severe and/or costly failure (i.e. requiring a rewind, or bearing or axle replacement), it is fixed due to the first cost bias. Another option, used perhaps less frequently, is to compare the lifecycle costs of the standard and efficient motors (i.e. investments, energy and maintenance costs for both alternatives).

In the specific case of motors presenting a high probability of failure, the analysis should consider not only the immediate fixing costs but also the net present value of the costs incurred on from successive breakdowns by both, the risky and the new motors weighted by the failure probability.

The corresponding failure probability curve was developed for a set of motors installed in the concentrator area of the main copper mining company in Chile and it was suggested that the company adopts this analysis tool as a guide for motor replacement.

Introduction

Inefficient motor replacement is not always profitable for industrial and mining companies. But, on the other hand, first cost bias avoids the introduction of efficient motor to replace the obsolete, unreliable and/or inefficient ones, even in cases when this could be the best choice. A methodology was developed to include, in the replacement analysis, the economic costs of maintaining in operation motors presenting a high probability of failure[1]. The methodology was applied to the main Chilean copper company, which has a consumption of 5.000 GWh/year, motors consuming 70% of that total.

Electric Motor Reliability Consideration on Mining Projects

During the recent years, electric motors reliability had been considered as a main subject for the operators of significant mining systems [2], as the concentrator[1], basically because:

- New milling equipment (mainly SAG (i.e. semiautogenous) mills) are significally bigger that conventional ones. As a result of this trend two milling plants having the same capacity but built with 20 years difference have a substantially different number of motors, as is shown in the following table.
- Flotation systems are increasing permanently their size and efficiency, then production is concentrated in a reduced number of equipments.
- Reliability becomes a central variable for modern milling systems, because a failure results in significant production loss, being the parallel lines, under this concept, the exception.

Table 1. Power and motor quantity strructure in two types of concerntration plants

	Conventional milling		Semiautogenous milling	
Power class HP	Number of motors	Power installed (HP)	Number of motors	Power installed (HP)
0-60	1.079	14.185	152	2.234
75-200	177	19.750	46	7.151
210-3.800	68	45.402	39	31.176
7.500-20.000	0	0	3	31.000
TOTAL	1.324	79.337	240	71.561

Note: The conventional milling plant produces 33.500 tonnes/day and the semi autogenous milling plant produces 30.400 tonnes/day

Reliability Rates, Histograms and Economic Evaluation of the Reliability

The development of a probabilistic model entails the knowledge of the failure statistic for a homogeneous group of motors, that is to say, motors subjected to similar environmental and mechanical stress. With this information in mind a fail-

[1] The analysis was focused on this area due to the importance of the concentrator's energy consumption used in motors.

ure histogram is built for this group of motors and through a mathematical process we estimate the parameters of a probability distribution function.

Statistical Analysis of the Motors Reliability. Histograms of Motors

The Weibull distribution is an adequate function to represent the equipment rate of failure. The corresponding analytical expressions are shown below.

- Probability's density function:

$$f(x)=\frac{\alpha}{\beta}\left(\frac{x}{\beta}\right)^{(\alpha-1)}e^{-\left(\frac{x}{\beta}\right)^{\alpha}} \qquad x\geq 0$$

- Distribution's function:

$$F(x)=1-e^{-\left(\frac{x}{\beta}\right)^{\alpha}}$$

The average period of time between failures is calculated through the following expression:

$$\bar{X}=\beta * \Gamma\left(1+\frac{1}{\alpha}\right)$$

where:

Γ : Gamma function

α, β : Weibull's distribution parameters

Two statistical parameters (α, β), associated to the Weibull's distribution, allows us to take into consideration the motors maintenance history in order to reduce system failures in a critical production area.

Histograms of Reliable and non Reliable Motors

The following graphic shows the distribution function for obsolete or non reliable motors and for reliable ones. Non reliable motors are those ones incorrectly designed, installed, specified or simply badly protected, these kind of motors should be replaced because of their high probability of failure. On the other hand, the reliable motors start to fail at the third year (even if this could occur with a small probability) and at that moment they should be submitted to a corrective maintenance in order to recuperate the original reliability for the next three years. The corrective maintenance techniques should be advantageously replaced by predictive or proactive maintenance techniques, through on-site and on-line measurements.

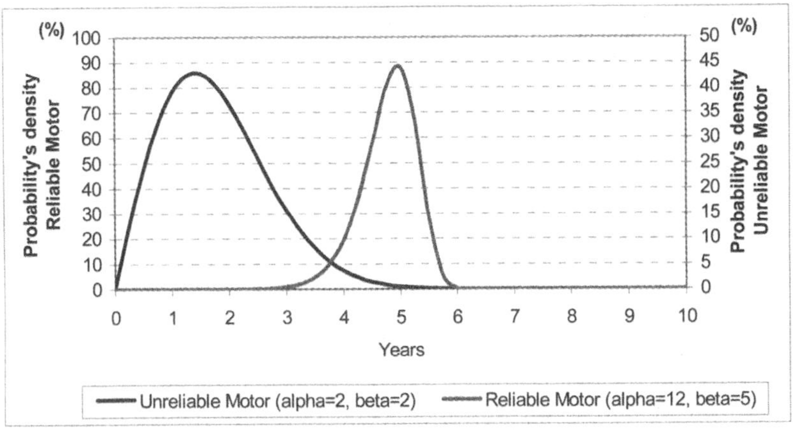

Graphic 1. Frequency distribution of failures for a representative sample of a set of reliable and non reliable motors.

When the parameters alpha and beta are lower than 12 and 5, respectively, it would be necessary to invest in order to increase it. Those investments are not necessarily associated to the replacement of one motor by another one, because it is possible that other reasons may explain the failures. In principle, the investment should be economically evaluated considering the costs associated to the energy consumption, motor repairs, and the like.

Reliability Economic Evaluation

The economic impacts of an unexpected motor failure or, even more complex, the impacts of having a non reliable population of motors, is difficult to evaluate. In this specific case, it was evaluated the costs of successive repairs of unreliable motors in relation with the costs of keeping a reliable population of motors. Even if the evaluation done is casuistic, in general, it can be asserted that it is more convenient to keep a reliable population of motors (investing from the very beginning in those kind of motors) instead of keeping untrusty motors subject to a high probability of failure (which should be repaired annually or even more frequently, anyway).

In function of the mentioned histogram was estimated the percentage of failure for the unreliable motors, the following equation allows to calculate the probability considered in the Table 2:

$$PF = 100\frac{\alpha}{\beta^{\alpha}} e^{-\frac{1}{\beta^{\alpha}}} = 38{,}9\% \quad si\ \alpha = 2\ y\ \beta = 2$$

The table included below shows the evaluation of the maintenance costs of the untrusty motors, considering an evaluation period of 4 years.

Table 2. Present value for the fixing costs (curep) of unreliable motors for the project evaluation period (years)

Variable		Year 0	Year 1	Year 2	Year 3	Year 4
CREPPC	Unitary repair cost[1]		1	1	1	1
PF	Annual failure [%]		38,9	38,9	38,9	38,9
CUREP	Repair costs' present value[2]	1,23				

Note:
1.For simplicity, repair costs have been considered equal to 1 and unchangeables each year.
2.Discount rate considered: 10%

The total absolute value of the successive repairs will be obtained for the unreliable motor product between the suggested unitary present value of the corresponding repairs (1,23) and the effective value of those maintenance costs.

The next table shows the same calculation, but for a population of reliable motors.

Table 3. Present value for the repair costs (curep)of unreliable motors for the project evaluation horizon (years)

Variable		Year 0	Year 1	Year 2	Year 3	Year 4
CREPPC	Unitary repair cost		1	1	1	1
PF	Annual failure probability [%]		0,0	0,0	0,9	19,2
CUREP	Repair costs' present value	0,14				

The above table shows that the reliable motors would fail for the first time at the fourth year and only 19% of the motors, as can be deducted from the graphic 2. To calculate the percentage of fails for this kind of motors, it is suggested to use the following equation:

$$PF = 100\frac{\alpha}{\beta}\left(\frac{year}{\beta}\right)^{\alpha-1} e^{\left(\frac{year}{\beta}\right)^{\alpha}} = 19{,}2\% \quad if\ \alpha = 12\ y\ \beta = 5\,and\ year = 4.$$

In consequence, the total repair costs for reliable motors are sustantially lower than those of non reliable motors. Finally, in any mayor repair it should be consider also the expenses associated to dismantle, transport and mount the damaged motor. The next table shown suggested values in function of the motor power.

Table 4. Costs of motor unmounting, transport and mounting (CDTM)

Power range	Lower than 26 HP	26 hp to 100 HP	101 hp to 350 HP
CDTM [US$]	424	687	2.223

Then, to evaluate the replacement of an unreliable motor by a new one, specifically an energy efficient motor, it should be considered the differential investment, the present value of the differential of the maintenance costs (including the CDTM costs) and the costs of the energy savings for the evaluation period .

As an application of this methodology, let's consider two frequent options: (1) to rewind a 30 HP unreliable motor, 85% efficiency, 70% load factor and operating 7.000 hours/year or (2) to buy a new efficient motor. The new motor cost plus the mounting costs is US$ 2.481, has an efficiency of 93,5% and work under the same conditions as the existing one. The energy cost is 0,045 US$/kWh. The rewinding plus unmounting, transport and mounting costs are estimated in US$ 1.219. The net earning benefit of replacing the motor instead of repairing it is estimated in US$ 1.706, as is shown in the following table.

Table 5. results of the methodology application to a 30 HP motor

Years	0	1	2	3	4
Differential investment [US$]	-1.794	0	0	0	0
Maintenance savings [US$]	532	474	474	460	172
Energy savings [US$]	0	531	531	531	531
Net benefits [US$]	**1.706**				
TIR [%]	**52%**				

Note: It was considered 10% as a discount rate.

Conclusion

In the case of motors presenting a high probability of failure, the replacement economic analysis should consider not only the immediate fixing costs but also the net present value of the costs incurred on from successive breakdowns for both, the risky and the new motors, weighted by the estimated percentage of failure.

References

[1] Kyung S. Park, "Optimal number of minimal repairs before replacement", IEEE transactions on reliability, vol.R-28, N° 2, June 1979

[2] Motor reliability working group, "Report of large motor reliability survey of industrial and commercial installations, Part I" IEEE transactions on industry applications, vol. IA-21, N°4, July/August 1985

[3] Brethauer et al, "The impact of efficiency on the economics of new motor purchase, motor repair and motor replacement", IEEE transactions on industrial applications. Vol.30, November/December 1994

BE$T, an Energy Savings Tool

John Malinowski

Baldor Electric Company, P.O. Box 2400, Fort Smith, Arkansas 72902 USA

Abstract

A new computer software tool has been created allowing the energy user to easily survey motors in his plant and record the inventory by means of a PDA (Palm Pilot or equivalent). The tool allows the user to easily perform a quick calculation, automatically selecting equivalent higher efficiency motors and calculating the payback time for the investment compared to energy savings. The tool is unique because of its ease of use and flexibility.

Why a Software Tool?

The need for a simple, yet flexible software tool are:

- Decision makers can understand savings if they are related to an investment with a favorable payback.
- Manual calculations are difficult and usually can be done only one motor at a time.
- BE$T automatically matches the old motor to a current NEMA Premium™ design and enters that data for comparison.
- If the efficiency of the old motor is unknown, BE$T defines that motor's efficiency as the average for that rating from the US Dept. of Energy survey.

International Usage

The software may be set up specifying local currency. Presently only dollars are available. Other currencies will be released later in 2002). The user's frequency (50 or 60 Hz) and efficiency preferences (NEMA Premium™ or European Union - CEMEP) may also be specified. These global settings will ensure that the calculations and motor selections will produce results that require no further calculations or adjustments.

Customer preferences

The user may select preferences: currency, electric rate, hertz, HP or kW, preferred energy standard (EPAct/NEMA Premium™ or CEMEP eff2/eff1), etc. Eventually, language could also be specified in the preferences.

Data Entry

BE$T has several ways to enter motor data:

- Direct entry to desktop or notebook PC
- Gather data and enter to a PalmPilot™ PDA (or compatible) and HotSync to PC
- Import motor data from an Excel spreadsheet

A simple drop-down menu (table 1) in the PDA allows the user to identify each motor's characteristics for horsepower/kW, speed, enclosure, application, voltage, efficiency, current (nameplate amps and measures on application), annual hours of operation, percent of full load, phase and mounting. A section is available for notes and asset or location numbers often associated with motors. If the motor is controlled by an adjustable speed drive may also be noted. If multiple motors of the same rating are used, quantity may be shown. Once the user has completed the plant survey, the data may be transferred to a PC to complete the calculations. If a PDA is not available, the survey data may be directly entered into the PC.

Table 1. Drop down menu for data entry

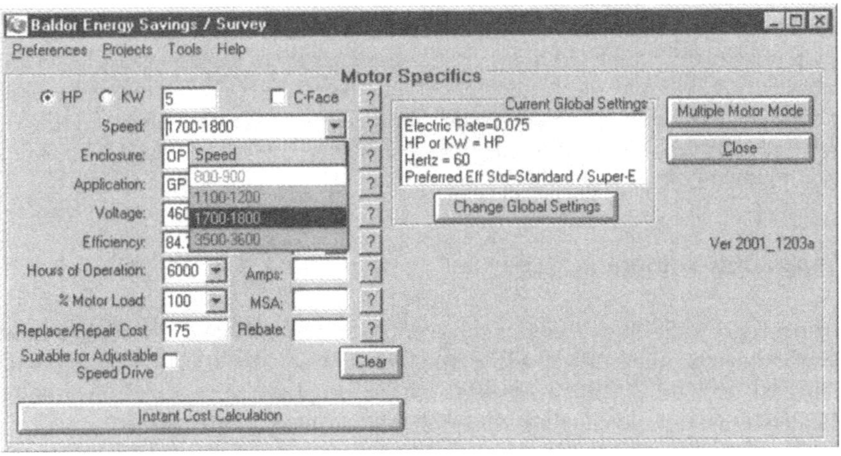

Efficiency levels

For the US market, the existing motor should be compared to equivalent models with EPAct and NEMA Premium™ efficiency levels. If the user does not know the efficiency level for his motor, a value equivalent the US Department of Energy average efficiency for that rating will be assigned. For IEC metric motors, European Union – CEMEP effl and eff2 motor preferences may be selected and used for comparison.

Calculations

The user may select to calculate savings on an individual motor or a complete plant full of motors, the BE$T software can handle either. On a multiple motor mode, more than 10,000 motor descriptions are possible for input with multiple quantities of each motor. Annual energy savings are calculated by the following formulae:

Single motor: (HP * .746 * %ML * Hours * $/kWh) / Efficiency

Comparing two motors:

(HP * .746 * %ML * Hours * $/kWh) / {1/(Efficiency$_{new}$– Efficiency$_{old}$)}

Single motor mode

The data for the single motor is usually entered directly to the desktop, but may be entered from a PDA via HotSync. In the single motor mode, an instant cost calculation is available that shows the annual electricity cost, kW/h used and motor efficiency for the motor being evaluated compared to an EPAct and NEMA Premium™ (eff2/effl) motor. Annual energy cost and kW/h savings for each is noted, along with the payback time.

Multiple motors mode

Data entry for multiple motors is through the same method as the single motor mode when using either a PDA or the desktop. Additionally, a list of motors may be imported from an Excel spreadsheet into the program. Multiple motor ratings are tabulated into a project listing, which may be scrolled for viewing.

Upon completion of the data entry, a project cost analysis is calculated. Based on the motors currently installed, the program selected the best motors for payback and reports on the number of upgrades available, total investment cost, total potential savings per year and the number of months to payback the total. A detailed report showing the savings data for each motor rating or a summary report showing the totals are available for viewing or print (table 2).

Table 2. Report summary for project and savings available

Asset #	Status	Qty	Replacement Standard-E catalog	Replacement Super-E catalog	Payback Months Standard-E	Payback Months Super-E	Repair Cost	Rebate	Annual Cost Your Motor
Line 2	Processed	1	WDM3615T	EWDM3615T	8	10	0	0	1982
recirc pumps	Processed	10	CP3587T-4	ECP3587T-4	12	14	0	0	12207
Line 3	Processed	1	CP3665T-4	ECP3665T-4	17	17	450	0	2022
recirc pumps	Processed	5	JSM3155	EJMM3155T	15	20	200	0	4476
Line 1	Processed	5	M3714T	EM3714T	17	20	375	0	19338
Test stand	Partial	1		ECP82276T-4		42	0	0	4340
steam cleaners	Processed	15	L1408T	EL1408T	45	22	200	0	20142
portable fans	Processed	100	L1203	EL1203	73	26	0	0	24506
Hazmat pump	Processed	1	M7054T-I	EM7054T	100	107	0	0	3832
fan #3	Processed	1	M3156T	EM3156T	136	41	125	100	628

Savings	Standard-E	Super-E
Upgrades Available	9	10
Total Investment Cost	30630	39968
Total Potential Savings/Year	9518	18297
Total Payback in Months	37	25

Payback time

An average selling price is assigned for purpose of calculating the payback time for the replacement of the higher efficiency motors. Payback time is the time it takes to pay for the motor using the savings from the energy that motor consumes. If the selected motor was a candidate for repair, the repair charges can be indicated. Often, electric utility companies offer rebate incentives that may be figured into the payback calculation. Repair costs and rebates are subtracted from the purchase price of the replacement motor in the calculation.

Adjustable Speed Drives

If motors are currently using or may be candidates for adjustable speed drives, these may be indicated at the time of data entry. Energy savings is only realized from an adjustable speed drive if the motor is driving a variable torque load, and the motor is operated below its base speed.

As the speed on a variable torque load is changed, the HP/kW requirements change by the cube of the speed change. At 80% of base speed, the HP/kW requirements are reduced to about 50% (see table 3). In a typical fan application using a damper or a pump using a valve for flow control, HP/kW requirements change very little as flow is reduced. Significant energy reductions are now possible because of the lower power requirements. Based on their operating profile, the added energy savings can be calculated (see table 4).

Table 3. Typical HP/kW requirements for variable torque loads

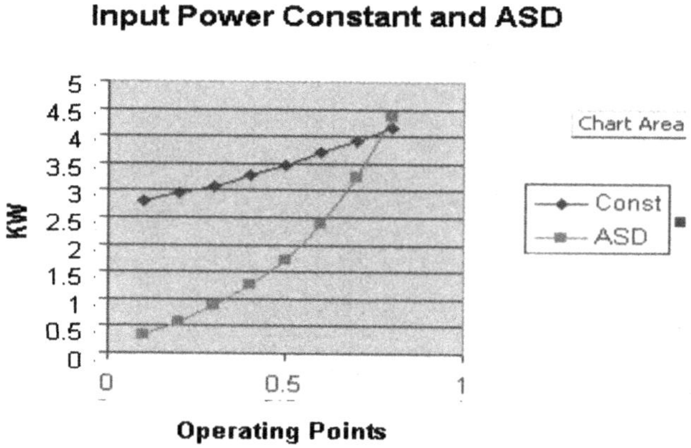

Table 4. Load profile from adjustable speed drive on variable torque load

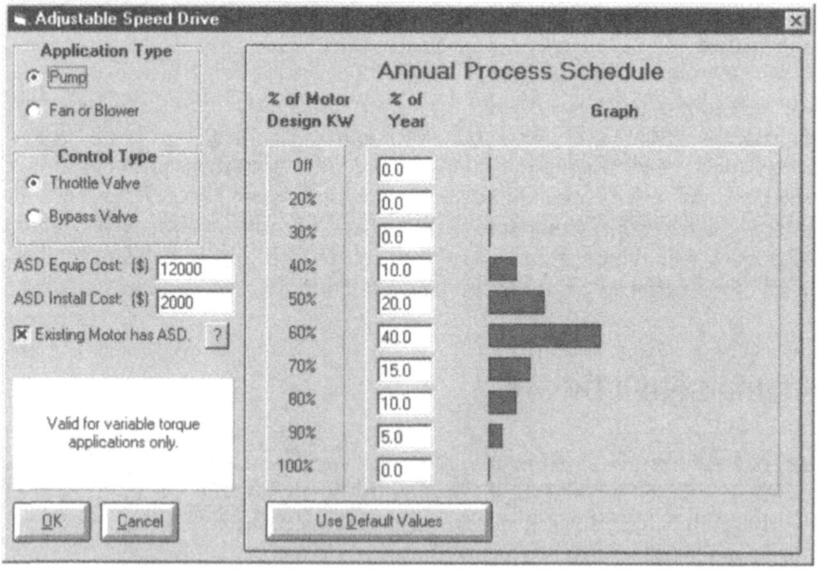

Applications with variable torque loads that are the best candidates for energy savings using adjustable speed drives are shown in table 5. These are also the most common types of applications consuming industrial energy.

Table 5. Applications best suited for ASD energy savings and percent of total electricity used by application

Application	Energy savings with ASD	% of total US electricity used by motors in this application
Air conditioning	20 – 25%	11.4%
Compressors	20 – 25%	15.8%
Central refrigeration	25 – 35%	6.7%
Blowers and fans	30 – 35%	13.7%
Pumps	30 – 50%	24.8%

Source: Wisconsin Center for Demand-Side Research, E Source 1999

Future Software Enhancements

A master database from various motor manufacturers could be used for worldwide comparison. Standardization of efficiency measurements will be a challenge and conversion will need to be performed for comparing results based on IEEE 112 and IEC 34-2. We hope that IEEE 112 will become the standard because all motor losses are taken into the efficiency measurement; no part is calculated or assumed. As this software tool is further developed, we hope it may be used universally for energy calculations.

Supporting the Effective and Efficient Reduction of Energy Consumption in Electrical Drive Systems

Ronald Tanner[1], Thomas Keller[2], Reinhard Riedl[2]

[1] Semafor Informatik & Energie AG, CH-5047 Basel, Switzerland

[2] University of Zurich, Department of Computer Science, CH-8000 Zurich, Switzerland

1 Introduction

Despite its increasing recognition in general public as well as by decision makers and authorities reducing energy consumption in electrical drive systems still is labor intensive, requires qualified expertise and sophisticated technical knowledge covering different engineering domains. Very often there is a variety of interfering constraints which have to be considered: environment conditions, performance requirements such as allowable maximum or minimum values of speed, torque and current, reliability and total life-cycle cost, just to mention a few. In many cases this is not a trivial task. And last but not least: its effort has to be taken into account when deciding about the feasibility of energy saving actions. Exploiting the saving potential of electrical drive systems is obviously closely related to the effort it takes for its identification.

This paper presents MOCA, a new software application system aiming to efficiently support the assessment process defined by the European Motor Challenge Programme.

2 MOCA's System Features

Adapted to the process described in the Motor Challenge documents MOCA provides the following features:

- browsing through multi-vendor drive component catalogs, searching and selecting components and comparing their properties,
- creating and modifying hierarchical structured drive system inventories including system topologies, component types, maintenance and operating conditions,
- identifying and assessing the energy saving potential by considering best practice recommendations,
- defining an action plan rated and ordered by economic feasibility
- generating user-adaptable reports.

3 MOCA's System Architecture

As depicted in figure 1 MOCA consists of the following components:

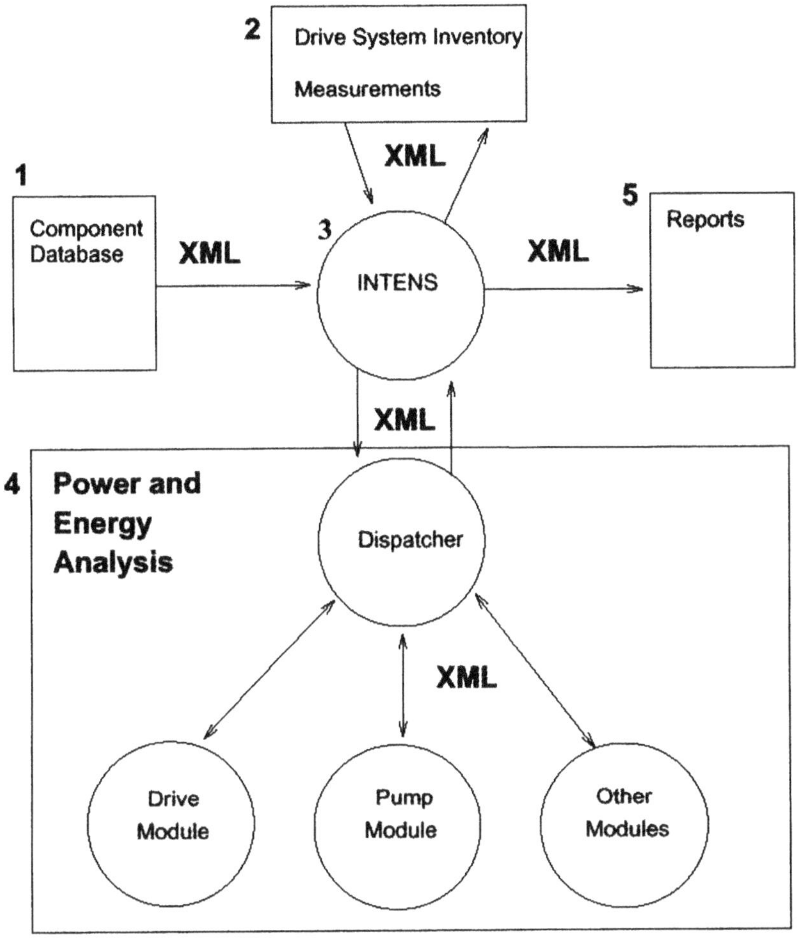

Fig. 1. Architecture

1. a component database that includes the catalog data of the drive system components (motors, inverters, gear boxes, pumps, fans etc.),
2. an inventory database of the motor driven systems and their associated load being used within the company,
3. INTENS, a fully configurable integration system with a graphical user interface that manages the project and inventory data and controls the execution and communication with external programs,
4. a power and energy analysis package,
5. a report generator that allows the creation of high-quality reports.

The chosen software architecture reflects the state of the art by incorporating well supported platform- and vendor-independent standards such as CORBA and XML thus offering a flexible and easily extendable environment. In addition, a set of documented interfaces largely simplifies the integration of third party modules for specific end-user requirements.

3.1 Analysis Procedure

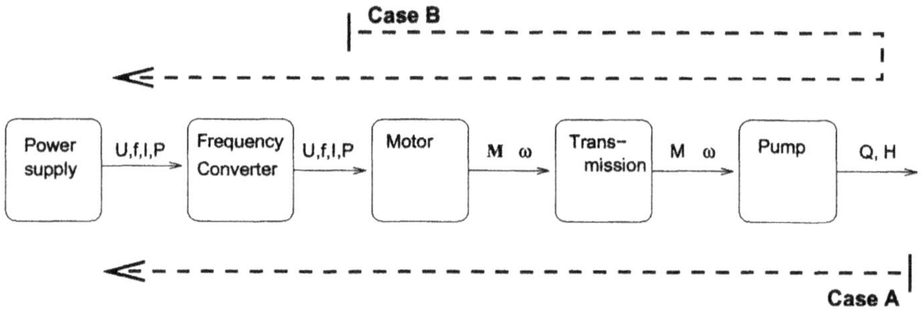

Fig. 2. Example of a drive system and possible directions of calculation

Figure 2 shows a drive system which consists of a variable speed drive, a motor, a transmission and a pump. For the analysis the following two cases are distinguished:

Case A: The operating condition of the end-use device can be measured. With this data and the name-plate data of the drive system components the total power and energy consumption can be calculated by simply traversing the drive system chain back to the power supply.

Case B: If the measurement at the end-use device is not possible either due to lacking measurement equipment or to limited accessibility measured data can be given at any intermediate connection point. The calculation starts at the closest point to the end-use device with given measured data and runs to the end of the drive system chain and back to the power supply.

In either case the load of the end-use device is left unchanged for the evaluation of subsequent energy saving actions.

3.2 The User Interface

Figure 3 shows the main window, which contains several folders and the graphical editor to create and modify the drive system topologies. The first step in a Motor Challenge project is the inventory definition of all motor driven systems being in use. The inventory contains a tree-like structure with one process for each drive system. One or more processes can be grouped to a department and each facility can be divided into one or more departments. After having activated the push button "New Process" a new drive system can be created by selecting its components in the tree list and dragging them to the diagram window. Each component (i.e. motor, transmission, pump etc.) has one or two ports with specific types: electric, rotatable or hydraulic. The diagrammer only allows the connection of two ports with the same type.

Activating the component with the right mouse button pops up a dialog window showing its properties as well as a table of the measured data. The power and energy calculation as well as the evaluation of saving actions can only be done for drive systems with at least one set of measured data.

At this point the influence of several energy saving actions can be evaluated. For example replacing components, installing on-off- or variable speed control equipment. The results can then be used for the formulation of an action plan.

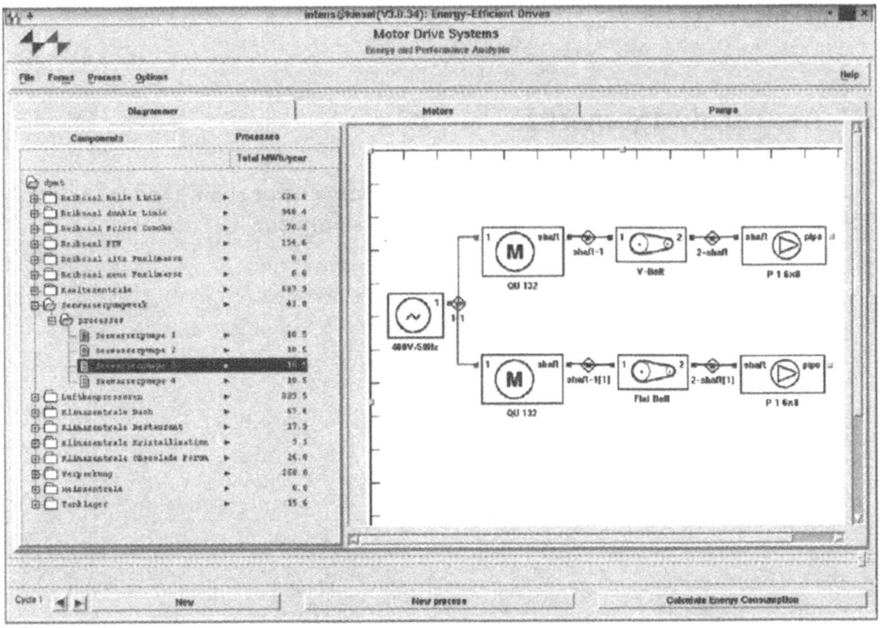

Fig. 3. Main window with graphical editor for drive systems

4 Status and Future Development

Currently the software is being used internally to support the Motor Challenge activities in Switzerland. It includes the following features:

- the graphical creation and modification of drive systems by dragging and dropping drive components from a tree list and connecting their ports

- the calculation of power and energy consumption of such drive systems consisting of converters, motors, gear boxes and pumps
- a motor database including the catalog data of several thousand motors
- the generation of Motor Challenge reports

For the first version of MOCA, which will be released in spring 2003, a range of improvements will be incorporated:

- completing the fan and pump module
- including some major fan and pump data in the component database
- additional searching and filtering functions for the simplified comparison of components and drive systems
- including a life-cycle-cost analysis module
- including best practice recommendations to support users in finding optimised solutions

5 Acknowledgements

The authors wish to thank the Swiss Federal Energy Office (BFE) and the Energy Office of the city of Basel (AUE) for funding this project.

The Energy Audit of Electric Motor Driven Systems

Flavio Conti

Dr. Ing. F. Conti, Energy Efficiency Consultant
Via Scirello 11, I – 21020 LUVINATE (VA) Italia
e-mail: flavio-conti@libero.it

1. Introduction

The growing importance of Electric Motor Driven Systems (EMODS) and the need to cope with Kyoto commitments require that national and international Energy Agencies launch wide energy efficiency programmes and set-up adequate tools for carrying them out. The share of electricity consumption associated to electric motor systems is very large: they consume today more than 40% of total electricity in the EU and forecasts indicate that this fraction will rise to 56% in the year 2010. Savings of more than 10 TWh/a at year 2015 have been quoted at last EEMODS'98 conference. (De Almeida, Fonseca, 1998)

The technologies used in Electric Motor Driven Systems are well established since many decades, although always new and more efficient products appear on the market. Design procedures are also very well defined since many years. But the need for making the use of electricity more rational requires a different approach and the development of more sophisticated and advanced energy analysis, management and auditing tools.

In real conditions an EMODS can be built up with various motor components, not always optimised, or recovered from other systems or simply chosen because they were the only available in the company's stock shelves. In old motor systems, information or design data are no longer existing or available, motor plates are missing, nobody remembers the motor specifications, the load may have changed with the time while keeping the same system, which in the meantime may become oversized. Every motor expert can mention many examples of very irrational layout of existing EEMODS.

It is therefore very often necessary, when entering particularly into a Small/Medium size Enterprise for energy efficiency upgrading, to have clear and effective guidelines for carrying out the audit of each EMODS. With motors using 66% of the electricity in a typical factory, they represent a tremendous place to start looking for energy savings. Because there are so many energy conservation opportunities (ECOs), (some hundreds on even a modest site), they are often overlooked simply because it is so hard to know where to start and assess them all.

The rational sequence in every action of energy audit of a system consists in the following steps:

1. Clear identification, description and modelling of the system under assessment
2. Identification of basic operational parameters and measurement
3. Comparison of performance data with those of Best Available Technologies and/or Reference Performance Values
4. Identification of specific Energy Efficiency Measures applicable for upgrading the system
5. Cost-effectiveness assessment of Energy Efficiency Measures set identified
6. Final Reporting to decision makers

This basic scheme must be tailored on the specific needs of the client. The Energy Audit theory foresees various schemes, depending on various targets that the client agrees with the auditor. A short discussion on possible audit types is also presented in this report.

The European Commission's Joint Research Centre had carried out, during recent years, two studies aimed at developing an Energy Audit tool for auditing EMODS. The availability of an up-to-date EMODS Audit tool will help to identify those motor system components which are likely to offer the best potential for energy savings, and then point to detailed information to evaluate the cost effectiveness of Energy Conservation Opportunities (ECOs) on each of them.

A summary of these two studies is presented in this report. The outcomes of these two studies can be the starting basis to set up a very effective audit scheme, to which the EuroDEEM software can be eventually linked for the selection of the most efficient components.

The outcome of the first study, after a short review of existing audit tools and software packages, presents the Electric Motor Audit scheme: this is split into Preliminary Screening Audit and Detailed Audit.

The second study has complemented the first one with an Industrial sectorial analysis, aimed at providing motor energy use reference values, to be used when such data are not available. Finally, guidelines for compiling the Final Audit Report were also developed.

2. Preliminary Audit

It is advisable to undertake an *Initial Screening* or *Preliminary Audit* of all electrical motor drive systems (EMODS) before starting a detailed study. By taking a quick *Walk Through* the plant/factory looking at motor systems, other energy saving opportunities may become obvious and provide greater energy saving opportunities.

Both a Preliminary Audit and a Walk Through Audit belong to the same class of audit, i.e. the so called *Scanning Energy Audit* [AUDIT SAVE Project, 1999].

While Walk-Trough Audit is targeted to SMI, Preliminary Audit applies to large Industries. Both types of audits are to point out areas, where energy saving possibilities exist (or may exist) and also to point out the most obvious saving measures which are normally "good housekeeping" and other no-or low-cost measures.

In this way, the result of a Preliminary Audit can be immediately useable by the client. The viable suggestions for a better housekeeping of company's facilities, with no or little investments, makes him aware of the energy saving potential obtainable in the company by upgrading all motor systems, and, maybe, also other types of systems. Therefore, he/she can be convinced to carry out a more detailed and complete energy audit of all company's facilities.

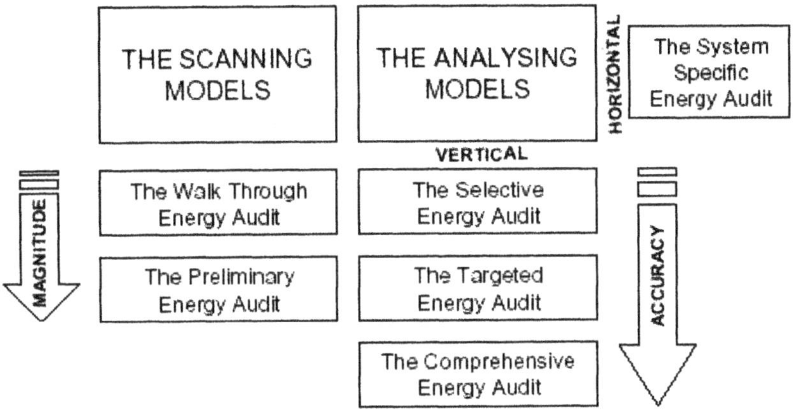

Fig. 1. The Basic Energy Audit Models

2.1 Step 1: Definition of Audit Type and EMODS Selection

a) **Audit Type:** The initial step of every Energy Audit is to define with the client (or facility management) the type of Audit wanted. Depending on the type chosen, the cost and time needed, the accuracy of results and many other items, it can vary considerably.

b) **Set Audit Criteria**: Once the audit type has been chosen, the audit criteria should be in agreement with the company's preferences, such as cost-effectiveness criteria, financial availability for energy retrofit actions, importance of production or service, type of contracting of eventual retrofit actions etc. If the energy improvement is made under a Performance-based Contract, it must be defined at this stage also the M&V Protocol to refer to.

c) **Select Motor Drive Systems to Audit:** Once chosen the type of Audit and the target area, the auditor can try to reduce the time for the audit to something manageable, picking up just the 50-100 motor systems on the site with the largest energy use, and to focus primarily the audit on them.

2.2 Step 2: Information and Data Gathering

a) **General Energy Information** the first type of information needed should be addressed to the facility overall energy uses and capacity installed. Specific indicators relating energy uses to production or service levels require also knowing such levels

b) **Gather of data of selected EMODS**: But the most important kind of data concerns the selected EMODS. In particular, the knowledge of the nameplate data, operational data, (type of loads, time schedules, load factors) are very important for the identification of possible Energy Conservation Opportunities.

c) **Previous Studies:** By looking at the results of previous studies (if existing) of motor use in the plant, very good clues can be given as to both where the big energy using motors are, and what the best ECOs are. This will greatly help in the next stage of the EMODS audit

In the case the audit type and related contract foresee to apply a M&V Protocol, then the information and data collected in this step can be part of the initial M&V activities needed to prepare a Pre-Installation Report.

2.3 Step 3: Identify and Select the Energy Conservation Opportunities

The several Electricity Conservation Opportunities have been organised and classified into 6 Energy Saving Areas:

1) Controls and Regulations
   - Switch Motor off when not needed
   - Install Speed Drives to match variable speed (Slow it down)
2) Reduce the motor losses
   - Improve motor efficiency
   - Improve the matching with the load
3) Reduce transmission losses.
4) Reduce losses in the power distribution
5) Improve the power quality
6) Reduce losses in the driven systems (pumps, fans, CAS, refrigerators, etc.)

It is not possible here, for sake of paper shortness, to comment ECOs of each energy saving areas. The preliminary audit proceeds from the most trivial and evident checks to more specialised ones.

The availability of a detailed database where each ECO is described according to a Standard Format can be of great help for the auditor. In fact, the assessment of an ECO involves several aspects. It is therefore important that none of them is omitted and forgotten when the whole information about the ECO is supplied.

The Standard Form has been structured according to the following issues:

a) Coding, Classification and Description
b) Cost Information

c) Decisional & Evaluation Procedure
d) Commercial And User Information

Obviously, if the Audit type is not comprehensive but limited to few target areas, also the Energy Saving Areas will be reduced, accordingly. This step can be considered as the conclusive one of the Preliminary Audit or a preparatory step for the successive Detailed Audit.

During this step, the Auditor should be able for each EMODS under investigation:

- to compile: a first list with all the ECOs to be implemented without any further investigation, because of their obvious and trivial convenience; in this list can be added, too, maintenance ECOs to be carried out subject to any equipment changes resulting from implementation of other ECOs
- to compile a second list with all the feasible ECOs but whose cost-effectiveness is not so evident and that require a further detailed audit step.
- ECOs with marginal economics and small saving potential: these also have to be further assessed and implemented, if favourable implementation strategy is found

2.4 Preliminary Audit (Intermediate- Pre-Installation) Report

At this point the auditor has got all elements sufficient to prepare a Preliminary Audit Report to be submitted to company's Direction for decision.
Even if based on a rough estimate of energy saving potential, the high cost-effectiveness and the evidence of their need are sufficient reasons for a prompt decision to implement the ECOs listed as "quick fixes" and related ECOs.

Since the payback of the suggested ECOs is so short and investment level so small, the estimate of the saving potential may be not very accurate, a greater accuracy being not needed.
At this stage, decisional issues to be debated with the company's decision-makers are:

- To implement the "quick fixes" suggested by the auditor and to put an end to the audit process or
- To further continue the Audit process to fully exploit and optimise the energy saving potential in the facility.

In this latter case, the Preliminary Audit Report can be considered as an Intermediate Report. In the case of use of an M&V Protocol, the Preliminary Report will include the so-called *Pre-Installation Report,* where Baseline situation is described. Each motor drive system, which will be object of an energy improvement, will be object of a careful description of the Pre-Installation condition. The report is compiled by the Auditor (or ESCO) and the customer must review and approve the report before any further step is made.

3. The Detailed Audit Process

A detailed audit will be necessary to identify the full extent of the energy saving opportunities. Some of the ECOs require a significant investment in new plant and this will need to be carefully verified and justified. The type of measurements to make depends on the need for the information - there is no point in collecting more information than is necessary. It is essential that data is collected that will help verify the savings potential and cost-effectiveness. Some basic measurements may be necessary.

4.2 Steps of an EMS Detailed Audit

The steps to be taken in the detailed audit are identified as follows.

a) Prepare a list of the motor systems where a detailed audit should be undertaken with related ECO's ID numbers for each.
b) Identify the data required to undertake the ECO assessment
c) If management staff cannot supply data concerning average load factors for various sizes of motors and applications, then the auditor can use reference values of the corresponding NACE industrial production (see next paragraph)
d) To each ECO apply the corresponding Audit Procedure (AP)and use the appropriate Measurement Technique (MT).
e) Calculate the energy savings and cost-effectiveness indicators (e.g. Internal Rate of Return and Payback periods)
f) Compile the Final List of ECO's for implementation, ranked and duly commented

4.2.1 Motor Audit Procedures

Each energy saving measure requires a specific *Audit Procedure*. With this wording it is intended the sequence of standard steps the auditor must carry out to assess the ECO under investigation and to be able to decide whether or not to recommend its implementation. During the execution of an Audit Procedure the auditor has to gather detailed data, by using a number of *Measurement Techniques* and relying on *Reference Values* obtainable from recent available *Industrial Sectorial Surveys*.

The description of as many as possible standardised Audit Procedures in a GuideBook can support and very much speed up the auditor's job.

4.2.2 Motor Measurement Techniques

Energy saving opportunities require data for motor systems that provide motor running time and the electrical power the motor is consuming. More precise data are required if power quality is an issue. The measurement of a EMODS parameters require one or more Measurement Technique (MT). Each MT should be described, with the types of instruments recommended for the measurement and the accuracy needed as a function of the kind of ECO to be assessed.

4.2.3 Reference Data from Industrial Surveys

During the energy audit of a motor driven system data concerning the average load factor of various motors are not always available. Reference values for motor load factors are very useful when making a rough estimate of the electricity consumption for electric motor systems where measurements are not performed. The knowledge of typical applicable ECOs and estimated energy saving potential can make the whole audit much easier and reduce the time needed to collect field data. As example of use, the results of surveys carried out in Denmark in the two industrial sectors are here mentioned (H. Hansen, August and Nov.1998): The sectors are: (1) the iron and metal industries (NACE Code 28) and (2) the food, beverage and tobacco industries (NACE Code 15).

For each of these sectors the following data are hereafter reported:

1) The electricity consumption of EMODS, with breakdown for end-uses and power categories expressed in percentages.
2) Reference values concerning average load factors for various sizes of motors and applications expressed in percentages.
3) Description of most common drive inefficiencies
4) Typical applicable ECOs and Estimated energy saving potential

The following figures show typical values for the iron and metal industry:

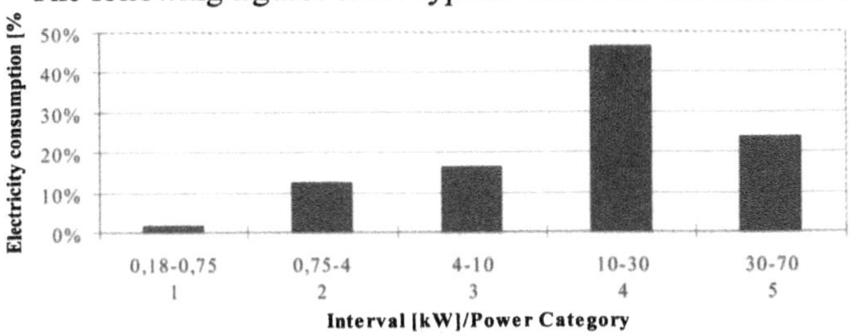

Fig. 2. Electricity consumption of motors within the iron & metal branch distributed on power categories

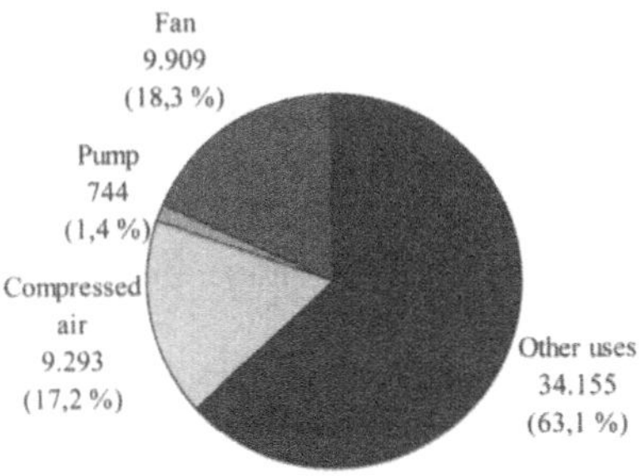

Fig. 3. Distribution of the EU-15 electricity consumption (GWh and %) for motors within the Iron & metal branch

Table 1. Reference values for motor load factors in iron and metal industries

No.	**1**	**2**	**3**	**4**	**5**	**6**	**7**	**8**
Power level [kW]	0.55	3	7.5	22	55	110	315	710
Use	**Load factor (LF) [%]**							
Other uses	20	30	35	45	55	-	-	-
Compressed air	60	65	70	75	80	-	-	-
Pump	55	60	65	70	80	-	-	-
Conveyor belt	20	30	35	40	50	-	-	-
Fan	55	60	65	70	80	-	-	-

From these figures an auditor auditing this type of industry can expect that nearly half of the electricity consumption is used for motors from 10 – 30 kW and that the distribution of the electricity consumption is split among 4 basic uses (fans, pumps, compressed air systems and "other uses). The analysis of the Load Factor table shows that there are significant problems within the iron and metal industries, primarily within the categories "other uses" and conveyor belts. Here, table indicates that the load factor varies between 20% and 55%, depending on the actual motor size. Proper sizing of motors in the categories "Conveyor Belts" and "other uses" can very likely achieve savings in the order of 20%.
Finally, industrial surveys indicate the most common drive inefficiencies of the branch. In the case of iron & metal industries the most frequent recommended ECOs are:

1. Major over-sized motors in the categories "other uses" and conveyor belts.
2. Limited use of High Efficient Motors. (HEMs).
3. Variable speed drives (frequency converters, etc.) are used to a limited degree only.
4. Technical possibilities with new motor technology are not utilised.
5. Service and maintenance of systems and plants are not performed regularly.

5. Final Reporting

From the content of the Final Report the decision-makers will decide about the actions to be taken and investments to make. Therefore, it depends very much on the ability of the auditor in presenting various solutions, saving and investment opportunities. Standard reporting tools can be sometimes used when the reporting is simple or very light. A computer tool can be used to produce the whole report and the appendix pages or tables of the report.

The Audit Report depends on the type of audit carried out, this in turn being related to the aim of the audit. In the case the Audit is part of a ESPC, the Audit Report must identify not only the energy improvements in the facility or in the EMODS, but also carefully record the energy and power uses "***before***" and "***after***" the installation action and all the "*adjustments*" due to the variation of the two situations.

The Final Report must include, besides the results of the Preliminary Report, also: the On-site Measurements, the assessment of Electricity Conservation Opportunities investigated, the suggested Action Plan, a description of the expected Post-Installation situation and overall savings.

6. Conclusion

The improvement of energy efficiency of motor drive systems both in industrial and service sectors can produce a large amount of savings. The European Commission and national energy agencies are paying growing attention to this energy "mine". Among support tools, information systems, motor component databases and audit guidebooks seems to be the most useful and effective. While large databases of induction motors have been set up, the need and the importance of exhaustive and comprehensive GuideBook for auditors seems to be neglected or underestimated, because electric motor theory is well known since long time. But assessment and evaluation procedures for a large number of energy saving measures need to beset up in a well-organised way. They must be connected with standard Measurements & Verification International Protocol, and all this knowledge collected in a Motor Audit GuideBook. In this way they can contribute considerably to disseminate and use correct practices of energy saving projects.

References

De Almeida, Fonseca, *Characterisation of EU Motor Use*, , page 143, Proceedings of EEMODS'98 International Conference, ISBN 3-540-67489-6

AUDIT SAVE –Project "*The Guidebook for Energy Audits, Programme Schemes and Administrative Procedures*", Final Report 1999, made by MOTIVA (Energy Information Centre, for Energy Efficiency and Renewable Energy Sources, Finland) IFE (Institute for Energy Efficiency, Norway) and C.R.E.S (Centre for Renewable Energy Sources, Greece).

Hans-Henrik Hansen, A survey of the stock of electric motors within the iron and metal industries. DEFU, Technical Report 397 E, August 1998

Hans-Henrik Hansen, A survey of the stock of electric motors within the food, beverage and tobacco industries. DEFU, Technical Report 411 E November 1998

The Software EuroDEEM as Support Tool for Auditors

Flavio Conti[1], Paolo Bertoldi[2], Vincent Berrutto[2], Stefano Cocchi[3]

[1] Via Scirello 11, I – 21020 LUVINATE (VA) Italia
E-mail: flavio-conti@libero.it

[2] European Commission, DG Joint Research Centre, 21020 Ispra (Varese), Italia
E-mail: paolo.bertodi@cec.eu.int, vincent.berrutto@cec.eu.int

[3] Via Einaudi 2, I – 21020 VARANO BORGHI (VA), Italia
E-mail: stefano.cocchi@libero.it

1. Introduction

The European Database for Electric Motor systems (EuroDEEM) was developed by the Joint Research Centre (JRC) on behalf of DG TREN. This activity, started in 1995, had the objective to develop a software tool for the promotion and selection of Efficient Electric Motors (EEM).

The first version of EuroDEEM contained only the motor selector database and was released in 1998 with about 3000 motor models available on the EU market. The motor data were provided directly from motor manufacturers and loaded by the JRC in the database. In 1998, it was decided to update and improve the software modules and to extend the software to other important motor system components such as Variable Speed Drives (VSDs), end-use devices (pumps, compressors, and fans) and other transmission and control devices. Development activities for the pump and VSD module started in 1999. A first demonstration version of the pump module was made available for comments by pump manufacturers. In year 2000, the motor database content was updated, raising the number of motors in the base up to 6000 with data supplied by 35 different European manufacturers.

During the period 1998-2002 the largest effort has been devoted to the dissemination and validation of EuroDEEM more than to its development. Studies on motor system audit procedures had also been carried out. New software modules containing these audit procedures will be one of next steps in EuroDEEM project development. Other improvements should concern debugging, I/O features, inventories of other motor components. A list of necessary improvements has been drawn which will be detailed hereafter.

2. Manufacturers Involvement

EuroDEEM's basic features were already reported at the previous EEMODS'99 Conference (Cocchi and Conti, 1999).
EuroDEEM is a software tool, aimed at helping the users improve the energy efficiency of their Electric Motor Driven Systems (EMODS). The original simple induction electric motor database of the first release of EuroDEEM has evolved toward a more comprehensive tool, targeted to the whole motor drive system, although there is still a long way to go.

This type of development cannot be carried out without a strong involvement of manufacturers for the data supply and the agreement on the data sets to be used in the database.

The assessment of the whole motor driven system implies that databases of various drive components coexist into EuroDEEM. Databases of Variable Speed Drives, coupling and transmission devices, end-use devices, such as pumps, fans, compressors, must be available together with the electric motor database. Catalogue or nameplate data of these drive components are not yet standardised, thus negotiations and discussion with manufacturers' associations are long processes.

The negotiations between the European Commission and Electric motor manufacturers in the second half of the nineties has led to the definition of motor efficiency classes (Eff I, Eff II and Eff III). The definition of what is efficient and what is not efficient is a fundamental step of an energy efficiency policy. Such definitions, unfortunately is not always so straightforward.

3. EuroDEEM as an Energy Efficiency Corporate Motor Policy

Each company should have its own database of motors and related components as basic information of corporate maintenance and management system. One of the most interesting features of EuroDEEM is to have such dedicated inventory integrated in its overall structure. Each EuroDEEM user can easily and quickly build up and update his corporate inventory by means of simple operations of drag & drop of the motor data, taken from EuroDEEM general motor database.

3.1 Motor Purchase and Maintenance Policy

Maintenance staff usually takes the replace/rewind decision without considering any energy efficiency reason. Too often the choice of the same brand and model for replacement avoids any kind of problem. Whenever a motor fails and it is replaced with the same type of motor instead of replacing it with a high efficiency motor, an improvement opportunity and lack of profitability is lost.

Similar considerations apply to rewind and maintenance practices inside the company. Very often, motors are let in operation until they fail. A good motor management should foresee what to do when motor fails or even decide an early replacement with a high efficiency motor, if this is cost-effective.

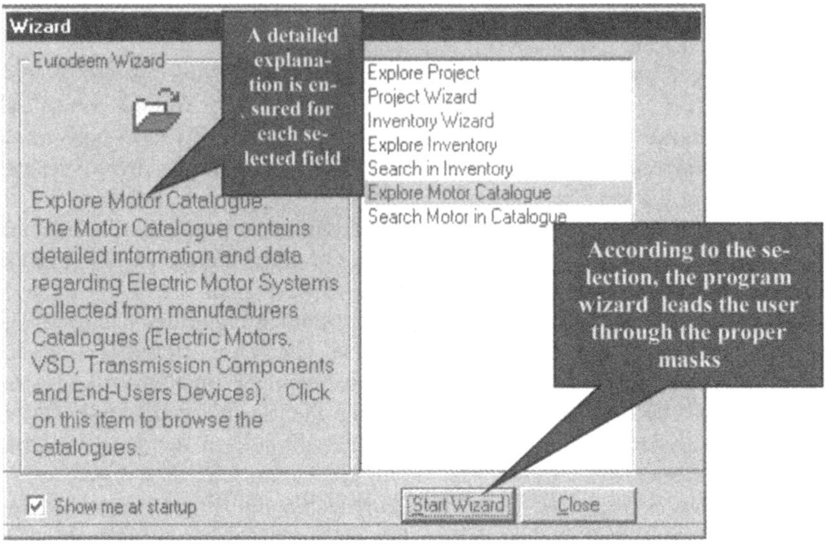

Fig. 1. EuroDEEM Screen with Wizard guided using modes

If the manager decides to rewind a burned motor, a careful choice of the shop must be made, because the quality of rewinding varies considerably. It is important that all rewinds, when cost-effective, be performed by reliable rewind shops that can guarantee that the efficiency of rewound motor is not less than 1% of the previous efficiency value.

The choices for a correct motor purchase can be easily carried out by using EuroDEEM as assessment tool, since it can sort out from all motors available on the market, those which best fit with various EMODS operating in the company's facilities. The fig. 1 shows the starting screen of EuroDEEM, where the query modes are highlighted, either using the *Wizard* or the Pull-down menu from *File* icon.

3.2 Company's Inventories

The user can create within EuroDEEM the company's own motor component inventory. In the EuroDEEM-2000 Version the only component concerns electric motors, but future versions will include other important components.

The company's motor inventory can be structured in 3 different levels (Company, Facilities, Departments), the higher directory including all lower subdirectories. Company's Inventory database can be expanded at Facility and down to Department level. Since facilities can be located in different regions and countries, with different electricity suppliers, EuroDEEM gives the possibility to enter different utility rates for each facility.

3.2.1 How to load Motors into the Inventory

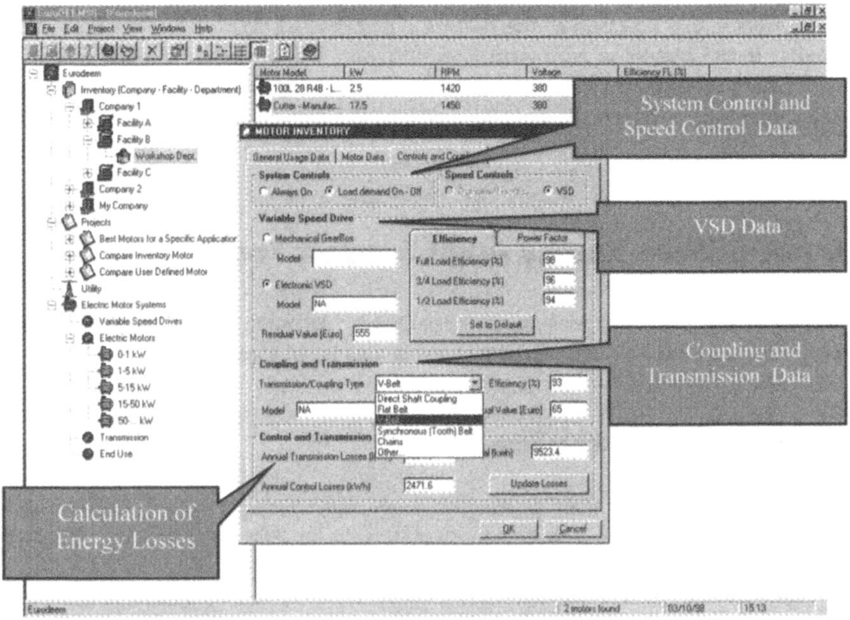

Fig. 2. Motor Inventory Data and System operation result

The loading of motor data into proper inventory location has been made as easy as possible. In the case the motor data are available in the EuroDEEM motor general catalogue (database), the user will simply drag and drop the motor icon into the desired department. If motor data is not in the base, data of a New Motor can be edited. After having entered the motor data, the user must specify where the motor is located in the facility, the load type and pattern it must drive and describe the system in which the motor is mounted.

3.2.2 How to assess and improve Inventory EMODS

The program calculates yearly operating hours, motor load factor, overall EMODS efficiency and energy losses, annual use of electricity and related expenditure.

The user can assess the saving potential of various Electricity Conservation Opportunities (ECOs) by entering various improved configurations of the EMODS. For instance, if the system is *Always On* this implies that the motor is using energy also when not required by the load and this reduces the overall system efficiency. The installation of a load-sensitive switch avoids the use of the motor when it is not needed and EuroDEEM displays the corresponding savings.

Another type of assessment concerns the Speed Control. The version 2K allows only the choice between *Dampers* and Variable Speed Drives (*VSD*). In the latter case the user must enter the data for the VSD, specifying whether it is a mechanical or an electronic device and entering the efficiency and power factor data. In a future version it should be possible to embed a function which can estimate the VSD efficiency as a function of the part load and duty cycle.

Finally, in the screen section with *Coupling and Transmission* heading, the user can enter data of different coupling mechanism and evaluate the savings by making the difference between the energy losses of the present state and the best solution.

4. EuroDEEM as Energy Audit Support Tool

As shown in the previous paragraph, EuroDEEM software is not only a valuable management tool but also a useful assessment and audit tool.

4.1 Motor Analysis and Optimisation

EuroDEEM can help the user in assessing and optimising a motor, defined as a *Project* (to be intended as a motor application and its operating mode). The following 4 options correspond to the most common needs of users:

a) Find best motor for a specific application
b) Find best solution for a user-defined motor
c) Find best motors of inventory replacements
d) Edit an existing motor project

A Wizard-procedure guides the user in entering data concerning (1) the load specifications the motor has to match, and (2) the operating schedule, power and speed percentages, control types and associated efficiencies, coupling and transmission choices and associated efficiencies and the utility rates. The input procedure differs slightly, depending on which of the 4 options above is chosen. In the case of motors whose data were already loaded in the inventory or in previously saved files, some input data masks can be skipped, otherwise for new motor projects all motor specifications must be entered by the user.

The outcome of the program search is a list of motors, which cope with *Project* requirements. Motors can be ranked according to increasing values of electricity use or energy cost. No motor price data are included into EuroDEEM, because real prices differ considerably from nominal prices from dealer to dealer. The user can extract from the list the motors he/she prefers, ask for price data and carry out a cost/benefit assessment with a standard Investment program. Such a procedure has not yet been programmed in EuroDEEM, although the cost-effectiveness routine is already inside the software.

4.2 Pump Analysis and Optimisation

Among end-use devices, pumps are not only the most important, but also the most difficult ones. Pumps in the EU use about 20% of the EMODS energy, corresponding to 160 TWh/a and 70 Mt CO_2. Their complexity is due not only to the huge number of applications (>80) but also because the efficiency of a pump cannot be derived by a single parameter, as is the case for electric motors. In fact, the pump efficiency depends on the flow (i.e. speed) and the head. It may happen that a pump with higher Best Efficiency point (BEP), when running in a given point of the plane Head-Flow, operates with lower efficiency of another pump with a lower BEP value.

The difficulty in finding a EU-wide and agreed definition of pump efficiency and pump data set, has caused delays in the development of EuroDEEM program.

In particular, as of today, no pump has yet been entered in the database.

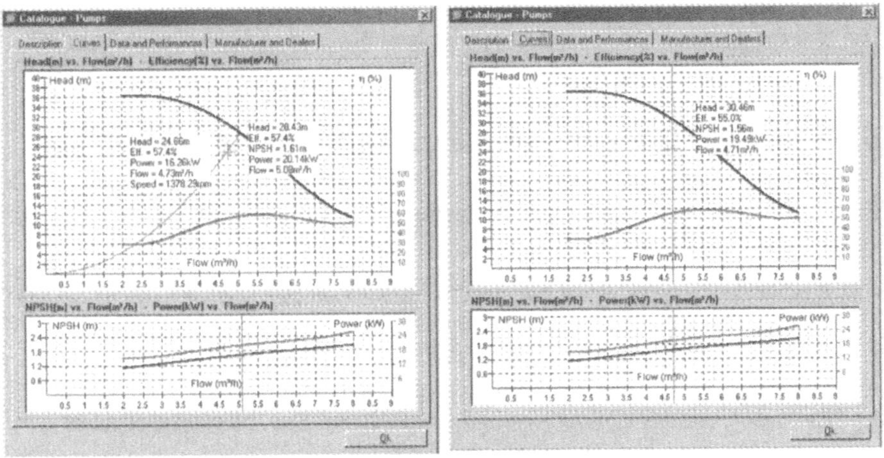

Fig. 3. a and b. Load curve and Pump characteristic curve

Three levels of pump selection have been identified and programmed:

1st Level: The user can select a pump simply specifying the manufacturer, the application, the Head and the Flow. Obviously, this level does not supply any help for energy optimisation.

2nd Level: The program searches for pumps, which can cope with operational needs specified by the user (e.g. Total Dynamic Head (TDH), Net Pressure Suction Head (NPSH), Flow, application, etc.). No energy analysis is carried out, but pumps can be sorted and ranked according to the distance of Actual or Desired Duty point from pump BEP.

3rd Level: the user must describe the system in which the pump is working, inputting the load curve as a function of the flow, the TDH and NPSH, and many other specifications. In this case EuroDEEM can calculate the power needed and highlight possible energy improvements.

EuroDEEM can identify and list the pumps, which have a working range compatible with the user load requirement. By clicking on a particular pump of the list a window is displayed and the user can look at specifications of the motor driving the pump. In a next window two multiple plots are displayed: in the upper one the pump working point is identified in the Head-Flow plane as the cross point of the working flow value with the circuit head. In the lower plot, pump power and NPSH as a function of the flow are quoted. Values of Power, Head, Flow, NPSH and efficiency of the working point are displayed. The user has the option to insert the load curve passing through the working point. If the working point does not lay on the pump characteristic curve, the program displays also the same set of values at the cross of this curve with the system curve.

From the example of Figures 3a and 3b it can be seen that for the same flow value, if a VSD is used the power drawn at desired working point is 16,26 kW. If the flow control is made by throttling the power required increases to 19,49 kW. With an easy click of the mouse the user can promptly assess the power control effect and the distance of the working point from the pump Best Efficiency Point.

5. Future Modification and Improvements of EuroDEEM

EuroDEEM software needs more refinements and further development. The main development issues concern:

- Inventory is limited to motor only: it must be enlarged also to other EMODS components (e.g. VSD, end-use devices)
- Import/Export features must be enhanced,
- Search option of motors according to efficiency classes
- Extension to other EMODS component and end-uses should be pursued;
- Cost-effectiveness calculations shall be facilitated;
- Its structure should be more oriented to identify a larger number of Energy Efficiency Measures, with links with EMODS Energy Audit outcomes.
- Comparisons of Pre and Post-Installation configurations must be made more evident to the user
- Finally, its structure should be more oriented to identify a larger number of Energy Efficiency Measures, with links with EMODS Energy Audit outcomes.

References

S. Cocchi and F. Conti, *The new European Database on Efficient Electric Motor Systems: EURODEEM-98*, pages498-513 of Proceedings of EEMODS'98 International Conference, - ISBN 3-540-67489-6

Energy Savings by means of Electrical Drives

Hugo Stadler

Dipl.Ing.(FH) Hugo Stadler, Head of Sales, Industrial Motors
Loher GmbH, Postfach 1164, D-94099 Ruhstorf, Germany
Tel.: 0049 8531 39284, Fax: 0049 8531 39543, E-mail: hugo.stadler@loher.de

Editors:
Dr. Herbert Auinger, Werner Blass, Dr. Martin Doppelbauer, Dirk Eggers,
Dr. Hans Funke, Berthold Göckel

Introduction

The manufactures of electrical drive systems play a key role in saving energy. Electrical drive systems account for two thirds of industrial power requirements and they are a central feature of automation systems. The economic potential for saving energy has been largely exhausted as far as the individual components of electrical drive systems are concerned. But there still remains a huge potential for saving energy by improving the design of the overall systems and their dimensioning. The purpose of this speech is to point out ways of exploiting this potential and thus counteract the wide-spread reluctance to take advantage of the possibilities on offer, something which companies have been slow to do for the most varied of reasons, such as short-term investment considerations. In addition to an objective description of energy-optimized systems, the potential for saving is indicated and inhibiting market barriers identified.

Saving Potential

Power Consumption of German Industry

In 1997, just over 18 per cent or 462 TWh (462 billion KWh) of the whole final energy consumption in Germany was accounted for by electricity. Out of this 462 TWh, industry consumed 42.5 per cent. Over 2/3 of the latter amount (133 TWh) was accounted for by electrical drives. The rest of industrial power consumption was accounted for by electrically-generated heat, technical galvanic processes and lighting.

In 1997, around 3.2 million low-voltage motors, 180 million fractional horsepower motors and 500 variable speed drives were sold in Germany alone.

The power-output spectrum ranges from a few watts to megawatt levels. There are no exact figures on the number of low-voltage motors in use but there are probably between 25 and 30 million installed units. The following estimates of energy-saving potential in the case of electrical drives only relate to industry.

Today, about 5 per cent (in terms of output) of the motors installed in German industry are operated with energy-saving electronic speed control. From an energy point of view, however, this would make sense for around 35 per cent of the drives. Given a total consumption of 133 TWh and an average energy saving of 40 per cent per drive unit, the total annual savings potential is 16 TWh, a saving which could be achieved by using electronic speed-control system alone. This is equivalent to 7 fossil-fuel power station units. Assuming an average price for industrial electricity of DM 0.15 per kWh, this would amount to DM 2.4 billion in terms of energy costs saved per year. A very good question can be asked here: What is stopping investors from taking advantage of this enormous potential for savings?

The same question can be asked in the case of energy-optimized motors. Energy-optimized means that the motors convert electrical energy into mechanical energy with as little loss as possible but still retain their necessary technical properties. By increasing the use of active material, the efficiency of standard motors can be raised, e.g. by about 8 percentage points in the case of 1 kW and by 1.5 percentage points in the case of 100 kW. Assuming that these measures make sense for half of the motors in use and achieve an average improvement in efficiency of 4 percentage points, the additional savings potential in industry amounts to 2.7 TWh (1.4%) or DM 400 million per year. This is approximately equivalent to a further fossil-fuel power station unit. (See Figure 1)

In addition to the economic advantages for the user, the application of energy-optimized drive systems is extremely significant for environmental policy. Given the savings potential in industry of 18.7 billion kWh per year and an average CO_2 emission of 0.59 kg of CO_2 per kWh (German energy mix) determined for the year 1997, the emission of 11 million metric tones of carbon dioxide could be avoided every year in Germany alone.

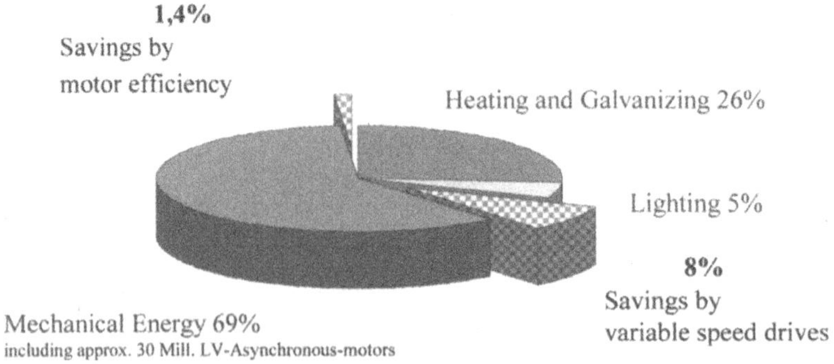

Fig. 1. Energy consumption and savings by German Industry (Source ZVEI/VDEW 1997 – 100%=194,6 TWh)

Minimizing Losses

Saving Potential by Motor Efficiency

Electric motors are electromagnetic energy converters whose function is based on the force exerted between electrical currents and magnetic fields – which are usually electrically excited as well. Due to their principle of functioning, electric motors have a relatively high efficiency. A typical value for an 11 kW standard motor is around 90 per cent and, for 100 kW, up to 94 per cent. Due to the laws of growth, the efficiency levels motors are higher than for smaller ones. Losses in an electric motor arise due to the following:

- current flows in the windings
- de-magnetization and eddy currents
- mechanical friction losses and
- parasitic effects.

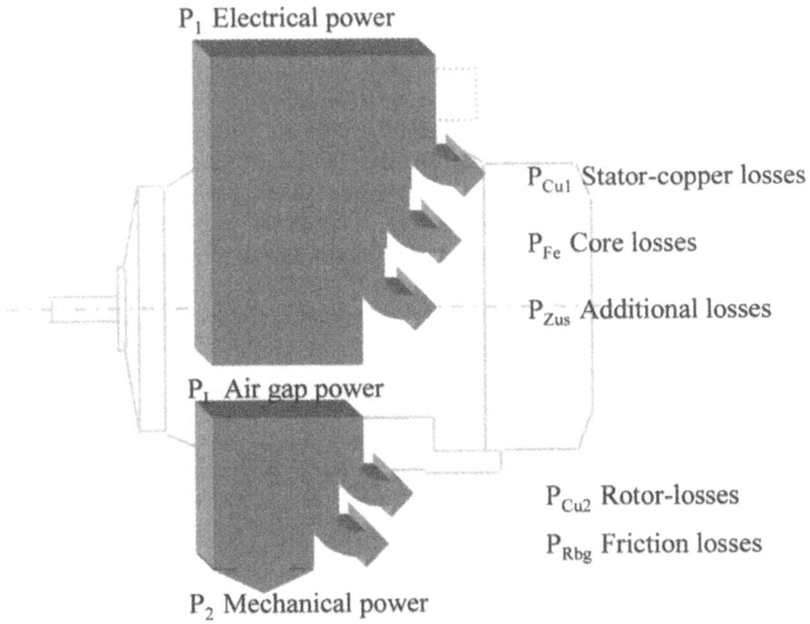

Fig. 2. Energy flow within an asynchronous motor

The art of designing motors is to find an optimum between the losses and the operating characteristics require. Depending on the type and engineering-principle of the electrical machine, any of the above-named factors can be influenced by designing to a greater or lesser degree. (See Figure 2)

Additional improvements entail higher costs which can certainly be justified according to the particular application. The example in the Figure 3 shows the number of operating hours after which the additional cost of an energy-saving motor is amortized, depending on the price for electricity.

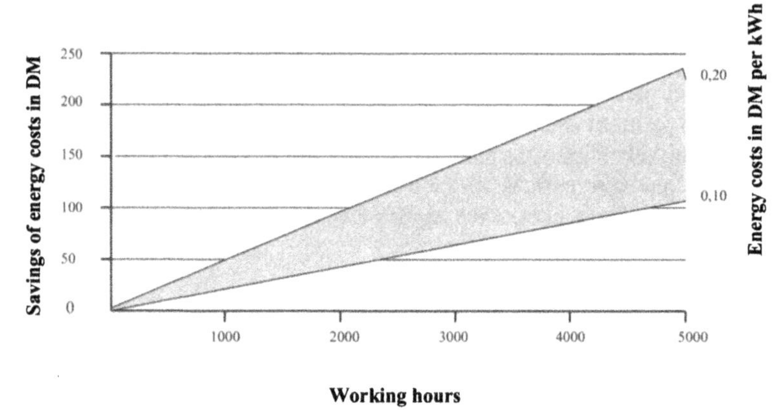

Fig. 3. Energy savings by motor efficiency

The present activities for increasing the efficiency of standard three-phase motors in the USA and Europe, three efficiency classes have been defined for 2 and 4-pole squirrel-cage motors in accordance with a voluntary agreement between the EU Commission and the European association of manufacturers, CEMEP. (See Figure 4)

In the case of variable speed drives, other motors such as permanent-magnet excited synchronous motors can be used as well. Due to the principle on which they are based, they do not have any rotor or exciter losses and are thus highly efficient.

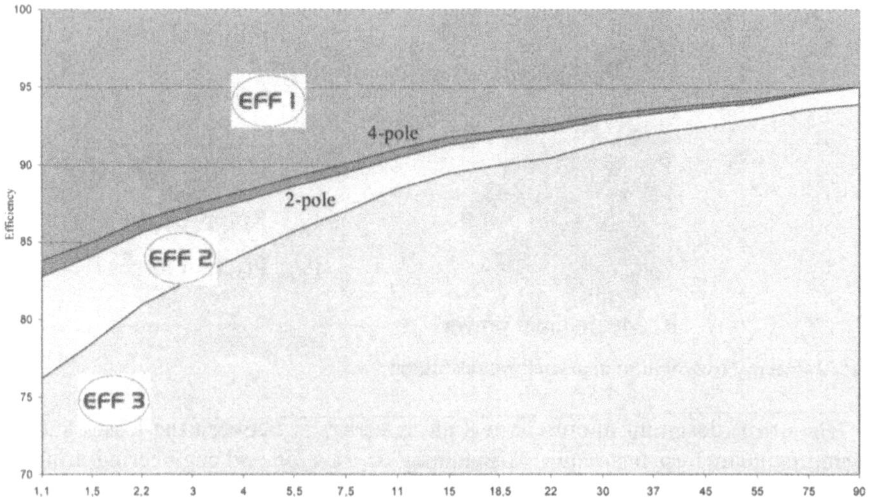

Fig. 4. Efficiency classes and marking of asynchronous motors

Saving Potential by Variable Speed Control

The public systems supply electrical energy in the form of AC voltage (single-phase) or three-phase voltage with a fixed frequency, usually 50 Hz. The frequency and voltage levels are changed in order to control the speed of the three-phase motor. This is done in the VSD by means of controllable semiconductor valves.

VSD are available for all power levels, from just a few watts up to several thousand kilowatts. These converters have been on the market as a standard product since the middle of the 70's. Control and information electronics were also improved decisively at the same time as power semiconductors. The use of microelectronics is opening up completely new markets both from a technical and economic point of view. The dynamic response of a drive with electronic speed control now meets the highest requirements. VSD are increasingly performing decentralized process-related automation tasks. The efficiency of today's type of VSD is typically above 96 per cent.

In industrial drive technology, two different groups can basically be distinguished. Driven machines which need an electronic speed-control system for process-technology reasons form one group. This group includes, for example, printing machines, machine tools and metering machines. The second group, to which pumps and fans belong, include driven machines which, in principle, can be operated without speed control as well. With regard to this second group, energy-wasting mechanical throttling systems are still being used today in many cases. Here, there is a great potential for saving energy with electronic speed-control systems.

For assessing the economic efficiency of a drive with a variable flow rate or with frequent alternation between acceleration and braking, the efficiency of the motor and other individual components is not decisive. Much more important is the selected control procedure. In comparison with conventional methods, electronic speed control can save between 20 and 70 per cent of energy costs. The additional investments for speed control are often already amortized within a few months. Nevertheless, conventional mechanical solutions with higher continuous energy costs or operating costs, respectively, are still being chosen, even today, for new installations. The main reason is that companies give priority to keeping purchasing costs down. A frequent cause of these misguided decisions is often that the responsibility for purchasing costs and operating costs is divided.

When pumps, fans or mills are being operated, the flow rates, pressure levels or quantity of material have to be changed according to the requirements of the technological processes. In the case of conventional drives, the motor runs at a constant speed. Quantity reductions required by the process are achieved by conventional control methods (throttle valves, bypass systems or inlet guide vanes) which can lead to very high energy losses. These losses can be considerably reduced by the use of modern drive systems, i.e. drives with electronic speed control.

Pumps play an important role here, the reason being that they account for more than half of industrial power consumption for drives. Many pumps are still being operated with the conventional methods named above. If the speed is electronically controlled, it is no longer necessary to reduce pressure by means of a governor valve and the high eddy losses are reduced to a minimum. (See Figure 5)

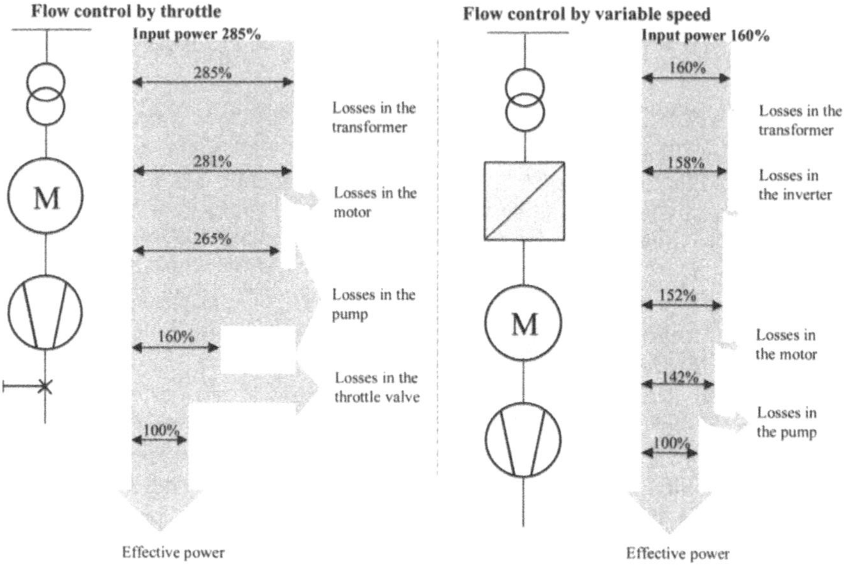

Fig. 5. Savings by variable speed drives: comparison of energy consumption for different pump applications

Why Energy-Saving Techniques Hardly Break Through ?

The Barriers

The reason why the known potential for saving energy is not being exploited have been investigated several times in recent years, from a scientific viewpoint as well. Science and practical experience have arrived at a series of shared explanations.

One of the main reasons is that buyers and users of drive components or whole machines generally belong to different departments within a company. Those responsible for making decisions in a purchasing department are primarily oriented to their purchasing budget and ignore possible savings in operating costs which can be made later. The reason is that such savings are of minor significance when the performance of the purchasing officer is being evaluated. In addition, investments in energy-saving measures are usually assigned a lower priority than other measures. Such expenditure does not help to maintain daily business. Neglecting such measures does not have any disadvantageous consequences for the present.

Furthermore, in the great majority of cases, the motor or the complete electrical drive does not go directly to the final customer but is supplied to an intermediary, usually a mechanical engineering company, for installation in other systems. For this intermediary as well, the energy consumption of the completed machine has lower priority in the beginning. His customers are mainly interested in the price of the machine and not in its energy costs.

Even if energy costs have been taken into account in considerations of economic efficiency when decisions are being made on investments, standard costing procedures, which are usually aligned to short project durations, neglect the investment in energy savings. The benefit of energy-saving measures, however, stretches over the whole lifetime of the equipment, usually for 10 to 15 years. The additional profits resulting from such measures are not apparent until after the calculated amortization period. They therefore have a subordinate influence on the decision-making process.

The psychological factor is also not to be underestimated. Whereas energy-saving arguments are convincing when new equipment is being purchased, they are usually resisted when old installations are involved. The customer does not want to take any unnecessary risks with new technology when it comes to the reliability and the operation of old installations. No businessman likes to interfere with a well functioning production plant when the only argument for doing so is to save energy. He will only do this out of technical necessity, either to increase production or the quality of his product.

How can These Barriers be Overcome?

The reasons for the low market acceptance of energy-saving techniques are varied and usually interrelated. Seen from an overall viewpoint, the greatest barrier is the lack of awareness surrounding this topic. This is where a strategy for overcoming these market barriers must be applied. The final customer, i.e. the buyer of a machine or system, must be sensitized to the topic of "energy consumption". This increases the pressure on the mechanical-engineering or systems companies to use the more expensive components with lower energy consumption.

In industry, 2/3 of electrical energy consumption is accounted for by drives. They are therefore of central importance when it comes to energy savings. The type of energy-saving method used depends on the application to a great extent. The greater the annual running time of the motor, the sooner the use of expensive energy-saving motors pays off. In relation to the total running time, over 97 % of the total costs can be accounted for by power consumption and only 3 % by purchase costs. Electronic speed control systems are used wherever the load is dependent on speed. Similar criteria apply to gears and power transmission.

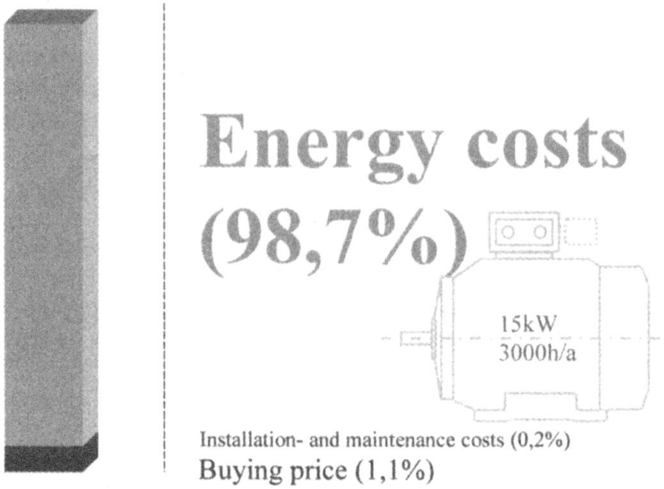

Fig. 6. Average Lifecycle costs of a standard asynchronous motor

Making up for lack of awareness and sensitizing mechanical-engineering companies and users must be the main objectives of a strategy for overcoming market barriers. Information brochures, publications in trade magazines and technical conferences on the topic of the technical and economic potential for saving energy are the instruments for achieving this objective. Freely available software programs with whose help drive tasks can be planned and improved by nearly all manufactures today as well as by independent institutes like Fachverband Elektrische Antriebe (ZVEI, Germany).

Energy Efficient High Voltage Induction Motors

Jakub Bernatt (jr), Maciej Bernatt

Research and Development Centre of Electrical Machines KOMEL,
Katowice, Poland, 188 Roździeńskiego street,
e-mail: info@komel.katowice.pl, http://www.komel.katowice.pl,
Fax: (0048 32) 2599948

1. Introduction

It is estimated that more than two thirds of electricity used by industry is utilised by electric motors, mainly induction squirrel cage motors, the real working horse of every industrial plant. These motors gained their primary position due to their excellent properties. These are: extreme simplicity of construction, high exploitation reliability, reasonable purchase price (equal to the energy cost consumed during the first two - three months of their usage) and relatively high efficiency when compared to the driven machines and apparatuses.

The problem of energy efficiency of low voltage small and medium size AC electric motors was discussed in many papers, at several conferences.

In US NEMA standard of efficiency level for 3-phase general-purpose induction motors up to 500 HP was established in mid nineties; in EU CEMEP efficiency classification Eff3, Eff2, Eff1, for 1.1 – 90 kW AC motors was agreed a few years later.

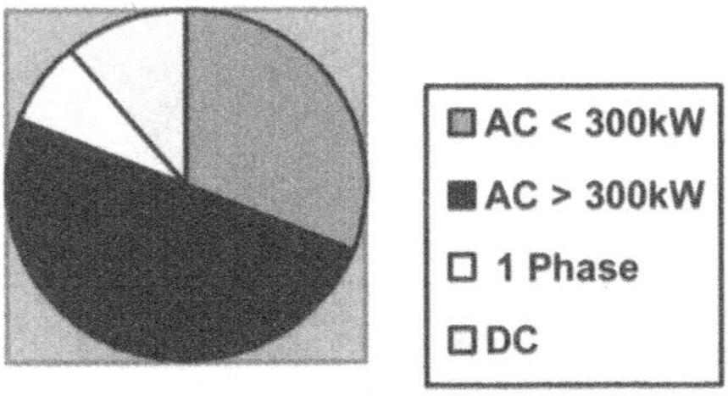

Fig. 1. Split of energy consumption by motor type

High voltage, large size AC motors refer to another issue. The motors are more specialist, there is physically less scope for improving their efficiency. But the large size, high voltage motors usually have very high duty cycles, 6000 or even more hours pa. They run with constant load, rather close to their full nominal load. Energy consumption by different type of electric motors can be shown graphically [1].

Since the invention of induction motors the main developing trend was to get their volume of active part, their mass and price as small as possible. Efficiency factor was regarded as being less important. Large size, high voltage motors built according to this principle usually have modular design, steel welded frame. The heat from active part is transferred by inner air circuit to the heat exchanger and than to outer air circuit (CACA type of design) - Fig. 2a. Up to the end of eighties TEFC motors with ribbed iron cast frames were built in power range up to 400 kW. In the nineties new trend in designing of large size high voltage motors emerged. Several companies started production of TEFC motors with ribbed cast iron frame up to 2600 kW - Fig. 2b. The heat dissipation of these motors is not so intensive as in CACA design; the iron core of large, ribbed TEFC motors should be less saturated, current density in the windings should be lower. Therefore the large, ribbed TEFC motors can be regarded as "intrinsic" high efficient.

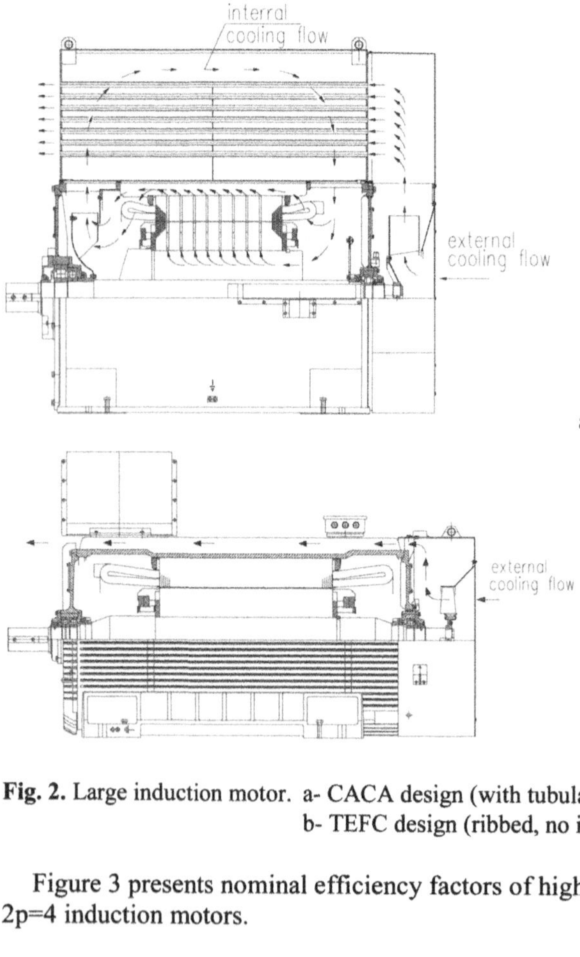

a)

b)

Fig. 2. Large induction motor. a- CACA design (with tubular heat exchanger);
b- TEFC design (ribbed, no internal cooling fan)

Figure 3 presents nominal efficiency factors of high voltage ribbed and modular 2p=4 induction motors.

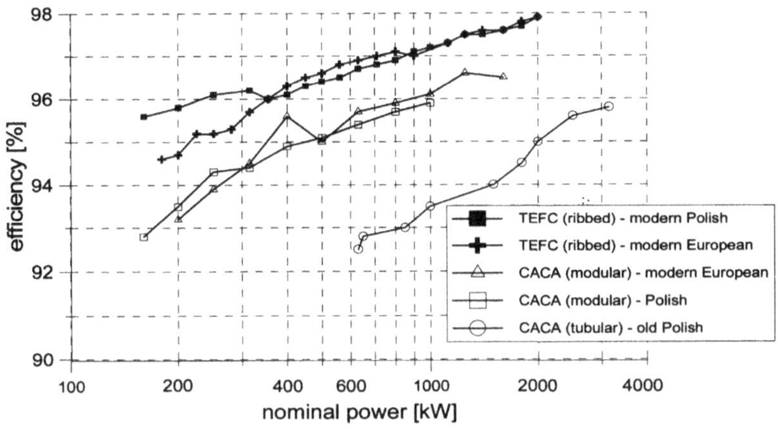

Fig. 3. Nominal efficiency of high voltage 4-pole motors

2. Current Status (Exploitation, Repair, Production)

In Eastern Europe countries most of high voltage motors output higher than 200 kW - used at present time at industry sites are the 20-30 years old products, technically outdated, manufactured years ago. Some were designed even in the sixties when such parameters as efficiency factor or noise emission level were not taken into account. Motors already used for several years are usually repaired and rewound.

During the repairs the efficiency factor of these motors is usually subsequently lowered. In Poland high voltage motors are repaired at several repair shops whose present 'working' potential significantly exceeds the demand. These shops remain in constant competition without any co-operation. None of them has the technological possibility to measure the efficiency factor of the repaired machine. The technology they apply in repair usually results in the lowering of the efficiency (due to lengthening of windings overhangs and increasing of iron loses). No repair shop takes action to increase the efficiency factor although in many cases it could be accomplished in the repair process. Only in Poland the outdated high voltage motors cause annual energy loses estimated at the level of 3900 GWh.

At present, one of the polish manufacturers (EMIT Company) have started manufacturing completely modern high voltage energy efficient motors in ribbed iron cast frames of output range 160 – 2000 kW, - Fig. 3. As far as the technical parameters (efficiency, noise level, exploitation reliability) and the price are concerned, they compete with modern foreign products. Motors have been designed in the Research and Development Centre of Electrical Machines KOMEL in Katowice. In the design process the ventilation circuit of the motors was optimised with the minimal noise and minimal mechanical losses taken as a criterion for the required cooling intensity (temperature rise of the winding as for the insulation class B – 80K, while windings insulation is of class F). The electromagnetic circuits were optimised with the minimal losses in the motor as a criterion for the set external dimensions of the active part of the motor. The iron core is made of a domestic low losses magnetic steel (P_{10}=1.5W/kg).

3. Motors Exchange – Cost Effective Way to Save Electricity

An exchange of old, disused, low efficiency motors by the modern energy saving motors (e.g. ribbed motors) will gain huge energy savings. The payback of exchanging an old high voltage motor by the modern one in Polish conditions was calculated to approx. 1 to 4 years. The examples of replacing of old motors by a modern ones are tabulated below.

Table 1. Payback time of exchanging old running motors by new, modern machines (4-pole motors)

Nominal output of the motor P [kW]	250	850	1800
old motor type nominal efficiency of old motor η_{old} [%]	SCDdm 114s 92.5	SZJr 134s 93.0	SZJre 144t 94.5
new (modern) motor type nominal efficiency of modern motor η_{new} [%]	Sh 355 H4C 96.1	Sh 500 H4A 97.1	Sh 560 H4B 97.7
diminished energy loses ΔP $\Delta P = P(\frac{1}{\eta_{NEW}} - \frac{1}{\eta_{OLD}})$.[kW]	10,12	38,59	62,38
annual energy savings when 6000 operating hours $E = \Delta P \cdot h$ [kWh]	60.720	231.540	374.280
value of saved electricity (1) $W = E \cdot C_{EN}$ [Euro]	4.554	17.365	28.071
average price of new (modern) motor C_M [Euro]	15.440	29.760	47.560
payback time of motor exchange $T = \frac{C_M}{W}$ [year]	**3.4**	**1.7**	**1.7**

(1) – price of electricity C_{EN} = 0,075 Euro/kWh (in Poland)

Table 2. Payback time of exchanging old damaged motors by new, modern machines (4-pole motors)

Nominal output of the motor P	[kW]	250	850	1800
average rewinding price(2) C_R	[Euro]	5600	8800	10200
payback time of motor exchange when 6000 hours p.a. $T = \frac{C_M - C_R}{W}$ [year]		2.1	1.2	1.3
payback time of motor exchange when 3000 hours p.a. $T = \frac{C_M - C_R}{0{,}5 \cdot W}$ [year]		4.3	2.4	2.7

(2) – in Poland

The replacing of the motor is particulary cost effective when the old motor is damaged and when it requires to be rewound or repaired. The table 2 shows the average rewound (winding exchanging) prices of the motors (in Poland), as well as payback time of replacing the damaged old motor by a new, modern one.

4. Goals Intended for Acquisition

- In the industrial drives operating at least 2000 hours per year replace the old, disused, low efficiency motors by the modern, energy saving ones.
- At the electric machines' repair shops introduce the repair technologies that:
  - would not worsen the efficiency factor or, in same cases
  - would increase the efficiency factor of the repaired motor.

5. Hindrances on the Way to Acquire the Intended Goals

The main hindrance is a human factor: routine behaviour (rewound instead of replace the failed motor) and lack of technical and economic awareness of the users of machines and electric drives. Users even do not know the losses and the cost of wasted energy in the motors they use; the value of the efficiency factor is not (and never has been) given on the nominal plates of the motors. Catalogues of the old motors showing this parameters are almost inaccessible now, so the users even if they wanted, could not estimate the losses of the motors they use.

As the result of their behaviour every old failed motor is send for repair although the cost in the case of stator's rewinding amounts to 30-50% of the price of a new motor. During the technologically traditional repair the efficiency of the motor is usually lowered.

Lack of financial means for purchasing new energy saving motors is a hindrance as well, but the main obstacle is just a human factor: routine behaviour and lack of economic awareness of the users.

6. Activities Proposed to Overcome the Barriers and to Achieve the Intended Goals

As the US EPACT proved, legislation is the most effective way to change market for high efficiency motors. But legislation, especially concerned high voltage, high power motors is unlikely at least for several years. Another kind of activities aimed to change routine behaviour and raise economical and technical awareness surrounding the topic should be carried out. The activities are:

- Elaboration and sending to the users of high power drives brochures with technical-economic data necessary to estimate the losses occurring in the motors used, with the full specification of efficiency factor of typical motors of old series compared with the efficiency of the currently available energy saving motors, (exemplary calculations of power losses in motors should be given in). Additionally, other benefits of modern high efficiency motors should be pointed out. Those are:

- Lower noise level
- Longer life and higher exploitation reliability
- Usually better power factor
- Less heat to be removed

These could be additional arguments in favour of replacing of the old motors. All these data should be placed in the Internet as well.

- Publications in trade and technical magazines papers concerned the same topics, with calculated examples for different kind of drives for various branches of industry.
- Organisation of technical seminars addressed to important users of machines and electric high voltage drives that would be dedicated to energy efficiency of this equipment. During these seminars the catalogues and the information leaflets concerned new energy saving motors and drives should be distributed, good energy saving stories from the industry should be presented. Technical seminars provide an excellent opportunity to address the competent proper audience. Research and Development Centre Komel has got a good experience in organizing these kind of seminars in pretty comfortable surroundings, away from the everyday pressure of work.
- Participation in technical fairs and exhibitions in form of specialised stands. This kind of promoting is not very effective, but provides good opportunity to gain publicity and demonstrate seriousness of a campaign.
- Addressing the producers of the motors with the proposal of presenting the nominal efficiency factor on the nominal plates of the motors, although neither the polish PN standards nor the IEC and EN standards require its placement.
- Elaborating special repair technologies for repair and service shops.

References

[1] Falkner H: Promoting Higher Efficiency Motors in Europe. The Role of the Copper Industry. ETSU AEA Technology, 2000

[2] Bernatt J., Bernatt M.: Energy Saving High Voltage Motors in Poland – From the Project to the Implementation , International Scientific Conference Energy Savings in Electrical Engineering, Warsaw 14-15.05.2001

New Brazilian DSM Program for Motor Driven System

George Alves Soares

National Program of Energy Conservation, PROCEL/ELETROBRAS
Rio de Janeiro, RJ, Brazil, E-mail georgesoares@eletrobras.gov.br

Abstract

This paper describes a new DSM Program for industrial motor driven systems, created by the Technical Committee for Efficiency of Energy Usage that was nominated by the Chamber for Management of the Energetic Crisis. It is being developed by the National Program for Energy Conservation – PROCEL. It aims to rationalise the usage of the motor systems in industry, considering their importance in that sector, as far as energy and costs are concerned. The expected results are the energy savings of 2 TWh per year implementing the program in the 2.000 biggest industries with the energy saved cost of US$ 13,5/MWh. This cost is lower than the one for expanding the generation system. It is also important to emphasise its innovative aspect as this program suggests structured actions aiming at rationalising complete and already installed motor driven systems.

Keywords: Motor driven systems, energy conservation, DSM Program

Introduction

During the electric energy crisis the possibility of a power cut motivated the whole society to rationalise electricity usage. At that moment, in May 2001, the Brazil's President created and enforced the Chamber for Management of Energetic Crisis – GCE. One of its assignments was the elaboration of the Strategic Emergency Plan for Electric Power. The aim of this Plan was to increase the offer of such electric power in order to secure the total required supply, at low risk of electric charge contingency. Therefore, either damages to the population or restrictions to the economic growth would be avoided, together with their undesirable impacts on employment and revenue and involving long and average run actions. Among the several different aspects to be approached, one of them referred to the development of compulsory programs for the rationalisation of energy usage. On July the 10th 2001, the President of the Chamber for Management of the Energy Crisis designated a Technical Committee for Efficient Energy Usage whose target was to establish measures to preserve and rationalise the use of electric power. Said Committee elaborated a document eventually consolidated as a program of measures and actions called Brazil Energy Plan – Energetic Efficiency. Such plan that was approved by the GCE on October 2nd, 2001 has the following targets: to get good results in saving energy both in the short and medium term and to promote a change in the market in terms of energy efficiency in the medium and long run.

The choice of the actions stated in that Plan was made by taking into consideration the actual results concerning saving energy in the short and average run. The energy crisis Brazil is facing and the ability of such actions to promote a change in the market in a larger scope have been kept in mind. That means that, though considering the seriousness of the present moment, it is important to realise that energetic efficiency cannot be tied to circumstantial questions only. It must be part of a national energetic policy as well.

This Plan was based on four basic axes: legislation, business environment, communication plan and specific programs. Among the specific programs, the one concerning efficient motor driven systems in industry stands out. Since the beginning of this year, PROCEL – National Program for Energy Conservation is responsible to run it. The operational aspects were improved and the approach through the state's industry federation was added to it.

The Importance of the Motor System in the Industry

In order to direct the industrial approach through different industrial sectors, a comprehensive survey on electromotive power usage in the Brazilian industry was developed. The aim of such survey was to identify the level of such system's involvement in it, both as far as the original consumption of energetic input is concerned as well as its cost.

The importance of electric power as energetic input within the industry is shown on Table 1, based on information from the 2000 National Energetic Balance (BEM) (based on 1999). This Table compares the total electric power in use in terms of consumption and cost on one hand with the total energetic input by industrial sector on the other. Table 1 also shows motor participation in the total electrical energy consumption. This study indicated which industrial segments are more cost sensitive to actions in motor driven systems.

Table 1. Cost and consumption shares of electricity related to total energetic input by industrial segment.

Industrial Sector	Motor electrical energy consumption related to total electricity consumption (%)	Electric energy consumption comparing to other energy inputs (%)	Cost of electrical energy related to the input total cost (%)
Cement	92,49	30,99	53,13
Pig iron and steel	29,27	25,52	46,36
Iron – alloy	13,50	73,92	92,69
Mining - / Pelletization	91,00	59,47	65,10
Non-iron and other metallurgical products	90,16	82,48	86,77
Chemistry	66,68	51,67	67,96
Food and drinks	75,35	25,73	75,08
Textile	87,87	80,03	84,74
Paper and Cellulose	88,10	40,22	69,65
Ceramics	77,70	22,50	40,36
Other industrial activities	74,93	78,31	83,55

Promotion to Energy Efficiency of Motor Driven Systems in the Industrial Sectors

Aim

This program aims at reducing any waste in the motor systems already in use in the Brazilian industry and also at motivating the replacement for more efficient equipment. As such, there will be negotiations with industrial sectors in order to offer their technical staff qualification. On the other hand, such industrial sectors will give a commitment that they will make the necessary technical analysis in their industrial activities in order to identify possibilities of energy saving in motor systems so that more attractive economical actions may be implemented.

The approach will be made through industrial unions (federations) and union leagues directed to the 2.000 major industrial plants and mills with an expected target of approximately 2 TWh/year.

Motivation

Efficient energetic measures in Brazil as in most countries aim at increasing motors' performance and at stimulating the buying of high efficient motors. Though important as they encourage the entrance in the market of more efficient motors such measures do not prevent working conditions of motor systems to be highly inefficient. As such, the effort made to incentive the use of efficient equipment is annihilated, considering that a simple replacement of motors do not reduce the major sources of energy waste such as the couplings, motors' loads and their control systems.

The optimisation of a motor system is complex and involves combining pieces of knowledge, which are usually isolated in Brazilian engineers' education. This is a strong knowledge obstacle that remains present in the structure of Brazil's industrial working staff.

It has been proved that the simple offer of diagnosis computer programs, job/professional training and courses are not enough to motivate the market to identify energy saving opportunities nor to implement them. Therefore, they do not generate the expected results. Generally speaking, technicians in industry are so concerned with keeping the production level that they do not analyse the motor system's efficiency and therefore neither the knowledge acquired nor the tools developed show the desirable intensity in usage. It goes without saying, however, that the importance of developing such tools is not being underestimated.

Among several international experiences, the one co-ordinated by the *Department of Energy – DOE* in the USA through a program called *Motor Challenge* is undoubtedly the best structured one to overcome such obstacles. Further information on that program's effectiveness and structure are available at the site www.oit.doe.gov/bestpractices/motors or in [3]. In one word, that program creates an information network directed to great consumers with a strong participation of manufacturers' distribution channels. In doing so, they try to involve manufacturers in a commitment to establish economically interesting measures. In exchange, they get training courses and didactical material.

Expected results

It amounts to 74,4% the total power energy installed in industrial systems for ventilation, compression and pumping. Several opportunities of energy saving were identified through studies made by PROCEL. Such saving involves motor loading coupling, motor oversizing, usage of electronic variable speed drives, among others. Diagnosis made by CEPEL in factories' compressed air systems indicated a potential for energy saving above 20 % just by acting on lowering the waste level in said systems and that without involving great investments. Such results can be justified because of the huge operational efficiency variation of the motor systems that ranges from 15% to 80%.

Using the American potential energy saving of 15% as a reference the Brazilian potential energy saving is of approximately 6,0 TWh. It was considered for such estimation that 8.000 major factories have a consumption of about 58 TWh in motor systems being 70% of them in ventilation, pumping and compression systems and that the total potential saving is of 15%. Based on that estimation it is hoped to get an energy saving of 2 TWh/year - 500 GWh/year in the first year and 2000 GWh/year accumulated in the second year by implementing this program in the 2.000 biggest factories plants.

Costs of this program concern the organisation of a qualifying network, which includes all the necessary educational material, from the follow-up, and support office to the report of successful cases. Such costs amount to R$ 6,2 millions. Other costs also involved concern the availability of R$ 400 millions to finance the implementation of said actions and of R$ 20 millions for the acquisition of the necessary instrumentation. The cost of saved energy is estimated in US$ 13,5 / MWh, considering 15 years of useful life span and an interest rate of 15% per annum.

This program foresees the organisation of a technically qualified staff involving Universities, Research Institutes, Equipment Manufacturers, Industry Unions, manufacturers and industry technicians aiming at minimising waste in motor systems including motors, ASD, starters, coupling, pumps, compressors, ventilators and exhausts. Figure 1 summarises the program's concept.

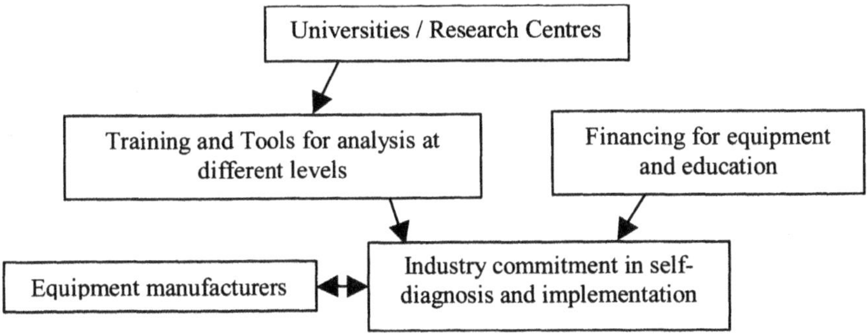

Fig. 1. Concept of the program for optimisation of industrial motor system

The structure of the program is based on four elements: a co-ordinating group, the program's bureau, the multipliers, and finally the agents and manufacturers. The co-ordinating group will be responsible for preparing the necessary material for gathering together a set of documents and computer programs. Such material will be the groundwork of the multipliers' training course. That group will also establish the market strategic approach, provide training, promote debate forums, and search partnership with other agents. The program's bureau will up-date and distribute the didactic material, do the following-up of actions and evaluate the impact of activities already carried out. The multipliers will either be highly experienced university teachers or consultants responsible for training the agents, for contacting local manufactures and for making diagnosis so that new successful cases may be elaborated. The agents will be factory engineers and managers who will train other engineers connected with the factory activities, identify obstacles to diagnosis making or to implementation of actions. Finally, manufacturers will provide the necessary information about the equipment and technical material and will co-operate in the speeding up of successful cases and in the diffusion of the program through their distribution channels.

Conclusion

This program denotes a significant enhancement in actions concerning energetic efficiency in motor driven systems in this country. Its innovative aspect is based on acting on minimising waste of energy in motor systems as a whole, thus involving motor, motor drives, coupling, and production procedures, and also including systems already in use. Its integral setting-up involving training, analysis tools, diagnosis, financing and implementation fulfils the gap that prevented effective energy saving before.

Last but not least, as an indirect result of the program, a technical reinforcement in the engineering curriculum in this country is expected. Considering that most of its multipliers are university teachers, there is a propensity that acquired multidisciplinary knowledge and experiences stand out as a new university subject, or in new researches or in a larger integration among the different branches of engineering

Bibliography

[1] MINISTÉRIO DAS MINAS E ENERGIA, Balanço Energético Nacional - 2000, Secretaria de Energia, Brasília, DF, 2000

[2] MINISTÉRIO DAS MINAS E ENERGIA, FDTE, Balanço da Energia Útil – 1995, Secretaria de Energia – MME, Brasília, DF, 1995

[3] SCHEIHING, P., ROSEMBERG, M., WEIL,S., Evaluating the Impact of a Government-Sponsored Motor-Driven System Efficiency: U.S. Experience, Energy Efficiency Improvements in Electric Motors and Drives. Berlin: Springer, 1999. p.395-414.

Assessment of Utilizing Energy Efficient Motor Drive Systems in the Egyptian Industrial and Agriculture Sectors

Mohab M. Hallouda

Electrical Power and Machines Department, Cairo University, Egypt.
E-Mail: hallouda@yahoo.com

1 Introduction

Replacing standard motors with efficient motors in new installations[1,2], using variable speed drives [3], and introducing power saving equipment, such as optimized efficiency soft starters [4,5], are leading technologies that are recommended to save energy and in many cases improve the performance of the drive system. One main disadvantage of these technologies is the initial higher cost of the new technology compared to the exclusive use of standard motors[1, 2]. This study addresses economical , environmental and performance issues related to the application of energy efficient motor drive systems in the Egyptian industrial and agriculture sectors. Such issues include energy savings, harmonic, cost analysis, custom duties and greenhouse gas reductions.

This paper presents the electric energy profile in Egypt with emphasis on industrial and agriculture sectors, the share of electric motors in energy consumption, the Egyptian market of motor drive system. This is followed by an analysis of two energy saving technologies based on cost, operating hours, efficiencies and electric energy tariffs. Recommendations for suitable applications, policies and type of technology are offered for the systems under consideration.

2 Energy Profile in Egypt

Electric energy consumption in Egypt has increased dramatically over the past three decades to satisfy the industrial, residential, agriculture and domestic needs. In 00/01, the Egyptian Electricity Holding Company (**EEHC**) generated 77,956 GWh of electrical energy, of that amount, 13,697 GWh (about 18%) is produced from hydropower plants. The rest is produced from thermal power plants, consuming more than 14,3 Mil.ton oil equivalent (MTOE). The installed capacity is 15,935 MW and the peak-recorded demand was 12,376 MW in 00/01. This is higher than 99/00 recorded peak demand by 5.5 %. Table 1 summarizes the electric energy production and consumption over the past 5 years.

Table 1. Electric energy capacity production and consumption (1997-2001)

	96/97	97/98	98/99	99/00	**00/01**
Installed Capacity (MW)	13303	13303	13935	14,582	**15,223**
Gross Production (MkWh)	57700	62336	68000	73,310	**77,956**
Hydro	12,000	12,222	15,300	14,659	**13,697**
Thermal	45700	50114	52700	58,628	**64,006**
Total Consum. (MkWh)**	49336	52977	56600	60,868	**64,330**

Primary energy production is based on three primary sources which are Oil with 2021 1000mt contributing 8.2%, Natural Gases with 14114 1000mt contributing 86% and Hydro energy with 15300 MkWh contributes 5.7%, finally Coal is a fourth minor participant with only 100 1000mt contributes 0.1 %.

In 200/2001 the Industrial sector electrical energy consumption reached 24,282 GWh, and the residential consumption 23,511 GWh, each comprising 37% of the total annual electrical energy consumption, other main consumption sectors are agriculture, public utility and governmental with 4, 11 and 5% consumption as shown in Figure 1. Table 2 shows the electric consumption distribution in industry based on the different industrial categories.

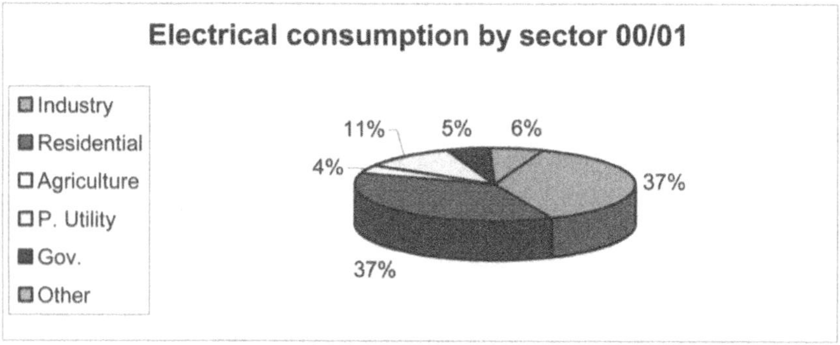

Fig. 1. Distribution of electric energy by sector

Table 2. Percentage of electric energy consumption in industry by type of industry

Sector	**Chemical**	**Metal**	**Textile**	**Food**	**Engineering.**	**Other**
% cons.	**30%**	**30%**	**20%**	**11%**	**5%**	**6%**

3 Electric Motors in the Egyptian market

Electric motors are the largest single piece of equipment that consumes electric energy in the Egyptian electric network; motors consume 60% of the industrial load, 23% of the residential load, 80% of the agriculture load and 50% of the public utility load [6]. This amounts to about 25500 GWh annually, which represents 39% of the total energy consumed, this figure does not include motors used in power stations which is about 1460 GWh.

The motors in the range up to 25 hp in the Egyptian market are dominated by local manufacturers. Because of the heavy custom duty protection, foreign companies focus on motors in the range above 25 hp. Local manufactured motors operate with efficiency within the international range up-to 15 hp then there is a difference of 2-4 % in favor of the European manufactured products.

Energy efficient motors are sold on a small scale due to the additional 15 to 25% increase in cost. Soft starters are also being introduced to the market basically to replace star delta starters, but not yet for energy saving intentions.

Electronic speed controllers often replace in-efficient throttling controls, in many other cases they replace two speed motors and result in higher efficiency and improved performance and comfort. The prices of electronic speed controllers are relatively high, at low range (<50 KW) cost is around 1000 LE per KW, and at medium range (50 KW to 200 KW) at 500 LE/KW and at high rang (>200 KW) at 300 LE/KW. The total annual market size of Electronic speed controllers is around 10,000 units, of which 80% are in the low range.

Egypt is an agriculture country with extensive activity based on direct irrigation it also has a large population which requires extensive water handling and treatment. Since the main and almost only source of water in Egypt is the river Nile, extensive pumping activity to handle and fully utilize the Nile water is considered necessary. The pumping energy consumed in agriculture pump stations and water handling government utility is about 6-7 % of the total energy consumed (5170 GWh). The efficiency of the pumping systems are relatively poor because of over sizing, poor energy management and limited variable speed drive applications.

Table 3 presents estimates of annual demand for 3-phase induction motors. Basically almost all the machinery used in the Egyptian industry is imported with built-in motors. Thus motors incorporated in original equipment manufacturers (OEM) machinery are the predominant motors installed in industry.

Table 3. Annual demand for polyphase AC motors (1997)

Range (kW)	Total Incl. OEM equipment	Non -OEM
> 0.55 - 7.5	200,000	50,000
> 7.5 - 37	30,000	7,500
> 37 - 75	8,000	2,000
>75	2,000	700

Table 4 summarizes electric motor production in 97/98 by public and private companies, and the top 6 categories of electric motors imports in 1998, the table is categorized according to the Central agency for public mobilization and statistics (CAPMAS) code.

Table 4. Electric motors production for public, private and investment sectors, and motor import in 1998 (value in 1000 LE)

Motors Type	Sector	Local 97/98		Item	Imports -1998	
		Qty.	Value		Qty	value
Single phase (1/3 HP)	Public	89453	7156	motors < 37.5 W	595728	10547
	Private	264785	34422	1-ph 92 to 552 W	251602	20198
	Total	354238	41578	1-ph > 552W	2124959	101640
Assembled	Public	19	8	3-ph 750W to 18.4 KW	37160	33756
1,000 RPM	Public	2255	902	3-ph 18.5 to 75 kW	10661	16629
3,000 RPM	Public	62	25	3-ph > 75 kW	2128	19244
3- phase	Private	1531	5605	Other types	223110	18500
Fan motors	Public	246318	52			
	Private	247849	16018			
	Total		16070			
1500 RPM	Public	5346	2138			

4 Assessment of Energy Efficient Technologies in Motor Drives for the Egyptian Market

Installation of efficient motor to replace standard motors, using power electronic converters to reshape the voltage and current wave forms are two of the main techniques offered in the Egyptian market for energy savings in motor driven systems, in the following sections an investigation of the two techniques for implementation in the Egyptian market is presented.

4.1 Efficient motors

The application of efficient motors in the Egyptian market is limited for two basic reasons, first, the lack of awareness concerning efficient motors, second, the price of efficient motors when compared to standard motors is 5 to 30% higher depending on the rating [1]. Regarding the first impediment, several activities have taken place over the past few years through governmental and international agencies to increase the awareness regarding the potential savings that can be achieved by using efficient motors [6], nevertheless the price of efficient motors hinders the success of its marketing.

In many cases government regulations are the driving vehicle for encouraging the use of efficient equipment [6]. As the price is a main obstacle to encourage the use of efficient motors, an effort to reduce custom duties on efficient motors is under way, the following presents the benefits of custom duty reduction on the price of efficient motors in the Egyptian market.

Since most local manufacturing is for motors below 18 kW, the custom duty for motors in this range is 30%, for higher ratings or special types it is 5%. Thus the custom duty reduction study focus on motors below 18 kW power, where reasonable reductions can be attractive[6]. Figure 2 and 3 shows motor imports below 18 kW in terms of value and number of units respectively over the years 96 to 2000.

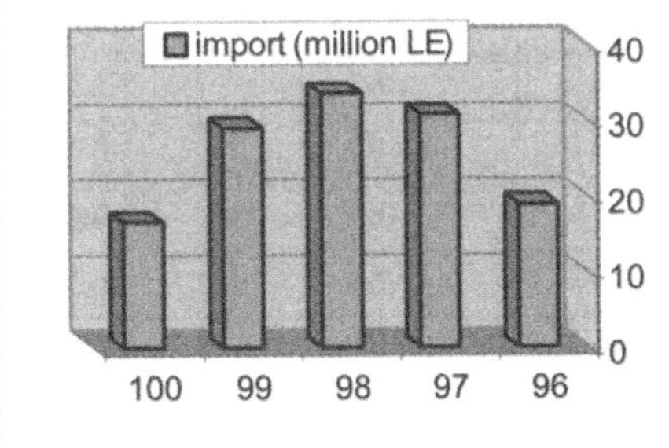

Fig. 2. Motor imports in value (L.E)

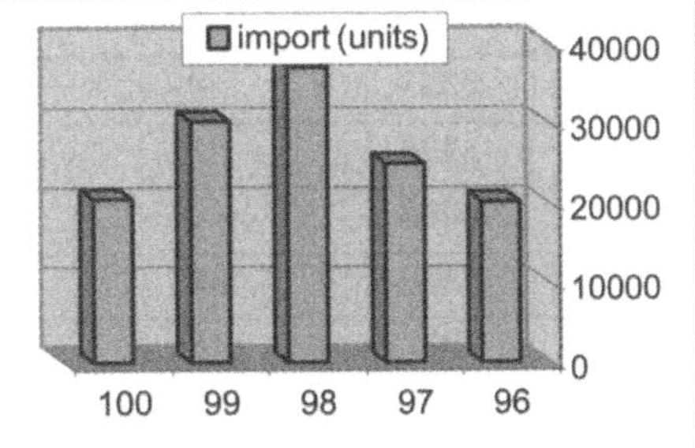

Fig. 3. No. of Units of motor imports

Analysis is performed for different motor ratings, and 8 hours daily operation. First the cost of each motor is calculated based on, the cost in dollars, the exchange rate per dollar, custom duties (30%), taxes (16%), and profit (20%), the cost of motor is:

$$\text{Cost of motor} = (\text{cost in } \$ \times 4)(1.0 + \text{custom duty})(1.6)(1.2) \qquad (1)$$

The energy cost/year (EC/Y), annual energy savings, payback period, and life cycle savings are calculated as follows.

$$\text{EC/Y} = \frac{0.746 \times \text{hp}}{\eta} \times \text{cost/KWh} \times \text{hours} \times \text{days/year} \qquad (2)$$

Where η is the efficiency of the motor from standard catalogs

$$\text{Annual savings} = \text{EC/Y efficiency - EC/Y standard motor} \qquad (3)$$

$$\text{Cost difference} = \text{cost of efficient motor - cost of standard motor} \qquad (4)$$

$$\text{Simple payback period (yrs)} = \frac{\text{cost difference}}{\text{annual energy savings}} \qquad (5)$$

$$\text{Life cycle savings} = \text{motor life} \times \text{annual energy savings} \qquad (6)$$

In table 5 standard and efficient motors having different ratings and 30% custom duty are analyzed, results of simple payback period (SPBP) ranges between 3 and 8 years, this relatively long payback period is due to the large initial cost difference. The SPBP in table 5 are not feasible for most customers. The analysis is repeated for the same motor types with the same assumptions (cost, efficiency, operating hours, electric tariff), but the custom duty is varied between 30% and 5%. Table 6 summarizes the results for SPBP obtained in each case. It can be concluded from the analysis that reducing the custom duties of efficient motors to 10% will have positive impact on the market to encourage industrial and agriculture customers to buy and install efficient motors. A different approach to consider could be to increase the custom duty rates of standard motors to 45 to 50%, while keeping the efficient motors rate at 30%.

Table 5. SPBP for efficient motors based on 8 hours operation, 0.153 LE/kWh

HP	Standard motor		Energy efficient motor		SPBP (Y)
	Cost ($)	Av. F.L Eff %	Cost ($)	Av. F.L Eff %	
3	235	80	264	86	2.92
5	329	82	431	87	7.67
7.5	408	84	538	88	8.44
10	516	85	650	89	6.67
15	677	86	864	90	6.35

Table 6. SPBP for efficient motors based on 0.153 LE/kwh, and different custom duty rates

	SPBP (y) for efficient motors different custom duties				
Hp\cd	5%	10%	15%	20%	30%
3	-ve	-ve	-ve	0.87	2.92
5	1.43	2.68	3.9	5.17	7.67
7.5	1.72	3.07	4.41	5.75	8.44
10	0.45	1.69	2.94	4.18	6.67
15	0.71	1.84	2.97	4.1	6.35

If all standard motors in the range under consideration are replaced with efficient motors the annual energy consumption savings based on average hp and average number of units will be approximately 23 million kWh of generation, this also results in a 20 MW reduction in demand power.

These amounts of energy saved corresponding to fuel savings (unburned at the power station) in the range of 5267 ton of oil equivalent fuel. The corresponding amounts of mitigated pollutants are in the range of 15801 ton of CO_2, 351 ton of SO_2, 19 ton of Nox, and 2.1 ton of CO.

4.2 Motor wave shaping

Electronic wave shaping is one of the recently offered technologies for improving motor efficiency. The commonly used soft starter with three pairs of anti-parallel SCR's has been modified for voltage and current wave shaping to operate at optimum efficiency [4, 5, 7]. This converter is useful when the motor is partially loaded, a case very common in Egypt due to motor over sizing, especially in pumps within the industry and agriculture sectors. To asses the performance of the converters a study is conduct on the three-phase types with 5 and 7.5 hp induction motors, to study energy consumption, savings, harmonics, power factor, and temperature rise. Figure 4 shows the laboratory set-up and sample waveform results.

The loss minimization is based on the principal that a light loaded induction motor operates at its highest efficiency when the flux (voltage) is reduced to achieve optimal trade off between copper and iron losses. This is achieved by blocking the voltage during a specified period synchronized with the zero crossing of the current as shown in figure 4.

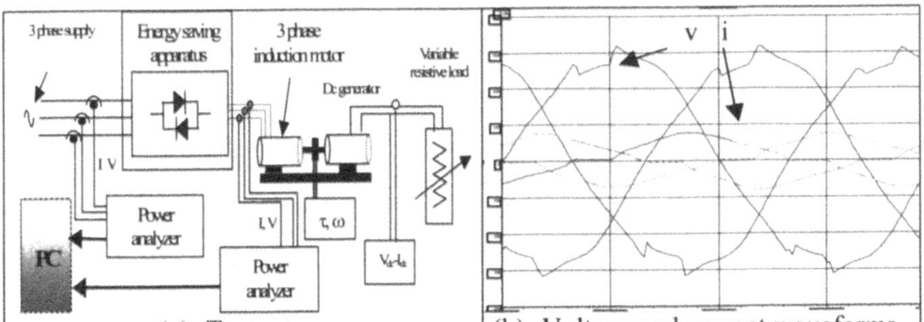

(a)- Test set-up (b)- Voltage and current waveforms

Fig. 4. Test set-up and sample waveforms of energy optimizing soft-starter

The normalized output power P_{nor}, the efficiency gain $\Delta\eta$ and the power saving P_{sav} are:

$$P_{nor} = \frac{\tau \times \omega}{P_{m,n}} \tag{7}$$

$$\Delta\eta = \eta_{sw} - \eta_{swo} \tag{8}$$

$$P_{sav} = \frac{P_{sw} - P_{swo}}{P_{swo}} \times 100\% \tag{9}$$

Where, τ is the torque in N.m, ω is the angular speed in rad/sec, $P_{m,n}$ is the rated motor power, η_{sw}, P_{sw} is the efficiency and input power of the system with the energy saving converter (soft-starter), η_{swo}, P_{swo} is the efficiency and input power of the system without the converter.

The results of the energy and performance analysis are summarized in figure 5, which shows the motor current, efficiencies, efficiency gains, percentage power savings, power factor and voltage and current total harmonic distortion for the system with and without the soft-starter.

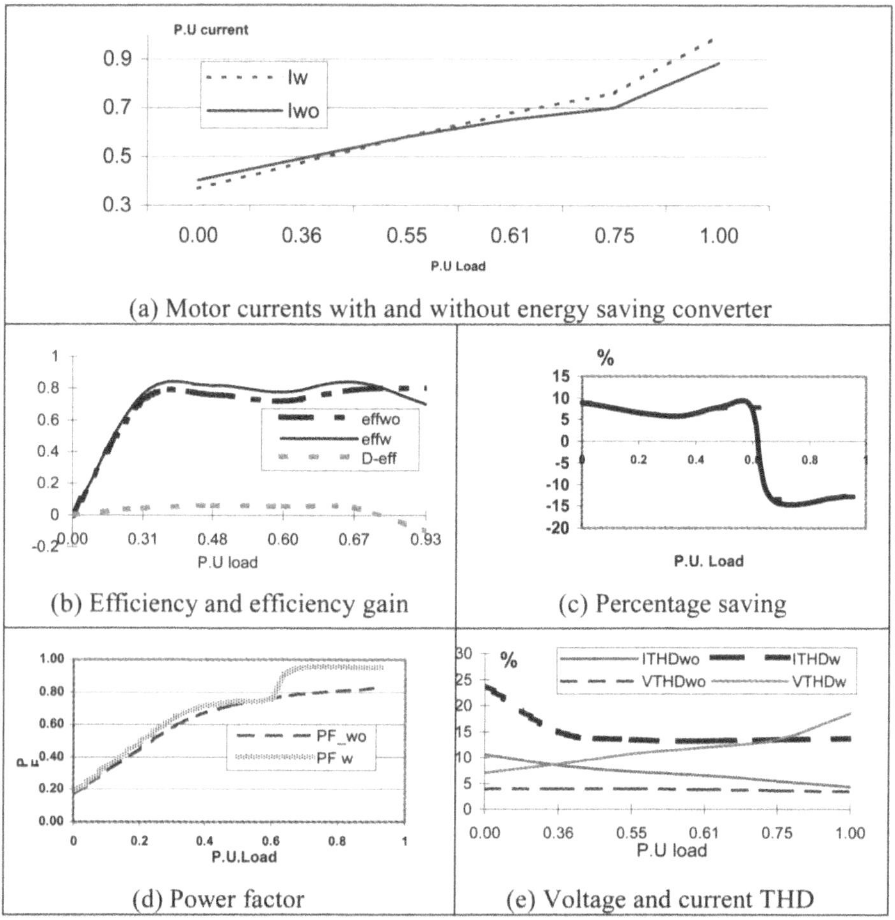

Fig. 5. Experimental results of the 3-phase energy optimizing soft-starter

The results in figure 5 (a), (b) and (c) show that the implementation of 3-phase energy saving soft-starter is feasible only when the motor is loaded below 60% for most of its operating cycle. Savings are around 10% from no load up to 60% loading, at full load the motor is not saving any energy, in fact it is consuming more energy and current than without the energy saving device. Power factor in figure 5(d) is slightly improved.

As the converter reduces the current to optimize the efficiency, harmonics especially the fifth and seventh become quit high. As show in figure 6(e), Voltage total harmonic distortion increase with loading and reach 18% at full load. While the current total harmonic distortion is high at no load and almost constant around 15.5% at normal loading. The voltage and current harmonics may cause problems for the motor drive system and other apparatus in the vicinity. Thus it is recommended to bypass the converter when motor is close to full load operation. Regarding temperature rise, no significant variation was observed between the system operating with and without the converter, this is assumed to be due to the balance between increase in harmonics verses decrease in current. Regarding speed drop, also no significant variation was observed, a slight drop in speed in some cases occurred.

5 Conclusion

The paper investigates the promotion of Efficient motors through changing governmental regulations and implementing soft starters to save energy for light loaded systems, especially pumps.

The main recommendation for efficient motors is to reduce custom duties to 10% on efficient motors while maintaining the custom duty rate of standard motors as it is, this will have a significant practical impact in encouraging the import of efficient motors. An alternative approach would be to keep the custom duty rates of efficient motors as it is at 30% and increase custom duties for standard motors to 48%.

For energy saving soft-starters, only applications with light loads are recommended with caution to harmonic effect and power standards. Bypass contactors are recommended when using such converters above three quarters load.

6 References

[1] J. Haataja, J. Pyrhoneon, "Improving 3-phase Induction Motors Efficiency in Europe, The Challenge for Manufactures ",IEE EMD97 conf., 1997. pp-190-194.

[2] S. Nadel, M. Shepard, S. Greenberg, G. Katz and A. De-Almeida "Energy efficiency Motors syst., HB on Technology, program and policy opportunities ",.ACEEE, 1992.

[3] H. N.Hickok, "Adjustable Speed- A tool for Saving Energy in Pumps, Fans, Blowers and Compressors" ,IEEE Trans. IA, vol 21, pp 124-136, 1985.

[4] F. Blaabjerg, J. Pederson, S. Rise, H. Hansen and A. Trzynadlowski ,"Can Soft-Starters Help Save Energy?", IEEE IA mag.. Sep/Oct, 1997. pp-56-66.

[5] N. mohan "Improvement in Energy efficiency of Induction Motors by means of Voltage Control", IEEE Trans. Power appar. And syst., vol-99, no 4, pp. 1466-1471, 1980.

[6] Egyptian Energy Service Consultancy Office, "Preliminary assessment for Energy Efficient Equipment Custom Duty Reduction in Egypt", Energy efficiency and Greenhouse Gas Reduction (EEIGGR) project, GEF, UNDP, September, 2001.

[7] Mahmoud Hakim, Mohab M. Hallouda, "Performance Evaluation of Power Planner Apparatus for Energy Savings", Energy efficiency and Greenhouse Gas Reduction (EEIGGR) project, GEF, UNDP, May, 2002.

European Policies and Programmes to Improve Energy Efficiency of Motor Systems

Paolo Bertoldi

European Commission DG JRC

1. Abstract

Electric motor driven systems account for the greatest part of industrial electricity consumption in Europe. Numerous studies carried out for the European Commission on individual component (motors, pumps, compressors) and on the consumption characterisation have shown the considerable potential for improvement of energy efficiency of these systems, and have recommended suitable policy actions.

A number of policies have succeeded in making improvements on the supply side for individual components of systems, such as the electric motor itself. For electric motor a EU wide classification scheme and labelling exist together with a voluntary agreement by motor manufacturers to substantially improve the efficiency of motor placed on the market and a comprehensive database EuroDEEM.

Following the recommendation of experts, the European Commission decided to that a concerted effort on the demand side could very usefully complement the efforts being spent on components and technologies. The European Commission has decided to launch "The Motor Challenge Programme" (MCP). This is a voluntary programme for motor systems users to agree to look at their system and to carry out within a specific time frame all the savings measure that are economic.

The essential elements of this new programme are:

- to raise awareness among industrial and service sector users of motor systems about the potential for energy saving (and money saving) measures;
- to create a European wide framework to encourage top level decision makers to make the implementation of these energy savings measures a management priority;
- to make available a wide range of information tools based on the EuroDEEM database, to aid users in optimally designing, purchasing, installing and operating motor driven systems;
- to get clear commitment by the company top management to carry out efficiency measures.

The publicity aspect of the programme would be used to convince top management of the usefulness of subscribing to the "MCP Guidelines", in some ways similar to the existing EU “GreenLight” Programme commitment. Because of the very wide variety of situations, this commitment would have to be open ended and flexible, a sort of "variable geometry" system, where each company, with aid from the Commission, would target those areas and measures most likely to be effective in its operations. The company will commit to carrying out these measures, and reporting on the results, within an agreed upon time period.

2. Introduction

Electric motor driven systems account for 69% of industrial electricity consumption in Europe. The base year for the characterisation of motor electricity use is 1998. Motor electricity consumption in the industrial and in the tertiary sectors in the EU in 1998 was responsible for 69% and 38% of the total electricity consumption, accounting for 590TWh and 190TWh respectively.

Figure 1 shows the share of motor electricity consumption by end use in the industrial and in the tertiary sector in the EU.

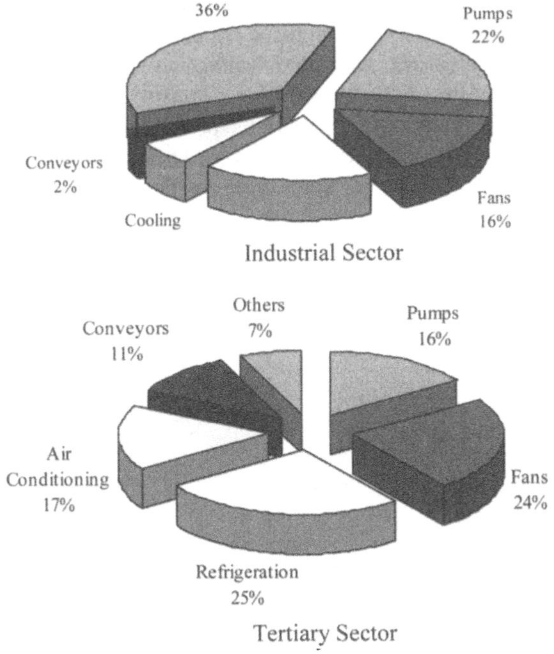

Fig. 1. Share of motor electricity consumption by end-use

Numerous studies and projects have shown the considerable potential for improvement of energy efficiency of these systems. The electricity savings potential are estimated for the year 2015. The annual average growth rates of the electricity consumption up to 2015, in the industrial and in the tertiary sector is assumed to be 1.2% and 1% respectively.

For the estimation of the motor electricity and carbon savings potential, the efficiency improvements considered, are the application of Energy-Efficient Motors (EEMs), Variable Speed Drives (VSDs), and energy efficient end-use devices and sytems (pumps, fans and air compressor systems). The total technical and economic electricity and CO_2 savings potential in Industry and in the Tertiary sector in 2015 are presented in Table 1:

Table 1. Total final technical and economic electricity and CO_2 savings potential in Industry and in the Tertiary sector by 2015

		TWh Savings by 2015	CO_2 Mtons Savings
Technical	Industry	107,1	42,9
	Tertiary	36,7	14,7
Economic	Industry	84,3	33,7
	Tertiary	24,5	9,8

The above estimated savings potential would be higher if other efficiency improvements would be included. The application of low cost efficiency measures which do not require sophisticated technology, such as improving maintenance practices, reducing waste, switching off the equipment when it is not being used (for example for the case of belt conveyors), would lead to large savings. These "House Keeping Measures", deserve to be strongly publicised among motor users. Losses in the power transmission system are often surprisingly large, but carefully selection and maintenance of drive-trains and their components are crucial for improving energy efficiency.

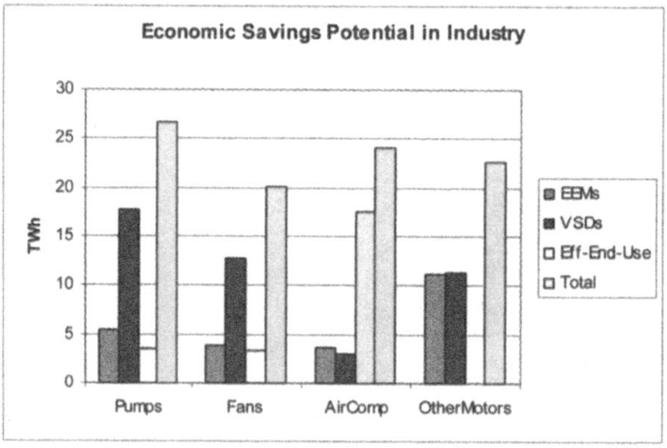

Fig. 2. Economic Savings potential in Industry by 2015

A number of EU policies and programmes have succeeded in making improvements on the supply side for individual components of systems, such as the electric motor itself.

3. Electric Motors

For electric motors a voluntary agreement was concluded in 1997 with motor manufacturers to substantially improve the efficiency of motor placed on the market and introduce a European classification scheme and labelling. The voluntary agreement has been signed by 35 companies. The target is to reduce the market penetration of EFF3 4-pole-motors in their joint sales by 30 percent by 2001 and by 50 percent by 2003. The target for of EFF3 2-pole-motors is a reduction in their joint sales of 50 percent by 2003. For 4-pole-motors the 30 percent reduction target has been achieved and the 50 percent reduction target will be achieved till end of 2003.

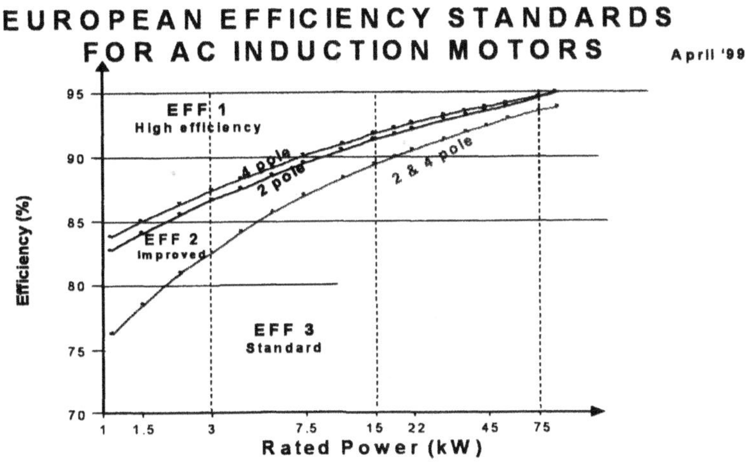

Fig. 3. The EU Motor classification scheme

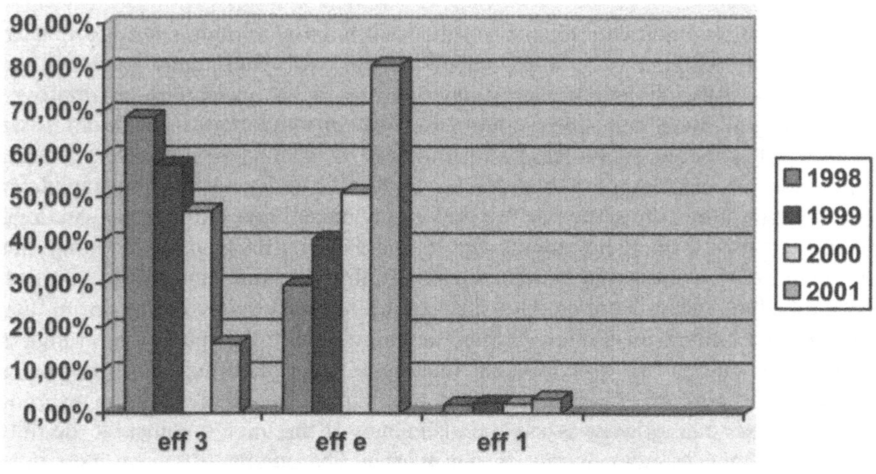

Fig. 4. Results of the EU/CEMEP Motor agreement

4. EuroDEEM

The European Commission has developed a European database for motor system, called EuroDEEM. This activity started in 1995 with the design of a tool for the promotion and selection of Energy Efficient Electric Motors (EEM). The scope of the database containing electric motor data was to make available an important information tool that allows users to easily carry out an evaluation of the best installation or replacement options, therefore helping the promotion of electricity efficiency. The EuroDEEM software package will permit users to select the most suitable electric motor for their purposes, evaluating energy and financial savings. EuroDEEM has been created to be a complete tool for very wide promotion and dissemination of information about energy efficiency in motor systems to a large range of end-users

EuroDEEM included in 1996 the Motor System Inventory DataBase for keeping track of all motor systems and electricity consumption in a Company. Utility data and tariff.

In 1998 it was decided to expand the database to other important motor system components such as Variable Speed Drives (VSDs), pumps, compressors, fans and other transmission and control devices. The first version of EuroDEEM containing only the motor selector database was completed and realised in 1998 with about 3000 motor models available on the EU market. The motor data are loaded directly from motor manufacturers.

Development activities for the pump and VSD module have started in 1999. A first Demo version of the pump module is available. In year 2000 a motor system audit procedures has been developed and it has integrated in EuroDEEM.

5. Pumps

The Commission attention has been focused initially on pumps for clean water. Following a comprehensive study and discussion with Europump experts it was agreed that although larger energy savings are to be made through improved system design, better selection of pumps is also very important In addition, it was agreed that labelling of pumps is not practical or useful to purchasers. Instead, information should be given to show the range of efficiencies of pumps available for different duties. In particular by drawing a "mean" line through the catalogue BEPs of pumps from many manufacturers, purchasers can see how the pump they are being offered compares to others available for the same duty. This is in itself very valuable. Manufacturers have also found it a useful way of comparing their products with others available. Through negotiation with Europump for a range of clean water pumps two procurement lines have been identified and a method to calculate the 'real' efficiency at any duty point has been agreed. With the resulting plots any user can quickly assess the efficiency of the quoted pumps at the duty point in relation to other pumps on the market. The pumps efficiency have been divided in three bands.

The Commission agrees that efforts should continue to be made by all the manufacturers to raise the general awareness of pump system design. The Europump/HI Life Cycle Cost Guide is particularly commended.

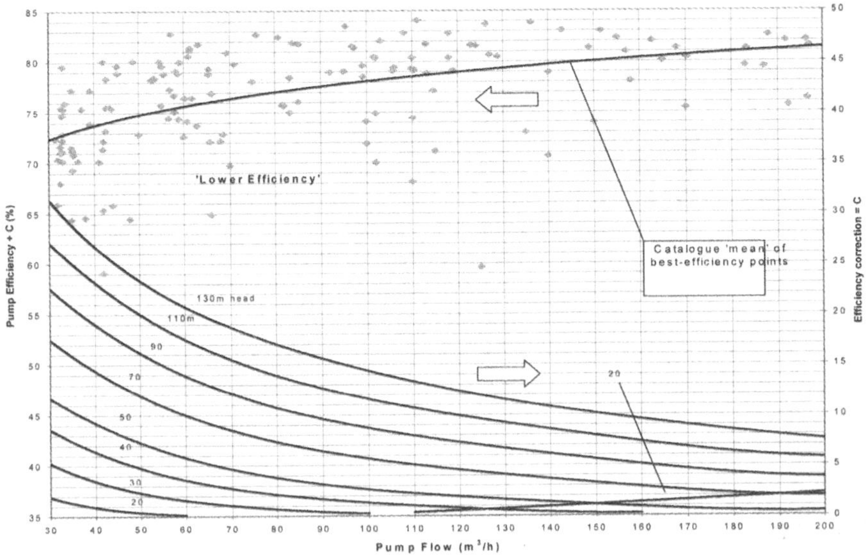

Fig. 5. Example of the pump procurement line

6. The Motor Challenge Programme

It appears that a concerted effort on the demand side could very usefully complement the efforts being spent on components and technologies. The essential thrust of such an effort would be to:

- raise awareness among industrial users of motor systems about the potential for energy saving (and money saving) measures;
- make available a wide range of information tools, to aid users in optimally designing, purchasing, installing and operating motor driven systems;
- create a European wide framework to encourage top level decision makers to make the implementation of these energy savings measures a management priority. This framework could adopt some of the successful elements of similar programmes, such as the European Union GreenLight programme, or the US DoE Compressed Air or Motor Challenges.

There are multiple reasons that explain why profitable (sometimes very profitable) energy savings measures are not put into practice in the private sector:

- Motor systems electricity consumption is "invisible" to top management, since it is most often a relatively small cost item for any one company.
- Electricity consumption in general, and motor systems consumption in particular, is usually treated as a general overhead item in company analytical accounting schemes. Thus reducing this cost item is not the responsibility of any particular manager.
- Measures to optimise the cost of equipment purchases, such as competitive bidding procedures, rarely take into account long term operating costs due to electricity consumption. Thus these cost cutting practices can be counterproductive in terms of reducing life cycle costs for electricity. This is particularly true since the optimal systems according to the electricity consumption criterion often require higher initial investment. Thus they are not even proposed by suppliers in competitive bidding procedures.
- Responsibility for potential optimisation measures is largely diffused among several management functions: Production, Maintenance, Purchasing, Finance. It is difficult to get high-level management agreement, cutting across departmental responsibilities, on a low priority item such as electricity consumption.

The “Electric Motor Driven Systems” considered in the MCP are the typical fluid handling application such as compressed air, pumping or ventilation. These applications have in common:

- an electric drive (consisting of a motor and perhaps an electronic motor controller) which converts electrical energy into mechanical energy in the form of a rotating shaft;
- a second conversion device (compressors, pumps or ventilators) which use the mechanical energy delivered by the drive to displace and/or compress a working fluid;
- a network through which the fluid circulates. (In compressed air systems, this network may terminate with an end use device which again transforms the mechanical energy in the air into some other type of service);
- some kind of control mechanism to adjust the output of the system to the needs of the application.

For the purposes of energy efficiency, it is essential to note that the overall efficiency of these fluid circulation applications depends of course on the efficiency of the drive and of the conversion device. It depends even more on the design and operation of the networks of which they are a part, and the inter-relationships between the components. For this reason, the MCP will mainly address the systems, and not only the individual components.

7. The Framework of the Motor Challenge Programme

The basic features of MCP are:

1. MCP is a voluntary programme managed by the Commission, as GreenLight;
2. Participating companies commit to continuous improvement of energy efficiency;
3. Top-management commits to energy efficiency;
4. MCP is open to all companies - especially SMEs;

The main responsibilities of MCP participants are:

- Compliance with all relevant environmental and energy legislation;
- Establish a management commitment to energy efficiency;
- Carry out selected cost-effective energy efficiency actions based on the initial energy review;

The first activity has been to establish the basic elements of MCP described in the MCP Guidelines. The purpose of the Guidelines is to define the nature of the commitment of companies, which choose to participate in MCP, and the requirements, which have to be fulfilled by participants. While the MCP must be sufficiently flexible to accommodate diverse situations, the general requirements of the approach must be sufficiently rigorous so that the commitment to the Guidelines has meaning. Thus, the commitment must contain clearly defined and verifiable actions, which the companies will carry out. These actions must be of such a nature, that they will lead to realising the bulk of profitable energy savings measures in plants and buildings of participating companies. The actions will include:

- inspection and audit procedures, to allow top level management to control energy consumption;
- integration into management procedures of those reporting and evaluation mechanisms necessary to verify dissemination of the MCP action within the company;
- a public Commitment, including internal communication of this Commitment;
- integration of energy consumption criteria into design and purchasing procedures (including, for instance, "Life Cycle Costing").
- Implementation of the identified and chosen cost-effective energy efficiency measures and investments.

The MCP is based on a number of "building blocks" covering: drives (motors, VSDs and transmission components), pumps, compressors, and fans.

The basic element of the building block will be to define the technical nature of an appropriate commitment for the specific type of end-use equipment. Because of the very heterogeneous nature of these systems, and the diversity of specific building and company installations, the "building block" must specify a general approach, consisting of verifiable actions, which when carried out lead to optimal system functioning.

The building block must define the notion of a "profitable energy savings measure". It is clear that this cannot be limited to lowering cost, but must include reliability and quality of service criteria. One important consideration is that the targeted types of systems (e.g.) are usually considered as "technical services" within a building or enterprise. Their failure, or a drop in quality of service, can have catastrophic results on the comfort and productivity of the building occupants. Thus, from a building management point of view, reliability and quality of service are the overriding criteria for judging the cost effectiveness of the service, rather than the actual cost of producing the service. For this reason, the MCP Guidelines will clearly state that profitable energy savings must maintain or improve reliability and quality.

8. The Motor Challenge Guidelines

The MCP Guideline will contain an overall framework for the "building blocks" already described. The framework will be modular, so as to permit the incorporation, in the future, of new building blocks (for instance on commercial refrigeration systems). The framework must be of an "à la carte" nature, that is to say that companies must be able to choose the elements or types of systems relevant to their operations or energy consumption. This in general means that a company will commit to examine those types of end-use devices, systems or buildings that are large energy users. Some companies might also choose a transversal approach focusing on different modules. Furthermore, the framework must be compatible with the range of approaches of the EU Member State Energy Efficiency programmes for commercial buildings, and the existing EU GreenLight programme.

In any case it is not be possible to specify quantitative requirements for energy savings (as is the case for the GreenLight programme), since the level of savings possible depends on the precise nature of each installation or building. Rather, the target for energy savings must be determined as a part of the audit process to which the company commits itself when signing on to MCP.

The Guidelines will define the process by which companies commit to MCP, define their specific company plan, carry out their plan and evaluate the results. Since the process is similar to other environmental and quality certification methods such as ISO 9000, ISO14000 and EMAS, care has been taken to use elements from these methods so as to simplify and reduce the cost of committing to the programmes.

The Guidelines will be accompanied by the following documents:

- awareness raising material to help top company deciders understand the purpose of the MCP Programme and the potential for energy savings. Special attention will be paid to the reliability and quality of service criteria;
- guidelines for the audit and implementation processes, including initial measurements and ex post evaluation procedures;
- lists of resources (co-operating equipment manufacturers, engineering consultants, software, documents and books, training material, list of possible financing mechanism, list of ESCOs operating in this field, etc.).

As it was experienced with the GreenLight Programme, initially there will an ongoing and permanent manner to improve the MCP Programme. Thus, it is to be expected that more than one working version of Guidelines will be issued during the course of the programme, leading to a consolidated version about one year after the launch of the programme. The first year will be used to test the overall MCP concept, and should contribute to improving and validating the MCP Guidelines.

9. The Motor Challenge procedures

Companies wishing to join MCP will proceed by the following 5 step process:

1. Carrying out the initial energy review (based on energy audit carried out by company staff or chosen contractor);
2. Formulating of an Action Plan, defining the scope and nature of the company's commitment (buildings and equipment covered);
3. Evaluation and approval of the Action Plan by the Commission;
4. Implementation of the Action Plan by the company, and annual reports to the Commission;
5. Renewal of Partner status by the Commission, upon review of the annual report.

9.1 Initial Energy Review

The MCP Guidelines and Modules (this contains for each building block the technical details, as audit procedures, list of energy saving option, and tips on how to save energy) will provide assistance to help companies in evaluating the potential benefits of joining MCP. Companies shall carry out an initial energy review to identify areas for energy efficiency improvements.

Where appropriate, companies may choose to be assisted by a third party, for instance an Energy Service Company (ESCO). The MCP website can provide lists of ESCOs who offer such a service. In some cases, National Energy Agencies may offer some type of technical assistance or financial support.

Alternatively, the company's staff may be qualified to carry out the initial energy review. An external audit and verification (as is the case for EMAS) is not a condition for submitting an application for Partner status.

All audits will include description and measurement of the initial state of the systems, so as to permit ex-post evaluation of the success of actions carried out. The audit recommendations will of course respect the "Reliability" and "Quality of service" clauses of the Guidelines.

9.2 Formulation of the Action Plan

The company wishing to participate in MCP formulates an Action Plan, defining scope and nature of its commitment and establishes an energy efficiency management system.

On the basis of the energy review in step 1, the company decides on an Action Plan. This plan must define the **scope** and **nature** of the company's commitment. The commitment will include a proposed reporting scheme.

The **scope** of the commitment defines the sites/buildings at which energy efficiency actions will be undertaken. The company will identify, within the chosen sites, the systems and devices to which the commitment applies. The commitment may cover all or some of the following systems (called “Modules” in MCP):

- lighting
- office equipment systems
- electric motor systems
- power and hot water generation systems
- HVAC systems (including heat-pumps to replace direct electric heating)
- Heat recovery systems
- Cooling systems (including water and food preservation systems)
- industrial cooking, ovens and laundry equipment
- Building insulation/shell and passive cooling

A Partner's commitment may be applicable to all European sites/buildings, to some sites/buildings, or may be limited to a single site/building[1].

- **Corporate Commitment**: a company will have “Corporate Partner” status if the commitment covers the major proportion of its energy consumption (i.e. more than 50% of consumption) at all of its European sites/buildings[2].

---

[1] According to the company choice of scope, the commitment will apply only to this specific site(s) or corporate level. The scope has to be clearly indicated in the proposed commitment. When a organisation joins MCP for a specific site (e.g. an enterprise joining only for plant XY in city ZZ), this has to be clearly indicated by the company in all the communication material relating to MCP, e.g. web-site, advertisement, etc. The Commission will also indicate the name of the site(s) in all its communication material e.g. web-site, catalogue, posters, etc. More sites/shops can be added to the company commitment and it is always possible to move from a site/shop partnership to a corporate partnership or vice-versa.

[2] This proportion will be applied in a flexible manner. A company may, for instance, choose to exclude a very large production system which represents specific engineering problems for which MCP tools are inappropriate. The partner should generally include subsidiaries in which the parent enterprise has a controlling share. The partner may choose to include or exclude other partially owned subsidiaries.

- **Site Commitment**: a company may choose one or several of its production sites or buildings. Within these sites, it may commit to treating one or several Modules.

The **nature** of the commitment defines the Action Plan that the company proposes to carry out. The Action Plan will include **specific actions,** which have been evaluated and decided upon (e.g. investments to upgrade or replace equipment, improved maintenance, etc.). The Modules provide guidance on typical actions that are recommended and/or expected by Partners.

In addition to these specific actions, Partners are required to establishes an **energy efficiency management scheme**, as described in the "Management Policies" Module. This Module proposes tools, which can help companies in making energy efficiency an element of management priorities, at every step of the life cycle of energy either by using equipment or systems: system design; choice of system components; installation of systems; ongoing operation and maintenance.

The reporting scheme should respect the recommendations and formats of the relevant "Module" documents, and be based on a suggested template. The Commission and National MCP Representatives (National Energy Agencies) may help potential Partners in defining the scope and nature of their commitment.

In the Action Plan, the company shall designate a person who would be responsible for ensuring the implementation of the plan and sending it to the European Commission. This person will:

- Verify that appropriate management tools and policies are created to implement the programme;
- Report to top-management on progress;
- Prepare the annual report to the Commission.

9.3 Acceptance of the Action Plan

Before the Commission grants Partner status to the applying company, it proceeds to the evaluation and approval of the Action Plan.

The proposed commitment (scope and nature of Action Plan, including reporting procedure, and the proposed management scheme) will be submitted to the Commission or MCP National Representatives (National Energy Agencies)) who will verify that the plan:

- Takes into account the essential recommendations of the relevant MCP Module documents;
- Retains the major part of technically and economically viable energy savings options identified by the initial energy review;
- Contains a satisfactory estimation of energy savings and overall reporting procedure.

The Commission will approve the Action Plan, otherwise it will explain to the applicant company its reasons for not approving.

Once the plan is approved, the Commission will grant MCP Partner status to the company, with all participation benefits.

9.4 Implementation of the Action Plan

The Partner carries out its Action Plan, implements the energy efficiency management scheme, and reports annually to the Commission on progress, in accordance with the reporting scheme proposed by the Partner in step 2. The Partner may be assisted in reporting by a third party, for instance an ESCO.

9.5 Commission Evaluates Annual Report and Renews Partner Status

The Commission will review the Partner's annual report, and upon approval, will renew Partner status. Fulfilment of Action Plan commitments may be verified. If the report is not approved, the Commission will explain the reasons to the applicant company. If the implementation of the company's Action Plan is notably weaker than agreed upon, or if the company does not honour its reporting commitments, the Commission reserves the right to terminate the company's participation in the MCP.

10. Benefits of Participating in the MCP

MCP Partners will receive advice and technical assistance from the Commission and from the MCP National Representatives (National Energy Agencies) in carrying out the initial energy review and in formulating and carrying out the MCP "Action Plan". Studies, carried out for Commission have demonstrated that there is a large number of cost-effective energy efficiency improvements available in commercial buildings and SMEs, for which the initial additional investments are repaid in a few years. By carrying out the Action Plan, companies will be able to:

- Maintain or improve both the reliability and the quality of service of the systems concerned;
- Realise the major portion of technically and economically feasible energy savings, therefore increasing the company's competitiveness. The plan must satisfy the criterion of economic efficiency, so that the allocation of scarce financial and human resources are justified from the viewpoint of the company;
- Contribute to EU's environmental objectives, in particular climate change

Furthermore, companies enhance their image in the marketplace by receiving public recognition for their efforts, through the MCP promotion campaign, aimed at raising public awareness of energy and environmental issues.

- Partners may use the MCP logo (draft version shown in figure 3). Partners may post signs, artwork of which is provided by the European Commission, publicising their energy saving actions and the contribution they are making to the environment.
- The list of MCP Partners, including a description of their specific contributions to the Programme will be publicised widely (brochure, internet, etc.). The Commission will send the list of Partners to Member State representatives in the SAVE Committee.
- Partners will be invited to provide documentation of a showcase installation that the European Commission can publicise, providing details on baseline energy use, what was done, and the resulting energy use after the upgrade.
- Company's relevant activities may be included in MCP promotional activities, such as Awards and the Catalogue.
- MCP Awards will be made to outstanding MCP Partnerships in several categories.

11. Conclusions

A very important market transformation has been achieved for motors. A new policy tools will soon be established for pumps. However the largest part of saving potential is in the systems approach and in the adoption of VSDs. To this end the Motor Challenge Programme directly aims to create the conditions for an energy efficiency commitment by top level management in industry. Experience in the American "Compressed Air Challenge" is that the original target of 15 to 20% energy savings will more than be met. It may reasonably be hoped that a broader scale European Programme would be equally successful.

The benefits of a successful Motor Challenge Programme would be very substantial. A conservative estimate would be 10% of industrial electricity use, i.e. about 70 TWh per year to be achieved after the five years life of the Programme.

12. References

[1] Bertoldi, P &. De Almeida A, "*Market transformation strategies for electric motors*". Proceeding of the ACEEE 1999 Summer Study on Energy Efficiency in Industry, .Washington, D.C.: American Council for an Energy-Efficient Economy

[2] De Almeida A, Bertoldi P & Leonhard W., "*Energy Efficiency Improvements in Electric Motors and Drives*" 1997, international conference proceedings, Lisbon November 1996, published by Springer Verlag, Heidelberg

[3] Bertoldi P, De Almeida A & Faulkner H., "*Energy Efficiency Improvements in Electric Motors and Drives*", 2000, international conference proceedings, London September 1999, published by Springer Verlag, Heidelberg

[4] University of Coimbra (Po), et al. 1996 *"Study for the Commission of the European Communities on Actions to Promote Energy Efficient Electric Motors in Europe"*, Final Report October 1996.

[5] University of Coimbra (Po), et al. 1999 *"Improving the penetration of Energy Efficient Motors and Drives"* Final Report May 1999.

[6] ETSU (UK) et al. 2000 *"Study on Improving the Energy Efficiency of Clean Water Pumps"*, Final Report December 2000.

[7] ADEME at al. 2000 *"Compressed Air Systems Market transformation Study"* Final Report December 2000.

[8] Fhi-ISI (GE) *"Market Study for Improving Energy Efficiency for Fans"*, Interim Report October 2000.

[9] Berrutto, V. & Bertoldi, P, "*European GrenLight Programme: a major initiative to reduce electricity consumption in non residential buildings*". Proceeding of the ACEEE 2000 Summer Study on Energy Efficiency in Buildings, .Washington, D.C.: American Council for an Energy-Efficient Economy

[10] The EuroDEEM database at: http://eurodeem.jrc.it

[11] The Motor Challenge programme at: http://energyefficiency.jrc.cec.eu.int/motorchallenge/index.htm

The Role of Electric Motors in Energy Planning in Slovenia

Tomaz Fatur

Jozef Stefan Institute – Energy Efficiency Centre, Jamova 39, SI – 1000 Ljubjana, Slovenia, Phone: +386 1 5885 278, Fax: + 386 1 5612 335

Energy Efficiency and Energy Planning

When discussing the role of energy efficient technologies we can not ignore main question how to evaluate these measures and how to implement and evaluate them within energy planning process. First question is how can we include several different options; on one hand we have to describe impact of energy efficiency regarding energy savings on the other hand we must integrate energy efficiency measures into wider energy planning framework. In this paper the role of energy efficiency measures for electric motors in industry have been presented as a result of integrated resource planning for energy demand and supply in Slovenia. The work has been done for Slovene Ministry responsible for energy and partially financed through EU funds.

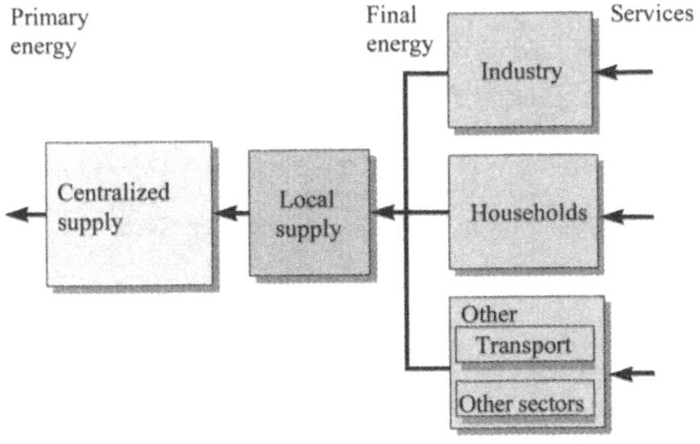

Fig. 1. Basic structure of reference energy system of Slovenia

We modeled Slovene energy model using model for integrated resource planning MESAP (Modular Energy System Analysis and Planning) with simulation module PlaNet (both developed by IER Stuttgart, Germany). Model can be used for long-term analysis of energy and environment for national, regional or local energy planning, and includes tools for demand analysis, DSM and simulation or optimization of energy supply. Model allows calculation of final energy demand; emissions and costs based on economic activities e.g. value added of Slovene economy.

Energy efficiency improvements in the field of industrial electric motors have been modeled as submodel in the reference energy model of Slovenia (Figure 1) where standard and improved technologies (high efficiency motors vs. standard motors and VSD standard "no VSD") have been introduced with two different efficiency rates and market penetration of technology. Two scenarios of economic growth have been used to describe reference (Minus) and intensive (Plus) development of economic growth. Three consistent strategies for each scenario have been then defined to describe level of intensity for implementation of energy efficient measures: reference, moderate and intensive. Besides, typically 0,5% efficiency improvements p.a. is expected in industry. Figure 2 describes major scenario and strategy framework for energy planning until year 2002.

BASE DATA			
Planning Horizon	1994 (base year), 2000, 2010, 2020		
Discount Rate	10 %		
Population Growth	0 %		
Imported Fuel Prices	Gas 11%, coal 4% *		
SCENARIO PARAMETERS			
	MINUS	PLUS	
GDP*	121 %	341 %	
VA and Physical Product	separated by activities	separated by activities	
Income Per Capita*	76 %	302 %	
Living Area*	11 %	18 %	
House Stock Number*	7 %	13 %	
STRATEGY PARAMETERS (for both scenarios)			
	Reference	Moderate	Intensive
Abbr.	MREF=Minus Ref PREF=Plus Ref	MMOD PMOD	MINT PINT
Financial support	0 %	5 %	10 %
Potential activation in industry	BaU	moderate	fast
Coal production		-20 %	-40 %
Energy prices growth in Slovenia	Ref. growth	Ref. Grow+10%	Ref. Grow+20%
CHP growth	7%	12%	16%

* Growth percentage in period 1994 in 2020.

Fig. 2. Data, scenario and strategy table

Modeling of Electric Motors

In this paper we focus only on industrial electric motors. Industry reference energy system connects services e.g. physical product and value added (on output side) to final energy (on input side). Vertical connections (Figure 3) describe commodities (services, energies) and include different activities, such as final energy, primary energy, emissions, costs, value added, whereas horizontal connections describe relations between these commodities and are presented as processes (electric motors, boilers etc.). Flows are presented as linear formulas.

Electric motors have been modeled as three groups of processes: motor electricity intensity, high efficiency (electric) motors (HEM) and variable speed drives technology (VSD). Electricity intensity describes needed mechanical energy to produce defined quantity of physical product in separate industrial sectors. Energy is distributed to motors according to market share of electric motors (AC motors rating below or over 100 kW, other motors). Energy efficient technologies include high efficiency motors and those drives, where VSD is applicable (turbo machines or 60% of AC motors population). Data for this submodel is shown in Figure 3.

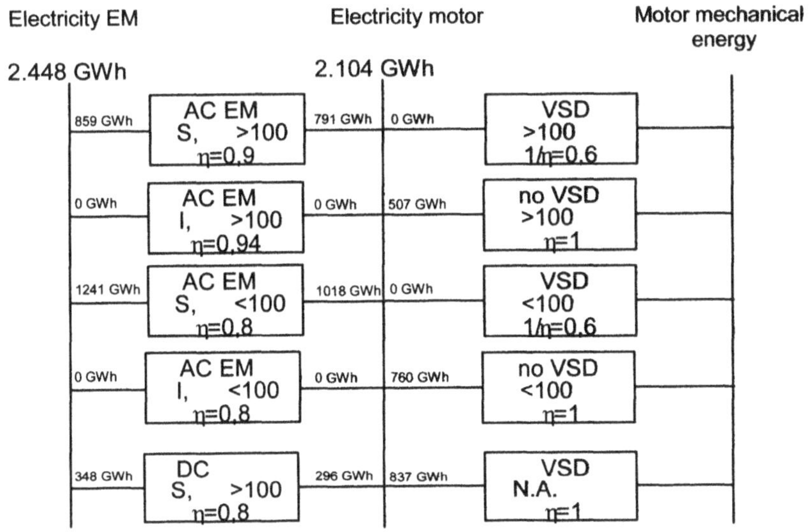

Fig. 3. Data in submodel Electric Motors, regarding total energy motor consumption (AC EM=asynchronous current electric motors, S=standard technology, I=improved technology, energy efficient technology)

Main features in the model can be grouped into four typical patterns:

- Submodel includes electric motors and VSD,
- Submodel uses different efficiencies for standard and improved technologies,
- Submodel uses time dependent market penetration proportions of individual technologies (standard vs. improved), and
- Submodel divides energy use for electric motors according to industrial classification.

Motor submodel is using a simple but effective transfer of market shares between standard and improved (energy efficient) technology (Figure 4). If investments into energy efficiency results in increased use of improved technology then market share of standard technology decreases in time and vice versa. An additional calculation has been performed to quantify these changes of market shares for pairs of motor technologies (AC EM < 100 kW, standard and improved) based on decision mechanism, in this case intern rate of return, electricity prices, difference in operational costs between standard and improved technology etc. Naturally, market share patterns are different for different scenarios and strategies.

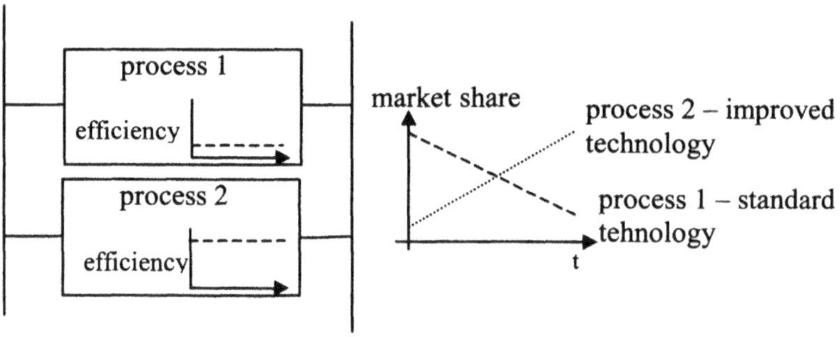

Fig. 4. Change of market shares of two processes

Impact Analysis of Efficient Motor Technologies

Based on the model structure, reference energy system, parameters of different scenarios (minus, plus) and strategies (reference, moderate and intensive) calculation of model flows can be performed for certain time periods. This allows us to get model estimation for energy demand forecast dependant on future economic developments and influencing circumstances. If we observe closely submodel for electric motors (within industry model) a good estimation of electricity demand trends for motors, based on interest for introducing energy efficiency measures in motors and drives in industry can be made. Also changing the characteristics of the model can influence energy demand for industry. Parallel to flow calculation, emissions and costs are calculated too.

Electricity demand for electric motors in industry will grow according to Scenario Plus from 2.448 GWh in 1994 to 4.618 GWh in 2020, representing 89% increase. Moderate strategy will bring 62% increase, and intensive strategy (intensive implementation of energy efficient measures) will increase energy demand for only 50%. In Scenario Minus electricity demand for electric motors will increase to 3.321 GWh, representing 36% increase, where moderate strategy brings 16% and intensive only 8% increase.

Based on these results costs of reduced energy can be calculated through investment costs into efficient motor technologies. And with known investment

DSM measures and other support mechanisms can be planned and observed. Using average production of electricity in Slovenia 0,5 kg CO_2 is emitted for each kWh used by end-user. Therefore electric motors in industry yearly indirectly produce around 1,2 mio t CO_2 and every saved kWh results in lower CO_2 emissions.

Since model allows changing influencing parameters, adaptation to different conditions can be performed, e.g. increased environmental standards. Model was so far only used to calculate CO_2 emissions based on proposed energy efficiency improvements, but also reverse procedure is possible. We can produce answers to questions like: What investments in electric motors do we need to reduce CO_2 emissions by certain percentage? A catalogue of possible improvements can be made using same principle for other technologies.

Campaign for Energy Efficient Motors in Slovenia – DSM Programme Proposal

With new Energy Law (October 1999), Article 67, DSM has become an option of promoting of energy saving options. The distribution network operator is obliged to perform DSM activities that can be financed through network fee (if confirmed by the Ministry responsible for energy). Practical application of the Article will be outlined by the end of this year. Energy Efficient Motor Programme is one of the options that show potential interest of interested parties. Work on promotion of High Efficient Motors (HEM) will be performed in the scope of 5. FP research programme, preparing Slovene industrial consumers to purchase HEM.

Analysis of energy efficiency possibilities is done in four steps:

- Programme outline needs pre-check, detailed economic analysis and preparation of DSM measure, suitable for chosen target groups
- Programme execution includes operational planning, pilot projects for technology assessment and market penetration activity,
- Programme monitoring assures that goals are achieved; as planned and in accordance with expectations,
- Programme evaluation must describe DSM Programme benefits regarding economic, environmental and social impacts.

The goal of DSM Programme for HEM is to increase number of HEM in Slovenia by 10% per year. Around 30.000 motors are sold within Slovenia each year and estimation of 10% of motor population equals 3.000 motors sold p.a. HEM will be defined on a "YES" list defining minimum efficiency levels of HEM. Major target group are old motors (some 10% of motor population is changed every year who need to be replaced by new motor) and new applications where new motors are compared to HEM with less pressure to reduce initial costs.

The tools and instruments in such DSM programme will vary according to the target group to which this programme is addressed. Promotional activities will be supported by already mentioned "YES" list will base on CEMEP and EU motor efficiency levels (also potential financial support). Main advantage of the list is transparent rules and compliance to EU agreements or standards. Financial support will be given to cover some 50% of increase costs of HEM – typically 20% higher costs of HEM will give consumer approximately 10% of total motor costs. System of subsidies in known in Slovenia for years when applying for subsidies for high efficient windows and insulation – application is simple and offers many end users to compete for the funds.

Such programme will be targeted to several target groups: end users, motor producers and resellers, engineering companies, OEMs, distribution companies (to actively participate in DSM programme), industry associations, and other actors. The programme would yield energy/emission savings of approximate 3% on average, resulting in more than 500 t CO_2 reduction per year.

Literature

[1] Tomsic Mihael, Urbancic Andreja, Damir Stanicic, Merse Stane, Fatur Tomaz, Jozef Stefan Institute, Integrated resource planning for the rational use of energy in Slovenia, 1996-2000.

[2] IER – MESAP-manual, version 3.1, 1998

[3] ETSU, Jozef Stefan Institute, Energy Conservation Strategy for Slovenia, 1997

[4] Fatur Tomaz, Solinc Hinko, Energy conservation with motors and drives, 1998

[5] Danish Motor Campaign, WWW source

[6] Stanicic Damir, Fatur Tomaz, Selan Boris, Jozef Stefan Institute, BOC Ireland, ETSU UK, ZRMK Slovenia, Demonstration projects for energy efficient investments in buildings and industry, 1997

The China Motor Systems Energy Conservation Program: Establishing the Foundation for Systems Energy Efficiency

Aimee McKane[1], Zou Guijin[2], Robert Williams[3], Steven Nadel[4], Vestal Tutterow[1]

[1] Lawrence Berkeley National Laboratory

[2] China Energy Conservation Investment Corporation

[3] United Nations Industrial Development Organization

[4] American Council for an Energy-Efficient Economy

Abstract

Industrial electric motor systems consume more than 600 billion kWh annually, accounting for more than 50% of China's electricity use. There are large opportunities to improve the efficiency of motor systems. Electric motors in China are approximately 2-4% less efficient on average than motors in the U.S. and Canada. Fans and pumps in China are approximately 3-5% less efficient than in developed countries. More optimized design, including appropriate sizing and use of speed control strategies, can reduce energy use by 20% or more in many motor-driven system applications. Unfortunately, few Chinese enterprises use or even know about these energy-saving practices. Opportunities for motor system improvements are probably greater in China than in the U.S. or Europe.

In response to this opportunity, China has established the *China Motor Systems Energy Conservation Program* in cooperation with the United Nations Industrial Development Organization and the US Department of Energy, and the Energy Foundation. This paper will describe the overall structure of the program. Elements include work to develop minimum efficiency standards for motors, a voluntary "green motor" labeling program for high-efficiency motors, efforts to develop and promote motor system management guidelines, and an intensive training, technical assistance and financing program to promote optimization of key motor systems in two pilot provinces, Shanghai and Jiangsu.

This year, an international team of experts is working with twenty Chinese experts in a series "train the trainer" sessions designed to assist the Chinese experts integrate a systems approach into their work in Chinese industry. We will focus on progress to date including plant assessment techniques and significant considerations in adapting market-based voluntary program techniques to the Chinese business environment.

Background

The China Motor System Energy Conservation Program is a direct outgrowth of activities undertaken by the China State Development and Planning Commission (SDPC) and the United States Department of Energy (USDOE) since 1997 as the result of a China-U.S. bilateral cooperation agreement. The Energy Conservation Law, put into place by the Chinese government in 1999, includes optimization of motor systems as a specific policy objective. The Chinese government has demonstrated a strong interest in improving industrial motor systems through its financial and in-kind support of workshops, industrial site visits, and other activities already completed under this agreement. Chinese officials view optimization of motor systems as one of the few activities that can yield near term benefits to Chinese industrial enterprises such as reduced operating costs and increased system reliability. These benefits are extremely important to industries trying to manage a difficult transition to open competition. Improved equipment design, more optimized system integration, and improved operations and maintenance practices can reduce motor system energy use by 20% or more, providing substantial energy and emissions savings (e.g., CO2 and SOx) while reducing factory operating costs and contributing to the economic viability of the factory. If 50% of electric motor systems in China are better optimized to achieve 20% average energy savings (a reasonable long-term goal), Chinese enterprises will save the equivalent of more than US$4 billion annually and carbon emissions will decline by more than 25 MMT annually.

Program Overview

The purpose of the China Motor System Energy Conservation Program is twofold: to develop a variety of education materials, analysis tools and standards for promoting motor system optimization in China and to put a local infrastructure in place in two provinces (Shanghai and Jiangsu) to promote greater efficiency in industrial motor driven systems. This is an important first step in a planned ten-year effort by the Chinese government to establish and train a network of motor system optimization experts throughout China, and to use these experts to assist individual factories to implement motor system improvement projects.

Primary funding for the pilot program is being provided by the United National Foundation (UNF) with substantial in-kind contributions from USDOE and SDPC. Additional funding is being made available from the Energy Foundation. The United Nations Industrial Development Organizations (UNIDO) is administering the program on behalf of UNF. The intent is to use lessons learned in the pilot phase of this project to develop an integrated program model that can serve as the foundation for a national program.Shanghai and Jiangsu were selected for the pilot program because they expressed strong interest, included a significant industrial base, have organizational support, and have a small cadre of energy professionals who can be trained on the technical specifics of motor system optimization.

Program objectives for the project period of 3 years, include the following activities:

1. Develop a series of educational materials and application tools (e.g., software) to assist motor system experts and factory engineers to assess and better optimize motor systems;
2. Revise existing Chinese national standards on the economic operation of motors, fans and pumps;
3. Train 10-16 motor system optimization experts who will work in Shanghai and Jiangsu;
4. Implement and evaluate a series of 8 to 12 motor system improvement demonstration projects in different industrial sectors in Shanghai and Jiangsu and prepare case studies of these projects;
5. Train ~400 Shanghai and Jiangsu factory managers and engineers and provide technical assessments to at least 32 Shanghai and Jiangsu factories to assist them to identify and undertake their own motor system optimization projects;
6. Evaluate the different project components and assess how the project can be improved; and
7. Assist the Chinese government to plan for and lay the groundwork for a major national motor system improvement program.

Program Elements

A primary subcontractor, the China Energy Conservation Investment Corporation (CECIC) was selected by UNIDO in August 2001 through a competitive bidding process. CECIC is responsible for overall management of program implementation activities, including contracting with local agencies in Jiangsu and Shanghai. These subcontracts were signed in October 2002. A substantial evaluation component is also included to determine the effectiveness of the pilot programs. Evaluation is the responsibility of a Project Management Office (PMO) established by the SDPC. Preliminary evaluation results are anticipated in early 2003. The pilots will be completed by late 2004.

UNIDO selected a Chief Technical Advisor (CTA) for the project in March 2001 to assist UNIDO in forming an International Team of Experts, selecting the Chinese experts to be trained, and to provide assistance to CECIC in developing an implementation framework for project management, reporting, and fiscal management.

An International Team of Experts was assembled by August 2001 to develop educational materials and application tools to assist motor system experts and factory engineers to assess and better optimize motor systems. This team is responsible for developing educational materials and application tools to assist Chinese motor system experts using their own material and information from existing motor system programs, most notably in the U.S. and the U.K. Topic areas will include: motors and drives, pumping systems, compressed air systems, and fan systems. Post-training technical support will also be provided for plant assessments, project development, and case study preparation.

The ultimate success of the program is dependent on the ability of the Chinese experts to understand and apply the systems approach in conducting plant assessments, develop projects, and train factory personnel on the benefits of undertaking additional projects using the systems approach. In addition, it is anticipated that some of these experts will form a core group that will be prepared to train other experts as part of a future national program. The pilot program has a fixed budget for 12 weeks of intensive training of the Chinese experts over a 10-month period by the International Team. For these reasons, selection of the Chinese experts was a critical element. CECIC, the Shanghai and Jiangsu Energy Conservation Centers, and the CTA finalized selection of 22 experts in a series of interviews with the candidates in December 2001, based on a set of criteria approved by the International Team. The selected experts are a mix of Energy Center staff and engineers from key industries in the two provinces. A group of 22 was selected to allow for attrition while still meeting the program target of 10-12 trained Chinese system optimization experts.

International team of experts

Program Administrator
Robert Williams, UNIDO

Chief Technical Advisor
Aimee McKane, Lawrence Berkeley National Laboratory

Motors & Drives Experts
Hugh Falkner, AEA Technology
Vestal Tutterow, Lawrence Berkeley National Laboratory

Pumping System Experts
Steven Bolles, Process Energy Solutions
Gunnar Hovstadius, ITT Fluid Technologies

Fan System Expert
Ronald Wroblewski, Productive Energy Solutions

Compressed Air System Experts
Wayne Perry, Kaeser Compressors
Tom Taranto, Pneumatech/ConservAIR

Chinese project management and implementation

Project Management Office
Dai Lin, ERI

China Energy Conservation Investment Corporation
Zou Guijin

Shanghai Energy Conservation Service Center
Ye Wen Baio

Jiangsu Energy Conservation Training Center
Li Yuqi

Program Activities

The pilot Program officially started with an inception meeting in Nanjing, China in December 2001. At this time, the training schedule through January 2003 was fleshed out in detail and the CTA provided financial management training. The period from January – March 2002 was used to develop and translate pump system training materials, select and order measurement equipment kits in consultation with the International Team, and identify factory sites for the first training session. The program focus through January 2003 is on building an infrastructure of trained system optimization experts in Shanghai and Jiangsu provinces. This is being accomplished by a series of three training sessions conducted by the International Team.

The training developed by the International Team for the Chinese experts combines classroom instruction, hands-on practice with measurement instruments, and onsite assessment of motor-driven systems at selected Chinese factories. Despite the considerable challenges resulting from differences in language and industrial practices as well as the physical difficulties inherent in conducting onsite training at factories, this model of instruction is beginning to produce results.

In April, the first train-the-trainer session on pumping systems was held in April 2002 in Shanghai. A meeting of the instructors, CTA, CECIC, and the senior members of each Energy Center was held immediately after completion of the training to review and discuss the students' written evaluations and instructors' observations. The "lessons learned" from these discussions were incorporated into the written report prepared after each training session and taken into consideration in preparing for the next training session.

In May 2002, training materials were prepared and factory sites selected for the motor/drive and fan system training (classroom and onsite training on measurement and plant assessment techniques). In addition, the International Team developed pumping system materials for use by the Chinese experts providing training to technical and managerial staff at factory sites.

In June 2002, the second train-the-trainer session on motors/drives and fan systems was held in Jiangsu province (Nanjing). The Pump System Experts returned at the end of the training to observe and assist the Chinese experts with measurement techniques and plant assessment reporting. At the request of the Chinese experts, additional training was given on project development and financing issues. The Pump Experts also observed the Chinese experts conducting their first factory training sessions on pumping systems and the benefits of implementing energy efficiency projects. Again, the experts training was followed by a discussion of lessons learned and preparation of a written report.

During July and August 2002, the Compressed Air Experts developed training materials and CECIC had them translated. Final preparations were made for measurement equipment purchases and factories for onsite training. In addition, the Motor/Drive and Fan System Experts began preparation of the materials for the Chinese experts to use in factory training. The third train-the-trainer session on compressed air systems is being held in Shanghai in October.

The Motor/Drive and Fan System Experts will be returning in November 2002 and the Compressed Air System Experts will be returning to Nanjing and Shanghai in January 2003. The purpose of these trips will be to observe factory training by Chinese experts and to provide additional technical assistance on measurement and plant assessment techniques. The International Team will also be available via email for several months to review plant assessment reports and case study descriptions.

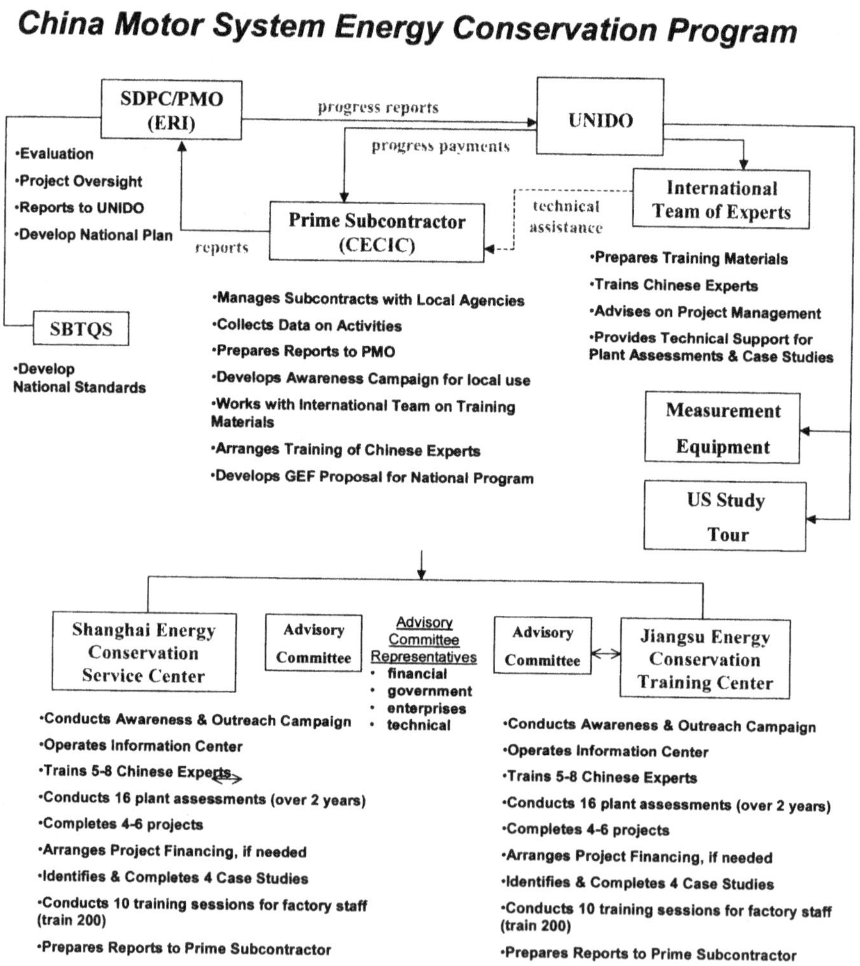

Fig. 1. Program elements

Lessons Learned

Cultural differences between Western and Chinese students surfaced during classroom training. Chinese students are accustomed to lecture-style instruction and require encouragement to interact with instructors and each other. Also, the Chinese experts are receiving training on all aspects of systems optimization for a pumps, fans, motor/drives, and compressed air. The next step for the Chinese experts will be to start thinking about areas of specialization, by system type and also by function. For example, in the US and the UK, some individuals become very skilled in the use of measurement instrumentation, while others become very skilled in determining what to measure and how to interpret the data collected. We expect that this will occur as individual strengths surface during the training. Intensive practice is essential to developing facility in the use of complex measurement equipment.

Another issue that surfaced is the need for standardization and quality control in plant assessments and report development. These could be accomplished by: standard report formats; plant wide general assessment approach (utility and facility information reporting); consistent use of standardized units for pressure and flow; standard spreadsheets for calculations; instrumentation procedures; and review of reports by senior members.

Finally, although the Chinese experts are making very good progress in developing system assessment skills, there is an ongoing need for technical assistance from the international experts as the Chinese experts begin to apply their system assessment skills in plants.

Next Steps

The Chinese experts have already begun conducting plant assessments and developing projects. In 2003, the primary program focus will be on pilot program objectives to: conduct 16 plant assessments in each province, complete 4-6 projects, develop case studies on these projects, and conduct training for 200 factory representatives in each province. A US study tour is also planned for mid-2003.

Development of national standards for the efficient operation of motors, fans, and pumps has begun and will be completed with the cooperation of the State Bureau of Quality and Technical Supervision in 2003.

An interim program evaluation will be completed during the first quarter of 2003 and the results from this evaluation will be used as an input to improving the pilot and planning a national program. UNIDO, USDOE, and the Chinese government will work to secure sources of international funds for a national program.

The objectives for the national program are to: continue working on refining the skills of the Chinese experts who have participated in the pilot program; expand the program to additional provinces with the assistance of these Chinese experts and the International Team; and gradually transition over several years to a training program for system optimization experts that is offered by a core group of Chinese experts who have developed their skills as the result of the pilot program. It is anticipated that training under the national program will be extended to the design institutes to integrate system optimization techniques into new system design.

In the final year of the pilot program (2004), a comprehensive program evaluation will be completed. It is anticipated that the pilot program model will be refined and applied in other countries with an emerging industrial sector.

References

McKane, Aimee, Vestal Tutterow, Gunnar Hovstadius and Steven Bolles. 2002. China Motor System Energy Conservation Program, Report to UNIDO on the Second Mission of the CTA.

McKane, Aimee, Ron Wroblewski, Hugh Falkner, Gunnar Hovstadius and Steven Bolles. 2002. China Motor System Energy Conservation Program, Report to UNIDO on the Third Mission of the CTA.

Nadel, Steve, Wang Wanxing, Peter Liu, and Aimee McKane, 2001. The China Motor Systems Energy Conservation Program: A Major National Initiative to Reduce Motor System Energy Use in China.

United Nations Industrial Development Organization. 2001. China Motor System Energy Conservation Program, Project Document.

A Statistical Analysis of the Induction Motors Produced by Italian Electromechanical Companies on the Base of the Efficiency Frame Proposed by the CEMEP Agreement

Giuliano Bergamasco, Aldo Boglietti, Andrea Cavagnino,

Politecnico di Torino, Dipartimento di Ingegneria Elettrica, Italy

Introduction

The reduction of energy consumption through an increase of electrical system energy efficiency is now an important target for every countries. This choice allows to reduce the CO_2 emission and other pollution sources due to electrical energy production. Making reference to the European Union, typically 60%-80% of energy absorbed by industrial sectors and 35% of energy absorbed by commercial sectors are due to electrical motors. In the industrial field, a motor consumes an annual quantity of electricity which approximately corresponds to 5 time its purchase price, throughout its whole life of around 10-12 years [1]. The electrical motors are by far the most important type of load. They are used in all the sectors in a wide range of applications, such as: fans, compressors, pumps, conveyors, mills, winders, elevators, transports, home appliances and office equipments. Since the motors are the larger users of electrical energy, even small efficiency increase will produce very large energy saving in the European Union. These concepts are not new, but their importance has assumed an important role across the end of the seventies and the start of the eighties, with the well known energetic crises. Even if the Italian people remember just the "by foot Sunday", those years saw the blossom of national proposals linked to a much more rational use of the energetic resources, with a particular reference to every type of energy saving. Making reference to the Italian approach, it is important to underline the "Progetti Finalizzati Energetica I and II" which put together the competencies of the academic and industrial people with very interesting results. At the same time the need of better energetic characterization was another important problem to be solved. A critical approach to the procedures used to determine the motor losses and efficiency was adopted and, in some cases, new methodologies were proposed in order to guarantee the accuracy and the validity of the measures. The first eighties were an important international jump platform for solving the problems concerning the high efficiency electric motor. The obtained results were a reference point for the future research activities on this field, but few years after the energy crises solution, the interest on the energy saving went down very rapidly. This is not the seat for analyzing this failure, but it is important underline that the interest on the energetic problems is again rose after the 1997 Kyoto Conference. In Europe, the research programs "SAVE I" and "SAVE II" and other national enterprises, linked to the European Union targets concerning the reduction of the CO_2 emissions, have given new impulse to the energy saving researches. Energy efficiency is a key element of the energy policy for the European Union, since it improves the efficiency of the economy, increases energy supply security and decreases harmful emissions due to the electricity generation. The EPACT (Energy Policy Act) [2] in the USA and the agreement between the European Union and the CEMEP (European Committee of Manufacturers of Electrical Machines and Power Electronics) [3] in Europe, have given a first answer to the energetic problem. For what concerns the CEMEP, the European Union guarantees to the producers much more information opportunities to the users, in particular for the use of high efficiency motor. On the other hand, the producers have to increase the realization of higher efficiency motors and to encourage their sale, reducing at the same time the production of the old motors with lower efficiency. The involved power motor range is from 1.1 kW up to 90 kW. The CEMEP protocol has introduced a revolution in the reference efficiencies of the industrial motors with a new classification based on motor efficiency. The industrial motors (types IP54 and IP55) will have to be reclassified in three efficiency bands. The lower class (labeled Eff3) involves the "standard efficiency" motors (motors now in production); the medium class (labeled Eff2) involves the "energy efficiency" motors, while the higher class (labeled Eff1) involves the "high efficiency" motors. The Effl and Eff2 values proposed for 2 and 4 poles motors are much more high with respect to the typical Eff3 ones, in particular for the small motors in Effl class. In Fig 1 the Effl and Eff2 limits are reported for 4 poles motors.

Statistical analysis of the induction motors produced by Italian companies

The Italian electromechanical industry is split up in small and medium companies with limited resources devoted to research and development programs. As a consequence, the new efficiency standard proposed by the CEMEP represent a very hard target to be reached by the Italian electromechanical companies which could be crushed by some giant European companies, in a market interested just on a low cost product. After several years from the CEMEP protocol, a statistical analysis of the industrial motors Italian companies is an interesting approach to understand how their products are close to the new efficiency frame. Since Italy is one of the world most industrialized countries the validity of this work is wider than the Italian border. The data collection has been made using direct contacts with the companies or using data sheets available on the web. In order to have homogeneous data with the Italian producer, only electromechanical companies with Italian capital have been considered. However the results obtained by some of the most important European companies, whose are present in the Italian marked, have been taken into account and used as reference points. For shortness problems just the results got on 4 poles motors are reported and discussed the considerations have to be considered valid for 2 poles motors too. The considered Italian companies have been sixteen while the European ones have been five.

Analysis of the obtained results

In Fig.2 the motor efficiency for the 16 Italian companies are reported using the CEMEP protocol limits as reference frame. It is well evident as a reduced number of motor are able to satisfy the Eff2 limit. In particular, the rated powers lower than 15 kW show the higher discrepancies with the Eff2 lower limit. A better situation is found for rated power higher than 15 kW where it is possible to find more motors with efficiency good for an Eff2 labeling. In Fig.3 the comparison between the maximum, average and minimum efficiency is reported using again the CEMEP limit as reference frame. From this picture it is evident as the average efficiency for rated power motors higher than 20 kW are corresponding to the Eff2 lower limit. The Eff1 lower limit are well far both from the maximum and the average efficiency. In Fig. 4 the efficiency standard deviation for all the rated power is reported. Also in this case, the rated power motors lower than 15 kW show an higher standard deviation putting in evidence that in this power range the efficiency dispersion is particularly significant. In other word, small motors with a large spread of the efficiency values are available on the market. From the energy saving point of view, this consideration has to be taken into account because the number of the installed small motors is very high [5]. In Fig. 5 the absolute differences between the average efficiency and the lower Eff1 and Eff2 limits are reported. It is well evident that the gap to be reduced for the Eff1 labeling is extremely heavy and a big job has to done by the Italian companies for getting this target. The gap to be filled for the Eff2 labeling is lower, even if for rated power lower than 18 kW ÷ 20 kW it is again significant. In Fig. 6 the motor efficiency of the motor produced by the most important European companies are reported together the Eff1 and Eff2 efficiency curves. It is well evident that all the motors are included in the Eff2 limit and a non negligible motors number has an efficiency higher than the Eff1 limit.

General considerations on the obtained results

In the previous section an aseptic description of the obtained results has been reported. It is important now to develop a critical analysis of the results, taking into account technical, economical and political references. A direct comparison between Fig.6 and Fig.2 highlights the difference between the Italian and the big

European company products, with an important gap to be filled. Induction motors cannot be considered an high tech product. All the processes involved in the induction motor design and the production are well known and in many cases, the same motor sheet shapes are used by different companies for the same motor size. As a consequence, just the magnetic material quality, the winding type and the core length are the free choice during the motor design. Few big punching firms have in their hands a big percentage of the sheets production, because only important electromechanical companies has the financial power to develop new designs by themselves. As an example, the cost of new punch tool is very high and the use of magnetic sheets outsource productions is more convenient from the economical point of view. From the Italian customers point of view, the induction motor choice is based on the purchase cost even if the company reliability is taken into consideration. The motor efficiency and the consequent energy save and reduction of the energy bill is not very often considered as a viable option. As a consequence, the motor market is looking for the best compromise between cost and reliability. It is important to highlight that the Italian companies do not have in their catalogs high efficiency motors and there are not reported references to the CEMEP protocol. On the contrary, the most important European companies have in their catalog high efficiency motors and a presentation of the CEMEP protocol with interesting discussions. In some cases the European companies have in production both standard and high efficiency motors. These results can be considered as an alarm bell which shows that an important industrial country like Italy is not still susceptible to use of higher efficiency machines. Obviously, it is not possible to generalize these considerations to other countries of the European Community, but the possibilities to find similar results could not be so remote. This marked behavior points out that in Italy there is not still a culture to the energy saving and this culture can be promoted only by means of government political acts and probably these considerations can be valid for other EU countries with similar economic and political system. Several approach to the energy save promotions can be adopted by the interested governments, but probably the best spring to move the industrial market is the possibility to have an immediate economical saving. A significant example is the reduction of the taxes if a high efficiency motor is bought instead of a traditional one. In Italy this type of action had an impressive impact on the Italian car fleet renewal when a government contribution was available if a old car using leaded gas was changed with a new one using unleaded gas. In other words the moneys move the world and with the moneys it is possible to introduce the energy saving culture in the economic and industrial systems.

Conclusions

In the paper the results concerning a statistical analysis on the efficiency of induction motors produced by Italian companies is presented. In addition the same results have been compared with the products of the biggest European electromechanical companies. The analysis has shown that the Italian products are not still competitive from the efficiency point of view on the European market. Furthermore, it is possible to conclude that, at the present time, in Italy there is not a sufficient culture of the energy saving. Since Italy is one of the world most industrialized countries the validity of this work is wider than the Italian border, in particular if countries without big electromechanical companies are taken into account. A critical analysis has been developed on the based of technical, economical and political point of view.

References

[1] A.T.de Almeida, P.Fonseca, "Characterization of the electricity use in European Union and savings potential in 2010", Energy Efficiency Improvements in Electric Motors and Drives, Ed. Springer-Verlag, ISBN 3-540-63068-6.

[2] Balducci, "EPACT Legislation- The United States experience of minimum efficiency standards for induction motor". Energy Efficiency Improvements in Electric Motors and Drives, Ed. Springer-Verlag, ISBN 3-540-67489-6.

[3] P. Bertoldi, G. Kuehneund, "The european negotiated agreement to improve motor efficiency", Energy Efficiency Improvements in Electric Motors and Drives, Ed. Springer-Verlag, ISBN 3-540-67489-6.

[4] A. Boglietti, M. Lazzari, F. Profumo, F. Villata, "About the selection of induction motors samples having representative character for energetic considerations", ICEM 86, Munchen 8-10 September 1986, GERMANY.

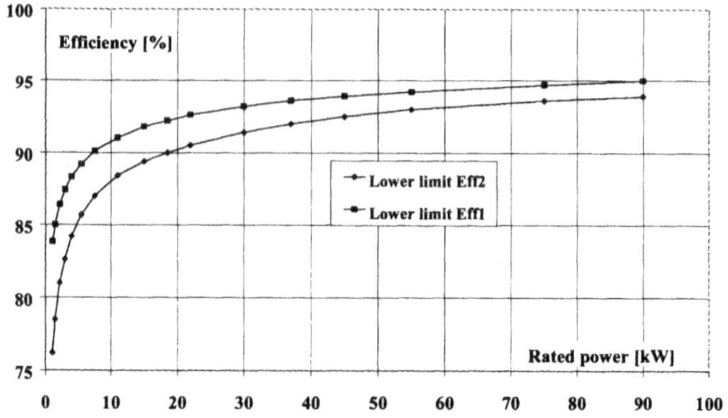

Fig. 1. Lower efficiency limits for Effl and Eff2 efficiency classification

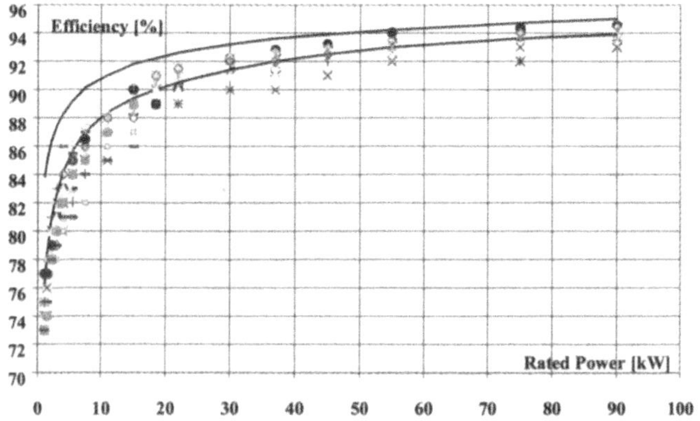

Fig. 2. Motors efficiency for the 16 Italian companies

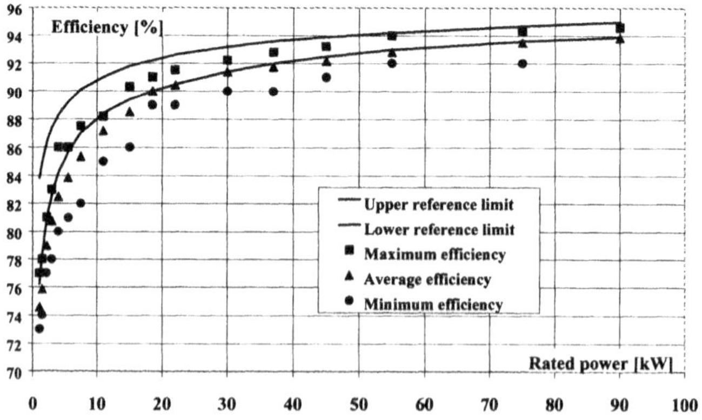

Fig. 3. Maximum, minimum and average efficiency for the 16 Italian companies

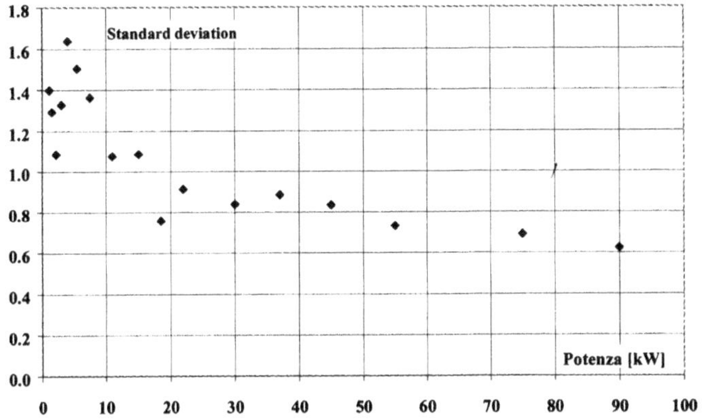

Fig. 4. Standard deviation for the Italian companies efficiency data

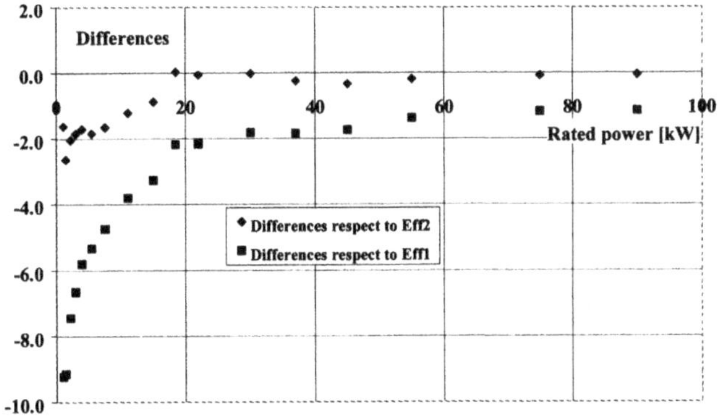

Fig. 5. Differences between the Italian motors average efficiency and the Eff1 and Eff2 lower limits

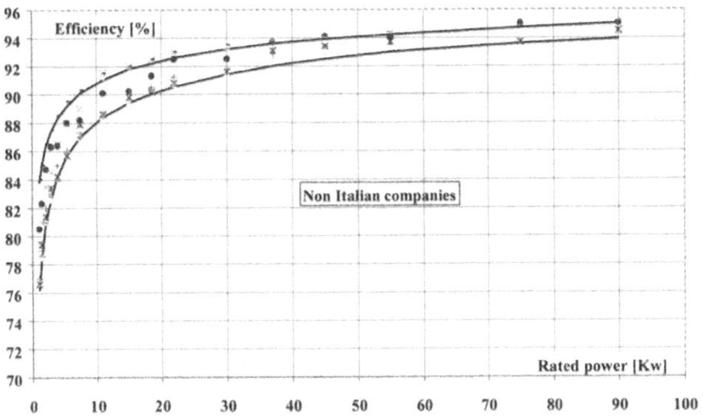

Fig. 6. Motor efficiency for 5 non Italian companies presented in the Italian market

Investment Grade Motors

Robert B. Boteler

Emerson Motor Technologies
8100 West Florissant Av.
St Louis, MO 63136

Abstract

This paper will chronicle the evolution of industrial motor efficiency in the United States beginning with the Energy Policy Act of 1992 through to the most current strategies. The paper will summarize and quantify the Energy Policy Act and its relative impact on motor sales in the United States. The paper will provide insight into the development of the latest NEMA Premium™ motor standard including how the motor manufacturers arrived at the efficiency levels used in the memo of understanding.

The paper will discuss the progress that U.S. Manufacturers have made in the past year promoting NEMA Premium, including gains in penetration. The paper will cover issues of used or rebuilt motors and the impact of motor repair that prolong the life of low efficiency motors, increasing energy demand.

The paper will document the energy cost savings available through the application of NEMA Premium motors. A particular focus will be made on the cost savings opportunity NEMA Premium motors provide and how the process of promoting these motors has moved from the factory floor to the boardroom. The paper will detail the steps needed to move from a single motor decision to a complete factory review, resulting in significant cost savings that rival industrial plant cost reductions in any other area of business such as labor and raw material.

Part One
Review of Energy Policy Act

The Energy Policy Act of 1992 [Epact92] established a uniform motor efficiency standard in the United States. The U.S. Congress worked directly with NEMA and the motor manufacturers to craft a law that all parties could support. Epact92 not only defined a minimum level of motor efficiency, it also recognized

specific test standards, established labeling and defined the product covered by the law.
Final rule 10 CFR Part 431

1) Efficiency
Efficiency levels were originally derived from the NEMA motor generator standard [MG-1] table 12-10.

2) Test Methods
Epact92 relies on two existing test methods: IEEE112 method B and CSA 390. Both of these similar test methods have been used by American and Canadian industry for many years. It is believed that they provide consistent and accurate measures of motor efficiency.

3) Labeling
Required labeling falls into two distinct areas. First, each motor that meets the definition of a "covered product" must have the NEMA Nominal Efficiency Number on the permanent name plate. Secondly, each covered motor must include the Department of Energy Manufacturers Compliance Number on each name plate.

4) Covered Product
Epact92 defines covered product as any general purpose "T" frame poly phase AC motor from 1 through 200 horse power. Covered motors are 2,4 or 6 pole design "A" or "B" that are usable on 60hz, 230/460 volts. The law exempts definite and special purpose motors. It is estimated that Epact92 impacts close to 75 percent of the 1-200 horsepower motors sold in the U.S. today.

Part Two
NEMA Premium

The motor manufacturers, along with utility groups, recognized there was an additional opportunity to save energy by expanding the scope of products offered, as well as stretching the levels of efficiency to an even higher level. The result of this exercise was the creation of a new standard known as, NEMA Premium.

The lack of a premium motor definition had begun to cause confusion in the market place. End users, OEMs and specifying/engineering firms all had different ideas about premium efficient motor performance. In addition, power utilities developed rebate programs that called out a wide range of motor efficiency, further confusing the market.

In the fall of 2000, NEMA's Energy Management Taskforce began work to draft a new premium motor standard. The member NEMA companies were determined to develop a motor efficiency standard that provided motor users with an optimum efficiency level, while maintaining the highest degree of interchangeability possible. The new NEMA standard is now published in MG1 as tables 12-11 and 12-12. NEMA Premium has been trademarked by NEMA and may only be used by those companies that sign a memo of understanding agreeing to its use and product application.

Product Range

NEMA members also believed there was an additional energy savings opportunity to be had if the range of NEMA Premium products were expanded beyond the limits of the Epact product definition. First, NEMA expanded the type of product from general purpose to also include definite and special purpose motors as described in MG1. This allows manufacturers to apply the new NEMA premium levels to a wide range of motors including vertical pump and close couple pump motors, as well as a whole universe of specially design motors for unique applications. By adding these additional categories, the Department of Energy estimated the impact of NEMA Premium to be the equivalent of removing 16 million cars from American highways over the next ten years.

The NEMA standard has also expanded the scope of the product to include motors beyond the 200 horsepower limit of Epact to 500 horse power. Unlike the Epact standard, NEMA Premium not only covers 230/460 volt machines, it also covers 2300/4160 volt motors.

DOE Savings Estimates

The US Department of Energy has reviewed the NEMA Premium motors' potential energy savings over a ten-year period. The DOE has concluded that motors having a 20 to 30-year service life can significantly save end users money by reducing operating cost as well as increasing up-time through improved reliability. Because an industrial motor can consume 4-6 times its original cost in energy each year, the saving of even 3 to 4 percent is significant over the motor's life. DOE estimates that NEMA Premium, when applied by American industry, can reduce 80 million tons of carbon and 5,800 giggawatt/hours over ten years or the equivalent of 16 million cars.

Part Three
The Promotion of NEMA Premium to the Market Place

The recognition of NEMA Premium by CEE, Copper Development Association, EASA and other industry associations has added awareness to an ever-increasing presence in the minds of motor users.

NEMA Premium efficiency motor sales have reached levels between 11.5 percent for 1-5 HP motors to 17 percent for 100-200 hp motors.. Perhaps, as a result of the slower economy or "energy fatigue," motor users appear to have reached this level and stalled.

Slower economic periods force motor users to cost reductions and postponement of upgrades and capital expenditures.

Previous efficiency upgrade and retrofit programs have focused on the maintenance or plant engineering level personnel within the industrial users' organization. This focus has brought us to the level of success we have today 11.5-17 percent premium motor sales.

A fundamental change must occur in the premium motor strategy for industry penetration levels to rise above their current level. This change involves effectively communicating plant-wide savings potential to the financial management of industrial companies. The challenge that we as manufacturers and environmentally conscious individuals have is to raise the understanding "return on investment" potential to the business manager without losing upper level managers in our technical evaluation processes.

We must provide a clear, easy to understand goal and process that can be uniformly communicated and implemented by businesses both large and small, drawing consistently on the NEMA Premium standard.

Historically, we have built cases for premium motor purchase based on the evaluation of specific units and application within a plant, operating unit.

This process is labor intense and has resulted in limited success. This process also creates multiple opportunities to revert to lesser efficient products.

In order to impact the greatest number of users on a plant-wide perspective, the savings opportunity and ROI [return on investment] will need to be elevated to a "business decision" that a business can review and allocate resources and funds to based on an overall savings potential. Industrial plants routinely examine their costs in hopes of finding those areas where reductions can be made, while maintaining overall product quality and production needs. Raw materials, labor and SG and A are always subject to scrutiny. Yet "fixed expenses" such as energy costs are all to often seen a "necessary evils" that become lost on the burden line within a plant's monthly costs.

The challenge we face is to successfully convey the return on investment potential in terms that other business decisions are made. We must create a process that clearly shows the value in an end user's AR [appropriation request] format. Whether large or small, financial managers or officers, understand fundamental financial data, based on plant level research that supports these actions.

With the advent of the NEMA Premium efficiency standard and the support of Motor Decisions Matter, including the previously mentioned associations, an element of consistency has been introduced that allows multiple organizations and individuals to capitalize on this single assignment. Working together in the creation of a model that defines this business plan will raise the NEMA Premium penetration to the next level.

Part Four
The Steps to Evaluating Motor Decisions and Implementing a Proactive Motor Business Plan

Step one
Determine a Plant-Wide Energy "Footprint"

Using an energy evaluation is the first step in developing the ROI in preparation for the AR. The plant or facility motor population must be audited or sampled to define the potential. This is the process that often stops motor system upgrade before it begins. The cost to effectively perform an audit can be substantial depending on plant size, motor population and difficulty to access. It is here that the introduction of Department of Energy case histories can be used to provide a basis of savings and resources.

Various resources such as the local EASA shop, ESCO or motor manufacturer can be of assistance. The outcome of this process must include plant-wide savings potential. A single spread sheet that captures total population, energy costs including peak demand charges, and existing motor efficiency data or age, new motor net cost and installation cost will provide the basis for evaluation.

Step Two
Plant Evaluation

Once the plant has gathered the energy footprint data, the information will be run through the energy payback calculator to classify the motor population. The purpose of this classification is to further simplify the decision-making process when presented to the financial team. The classification allows a tiered approached to the AR-making-financial- decision to occur in a logical progression based on three levels of payback. Level one, less than two years payback, level two greater than two to five years payback and level three greater than five years until payback is achieved. Once we have made this calculation and placed units into one of the three categories we can document our motor business plan.

Step Three
Prepare an Appropriation Request

Once the levels have been established and the units identified, an action plan can be implemented based on the ROI goal of the individual company. Each of the three motor levels should be reviewed separately to determine the ROI for the group. For example, if 200 motors in level one will cost $175,000 including installation and save $72,000 per year, the ROI is 20.2 percent. If the motors in level two will cost $100,000 including installation and save $30,000 per year, the ROI is

10.6 percent. Level three, likely being the least used motors in operation, cost maybe $65,000 including installation, while saving only $8,000 per year, would have a negative ROI.

The financial officer or operations manager needs to be provided with two or three distinct alternatives to premium motor usage. By dividing the plant or facility motor population into the three groups we can make a case and prioritize each of the three categories based on traditional financial evaluation methods, doing an appropriations request for each category. Keeping our goal in focus and allowing the financial management to compare the premium motor saving program to other cost reductions such as labor or material savings is absolutely necessary to improve the odds of project success.

In most industrial companies an ROI of 20 percent or more are considered good investments. One large motor [100 HP or greater] can save the equivalent of a 2 percent labor reduction per one worker. A plant with several thousand connected horse power can replace old, low efficient motors and reduce operating costs comparable to a hundred of the plant workers taking a salary decrease.

Summary

Motor efficiency must be moved from the shop floor to the business plan. Plant-wide evaluations of motor populations can show cost reduction opportunities that rival most other cost savings available to modern processing or manufacturing facilities. The challenge is to provide users with the insight to evaluate and make good financial decisions with consistent application of the new NEMA Premium motor products.

Life Cycle Cost Guidelines www.lcc-guidelines.com

Gudrun Bågstam

VI Association of Sweden Engineering Industries,
P.O. 5510, SE-114 85 Stockholm, Sweden,
Ph. +46 8 782 08 00, Fax +46 8 782 09 80,
E-mail gudrun.bagstam@vi.se

Introduction

This paper will deal with two resent developments in the LCC- area. Namely a new LCC website but also a new tool to help with procurement and quotation work with LCC as the basic evaluation criteria. This new tool is named "Kalkylera med LCCenergi" which means "Calculate/evaluate with LCCenergy". This is aimed to be a Swedish Engineering standard and is presently only available in Swedish.

1. The LCC Website www.lcc-guidelines.com

This new site was designed during the spring of this year and was launched in June 2002. The introduction theme is "Welcome to the world of LCC" and two key questions are highlighted.

Do you want to improve your performance ?

Do you care about the environment ?

The website has an impressive content with several buttons with direct access to important and interesting areas. Below a few of them are described in short.

What is LCC?

LCC stands for "Life Cycle Cost" and is the total cost for an equipment over it´s lifetime from time of installation to decommissioning.

Think *systems* not components, this is vital to be successful in your work. The cost can simply be added together year by year to a total value or ; one can take inflation and interest rate into account and refer to a specific year. Normally the year when the equipment is installed. When comparing different alternatives of investment it is convenient to compare the different LCC- values. Other used methods are "Pay-Off" in years and "Return On Investment" in %. However for both these methods you also need to calculate the difference in cost over the equipment life, even if you often only look at one year to make it simple. The advantage with LCC is that the full cost is focused and the different cost elements are easily compared.

It is easy to see how large the initial investment cost is in comparison with other yearly cost elements. The normally used LCC- model to allow analyses and comparisons of systems are:

LCC = the sum of (Initial + Installation + Energy + Operating + Maintenance + Down time + Environmental + Decommissioning costs)

How to Calculate

The costs estimated for the various elements making up the total life cycle costs need to be aggregated to allow a comparison of the designs being considered. This is best done by means of a Tabulation/Form, which identifies each item and asks for a value to be inserted. Where no value is entered an explanatory comment should be added. The estimated costs can then be totalled to give the LCC values for comparison, and attention will be drawn to qualitative comparisons as well.

Calculating Life Cycle Costs (LCC)

Following is a brief LCC overview:

LCC = the sum of ($C_{ic} + C_{in} + C_e + C_o + C_m + C_s + C_{env} + C_d$)

- C = cost element
- ic = initial cost, purchase price of equipment.
- in = installation and commissioning cost (including training)
- e = energy costs (predicted cost for system operation)
- o = operation cost (labor cost of normal system supervision)
- m = maintenance and repair cost (routine and predicted repairs)
- s = down time cost (loss of production)
- env = environmental cost (contamination from equipment)
- d = decommissioning/disposal cost (including restoration of the local environment and disposal of auxiliary services)

There are also financial factors to take into consideration if you choose to discount the costs to a certain year. These include:

- Present energy price
- Expected annual energy price increase (inflation) during the life time
- Discount rate
- Interest rate
- Expected equipment life (calculation period)

In the energy prices used, the influence of fixed charges, power charges, penalty charges for reactive power demand, and so forth, must be weighted if possible. Corresponding factors must also be considered for energy forms other than electricity.

In addition, the user must decide which costs to include, such as, maintenance, down time, environmental, disposal, and other important costs.

Procurement

The most convenient time to influence the life time cost (LCC) is when the system is purchased. This is true whether a replacement, upgrade or new system is desired. Think ***systems*** not components, this is vital to be successful in your design work and the ability to achieve a low life time cost.

To be sure to chose the best alternative it is recommended to base the evaluation on total LCC.

Different general conditions for procurement of equipment can apply. This can vary from a simple product procurement to a turn key installation.

In Europe ORGALIME has published GENERAL CONDITIONS like ORGALIME S2000 for product deliveries and SE94 also including erection [1].

There are other general conditions specific to that country which equipment suppliers, contractors & purchasers are applying to procurement practices.

Showcases

It is the intention that the showcases will highlight the value of savings possible by applying the LCC-approach described on the website.

Showcases are taken from a wide range of industries and applications, but it will be seen that the method of achieving the saving is, of course, usable outside the industry of the showcase. Some guidance on the wider usage of the cost saving techniques is presented with each showcase.

Most of the cases are from existing applications. It does not have to be **new** installations to achieve big savings on energy and maintenance.

The following typical cases could be suggested , however it should be remembered that all cases are unique;

- New installation –few limitations.
- Replacement installation – limitation due to surroundings.
- Break down. Repair or ?
- Replace the existing installation with one with better efficiency.
- Match the system to the true need. Many are oversized.

Discussion Forum

Here you can discuss LCC topics with others and get valuable information.

LCC-Guidelines – The EU Project

The work on developing the Guidelines has been carried out in the EU SAVE project XVII/4.1031/Z/98-037," LCC-based Guidelines on Procurement of Energy-Intensive Equipment in Industry."

The initiative to the LCC Guidelines was taken by the Association of Swedish Engineering Industries, and supported by Swedish National Energy Administration. The objectives were to develop European guidelines for procurement and to promote the application and implementation of the guidelines among manufacturers, installers and end users in the European Union.

Those involved in the SAVE project have been:

- VI, the Association of Swedish Engineering Industries, co-ordinator
- AREA, Air Conditioning & Refrigeration European Association
- EUROPUMP, European Association of Pump Manufacturers
- EUROVENT/CECOMAF, European Committee of Air Handling and Refrigeration Equipment Industries

A joint management group, JMG, with representatives from the partner organisations, directed the development work carried out by specialists and working groups from the partner organisations. A main document with separate documents for each particular equipment has been prepared. This work was carried out by participating associations with a strong involvement of manufacturers.

Members of the JMG were:
Gudrun Bågstam, VI, chairman
Lars Nordell, AREA
Lars Frenning, EUROPUMP
Sule Becirspahic, EUROVENT/CECOMAF

Guestbook

Here you can register and get information about news in the LCC-area.

LCC – How to Sell it

Do you need to convince some people how great the LCC approach is?

- Here you can download ten pictures.
- Hope this will help you in your selling !

2. "Calculate/Evaluate with LCCenergy"

This new Swedish calculation tool can be found on the website www.industrilitteratur.se. It is split into different modules and a few of them are:

- Lightning systems
- Air handling
- Cooling systems
- Transformers
- Motors
- Pumping systems
- Compressed air systems

With this tool you can easily do LCC calculations/evaluations and take the most important cost elements into consideration; like initial cost, installation & commissioning, energy and maintenance costs.

The "LCCenergy" web site based tool is available in Swedish only.

References

[1] http://www.orgalime.org/shortpub.htm

Specifying Energy Efficient Motors in Industry Standards

John Malinowski

Baldor Electric Company, P.O. Box 2400, Fort Smith, Arkansas 72902 USA

Abstract

Despite the promotions of energy-conscious organizations, premium efficient motors are not being specified by industry nor included in industry specifications. We propose that manufacturers of motors and drives, consultants and users who serve on industry standards committees make an effort to revise standards to include efficiency levels and call attention to life cycle costs relative to motors and drives.

US Energy Usage

In the United States, electric motors account for 52.5% of the total of all electricity used and 70% of electricity used by industry. Even with this fact, it seems to be taking a long time for industry to upgrade their older motors to premium efficient levels. Compared to the mid-2000 energy crisis and rolling blackouts, today's low energy costs have given US industry temporary relief from high electricity costs. But even with today's lower energy costs, upgrading motors can contribute to reducing plant downtime while saving electricity. Studies have shown that 28-42% of all US motor input energy can be saved using good electrical practices available with existing technologies.

Voluntary programs do not seen to be working well, except in areas where rebates from the electric utilities are offered to help defray the cost of the premium motors. Utilities in the Northeast and northwest US are offering rebates on motors and drives.

Industry Standards

Some motor standards used in industry mandate efficiency levels above the EPAct minimum efficiencies. An example of this would be IEEE 841-2001 "IEEE

Standard for Petroleum and Chemical Industry – Severe Duty Totally Enclosed Fan-Cooled (TEFC) Squirrel Cage Induction Motors – Up to and Including 370 kW (500 hp)".

The most recent IEEE 841 revision of 2001 (that replaced a 1994 standard) changed the minimum efficiency standards from EPAct levels to EPAct plus one efficiency level from the NEMA standards. We believe the efficiency level will be upgraded to NEMA Premium™ at the next revision. The IEEE 841 standard is used by the petro-chem industry as well as pulp and paper.

Many other industry standards make attempts at specifying motors with high efficiency. The API (American Petroleum Institute) Standard 541 "Form-Wound Squirrel Cage Induction Motors – 250 Horsepower and Larger" is another standard used in the petro-chem industry for above-NEMA sized motors. It has no efficiency standards listed, but primarily focuses on defining performance standards (electrical and mechanical) that lead to a robust, low-maintenance motor. The standard should also evaluate the motor's life cycle cost.

The Cooling Technology Industry has published "Cooling Tower Manual, Chapter 11, Electrical Components for Cooling Towers" of January 1984. The standard states that motors purchased with packaged systems should be evaluated for efficiency and cost of operation. No minimum efficiencies are given, only references to calculating life cycle costs. This CTI standard is presently under revision and adding NEMA Premium™ efficiency levels and EPAct as a minimum has been proposed.

To the authors' knowledge, most industries do not have a standard that specifies a motor suitable for their typical applications. The EPAct minimum efficiencies are utilized for applicable motors. Many non-standard motors are used for pumps across all industries and these do not need to comply with EPAct. Also motors with C-face mountings attached to gearboxes as used in material handling are not covered by EPAct. Therefore, very low efficiency motors are allowed if the user is not looking at life cycle costs.

In many electricity dependent industries, it is possible to separate the motor usage from other electrical consumption. Figure 1 illustrates many major industrial segments.

Electrical energy usage by industry

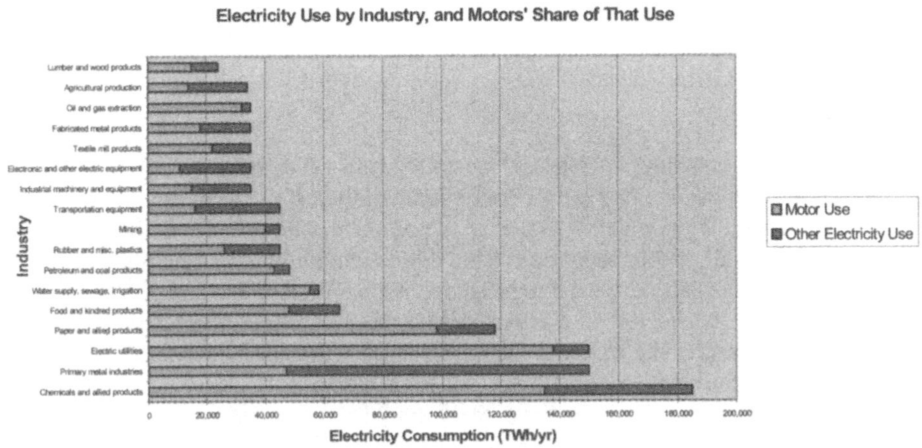

Fig. 1. Sources: EIA 1997 for motor energy use; U.S. Census Bureau 1996b for electricity use; utility motor energy use based on E Source 1999; utility electric energy use form EIA 1995; and agricultural, mining, oil/gas extraction, and water supply data from Xenergy 1998. From Energy-Efficient Motor Systems: A Handbook on Technology, Program, and Policy Opportunities – Second Edition by Steven Nadel, R. Neal Elliott, et al.

User industries

If we look into process industries, we find their motor selection to be more controlled. Petro-chem companies look toward IEEE 841-2001 and API 541 for their motor specifications. In addition, most oil companies also have their own proprietary motor specifications that may further refine the IEEE or API standard. Most pulp and paper companies also follow this same practice.

Consulting and specifying engineers

Consulting and specifying engineers (CSE) are most concerned with mechanical and performance issues. Their specifications are often conglomerations of specs written over the years. Motor manufacturers no longer in business are sometimes listed as approved. Components and processes that are not used in the industry are often specified. Many of these CSE specifications are in dire need of rewriting, addressing efficiency and life cycle costs.

Consulting engineers who write proposals for clients should be a target for addressing energy efficient motors and drives in their specifications. Some consulting firms are not using sound practices for sizing motors and waste energy from over-sizing the motors. Consulting engineers are often members of standards committees and may be influenced by the energy-conscious delegates who are also on those same committees.

OEMs

As a manufacturer of motors, most OEM specifications we see normally deal with the mechanical and performance issues relating to the interface of the motor to the driven equipment. Efficiency is usually not addressed. Often the OEM is concerned with the selling price of the equipment and not its operating cost. Since the motor is usually the most expensive component, a lower-cost and less efficient motor may be selected so a lower selling price can be achieved. Life cycle cost wasn't taken into the sale and the user may pay unnecessary energy costs for the 10-20 year life of the motor.

The initial purchase price of an electric motor is generally only two to four percent of its life operating cost. Most users are not aware of this. Many equipment manufacturers are selling their equipment in a very competitive market and often do not differentiate their equipment by cost of operation. The purchase price is usually the only issue. Adding higher efficiency motors and drives would give added value to the equipment and ensure a lower operating cost over the equipment life.

Benefits Beyond Energy Savings

We shouldn't totally dwell on energy savings. Most facilities are more concerned about eliminating downtime than on their energy bill. When a motor fails and the flow of widgets to the shipping departments stops, attention gets called to that motor and getting the line moving again.

Premium efficiency motors are simply better motors than older standard efficient designs. Lets look at some of the benefits of these premium efficient motors. To achieve their high efficiency, motors have lower losses.

- Motors operate cooler, motor insulation and bearings last longer. For every 10°C cooler, the motor insulation life is doubled.
- Better components and tighter tolerances result in better balance and longer bearing life.
- With a lower temperature rise, premium motors are better suited for adjustable speed operation using a PWM supply.
- Laminations made of high-grade steel that are capable of being run through a temperature-controlled bake-out oven for rewind at some point in the future. Lower grade steels can degrade and reduce the motor's efficiency after rewind.
- Lower temperature rise means that a smaller fan can be used resulting in quieter operation.

- Many premium efficiency motors use cast iron frames and endplates, making the motor structure more rigid and helping to reduce vibration.
- Cast iron frames bases are machined flatter than a stamped steel base, resulting in easier alignment and longer bearing life.
- Cast housings have cooling fins and generally dissipate heat more effectively than a steel band motor.
- Many cast iron motors are offered as "severe duty" models with extra corrosion proofing (inside and outside), epoxy paint, neoprene gaskets, grease fittings and reliefs, corrosion-resistant hardware and other features to extend life.
- Motors with cast iron housings and endplates resist corrosion better compared to rolled-steel band or aluminum frame motors with aluminum endplates.

Call to Action

Industry standards cover those maintained by IEC and IEEE as well as industry-specific organizations. Applications would include pumps, compressors, fans, cooling towers, elevators and many other energy intensive industrial applications.

As energy delegates, we are often members of committees that write and revise industry standards. We should strive to write aggressive minimum efficiency levels into the various standards. A grassroots movement to add efficiency minimums to industrial machinery and process specifications should be initiated. A symposium on industrial standards could address the various standards that could be improved by adding minimum efficiency levels of CEMEP eff1 and NEMA Premium™.

References

[1] NEMA MG1, 1998 "Motors and Generators"
[2] IEC Std 60034, August 1999, "Rotating Electrical Machines"
[3] IEEE Std 841-2001, "IEEE Standard for Petroleum and Chemical Industry – Severe Duty Totally Enclosed Fan-Cooled (TEFC) Squirrel Cage Induction Motors – Up to and Including 370 kW (500 hp)"
[4] API Std 541, Third Edition, April 1995, "Form-Wound Squirrel Cage Induction Motors – 250 Horsepower and Larger"
[5] Cooling Technology Institute, Houston, TX, "Cooling Tower Manual, Chapter 11, Electrical Components for Cooling Towers", January 1984
[6] "Energy-Efficient Motor Systems: A Handbook on Technology, Program, and Policy Opportunities", Second Edition 2002, Steven Nadel, R. Neal Elliott, Michael Shepard, Steve Greenberg, Gail Katz and Anibal T. de Almeida.

Introducing a Minimum Efficiency Standard for Electric Motors — Experiences in Australia and New Zealand

David Cogan

Standards Engineer, Energy Efficiency and Conservation Authority, New Zealand

Introduction

This paper recounts the experience of the author in developing and introducing mandatory requirements for electric motor efficiency for New Zealand and Australia. Problems encountered are described and recommendations made for those contemplating similar regulatory measures. There is emphasis on the need to take account of international developments.

Background

New Zealand and Australian agencies charged with improving national energy efficiency both identified squirrel cage induction motors as a likely subject for a mandatory minimum efficiency standard round about 1993. This was based on the statistic that around 70% of the electricity used by industry was used in motors and motor drives. Even though an improvement in the motor efficiency itself represents only a small part of the total potential savings, it is relatively easy to achieve and is still worthwhile. In New Zealand a study by Energetics and George Wilkenfeld & Associates [1] in 1994 demonstrated clear economic and energy benefits for a MEPS that eliminated the worst performing 20% of the motors then being sold. Further consultation by the then Department of Primary Industries and Energy in Australia indicated industry support for a more stringent level, dubbed the "40% MEPS".

Development of Standards

In Australia energy efficiency legislation is a function of the state and territory governments, and so in order to co-ordinate the technical aspects, these are contained within national standards. In New Zealand, the Energy Efficiency and Conservation Act enables the production of regulations, and makes specific mention of standards as a vehicle to carry the technical requirements. This is partly to ensure a degree of participation by interested parties in the regulations via the standards consensus process, and partly to make the regulations themselves more streamlined and less liable to frequent alteration. An additional reason is that standards are increasingly international in character, and basing regulations on an international standard reduces the number of trade barrier issues while also reducing the compliance costs faced by manufacturers and exporters.

Australia and New Zealand have a formal agreement to develop joint standards [2] as the first option, and the first joint standard produced was that dealing with the energy efficiency labeling of domestic refrigerators. It was therefore assumed that the production of the standard for motor efficiency would be a joint project.

The joint Australian and New Zealand energy efficiency standards are generally produced in two parts. One part deals with the measurement of energy performance, while the other part contains regulatory requirements, such as the performance that each item must meet (when tested in accordance with the first part) or how its energy performance must be labeled. It is the second part that is actually referenced by regulations, and so the energy efficiency regulatory authorities have a greater say in what is contained in that part than would be the case for most other types of standard.

Measurement standard

The existing Australian Standard for measuring electric was based on IEC 34-2. Most suppliers of motors to the Australian and New Zealand markets tested to that standard, as would be expected in most countries with a 50 Hz mains supply. However, at that time — late 1996/early 1997 — the question of the suitability of IEC 34-2 for determining efficiency for regulatory purposes was being questioned. Discussions at the 1996 Conference in Lisbon [3], [4] had led to a start being made on producing an IEC Standard that would be more akin to IEEE 112. It was decided that, for the sake of test repeatability and longer term consistency, the new IEC Standard, the future IEC 61972, would be adopted. But, just in case the IEC project took longer than planned, work started on producing a local version that aimed to anticipate the international standard. The result was AS/NZS 1359.102.3 [5], which was published in 2000.

It turned out that IEC 61972 was not published by the time the testing standard was required to support the regulation, so it is the joint Australian/New Zealand standard that is referenced as the measurement method.

Performance standard

Technical aspects of regulatory requirements are contained in the joint Standard AS/NZS 1359.5:2000 [7].

It was intended to select the minimum efficiency values on the basis of excluding a portion of the motors from the market. However, when drafting the standard, it was found that for 2-pole and 4-pole motors, the intended values were similar to the CEMEP "eff 2" minimum values. Since most motors supplied to the Australian and New Zealand markets come from manufacturers who also supply the European market, it was decided that, rather than introducing a new and different set of requirements, the "eff 2" level would be adopted as the minimum efficiency level.

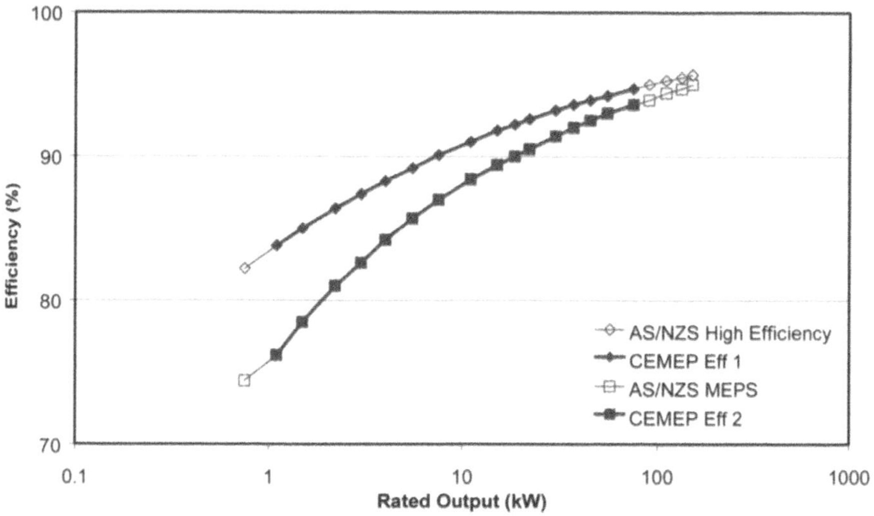

Fig. 1. Minimum and High Efficiency Values Compared with CEMEP Eff Values (when tested to IEC 34-2 or similar)

As may be seen from Fig. 1, the range covered by the Australian/New Zealand standard is greater than that covered by the CEMEP requirements, and range from 0.73 kW (1 HP) up to, but not including, 185 kW (which effectively means 200 HP). In addition, the Australian/New Zealand standard includes requirements for 6 pole and 8 pole motors, also over the same range of outputs.

The adoption of the CEMEP "eff 2" values as minimum efficiency requirements provided an opportunity to use the "eff 1" curve (and equivalent curves for 6-pole and 8-pole motors) to define motors that may claim to be "high efficiency" or similar. In the past, it had been found that some manufacturers' "high efficiency" ranges were less efficient than other manufacturers' "normal" ranges [8].

While the specified measurement standard is technically equivalent to IEEE 112, it is recognized that most motors on the market in Australia and New Zealand are still tested to IEC 34-2 or one of its equivalents. Therefore AS/NZS 1359.5 contains separate sets of tables for each test method. These tables take account of the differences in assumed winding temperature as well as in the different treatment of stray losses. Testing of individual motors to the different test procedures indicate that the two sets of tables are equivalent.

Introduction of Regulations

Both New Zealand and Australia have now introduced their regulations in respect of minimum motor efficiency. The two regimes are by no means the same. Australia has a system of registration, with deregistration being a sanction, and the subsequent sale of a deregistered model of motor being the offence. The New Zealand regulatory authority maintains a list of motors and their efficiencies, but the offence is selling a non-complying motor. However, as the technical requirements are contained within joint standards, the regulatory regimes are compatible, and a motor that meets one country's requirements can be sold with confidence in then other country. The regulatory authorities also co-operate on technical issues, and tend to refer them to the same person with specialist knowledge.

Issues

The Standard covers three-phase cage induction motors with a voltage of up to 1100 volts. Certain exclusions are listed; these are:

- Submersible motors specifically designed to operate immersed in a liquid
- Motors that are integral with, and not separable from, a driven unit
- Multi speed motors
- Motors for use only for short-time duty cycle (i.e. S2 duty rating)

And, in addition, a clause was added to provide some flexibility and to deal with obscure or unforeseen cases:

- Motors that have been granted exemption by the relevant Australia/New Zealand regulatory authority due to their application placing restraints on the motor dimensions or other key design aspects.

The regulations apply to new motors. They do not apply to rewound motors, except where there are claims made about the performance of a rewound motor. If the seller claims an efficiency or an efficiency class, the motor must meet it.

In practice, there have been fewer applications for special exemptions than was expected. Only two cases have been brought up.

One exemption was sought for a range of gear motors on the basis that they were made specifically to fit a specific range of gearboxes. It was ruled that the motors were separable, and that fitting the gearboxes did not place restraints on key design aspects. However, it was found that the motors in question had been made to have an S3 frequent start rating by designing them with low inertia rotors. This can adversely affect efficiency. Calculation showed that for most of the motors in question the energy lost due to comparatively low efficiency was less than the savings in rotor kinetic energy involved in the frequent starting. For such motors, an exemption was granted.

Another application was made for exemption for a range of Ex(d) motors. In this case, it was considered that the need for the special flameproof enclosure did not affect the ability to produce an efficient motor. But such motors with a low temperature classification, say T4, T5 or T6, could possibly have their design affected by the need to keep the motor cool. At the time of writing, the applicant has not furnished the regulators with information to allow them to make a decision on that point.

Introducing MEPS in Other Countries

The experience gained from the introduction of motor MEPS provides some pointers to other countries who may be considering a similar exercise.

Selection of efficiency values for a MEPS

Those countries that were among the first to introduce a MEPS for electric motors undertook an analysis of stringency levels, the proportion of motors that would be affected, costs and resulting benefits to set the levels of efficiency. Such a process is becoming less appropriate; it is more effective to investigate the sets of levels used elsewhere, particularly by trading partners, and to adopt one of those. Up-to-date information about energy efficiency requirements can be obtained from sources such as the APEC Energy Standards Information System [9]. This system will, once fully established, provide active updates of developments being undertaken by member economies in the area of energy efficiency standards.

Choice of measurement standard

While it will take a while for the new IEC 61972 to be fully adopted by those currently using IEC 34-2, it will happen, and so there is merit in specifying the required efficiency levels are to be measured according to the new standard.

It is possible to calculate the equivalent efficiency figures when different test procedures may be used. A slightly cruder calculation can be carried out using the different sets of tables in AS/NZS 1359.5.

Use of a performance standard for technical requirements

The experience (not just with motors) in Australia and New Zealand of using Standards to provide the technical requirements of energy efficiency regulations has proved to be successful in giving all stakeholders some effective say on technical aspects, and in providing a mechanism for aligning requirements between jurisdictions. It is far easier to share or to adopt a Standard than it is to share regulations.

Conclusions

The use of a test standard and a performance standard to provide the technical requirements for motor efficiency regulations results in an effective regime.

In developing mandatory motor efficiency requirements, note should be taken of developments elsewhere, and the adoption of a standard in use elsewhere reduces problems for regulators and manufacturers.

References, notes and acknowledgements

[1] Energetics Pty Ltd and George Wilkenfeld & Associates, *The Feasibility of Introducing Minimum Energy Performance Standards (MEPS) into New Zealand*, produced for EECA, Wellington, NZ, December 1994.

[2] For further details, see David Cogan, *Harmonization of Standards — The Australian and New Zealand Experience*, Regional Symposium on Energy Efficiency Standards and Labeling, Bangkok, May 2001

[3] R.G.Bartheld and J.A.Kline, *Comparative Efficiency Measurements IEC 34-2 vs IEEE 112*, Energy Efficiency Improv. in Electric Motors and Drives, Lisbon 1996

[4] C.N.Glew, *Efficiency Measurement Testing Standards — Stray Loss, the Key to Efficiency Determination*, Energy Efficiency Improvements in Electric Motors and Drives, Lisbon 1996

[5] *AS/NZS 1359.102.3:2000, Methods for determining losses and efficiency— Three-phase cage induction motors*, Standards Australia/Standards New Zealand, 2000

[6] Neville Glew, *The Challenge to Define the New IEC Test Method*, EEMODS'99, London 1999

[7] *AS/NZS 1359.5:2000 Rotating electrical machines— General requirements, Part 5: Three-phase cage induction motors—High efficiency and minimum energy performance standards requirements*, Standards Australia/Standards New Zealand, 2000

[8] David Cogan, *Energy Efficiency of Electric Motors and Drives: Australian and New Zealand Programmes*, Energy Efficiency Improvements in Electric Motors and Drives, Lisbon 1996

[9] http://www.apec-esis.org

The author wishes to thank colleagues involved with the APEC Energy Standards Information System (http:/www.apec-esis.org) and the International Copper Association, in particular John Mollet, for their support and assistance with producing and presenting this paper. Frank Duff and Nat Krishnan of Standards Australia did sterling work producing the joint standards featured in this paper. Thanks also to Denise Knight of the IIEC for pointing me in the right direction.

Minimum Efficiency Standards for Electric Motors (MEPS) – Future Direction of Regulation in Australia

Tony Marker

Dr. T. Marker, Manager, Energy Efficiency, Australian Greenhouse Office, GPO Box 621, Canberra, ACT, 2601 Australia

Abstract

In October 2001, Australia introduced mandatory requirements for electric motor efficiency upon all products manufactured in and imported into the country.

While an improvement in motor efficiency represents only a small proportion of the energy saving potential connected with motors, it is both technically possible and cost effective in Australia. The 2001 MEPS in Australia are projected to save 4,100 GWh over 15 years (the effective life of the product), abate 0.33Mt pa of greenhouse gasses by 2010 and save Australia $AU 70 million (Euro 39 million).

Australian government officials, however, believe that further improvements can be made. Australian Standards specify both the MEPS level and the High Efficiency levels (which establish a marketing standard for motors sold as "superior").

In 1999, Australian jurisdictions agreed a policy of "matching" the best regulatory standards imposed by our trading partners, subject to public consultation and economic analyses demonstrating the benefit of regulation. Australia imports the majority of its electric motors from the major economies in Europe, Asia and North America. As a "taker" of technologies, our approach seeks to avoid the costly, sometimes divisive technical debates about MEPS in favour of a pragmatic approach of following the most stringent minimum regulatory standard adopted in a major economy. Australian governments thereby avoid imposing substantial research and development costs on suppliers in Australia and our MEPS debates are about when (not if) new standards should commence.

This policy obviously encourages Australian regulators to examine and anticipate developments overseas. The 2001 MEPS in the Australian Standard matched the efficiency level 2 in Europe while the High Efficiency levels are equivalent to the MEPS level already imposed in the USA (similar to efficiency level 1 in Europe).

Australian governments will commence a process in 2002 that will develop the next round of MEPS levels. By 2003, the relevant Australian Standard AS/NZS 1359.5 will carry the new MEPS and High Efficiency levels that will come into effect at a date set by Ministers, probably 1 January 2006.

The opportunity exists for Australia to move the MEPS target from matching the minimum of Europe in our existing standard to the minimum of North America in our next standard. As Australian governments undertake to fix MEPS for at least four years (to allow industry certainty to write off development costs), the challenge in 2002 or 2003 is to anticipate the next, more stringent MEPS level in a major economy that may commence sometime in 2004 to 2010. This level would be matched in the new standard as the High Efficiency level.

The Australian Context

Introduction of Electric Motor MEPS 2001

At EEMODS 1999, I presented a paper on "Australia's National Greenhouse Strategy and Electric Motor Efficiency".[1] In that paper the proposal and process to introduce minimum energy performance standards (MEPS) for electric motors was discussed. It was anticipated that MEPS would be introduced from 2001 to apply to three-phase induction motors with a rating of 0.75 to 150kW.

In October 2001, Australia introduced mandatory energy efficiency requirements for electric motors. The mandatory energy efficiency requirements were set down in joint Australian and New Zealand Standards to which implementing legislation referred. The mandatory minimum efficiency levels applied to three-phase motors from 0.73kW up to185kW. The Australian/New Zealand Standard also established requirements for minimum levels for motors to be marketed as "high efficiency" motors. After extensive consultation with the Australian motor industry, the 2001 levels for minimum energy efficiency were based on the European "eff 2" grade, and the "high efficiency" level equated to the European "eff 1" level. The aim of having a definition of "high efficiency" set out in the Standard was to codify a consistent approach permissible for use in advertising and equipment labelling.

Expectations of Early MEPS Review

In the consultation processes and Regulatory Impact Analysis for the introduction of MEPS in 2001, the intention to re-visit the stringency of MEPS levels was signaled to all stakeholders with a target date of 2005. The AGO, proceeding to a new "Timetable for MEPS Process" commenced actions in 2002 to accomplish a review of the stringency levels in relation to world's best regulatory practice.

Government Policy for Standards

Industry Consultations & Notifications

Industry involvement in proposals for MEPS on additional equipment types or proposals for increasing MEPS stringency levels is sought at the earliest possible stage. Usually, a discussion paper reporting on studies into the feasibility of the MEPS proposal is published by the National Appliance and Equipment Energy Efficiency Committee (NAEEEC) and widely circulated with a two-month public comment period.

Subject to no substantial dissenting comments, steering committees are convened by the AGO on behalf of NAEEEC, drawing membership from a wide variety of interested groups, industry representatives, academia, and the interested public. The steering committees may have technical working groups and develop the MEPS proposal having regard to the consensus view. Their input is fed into the Standards Australia process and into the economic analysis and regulatory impact assessment process over a nominal one-year period.

Revised MEPS Policy

The approach in Australia for a process for introduction of MEPS has been modified following a Ministerial Council decision. The current policy is that MEPS are to be based on ensuring more effective engagement of all parties in a MEPS debate in order to deliver more certain outcomes, and that MEPS levels should be set not at a "no regrets" level, but at a level that matches "world's best regulatory practice". The policy approach has the following three elements.

Matching MEPS Best Regulatory Practice

Under the revised policy for MEPS in Australia, the most stringent energy performance standard in place elsewhere in the world will be adopted, provided that it meets other government policy requirements and provided that regulatory impact analysis indicates overall economic benefits.

Introduction of a MEPS Timetable

Government ministers have agreed upon a 'target timeframe' for the introduction of MEPS, to provide some degree of certainty to the process and give industry an appropriate notice period to undertake any necessary modifications to product designs and/or production procedures. While this proposed timeframe is indicative only and flexible enough to take into account specific circumstances that may arise, it creates reasonable expectations amongst all parties of the time that MEPS will commence. The AGO uses the timetable as a performance indicator, spurring all stakeholders to work toward positive outcomes within the allocated times.

The timetable published outlines stages in the development of MEPS with the aim to reduce development periods to between three to five years (compared with the seven to ten taken during the 1990's).

Developing a Consensus Processes

In addition to these policy improvements, Australia is embracing a more collaborative approach to MEPS negotiations. A range of procedural improvements have been agreed to better engage stakeholders, improve transparency and deliver more certainty to the process.

Future Standard for Australia

Prospects of 2001 Motors MEPS Revision

The AGO commissioned a review of the 2001 MEPS level in early 2002. The study, (conducted by David Cogan, Standards Engineer, Energy Efficiency & Conservation Authority, New Zealand)[(2)] compared the 2001 MEPS level with the standards in place or planned for North America, Europe, Thailand and Taiwan. This revealed that, for two pole and four pole motors representing the majority of the existing stock, the 'world's best practice' in MEPS regulation is represented by the North American Standards. These levels were assessed as approximately equal to the existing Australian "high efficiency' standard.

The conclusions of the report were that:

- Increasing the MEPS level to the 2001 High Efficiency level for two and four pole motors would produce worthwhile energy savings of about 25 GWh per year;
- This increase in MEPS level would match World's Best Practice.
- Corresponding greenhouse gas emission savings would be 20 kt per year accumulating, for a savings of up to 200 kt in the 10th year from introduction;
- Benefits of including six pole and slower motors in the exercise would be smaller and more marginal, requiring further analysis of actual benefit;
- Introduction of a "premium efficiency" standard would provide an incentive for further improvements above the new level for minimum efficiency; and
- Specification of values for High Efficiency and Premium Efficiency could be introduced for motors with outputs up to 550 kW to encourage production of motors with greater efficiency in the larger sizes.

The report's conclusions, if adopted by Australian governments, will be communicated to industry and other stakeholders as an intention to re-set the MEPS and High Efficiency levels in three years.

Drafting of Proposed Revised Standards

To revise the current MEPS, the Australian Standard must be revised to set out the more stringent efficiency levels. This process would commence with development and publication of a draft revised Standard, expected by early 2003. The draft revised Standard is then circulated for comments and proceeds through other consultation processes before finalisation. Concurrently, the AGO broadly announces the proposed revised MEPS level set out in the revised Standard to all known stakeholders to give industry adequate notice.

Potential Savings from MEPS Revision

Initial analysis indicates the projected savings from the proposal to revise motors MEPS are worthwhile. The accumulating annual savings in electricity and expected increases in purchase costs are set out in Table 1. The relatively low savings expected for six pole motors, and associated relatively high costs indicate that further specific investigation of the feasibility of whether to include or to exclude six pole motors in the revised Standard is required.

Graphically, accumulating savings versus costs are shown in Figure 1.

Table 1. Revised MEPS levels – Expected accumulating annual savings vs. costs[3]

Motor speed	Annual savings from each year's purchase (GWh)	Total premium paid ($AU)
2 pole	9.7 GWh	$ 2.7m
4 pole	15.5 GWh	$ 4.4m
6 pole	0.8 GWh	$ 0.4m
Total	25.0 GWh	$ 7.5m

Opportunities to Match USA Minimums vs EU Minimums

The challenge for Australia in this MEPS revision process is to decide to which stringency level to revise the Standard. There exists both the opportunity and risk of matching either the current USA MEPS minimum levels, or the European Union "eff 1" levels. The issue could be resolved by agreement on the new IEC Standard. Revised "high efficiency" levels are also required.

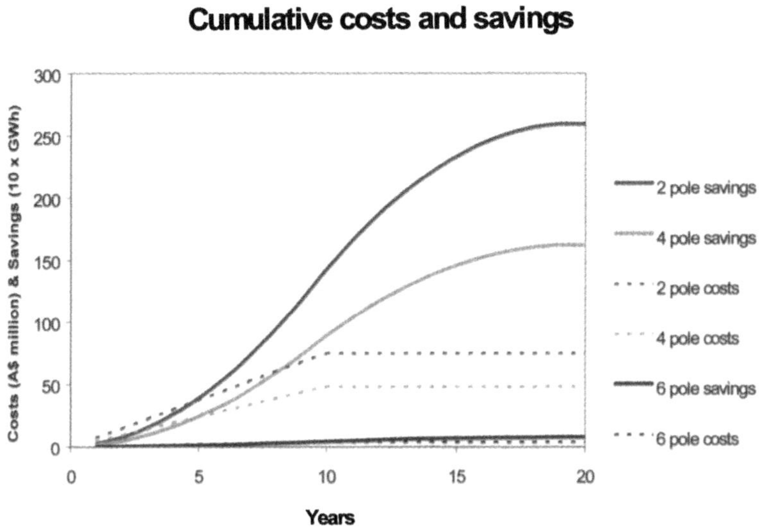

Fig. 1. Overall savings and costs predicted for ten years[4]

It is recognised that there is a challenge in anticipating a future more stringent MEPS level in the world that might be introduced during the period 2002-2012. Also, the question of possibly excluding six pole motors from the higher standard requires further economic analysis.

Next MEPS Levels the Limit?

It is possible that the current revision of the 2001 motors MEPS levels may be the final one. Current expectations are that further minimum efficiency mandates will result in diminishing returns for anticipated savings, making justifications on benefit/cost analyses unachievable. Unless there occurs a quantum leap in motor efficiency technology, it is thought that a second round of stringency increases may not be possible. Alternatively, future increases in the cost of electricity may lead to more stringent standards being cost effective.

In either case, the prospects for further energy efficiency gains in the motors and drives arena lie mainly in promotion of 'energy efficiency best practice' – improving the application of motors to driven equipment. It is clear that substantial gains in overall efficiency in the economy may be achieved by focusing attention on how motors are utilised with pumps, fans, compressors and other industrial equipment.

Conclusions

Australia introduced MEPS on three-phase electric motors in October 2001. The intention was signalled to industry in the consultation process leading to the introduction of MEPS that the stringency levels would be revisited with a view to increasing the MEPS level in four years.

Australia commenced the process of establishing a revised MEPS level in 2002. The processes to revise the Australian Standard and to select the appropriate MEPS level are underway with a view to introducing the new levels by January 2006, allowing a four-year notice period for industry to adapt.

Challenges and opportunities are present in choosing the correct MEPS level for the Australian economy. Based on current reviews of opportunities in reviewing the initial 2001 MEPS level, it is estimated that substantial worthwhile savings may be realised from the raising of minimum efficiency levels to something near the 2001 'high efficiency' level, and introducing a new 'premium' class motor standard to provide a 'best practice' incentive.

Due to diminishing prospects for achieving approval of further increases in MEPS stringency levels in 2010, the question arises as to whether this round of revision to motor MEPS in Australia will be the last one. Opportunities are thought to exist for further energy efficiency gains primarily from 'energy efficiency best practice' programs, and attention to optimum application and use of motors with driven equipment.

References

[1] Marker, Tony. "Australia's National Greenhouse Strategy and Electric Motor Effciency", in Springer-Verlag, Berlin, 2000, Energy Efficiency Improvements in Electric Motors and Drives (EEMODS '99), pg. 440.

[2] Cogan, David. 2002. "A Review of the Stringency Levels for the Mandatory Minimum Energy of Three-phase Cage Induction Motors in Australia – a report to the National Appliance and Energy Efficiency Committee. 30 pgs. (unpublished).

[3] Cogan. 2002. pg 15.

[4] Cogan. 2002. pg 15.

Acknowledgements

The author would like to thank the following for contributing to this paper:

Shane Holt, Australian Greenhouse Office, Canberra
James Weston, Australian Greenhouse Office, Canberra
Lloyd Harrington, Energy Efficient Strategies, Melbourne

Motors and Drives Energy Efficiency Increase in Industrial and Civil Applications: the Research Approach by a Pool of Italian Universities

Nicola Bianchi[1], Aldo Boglietti[2], Antonino Di Gerlando[3], Marco Villani[4]

[1] Università di Padova, Via Gradenigo 6a, 35131 Padova, Italia

[2] Politecnico di Torino, Corso Duca degli Abruzzi 24, 10129 Torino, Italia

[3] Politecnico di Milano, Piazza Leonardo da Vinci 32, 20133 Milano, Italia

[4] Università di L'Aquila, Monteluco di Roio, 67040 L'Aquila, Italia.

Introduction

The energetic problems has again risen after the 1997 Kyoto Conference. In Europe energy efficiency is a key element of the European Union energy policy, since it improves the productivity of the economy, increases energy supply security and decreases harmful emissions due to the electricity generations. The agreement between the European Union and the CEMEP (European Committee of Manufacturers of Electrical Machines and Power Electronics) [1] has given a first answer to the energetic problem. For what concerns the CEMEP, from the producers point of view the European Union guarantees much more information opportunities to the users in particular for the use of high efficiency motors. On the other hand, the producers have to increase the fulfillment of higher efficiency motors and to encourage their sale, reducing, at the same time, the production of the old motors with lower efficiency. The targets imposed by the CEMEP protocol will require a very hard effort by the electromechanical Italian companies, from the research and development point of view. About the efficiency evaluation for electric householder appliances with refrigerator system, it is possible to find accurate indications in the European Union laws now in force (Document 39610057 - CE directive 96/57) [2], [3]. This law highlights that the electric energy consumed in householder refrigerator systems is a non negligible percentage with respect to the total electrical energy consumed in European Union civil applications. The European Council decision (89/364/CEE) of the 5 June 1989, regarding the necessity to start with a common program devoted to a better use of the electrical energy has confirmed that one of the European Union target is the environment and its protection and quality. In particular, an efficiency increase of the refrigerator apparatus and a solicitation to the customers for buying higher efficiency householder appliances have been highlighted. The countries, which have agreed to the European Union resolution, have adopted all the measures to guarantee that only civil refrigeration systems with an energy consumption equal or lower than a maximum one depending on the appliance category will be able to be introduced in the European market. On the basis of this situation, it is evident that the Italian electromechanical industry, which is split up in small and medium companies, could be crushed by the giant European companies. From these considerations, the academic world has to give new impulse to the applied research on the energetic problems in electrical motors and drives and a pool of Italian Universities has proposed to the "MIUR" (Italian Ministry for Education, University and Research) a research project within the framework of the "PRIN" Projects (National Relevant Research Projects).

The Italian "PRIN" Projects (Relevant Interest National Research Projects).

Every year the MIUR finances some research projects which are considered of relevant importance for the national interests. These research projects are presented by pool of Italian Universities, and the projects selection is made by means of a Guarantee Commission that involves anonymous national and international reviewers specialized in the several technical subjects. For these reasons the proposals have to be prepared both in Italian and in English language. On the basis of the reviewers selection, the Guarantee Commission prepares a priority list of the projects evaluated as positive and which can be supported from the financial point of view. The MIUR financial cover is up to the 70% of the proposed project total cost, while the other 30% has to be found by other financial resources. In particular, the financial support by industrial companies is considered in extreme positive way by the MIUR, because this fact shows a direct interest of the Italian industrial world on the research results. Each research project is developed by "Research Units" corresponding to the number of the Universities involved in the proposed Projects. Each Research Unit is coordinated by a Local Scientific Coordinator and all the Research Units are coordinated by a National Scientific Coordinator. The National Scientific Coordinator corresponds to the Local Scientific Coordinator of the University which has proposed the global project to the MIUR.

The research project presentation has to show and cover the following main sections:

- Italian and international state of art;
- Final targets to be reached;
- Developing phases of the project and related times;
- Partial targets for each phase;
- Financial cost for each phase;
- Human resources involved in the project;
- Financial resources to cover no lower then 30% of the project costs;
- Elements and criteria whose can be used to verify the obtained results.

To close this short presentation of the Italian PRIN Projects, it is interesting to highlight that in the 2001 the total amount of the financial resources available for the projects in the industrial engineering area have been of 13 million of €.

The 2001 PRIN Project "Energetic efficiency increase in electric motors and drives for industrial and civil applications"

In the 2001 a pool of Italian Universities have presented a research project entitled: "Energetic efficiency increase in electric motors and drives for industrial and civil applications" to be reviewed within the framework of the "2001 PRIN" for a total cost of 420000 €. The Project has been accepted by the MIUR with a high priority grade highlighting the sensitivity of the MIUR to the energy saving problems. The research program is two years long, starting at the beginning of 2002. The Universities involved in this research program are Politecnico di Milano, Politecnico di Torino, Università de L'Aquila, Università di Padova and they are cooditated by the Politecnico di Torino Unit. The Project total costs cover, in percentage, is reported in Table I.

Table 1.

Project Cost	Funds from MIUR	Funds from Industrial Companies	Funds from Involved Universities
420000 €	68 %	26 %	6 %

Table 1 shows that the financial support got from the industrial companies is high with a cover of the 26% of the total cost. This fact has been considered by the reviewers as a positive evaluation key for the global and the single parts of the proposed research. In addition, the immediate fall down of the research products have been considered an interesting choice key by the involved companies and a significant key for a correct evaluation of the project. The main target of the project is the development and the aggregation of the know-how, both from the theoretical and practical point of view, to be transferred to the electrical machines manufacturers interested on the production of higher efficiency motors and drives . In particular, not only theoretical results, but the immediate fall down on the companies, and the national energy saving has been mainly taken into account. It is important to underline that the research is based on the energetic problems concerning motors for industrial and household appliances at the same time. As a consequence, motors with rated power starting from some hundreds watts up to some hundreds kW are involved covering all the low voltage electrical machines. The obtained results will be particularly interesting because of their general validity in the low voltage systems.

Detailed analysis of the Project

The research main topics for the four Research Unit are reported hereafter:

TORINO UNIT- The aim of research is the efficiency improvement of low voltage three-phase induction motors in the range 0.75-55 kW, for industrial applications. The main objectives and the detailed activities are reported in the following:

- Review of the motor design for higher efficiency machines without change the punched magnetic sheet in use (No Tool Cost approach)
- experimental validation on new production technologies for increasing the motors efficiency
- Analysis of the energetic limits imposed by the new standards and the new protocols with particular reference to the European Union. Comparison of the Italian production position inside the frame previously analyzed.
- Analysis of the international countries standards for efficiency measurement methods on industrial motors.
- Research and analysis of low impact solutions on the productive process for the motor efficiency increase, such as variation of the rotor length, magnetic material influence, reduction of the additional losses. In order to verify the proposed approach, finalized prototypes will be built by the partner company.
- Review of the motor design criteria and analysis of the production technique for attaining the efficiencies imposed by the new standards and protocols. Comparison between the theoretical and experimental results
- Cost comparison of the suggested solutions with the production cost of standard motors.

L'AQUILA UNIT - The aim of research is the efficiency improvement of low voltage three-phase induction motors in the range 0.75-22 kW, for industrial applications. The main objectives and the detailed activities are reported in the following:

- Review of motors design when "premium steels" (low loss or/and high permeability) and copper pressure die-casting rotor cages are used;
- Analysis of real efficiency improvements when these innovative solutions are adopted.
- Evaluation by simulations of electrical steels and copper rotor cage influence on standard motor performance
- Results comparison with commercial motor ones in order to evaluate the achievable improvement due to the use of "premium steels" and copper cages only without affecting the design of standard motors
- Motor design optimization with low cost strategies without changing the stator and rotor shapes.
- Electrical steel selection and prototypes realization with rotor cages in aluminum and copper but without modifying the standard designs

- Analysis of the effects on magnetic properties of the used steel due to the high temperature and pressure during copper die-casting

MILANO UNIT – The aim of the research is to theoretically and experimentally analyze the high speed universal motor performances (energetic efficiency, commutation and brush wear conditions). The main objectives and the detailed activities are reported in the following:

- redesign of the universal motor, aimed to reduce energy consumption and brush wear;
- comparative analysis and, in case, choice of an alternative solution to the universal motor, for vacuum cleaners applications.
- Choice of a product with an important energetic impact (million devices), like vacuum cleaner applications.
- Analysis of the international standards concerning the energetic aspects of the chosen application and critical analysis of the test procedures and of the design criteria of the high speed universal motors.
- Evaluation of different solutions for the high speed motors.
- Development of a universal motor model able to analyze the motor behavior from the global point of view and as regards the commutation conditions, taking into account all the non linear effects.
- Experimental validation of the universal motor proposed model, using tests on commercial and modified motors.
- Redesign of the universal motor and comparison between the actual and the proposed solution, taking into account the restrictions due to costs, overall dimensions, reliability, efficiency.
- Realization of universal motor prototypes designed following the previous research activity in collaboration with the partner company.
- Comparison of the proposed solutions with the present ones from the economical point of view.

PADOVA UNIT – The aim of research is the energy improvement of a home appliance system including an electrical motor and, if necessary, the static converter.

- Analysis of the state of art for high efficiency motors used in electrical home appliances.
- Analysis of the magnetic and electric materials that better agree to the project.
- Development of a model of different types of motor for combination with electronic converter, including electromagnetic and thermal analysis. The model will be used for the optimization of the motor-compressor system.
- Comparison between simulation and experimental results using motors currently in production
- Review of the motor design criteria in order to obtain efficiency meeting the new standards and protocols. A comparison between several motor topologies

will be made and in particular, a comparison between the single-phase capacitor motor (constant speed of 3000 rpm) and a brushless motor with surface magnets supplied by a power converter (constant speed but different from 3000 rpm or with variable speed control).

- Definition of the control and simulation of the drive and system optimization for meeting the target of higher system efficiency.
- Comparison between the obtained performances and the required limits.
- Comparative evaluations of the proposed system with that adopted nowadays.
- Suggestions for line production and cost analysis

Conclusions

In this paper a research project proposed by a pool of Italian Universities on the energy efficiency improvement of electrical machines and drives has been presented. The research involves both industrial motors and motors for civil use. This choice allows to cover the range of low voltage motors starting from some hundreds watts up to some hundreds kW. The main aim of the project is to analyze and propose new design criteria and new driving modes for electrical machines, in order to increase the energetic efficiency of electromechanical systems in industrial and civil applications.

References

[1] P. Bertoldi, G. Kuehneund,"The european negotiated agreement to improve motor efficiency", Energy Efficiency Improvements in Electric Motors and Drives, Ed. Springer - Verlag, ISBN 3-540-67489-6.

[2] Commission of the European Communities, Communication from the Commission “Energy Efficiency in the European Community- Toward a strategy for the rational use of Energy. Brussels 29.04.1998 COM(1998) 246 final.

[3] Department of Energy, Sept. 17 1997: Policy statement for electrical motors covered under the energy policy and conservation act.

The Compressed Air Campaign “Druckluft effizient”: a Collaborative Action to Achieve Market Transformation in Germany

Peter Radgen

Fraunhofer ISI, Breslauerstr. 48, 76139 Karlsruhe, Germany,
E-mail: peter.radgen@isi.fhg.de

Introduction

Compressed air is an energy form offering an unrivalled range of applications and combining speed, power, precision and safe handling. These characteristics make compressed air irreplaceable in many cases. However, there are applications in which compressed air is in competition with other forms of energy such as electricity or hydraulics which are of interest. Here, the principle of economic efficiency demands a precise cost-benefit analysis. The relatively high costs of producing compressed air always have to be weighed against factors such as working speed, reliability, maintenance costs etc. The best available technology should be taken as a base. Compressed air applications have made tremendous progress in recent years with regard to energy efficiency.

However, a study made for the European Union [1] has show an economic saving potential of about 33 %. Compressed air systems actually consume about 80 TWh of electricity annually. These figure sounds quite attractive but as have been seen for many energy efficient technologies, these saving potentials remain often unused by customers, due to existing barriers. Typically this is more a question of behaviour and knowledge than a question of the availability of the technology. Stimulated by the favourable results of the EU study, the German Energy Agency (dena), the Fraunhofer Institute for Systems and Innovation Research (ISI, project co-ordination) and the German Manufacturers Association (VDMA) decided to develop and start a compressed air efficiency campaign in Germany. To get a wide recognition of the project, compressor manufacturers, manufacturers of air treatment systems and engineering and consulting firms have been invited to join forces. For the target of the campaign, Industry, the National Energy Agency, the manufacturers association and a Research Institute have joined their forces.

Finally 19 companies agreed to join the campaign and to co-finance the activities to be implemented. Financial support was also provide by the German Ministry of Economic Affairs (BMWi). Just in time to the Hanover Trade Fair in April

2001, the project group announced the start of the compressed air energy efficiency campaign called "Druckluft effizient".

Campaign Target

The campaign "Druckluft effizient" (efficient compressed air) aims to inform system operators in industry and commerce about the possibilities of cost and energy savings and to motivate them to exploit these potentials. The widest possible variety of ways to optimise technical compressed air application will be shown ranging from planning, generating, processing and distribution to financing possibilities.

The key message of the campaign is that the energy consumption of compressed air systems can be reduced by 1/3 using simple and economic measures. Lots of systems are not being operated optimally due to insufficient maintenance and service, but also due to faults in plant layout and design and mistakes made during procurement. The most common faults are leaks in the system which may result in additional electricity costs of several thousand Euro per year which often go unnoticed. The efficiency improvements will give a perfect win-win situation, as about ¾ of the total life cycle costs of a compressed air systems relates to energy consumption.

Campaign Program

The campaign has identified three main areas to work on: Information, Education and Strategic Planning. To have an easy opportunity for the identification of activities related to the campaign, a logo was developed for easy recognition, cf. Fig. 1.

Druckluft effizient

Fig. 1. The campaign logo

The main elements of the campaign are:

1. Information Platform on the Internet
2. Measuring Campaign
3. Compressed Air Benchmarking

4. Compressed Air Seminar
5. Compressed Air System Award
6. Compressed Air Contracting Guidelines
7. and many others.

Internet Platform

The different activities are all linked together, the Internet platform [2] is the main source of information for all project activities. Based on the experience, that in the field of compressed air systems their is not mainly a deficit on information available but instead the available information is hard to found or seems to be not from a neutral source, the project collects the information available, ranging from compressed air seminars over books on compressed air to public funding or technical regulations. In addition the available publications on case studies or reports related to compressed air systems in English and German have been collected and sorted by topic.

In addition additional material, the so called factsheets [3], have been developed. In two to six pages the main questions related to compressed air systems are explained, starting from the thermodynamics over compressors and air treatment up to the air distribution and tools.

An electronic newsletter and press archive, together with available photos to complement publications is also available. The newsletter gives a regulary update on the campaign and compressed air related issues.

As the site is mainly targeting the German market, most of the information is only provided in German. However, based on the strong request from foreign countries, we have started to translate and to publish parts of the information also in English on the Website.

Free Compressed Air System Audit

Many companies running a compressed air system don not know anything about it. Often energy consumption, compressed air production and required and produced air quality is not known. Also problems occurring from air leakage and from pressure drops are often not identified, not even speaking from the money wasted by wrongly planned, dimensioned and operated compressed air systems. To bring to light this potential, the free audit campaign has been set up, which allows to perform about 100 to 150 Audits. These audits will cover all sectors of industry and all sizes and types of plants, therefore allowing to have a consistent picture on saving potentials for everyone. Fig. 2 shows the distribution of applications by size of the plant and the sector of the compressed air user.

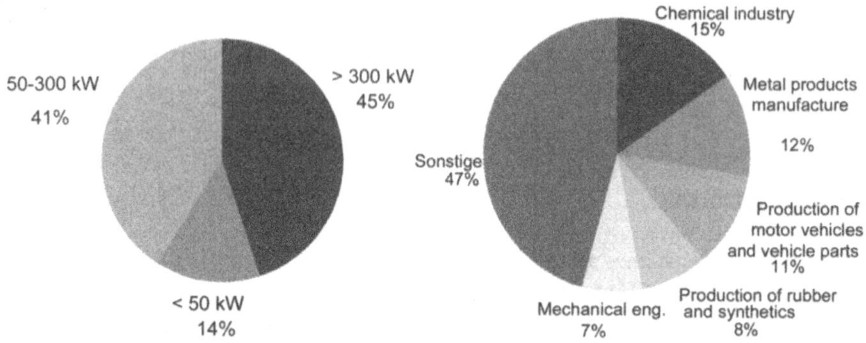

Fig. 2. Distribution of applications by size and sector (August 2002)

The results of the case study audits will be published, to inform compressed air users and to motivate them to follow with an compressed air audit at their own plant.

As a first result from the audit campaign it can be stated, that there is a great interest in receiving the free audit. During the first audits it became already clear, that in many cases the site visit of an compressed air expert without taking measurements typically already identified savings potential.

Compressed Air Benchmarking

An often encountered problem when analyzing a system is the question of comparison. How good is system is can theoretically calculated from the basic of thermodynamics. However the ideal solution can not be built. Therefore a real world comparison is necessary, which is well know under the Term Benchmarking. The campaign Druckluft effizient is actually preparing an internet based benchmarking tools for compressed air users, to give them the possibility to compare their system performance to that of their competitors or the whole industry. The data protection will be secured from the neutral scientific Fraunhofer ISI, who is responsible for the implementation and maintenance of the system. Each company will then be able to compare their Indicators in an anonymous way to the Indicators of other companies. This will enable the company to decide on measures to improve their system, if for e.g. the energy consumption indicator kWh/Nm^3 is far below the industrial or sector average.

Compressed Air System Award

A main problem when trying to improve the energy efficiency of compressed air systems is to convince the technical or the principal manager to look at the compressed air system. Therefore a compressed air system award has been set up to give a broad public recognition to a company, who has been successfully demonstrated who efficient compressed air systems can be operated, when all possible measures are realized, and who they have saved a large amount of money for their company. The application period for this years award is until October, 13th, 2002. The prize will be presented by a high level official at a special event.

Contracting Guide

During the last years, many companies have started to focus their activities on their core business. This have created the awareness for the outsourcing of some parts of the production and for the energy supply. Compressed Air can be sold like an utility such as electricity and gas, but when looking into details, a much more complex situation and contractual agreements have to be made. Often Contracting Companies are large or Contracting is offered by the local utility. Therefore the compressed air user should be carefully when making contracts. The aim of the contracting guide [4] is to inform the compressed air user about the advantages and disadvantages of different contracting systems. The guide will therefore help to ensure, that compressed air contracting will lead to an energy efficient supply with a maximum of profit for both partners.

Other Activities

Together with the described activities, additional work is done. For example a tool for calculating life cycle cost of compressors have been developed which will come online soon. The project group is also working in close co-operation with local authorities and industry association to promote the efficient use of compressed air. The project group is present at important trade fairs such as the Hanover Trade Fair and the Achema in Frankfurt, where many people from industry can be reached. However due to the broad information for the project itself and in addition from the 19 industrial partners have showed a good success.

A training course on the improvement of compressed air systems has been developed and can now be booked, either as external seminar or as internal seminar for a company alone. To facilitate the exchange of information between the manufacturers, compressed air users, consultants and R&D Institutions a compressed air conference was organized in cooperation with the association of the German Engineers.

The campaign is also offering students to make a bachelor or master thesis in the field of compressed air systems. Already two thesis have been supervised, one on the comparison of the Internet presentation of compressor manufacturers [5] and one on the acceptance for different compressed air pricing schemes by potential customers for compressed air outsourcing [6].

Conclusions

This article described the German compressed air campaign "Druckluft effizient" which has been started in April 2002 by dena, Fraunhofer ISI and VDMA. 19 Industrial Partner joined and support the campaign with the aim to achieve the significant energy savings which exists in compressed air systems. The large and cost effective saving potential has been identified by the European Study on compressed air systems, which has preceded the German Campaign.

The project group has also received strong support from the Federal Ministry of Economic Affairs (BMWi), which was important to insure a product neutral pre competition campaign, and secures the accompanying scientific evaluation of the project result. Based on the very positive feed back from compressed air user during the first project year, it can be stated, that the improvement of compressed air systems is an adequate target to achieve energy savings and increasing the competitiveness of the industry. The campaign will go on at least until 2004, therefore avoiding a flash in a pan. The energy savings achieved during the campaign will sustain and the effect of the campaign may even last longer than the campaign lifetime, as the awareness of the importance of energy efficiency in compressed air systems will be fixed in the minds of the managers.

We also recommend to use the experience of the German campaign to start similar campaigns in other parts of Europe or even on an European level. The national programs should therefore linked with European Programs such as the Motor Challenge Program, which is actually in its pilote phase.

References

[1] Radgen, P.; Blaustein, E.: Compressed Air Systems in the European Union Energy, Emissions, Savings Potential and Policy Actions. LOG_X Publisher, Stuttgart, 2001. (Online : www.isi.fhg.de/publikation/c-air/compressed-air.htm).

[2] Druckluft effizient: www.druckluft-effizient.de

[3] Factsheets Compressed Air (Different topics). Available at: www.druckluft-efficient.de/wirtschaft/fakten/fakten-dl.htm)

[4] Radgen, P.; Dudda, C.; Schmid, J.: Contracting, Finazierung, Betrieber Modelle. Leitfaden für die Anwendung bei Druckluftanlagen. Karlsruhe, Fraunhofer ISI, 2002. (Online: www.druckluft-effizient.de/wirtschaft/finanzen/finanzierung.htm).

[5] Poppe, N.: Marketingrelevante Bewertungskriterien von Internetauftritten – Benchmarking in der Investitionsgüterindustrie. Master Thesis, European Business School, Oestrich-Winckel and Fraunhofer ISI, Kalrsruhe, Germany, February, 2002.

[6] Fritz, M.: Cutting Edge Management of Utilities in Industrial Facility Management - UTILITIES AS A PRE-PRODUCT -. Master Thesis, FHS KufsteinTirol, Austria and Fraunhofer ISI, Karlsruhe, Germany, 2002.

[7] Bertoldi, P.; Blaustein, E.: MCP. The Motor Challenge Program. EEMODS 2002 proceedings, Springer Publisher, 2002. (Informations also on the Web at www.motorchallenge.de)

Public/Private Sector Cooperation to Promote Industrial Energy Efficiency: Allied Partners and the US Department of Energy

Aimee McKane[1], Chris Cockrill[2], Vestal Tutterow[1], Anthony Radspieler[1]

[1] Lawrence Berkeley National Laboratory

[2] US Department of Energy

Abstract

Since 1996, the US Department of Energy's Office of Industrial Technologies (USDOE) has been involved in a unique voluntary collaboration with industry called the Allied Partner program. Initially developed under the Motor Challenge program, the partnership concept continues as a central element of USDOE's BestPractices, which in 2001 integrated all of USDOE's near-term industrial program offerings including those in motors, compressed air, pump, fan, process heating and steam systems.

Partnerships are sought with end use industrial companies as well as equipment suppliers and manufacturers, utilities, consultants, and state agencies that have extensive existing relationships with industrial customers. Partners are neither paid nor charged a fee for participation. Since the inception of Allied Partners, the assumption has been that these relationships could serve as the foundation for conveying a system energy-efficiency message to many more industrial facilities than could be reached through a typical government-to-end-user program model. An independent evaluation of the Motor Challenge program, reported at the last EEMODS conference, attributed US $16.9 million or nearly 67% of the total annual program energy savings to the efforts of Allied Partners in the first three years of operation. A recent evaluation of the Compressed Air Challenge®, which grew out of the former Motor Challenge program, attribute additional energy savings from compressed air training alone at US $12.1 million per year.

Since the reorganization under BestPractices, the Allied Partner program has been reshaped to extend the impact of all BestPractices program activities. This new model is more ambitious than the former Motor Challenge program concerning the level of collaborative activities negotiated with Allied Partners. This paper will describe in detail two new types of program initiatives involving Allied Partners: Qualified Specialist Training and Energy Events.

The Qualified Specialist activity was conceived as a way of engaging the supply side of industry, consultants, and utilities to greatly increase use of decisionmaking software developed by USDOE to assist industrial facilities in assessing the energy efficiency of their energy-using systems. To date, USDOE has launched Qualified Specialist training with member companies of the Hydraulic Institute (HI) and with distributors and consultants associated with the Compressed Air Challenge. These activities train and qualify industry professionals to use and to train customers to use USDOE's Pumping System Assessment Tool (PSAT) and AIRMaster+ software programs, respectively. The industry experts provide a public benefit by greatly increasing customer access to the software and assessment techniques. Participating Specialists anticipate a business benefit by providing a valuable service to key customers that is associated with USDOE.

The Energy Event concept was developed in 2001 in cooperation with the California Energy Commission in response to the state's energy crisis and has been extended to other geographic areas during 2002. The three California events, named "Energy Solutions for California Industry", relied on Allied Partners to provide system-based solutions to industrial companies as both speakers and exhibitors. These one-day events developed a model for serious solutions-oriented format that avoids the typical trade show atmosphere through strict exhibitor guidelines, careful screening of speaker topics, and reliance on case studies to illustrate cost- and energy-saving opportunities from applying a systems approach. Future plans to use this activity model will be discussed as well as lessons learned from the California series.

Background

Beginning in 1992, US DOE Office of Industrial Technologies (USDOE) sought to design a program that promoted increased energy efficiency of motor systems and was responsive to industry needs. The Motor Challenge program began as the result of an industry roundtable discussion and evolved through a series of events that presented industrial end-users and the companies that serve them with a unique opportunity to share in and help shape the program. The result is a program "designed with industry for industry" that relies extensively on existing market forces to bring program messages to the industrial end-users.

The program was initiated in 1993-94 with three initial offerings: showcase demonstrations, MotorMaster software, and the Information Clearinghouse. As additional program offerings were developed and the ongoing dialogue with industry matured, a program structure emerged to deliver the program message within existing market mechanisms.

By 1996, a primary feature of the Motor Challenge program model was its reliance on partnerships with industry both to develop new program materials and to deliver this information to industrial customers. Two additional programs, Compressed Air Challenge and Steam Challenge, subsequently emerged based on elements of this program design. By 1999, other sectors, such as process heating, had expressed interest in developing their own program initiatives. Rather than continuing to create separate new programs, USDOE moved to integrate all existing and proposed program initiatives of this type under the program heading of BestPractices.

Existing markets for industrial equipment and services (both supply and demand) traditionally focus on components rather than systems. This piecemeal approach to industrial motor-driven and steam systems in the US typically results in less than optimal system operation, reliability, and efficiency. In addition, for industrial manufacturers production is their top priority, not energy efficiency. BestPractices, pursues an objective to develop best practice information and tools in cooperation with industry associations and energy efficiency organizations. These information and tools are then distributed either directly, or through Allied Partners (vendors, utility companies, end use companies and others) to end users of motor systems to promote a systems approach in the way these systems are managed, maintained, upgraded and improved.

The overall program design evolved over several years with substantial input from the industrial community. Throughout, program development has been guided by the following technical and program design principles:

- Promote a "systems" approach
  Industrial engineers have long known that careful matching of the elements of a plant system (in the case of motor systems – motors, controls, couplings, and process machinery) to the work to be performed yields far more savings than upgrading just the individual components. The Motor System Market Assessment found that over 71 percent of total potential savings came from systems-level measures such as improving the configuration and control schemes in pump, fan, and compressor systems. The practical procedures and the benefits of the system approach are stressed in program tools, publications, and case studies.
- Harness the business motivations of end-users, manufacturers, and vendors in disseminating technical information and promoting energy efficiency.
  The program has and continues to emphasize not only the energy savings associated with improved system efficiency, but other benefits of efficiency improvements such as increased control over production processes, reduced waste, and an improved production environment for workers. Emphasis is placed on demonstrating the links between energy efficiency, system reliability and lower costs of operation through case studies, plant wide assessments, and showcases.
- Recognize suppliers and users of energy-consuming industrial systems who "walk the walk" by applying a systems approach to new systems and retrofits.

Allied Partnerships are voluntary, based on shared interests, and have no fees associated with their formation. USDOE currently has more than 200 Allied Partners (APs) who distribute more than 10 times the amount of BestPractices information as is distributed through other program channels. An independent evaluation of the program completed in 2000 found that $16.9 million in annual energy savings or 67% of the total program savings was attributed to AP activity.

Building Partnerships

The first step in developing the Allied Partner effort was to identify why organizations would seek to partner with USDOE. For each type of partnership, USDOE must characterize the public benefit (in this case, energy efficiency) and work with companies to identify the intersecting private interests that have the potential to carry the desired actions forward.

Participant	Potential Contribution	Initial Motivation	Primary Drivers
Equipment Manufacturers & Distributors	Detailed technical & market information Customer influence	Brand enhancement Perceived threat to business	Sales Customer retention Increased profits Increased brand recognition
Specialized Consultants	System engineering Broad knowledge of applications/ problems	Increased visibility Legacy	Increased billable hours Recognition
End Users of Compressed Air Systems	Knowledge of specific applications & organizational dynamics	Alignment w/ green image May fix problem	Reliability Profitability Best Value Efficiency Recognition
Influencers: *Utilities* *Energy-efficiency organizations* *Government*	Market credibility Outreach mechanisms Perceived neutrality	Meets mission Highly leveraged Politically beneficial	Customer education Customer retention Energy savings Pollution prevention Recognition

Fig. 1. Partnership Motivation Analysis: Industrial Compressed Air Systems

Key Questions for Partnership Development:

- what is the potential contribution of each participant in the collaboration (why are they desirable partners)?

$ what is their initial motivation to join the collaboration?
$ what are their primary drivers?
$ what do they hope to gain from their participation?

As the result of the proposed partnership-

- What will take place to promote greater energy efficiency?
  - Enhanced existing activity- frequency, scope
  - New type of activity
  - New piece of information, software, training, etc.

If this question cannot be answered, the public benefit has not yet been identified.

Partnership Strategies

Over time, the approach to developing Allied Partnerships has evolved based on partnership activity and feedback. The current approach emphasizes targeted partnerships- those companies with multiple facilities and groups with a large industrial customer base, such as associations and supply chain companies. The objective is to reach as many industrial end users with each partnership contact through a "multiplier effect" as follows:

- one to one- USDOE agreement w/ association
- one to many- Association works w/ member companies
- many to many- Member companies work with customers or employees

In addition, Allied Partner agreements have a defined scope. They are individually negotiated for a period of one year and are renewable based on mutual agreement.

Applying the Allied Partner Model

An example of a market opportunity for partnership is illustrated by the findings of the Assessment of the Market for Compressed Air Efficiency Services, completed for USDOE, in cooperation with the Compressed Air Challenge in 2001. Phone interviews of 222 industrial compressed air users revealed the following:

- Only 9% identified controlling energy costs as a primary concern
- 71% identified **consistent, reliable compressed air supply** as a principle objective of system management
- 35% had experienced **unscheduled shutdowns** during the previous 12 months- 60% of these shutdowns were for 2 days or more
- 30% had service contracts- had no effect on incidence of shutdowns
- 75% of system operators had **no formal training** in compressed air system efficiency
- 57% had **taken no action** (include leak repair) to improve compressed air system efficiency in past 2 years

The Compressed Air Challenge® (CAC) is an outgrowth of work on Industry Partnerships for the USDOE Motor Challenge Program. The themes which ultimately led to this project were first identified in the April 1995 Roundtable on Market Transformation Strategies for Industrial Motor Systems breakout session on compressed air systems. The CAC, founded in 1997, was incorporated as a separate, not-for-profit organization that includes USDOE, equipment manufacturers and distributors, government agencies, non-government organizations, and utilities. In all the CAC counts 15 separate organizations as sponsoring members, all of which contribute both funding and time to the collaborative efforts. The purpose of the organization is to promote more efficient compressed air systems.

The CAC is a good example of the "one to one & one to many" approach of using a single point of contact, the collaboration, to reach thousands of industrial end users. The results of this collaboration to date are as follows:

- Published & distributed more than 4000 copies of Compressed Air Sourcebook
- Developed 1-Day Training "Fundamentals of Compressed Air Systems" & 2-day "Advanced Management of Compressed Air Systems"
- Qualified pool of 22 qualified CAC Fundamentals instructors; 10 qualified Advanced Instructors
- Trained more than 4000 end users, consultants, suppliers
- Website, more than 20 case studies, new projects

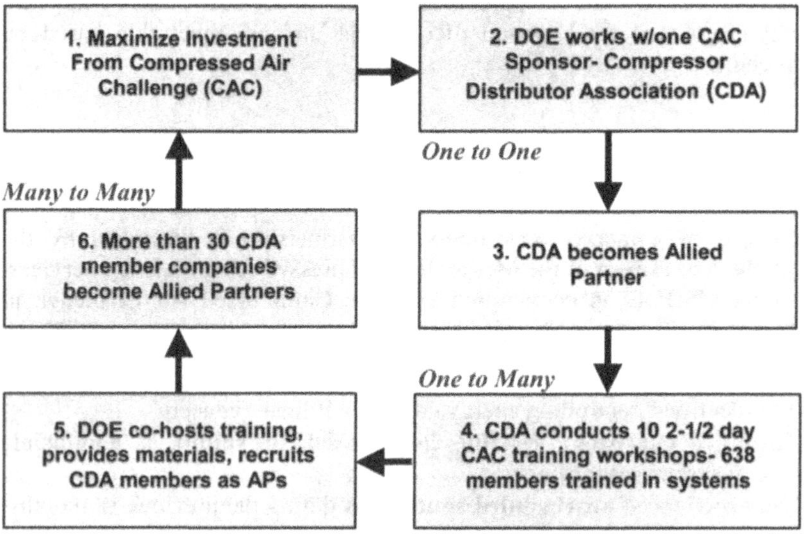

Fig. 2. Applying the Multiplier Effect to the Compressed Air Market

Increasing Use of USDoE Software Through Allied Partners

Since the development of MotorMaster+ in 1993, USDOE has worked with industry experts & key associations to develop software tools to help industrial end users identify and assess the relative importance of system improvement opportunities. The purpose of the software, which now includes MotorMaster+, Pump System Assessment Tool (PSAT), AIRMaster+, Process Heating Assessment Tool (PHAST), and the Steam System Scoping Tool, is to make it easier to identify the substantial energy efficiency opportunities that are frequently overlooked in industrial systems. Feedback from software users revealed that few people use the full capability of these tools as designed to identify system improvement opportunities. The primary barriers to effective use are 1) the time required to understand the tools and collect required input data and 2) the specialized systems expertise required to use the tools effectively.

Allied Partner Qualified Specialist Training

To substantially increase the use of USDOE software tools with and by industrial end users, USDOE worked with Allied Partner industrial trade associations and their member companies in developing a Qualified Specialist training and recognition program. This approach, which was initially conceived for PSAT in cooperation with the Hydraulic Institute and its member companies, is designed to build use of the assessment software into existing interactions between industrial customers and their suppliers. The Allied Partners, who are already highly skilled in their respective area of specialty (in this case, pumping systems) also become skilled, qualified users of the assessment software. There are several advantages to this approach. First, it addresses the two major barriers to increasing effective use of the software- the time required to understand the tool and technical expertise in systems required to make full use of the software's capabilities. The goal is to increase the number of professionals qualified to use each software tool from <5 to >50 and greater in a relatively short period (two years).

In addition, the process of developing the delivery program also encourages industry suppliers to provide detailed technical input to fine-tune the software, while building ownership in the final product. The long-term goal of USDOE is to develop a well-defined Qualified Specialist program element for each type of system assessment software and to seek out transfer of long term stewardship of Qualified Specialist programs to key industry associations. By involving these associations in the early stages of developing the software and the corresponding Qualified Specialist delivery mechanism, it is hoped that this transition can gradually evolve over a period of approximately five years from each Qualified Specialist program launch.

In order to engage industrial suppliers and consultants in the Qualified Specialist effort and sustain their interest in participating, USDOE offers the following benefits:

- Recognition as a skilled user of the software(PSAT, AIRMaster+, PHAST, etc) each Specialist gets a signed, numbered certificate;
- Listing on the DOE BestPractices Website & access to quantities of software;
- DOE refers callers and trainees interested in software to the website list of Specialists;
- May also become qualified to offer DOE software training classes to customers;
- Invitation to provide referrals for possible DOE case studies on energy-efficiency projects identified by using software;
- Invitation to participate in review of proposed changes to software; and
- Ongoing technical support.

It is important to note that the work with the Hydraulic Institute (HI), an industrial trade association for the pumping industry, and its member companies on the Qualified Specialist activity is based on an ongoing cooperative relationship with USDOE that began in 1993. To date, the Allied Partnership with HI and its member companies has resulted in the following activities:

- 1997- DOE and HI co-produce Energy Reduction Video Program
- 2000- DOE works with HI members and the Federal Energy Management Program (FEMP) to develop a FEMP Product Efficiency Recommendation How to Select an Energy-Efficient Centrifugal Pumping System
- 2000-2001- DOE works with HI and EUROPUMP to create Pump Life Cycle Costs: A Guide to LCC Analysis for Pumping Systems
- 2001- DOE publishes Executive Summary of the LCC Guide - available to pump users both in print and as a pdf on the BestPractices website
- 2001- HI and DOE launch the Qualified Pump System Specialist training using the Pump System Assessment Tool (PSAT)

The Qualified Pump System Specialist training includes 1-1/2 days of classroom and hands-on training followed by a written qualifying exam. To date, five classes have yielded 44 Qualified Pump System Specialists. In addition, five individuals have become Qualified PSAT Instructors and are offering pump system end user training.

The Qualified AIRMaster+ Specialist training is being offered in cooperation with the CAC. Again, this program element was developed in cooperation with industry suppliers and consultants within the context of a long term cooperative relationship. The training includes 2 days of classroom training on AIRMaster+ and measurement techniques, a practical exam on measurement techniques, a take home exam using AIRMaster+, and a 4-1/2 hour written qualifying exam. As with the Qualified Pump System Specialist exam, the AIRMaster+ Specialist exam is open book and designed to test the individual's ability to apply systems knowledge as well as use the software. To date, five classes have yielded 41 Qualified AIRMaster+ Specialists. These Specialists are being used as the primary avenue for delivering AIRMaster+ training to industrial end users through an awareness training module.

Energy Events

Another avenue for working with Allied Partners was developed in response to the needs of the energy crisis in California in 2001. The goal was to promote a "many to many" approach to increasing awareness of the systems approach by organizing events that provide opportunities for individual Allied Partners to meet while providing service to industrial customers. The desired results were twofold: to create new business opportunities for Allied Partners and strengthen the market for energy-efficiency services.

Three "Energy Solutions for California Industry" events were held in California in 2001 and early 2002. The model for the events was developed cooperatively by USDOE and California Energy Commission, in conjunction with an Advisory Committee of representative Allied Partners. Local Sponsors were sought to publicize events and provide topical speakers. Allied Partners were invited to exhibit and demonstrate DOE software. Care was taken to coordinate speakers and exhibits to present system solutions. Standard signage was used and product displays were not allowed. The idea was to keep the emphasis was on systems-oriented assistance to participants.

The three events attracted 563 participants from 338 companies. Six months later, 85% of participants interviewed stated that the event had a positive effect on the energy efficiency of their business. USDOE is evaluating how the model for these events can be applied in cooperation with other states and regions in the US.

Lessons Learned

Partnerships with industry (suppliers, utilities, large end users, utilities) can highly leverage the ability of the public sector to reach and influence the energy-consuming behavior of industry. These partnerships take time to develop and involve understanding the goals and agendas of all involved. Key partnerships that represent very large constituencies, such as trade associations, can take several years to mature to the point where they result in concrete outcomes. Properly managed, this infrastructure-building phase provides a foundation of ownership and trust essential to obtaining highly specialized technical assistance and broad-based organizational support for joint projects. The Qualified Specialist activities and the Energy Events are two examples of joint projects made possible through this type of partnership development.

References

[DOE] US Department of Energy. 1998. United States Industrial Electric Motor Systems Market Opportunities Assessment, Office of Industrial Technologies, December 1998, Washington, DC

[DOE] US Department of Energy. 2001. Assessment of the Market for Compressed Air Efficiency Services, Office of Industrial Technologies, August 2001, Washington, DC.

McKane, Aimee T, Chris Cockrill, and Vestal Tutterow. 2001. Making Industrial Energy Efficiency Mainstream and Profitable: Where Public Benefit and Private Interests Intersect. In Proceedings of the 2001 ACEEE Summer Study on Energy Efficiency for Industry, Tarrytown, NY, July 2001

McKane, Aimee T., Joseph P. Ghislain, and Karen Meadows. 1999.Compressed Air Challenge: Market Change from the Inside Out, Proceedings of the 1999 ACEEE Summer Study on Energy Efficiency in Industry, Saratoga Springs, 15-18 June 1999

XENERGY Inc. (2000): Evaluation of the Motor Challenge Program. US Department of Energy and Oak Ridge National Laboratory, Washington, DC (unpublished)

Actions to Promote VSDs

Aníbal T. Almeida[1], Fernando J. Ferreira[1], Dick Both[2]

[1] Dept. of Electrical Engineering, Univ. of Coimbra, 3030 Coimbra, Portugal
E-mail: aalmeida@isr.uc.pt, jeepisr@isr.uc.pt

[2] NOVEM - Utrecht, Catharijnesingel 59, Netherlands, PO Box 8242, 3503RE
E-mail: d.both@novem.nl

1 Introduction

In the previous studies[1], the application of Variable Speed Drives (VSDs) was identified as the motor systems technology having the most significant energy savings potential.

The loads in which the use of speed controls in electric drives can bring the largest energy savings are the fluid handling applications (pumps, compressors and fans) with variable flow requirements. Other applications which can benefit from the application of VSDs include conveyors, machine tools, lifts, centrifugal machines, etc..

In this paper the main results of the "VSDs for Electric Motor Systems" project are presented, which are the characterisation of current market of the VSDs, the estimation the potential energy savings through the use of VSDs and the identification of actions to promote VSDs.

2 Market Characterization

The presented characterisation of the VSDs current market in the European Union (EU) includes the number of units sold and sales value per power range (Fig. 1) and the average retail prices and total cost (including installation costs) of VSDs per kW for each power range (Fig. 2). The information was collected, through several sources (questionnaires, trade associations, large manufacturers, etc.) in each country of the study (Denmark, United Kingdom and Ireland, France, Germany and Austria, Netherlands, Portugal and Spain). These EU countries represent around 70% of the total EU VSD market, and the estimated average values were then extrapolated to the EU, based on previous SAVE studies and EU statistics. The base year for the market characterisation was 1998. Fig. 1 shows that the

[1] SAVE II project "Improving the Penetration of Energy Efficiency Motors and Drives".

VSDs market, in 1998, was dominated by low power drives in the range of 0.75 to 4 kW, representing about 76% of the total units sold in the considered countries. The number of VSD units sold in the EU in 1998 was 1 268 400, representing a total value of 930 400 000 Euros. Induction motors are by far the dominant type of motor used with VSDs, but other more advanced motor designs are entering the market, particularly in the low power range.

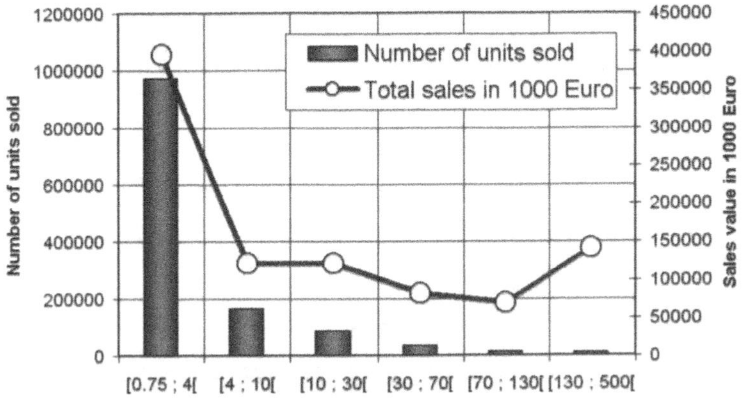

Fig. 1. Number of units sold in the EU and sales value per power range, in 1998.

As it can be seen in Fig. 2, the prices per kW decrease with the increase of the power, but more sharply in the low-medium power ranges. In fact, the prices per kW decrease until the 30 to 70 kW range, then they stabilize in the higher power ranges. The prices per unit have a more regular behaviour, increasing continuously with the VSDs power. Fig. 2 also shows the total cost (unit cost plus installation) per kW. The curve is influenced by the fact that the higher the power, the lower the installation costs per kW.

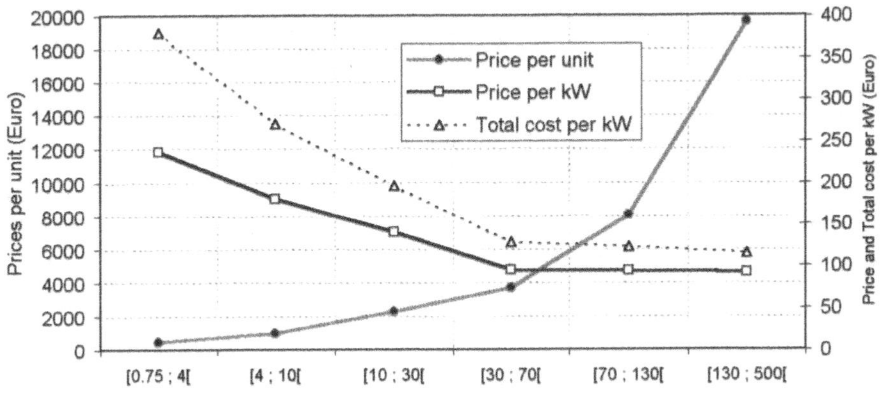

Fig. 2. Total cost per kW and average prices per unit and per kW, for each power range, in the EU.

3 Savings Potential

The estimated motor electricity consumption in the EU by 2015 is 721 TWh in Industry and 224 TWh in the tertiary sector. For the assessment of electricity savings potential with the application of VSDs, three different scenarios have been considered: the technical savings potential, economic savings potential assuming constant VSD prices, and the economic savings potential assuming a VSD price decrease of 5% per year. In general, VSDs are not cost-effective in the lower power ranges. Table 1 summarises the technical and economic savings potential in the industrial and in the tertiary sector with the application of VSDs.

Table 1. Estimated total electricity savings potential in TWh pa, by 2015.

	Economic Potential (TWh pa)		Technical Potential (TWh pa)
	Constant prices	5%/year price decrease	
Total Industry	39	43	62
Total Tertiary	8	11	22
Total	47	54	84

The identified electricity savings potential with the application of VSDs, by 2015, would translate into 19 Mton CO_2 savings (Economic savings potential with VSD constant prices), contributing to the goal of reducing the greenhouse gas emissions in the EU.

4 Actions to Promote VSDs

4.1 The market process and barriers

VSDs are typically considered as an extra component to 'common' systems or machines. They are acquired either directly by the end-user from the VSD supplier or through intermediate parties such as OEM and installers. At present both channels are used. A relatively small part of the potential end-users is willing to actively search for VSD related solutions; these typically are early market clients. VSDs have to move further to main market clients. Most of these will rely upon their traditional channels for motor driven system, such as system suppliers, OEMs and installers. For successful dissemination of VSDs consequently these parties will become more crucial . They will have to play an important role in actions to promote VSDs. Though VSDs face strong growth, the general impression is that applications are driven mainly by process-related considerations (extra throughput, process control advantages, quality, etc.). The relative growth in the market segments 'driven' by energy-efficiency advantages lag significantly behind in growth and maturity. The reasons are well known, e.g. low priority for energy efficiency, low visibility of benefits of VSDs in most energy-driven applications, and the market is dominated by lowest first cost decisions in competitive bidding procedures. Table 2 summarises the main barriers in the VSD market for the different market parties. The difference in market maturity between process- and energy-driven applications (that use the same technology) clearly indicates that the real barriers are not technology related.

Table 2. Overview of barriers in the VSD market.

Market Parties	Main barriers
End-users : industry	
In-house Utility manager (head of technical services department)	*Low visibility*: The benefits are not easily visible. In various industries the managers are not aware of the full application potential of VSDs e.g. in motion control applications, such as conveyors, centrifugal machines, etc. In buildings this is even more so (e.g. in HVAC controls, regenerative direct drive lifts, etc.). If aware, the benefits are not always easy to sell to the superior (lack of clear information on economic benefits).
	Low priority: Time is money. Priority is given to core business. Utility departments often are reduced in capacity over recent years. They have no time to assess life cycle cost, to gather and compare information, etc..
Plant manager/ production manager	*Higher initial cost*: The required extra investments compete with core business investments. The latter will be given priority.
	Perceived higher risks: Though decreasing, there are still some industries that perceive extra risk for safety and/or loss of production as a consequence of potential technical problems with VSDs.
End-users: buildings	
Investor/ project developer	*Low visibility*: The benefits are not easily visible and revert to other parties e.g. the users. The extra cost incurred by VSD are difficult to 'sell' to users that do not see the benefits.
User	*Low priority*: The user does not feel he has any influence on decisions with regard to VSDs.
Intermediate parties	
OEM	*Higher initial cost* (sales price): Fear to offer a higher price than the competitor.
	Higher initial cost (design and engineering cost): Extra design and engineering cost for OEMs in incorporating speed control.
	Extra expertise: Need for additional (electronics) expertise.
	Technical constraints: Though decreasing, some OEM have concern over interference, harmonics or need for additional cooling.
Installer	*Lower profit margin*: Little margin to be gained on inclusion of VSDs in sales package.
	Higher initial cost (installation/sales): Fear to offer a higher price than the competitor.
	Higher initial cost (engineering): Extra engineering cost for incorporating VSDs.
	Extra expertise: Need for additional (electronics) expertise.
Suppliers	
VSD supplier	*Marketing dependency*: Lack of direct contact with the main market end-users, in particular smaller end-users. In these segments distribution depends largely on OEMs and installers.
Influencing parties	
Advisors/consultants	*Low visibility*: Lack of 'independent' information on life cycle cost.
Energy suppliers/ electricity distributors	*Technical constraints*: Some concern over power quality problems. Here though VSDs may be both the 'victims' or the 'villains'.

4.2 Possible actions to promote VSD

The study identified that: a mix of measures is needed, directed especially to the end-user, to increase the importance and awareness of more efficient VSD systems, and to the intermediate parties, to improve the services offered to the clients (total solutions, that do not require extra effort from the end-user).

Table 3 summarises the various actions to promote VSDs and gives a brief indication on their cost-efficiency. Most actions though will be cost-effective especially when carried out simultaneously with others. The following considers combinations of actions with significant synergy. This should be treated with care though, given the difficulty in predicting effects. Most actions are system related and not solely for VSDs as such. The present market requires a system approach. Most energy benefits with VSDs also result from their integration in systems.

Table 3. Indicative summary of cost-efficiency of various actions in disseminating VSD.

Overview of actions				
The actions	**Cost**	**Likely cost-efficiency**	**Time to effect**	**VSD applications that (may) benefit**
Negotiated agreements				
- on energy efficiency	Medium	Limited	Medium	All
- on utilities	Medium	Good	Medium	All
Procurement/contests/awards	High	Medium		Only possible for some priority subsegments
Labelling/testing/standards:				
- for VSDs	High	Low	Medium	Not relevant
- for systems with VSD	High	Low	Long	Not considered feasible
Joint action of OEM sectors	Medium	Good	Medium	Priority segments
Information/training				
- decision support tools and databases	Medium	Good	Medium	Per application type
- guidelines, formats, cases	Limited	Good	Short	All
- training material	Limited	Med./Good	Short	All
- articles, PR, internet	Limited	Good	Short	All, mainly as support to other actions!
Technical demonstration projects	Medium	Limited	Medium	In present market little added value
Subsidies/fiscal incentives	Medium	Limited	Medium	To be considered if specific financial barriers occur with other actions
Negotiated agreements with:				
- VSD suppliers	Medium	Low	Medium	All
- OEM sectors	Medium	Medium	Medium	Priority segments
Outsourcing:				
- guidelines	Limited	Good	Short	
- case material	Limited	Good	Medium	

The study team distinguishes between three basic approaches to transform the VSD market. None of these will likely do the job alone; however they can be considered as extremes in which, depending on preferences of the policy makers, a balance should be found.

The 'awareness' approach - This aims to increase awareness with relevant parties. The assumption in this approach is that the main barrier is lack of information and know how. This approach could be adopted in two ways:

- as a sole action package of information and training, it would likely increase demand to some extent.
- as support to actions in other approaches, providing user-friendly information on application and economics.

The 'demand stimulation' approach - This approach focuses on increasing market demand. The core would obviously be negotiated agreements with end-users on utilities. This step is planned in the Motor Challenge Programme. End-users will only embark on such agreements if the EU offers extra actions, such as promotional activities that improve the reputation and image of the clients and supporting activities that ease implementation, such as audit tools, decision support tools, case materials, subsidised audits, etc.. This approach will likely be cost effective.

The 'improved services' approach - The focus in this approach is stimulating and facilitating system suppliers and installers to develop product or service packages that better suit present market demand. The suppliers have to shift from product seller towards provider of a total solution (integrating VSDs). These could range from integrated VSD-pump systems to more comprehensive 'built, own, operate and maintain' services for utilities.

5 Conclusions

In order to promote the large scale penetration of VSDs, the authors recommend a mix of the awareness and demand stimulation approach. The ongoing Motor Challenge Programme offers a good basis for integration of actions for dissemination of VSDs. It is recommended to add one or more pilot actions with the supply side in specific target segments that aim at enabling improved services and/or systems ('joint actions').

References

[1] ISR-University of Coimbra, "VSDs for Electric Motor Systems", European Commission, Directorate-General for Transport and Energy, SAVE II Programme 2000;

[2] ISR-University of Coimbra, Motors Study Group, "Actions to Promote Energy-Efficient Electric Motors", DGXVII, European Commission, October 1996;

[3] ISR-University of Coimbra, "Improving the Penetration of Energy-Efficient Motors and Drives", European Commission, DGTE, SAVE II Programme 2000;

Energy Saving: Technology Makes the Difference, Joint Forces Make Technology

Gianfranco Pozza[1], Federico Di Santo[2], Maurizio Russo[3]

[1] ITACO S.R.L., Via Riviera Berica, 40, I-36024 Ponte di Nanto, Italy
Ph. +39 0444639922, Fax +39 0444638213, E-mail progettazione@itaco.it

[2] REEL S.R.L., Via Riviera Berica, 42, I-36024 Ponte di Nanto, Italy
Ph. +39 0444730003, Fax +39 0444638213, E-mail ut@reel.it

[3] ELECTRO ADDA S.p.A. , Via Nazionale, 8, I-23883 Beverate di Brivio, Italy
Ph. +39 0395320621, Fax +39 0395321335, E-mail maurizio.russo@electroadda.com

Mankind has been inconsiderately exploiting our planet's resources for years , and industrial countries lately have made efforts to introduce machinery and devices able to reduce the energy consumption as far as possible, which is one of the main concerns regarding the resource exploiting.

Another aspect motivating the choice is the resulting economic advantage, as the costs for the electric energy production are steadily increasing.

We can say that the ambitious target for the years to come is to contain the exceeding energy consumption and thus reduce the pollution of the environment. As the restoring costs are laid upon those who exploit environment , to reduce pollution also means a definite cost reduction and consequently ENERGY AND COST SAVING.

To achieve such goal, a strong commitment and big efforts from a point of view of research and development are required , the only way being a technology of higher performances.

Electro Adda S.p.A. and **Reel S.r.l.** have decided to join forces and expertise to offer new solutions for the market . Thanks to their synergy a new company was founded that combines know-how and experience: **EARP S.p.A.**

The first ambitious target was to find new energy saving solutions.

EARP manufactures and distributes complex drive systems, where drive and motor are studied and developed as one system, although a separate application on heterogeneous systems is possible as well. The synergy has allowed developing advanced solutions in which the efficiency and performances of the single components are among the best available ones on the market. New ideas are constantly developed and soon transferred on highly technological products.

EARP - REEL - ELECTRO ADDA, together and individually choose setting up strategies aiming at saving or even producing electric energy, either directly or through recovery.

Among many applications we would like to mention some examples:

1. production of electric energy through wind power plants
2. engines test bench systems recovering the braking energy to the mains
3. centrifuge supply within sugar production lines

Single-Bladed Wind Power Plant - 350 kW

Foreword

The exploiting of the wind energy is an excellent example for environment-friendly technology. On the other side, the installation of wind power plants is rather expensive and requires much work. A wind power plant is located near the sea or the mountains, where the wind can be fully exploited. The installation is not always comfortable ,due to the need for bulky instruments and materials to build up several generating towers.

That was the reason why a project with a smaller, single-bladed tower of about 350 kW was started, installed inside several units to satisfy a medium consumption.

The single-bladed generator is easily assembled with a crane and consists of one hollow metal tower, 3 m high, having a cabin on top that contains the motor group and the electrical cabinet. The cabin turns following the wind at a 90° angle towards the wind direction and thus opposing maximum resistance and exploiting the force at a maximum level. When they are not running, the blade lies in a line with the wind direction.

Generator by EARP

EARP actually deals with the construction of electric devices for the conversion of wind energy into electric energy. Combining the resources ELECTRO ADDA (motors) and REEL (electronic drives), EARP is able to offer the whole control system by a single supplier.

As already mentioned, the generator has a single blade and generates up to 350kW. Several groups of generators are usually in stalled in parallel.

The motor manufactured by Electro Adda is an asynchronous one with wound rotor, provided with brushes that make the rotor windings available for the Reel inverter; the drive takes care of the mains synchronism and manages the energy recovery through the rotor supply, obviously aiming at obtaining the highest recovery level.

The rotor is wound for voltage 1850V so that the current is limited to 102 A. The inverter controls the slip in order to keep the voltage within the mentioned values. The system is able to recover electric energy through both the stator and rotor circuit optimising the efficiency.

The inverter by Reel is a regenerative drive connected between rotor and mains to control the stator flux and recover both the stator and rotor energy.

A pre-charge circuit supplies the intermediate circuit's capacitors step by step at POWER-ON. Once the dc link is supplied at nominal voltage 535 V (380V * 1.41 supply), the inverter on the mains side increases the value on the dc bus to 700V dc in order to control the section on the motor side. When the wind is strong enough, the previously synchronized stator (through the rotor circuit at 380V 50 Hz) , is directly connected to the mains. The inverter controls the slip by accurately handling the rotor voltage/frequency . Thus the conversion of wind energy into electric energy begins. The current recovered through the stator results from the stator-rotor slip processing, and its features strictly depend on the motor performances. The energy surplus of the rotor circuit is recovered to the dc link and recycled through the inverter on the mains side, that inverter modulates in mains synchronism at frequency 5kHz: the harmonic distortion (THD) of the recycled power never exceeds 4%; the related cos-phi value is near 1.

Test Bench Control Systems

Foreword

Since their introduction and until nowadays, most inverters have been provided with a diode or thyristor rectifier bridge on the supply section. Such solution offers several advantages such as strength, compact dimensions, low cost, etc. Nevertheless, some drawbacks are given, too.

A diode bridge generates high harmonic distortion on the supply and the output voltage can reach a maximum value equal to the supply voltage.

Besides, the energy can flows just in one direction from the mains to the motor and not in the opposite way.

When a PWM rectifier is used, the converter can operate in both directions; besides, the THD is sensibly improved and the cosφ can even be set. In fact, a 6 pulse rectifier along with a well dimensioned choke puts in a 30% higher current distortion (THD) while, when using an IGBT input bridge, the THD value can easily reach 4%.

Thus the effective current delivered by the mains decreases, which also means lower losses on cable and transformer.

The IGBT input bridge allows increasing the rectified voltage on the dc bus beyond the peak value of the mains voltage, so that a motor can be supplied with a higher nominal voltage and the output current values can thus be decreased.

Since the energy is recovered to the mains, it is no longer necessary to dissipate on a brake resistor in order to brake a load; that means that many applications such as cranes, lifts and elevators, centrifuges reach higher performances; besides, the system can be used for several more recent applications such as wind energy plants and engine or electric motor test benches .

A test bench system can be described as a "machine" able to absorb mechanical energy. Test benches find application in several fields: motor power measures, gear box or transmission tests, etc.

The most commonly used test benches are currently of three types, according to the related brake: *hydraulic, eddy current, dynamic test benches.*

Hydraulic Brake

Inside the hydraulic brake, the rotor is braked by water flowing inside rotor and stator that also provides cooling. The control is obtained by changing the water pressure inside the dynamometer. Main features of this system the low inertia and the fact that high power values compared to the size can be braked. Power values up to 1800 KW. Acting as brake only. Response time: hundreds of ms.

Eddy Current Brake

The rotor is braked through the magnetic interacting with the stator (principle of the Foucault currents). The generated heat is removed by the stator through the water cooling circuit. Typical for such brake is the control accuracy. It acts as brake only. Response time: values up to 100 ms.

Until today, such brake type is generally much used in the field of electric motor or engine test benches. On the other side, an eddy current brake causes higher energy consumption: taking an engine test bench as example, energy is required to run the motor under test (fuel or electric energy), supply the brake, remove the generated heat.

Dynamic Brake

The dynamic brake is usually a AC motor used as generator. "Dynamic" means basically that it can also run as motor. The mechanical energy of the motor being tested is transformed into electric energy that is recovered to the mains through a converter.

This is the system that allows exploiting the braking advantages at a maximum level, while reducing the waste of energy and making possible to save on the cooling system from a point of view of costs and space.

The cooperation Electro Adda-Reel made the construction of highly dynamic test bench systems possible. The performances obtained with the vector drives by Reel combined with special Electro Adda motors are: torque response = 2 ms from +100% to -100 % of the load; torque control accuracy 0,1% with external torque loop (torque flange or load cell), etc.

Electric Drives for Sugar Centrifuges

Foreword

Until the 16th century, the raw material for sugar production was the sugar can; in 1747 the beet was discovered for sugar production as well. Since then, the basic beet working has remained unchanged. Beets contain about 16-18% sugar; they are washed, cut and put into hot water flow. After that, the whole is purified with lime and concentrated by evaporation until a sugar concentration of about 50% is obtained. The mass is then cooked at 70° C , while micro-crystals are added acting as condensing agent that crystallizes the sugar when the liquid evaporates.

The centrifuges are the final step, extracting the more or less white sugar crystals from the thick sugar syrup. The mass goes then into the centrifuge drum with about 650 Kg load, then accelerated. The load of the cooked mass come from the tank through an adjustable, pneumatically controlled gate. The centrifuge force makes the sugar crystals deposit on the grid and the extracted liquid is recycled. The crystal washing step follows, through a pneumatically controlled gauge. After the low speed deceleration step (about 70 rpm) an oblique scraping ploughshare going into the drum removes the sugar layer.

The full centrifuge cycle lasts for an average of 3 minutes and follows the steps described below:

MATERIAL LOAD ACCELERATION	5 sec.
MATERIAL LOAD	18 sec.
ACCELERATION FOR CENTRIFUGE	50 sec.
CENTRIFUGATION	15 sec.
BRAKING TO UNLOAD	50 sec.
MATERIAL UNLOADING	45 sec.
TOTAL CYCLE TIME	183 sec.

The described application regards tow batteries of 5 650Kg ASEA centrifuges located in the plants of ERIDANIA S.p.A, in Russi near Ravenna.

Each centrifuge was previously driver through a Ward-Leonard system with a 55kW dc motor, which was replaced with a 75 kW 6-pole three-phase asynchronous motor.

The installed power needed to be increased in order to reduce the cycle time required by the Customer.

The centrifuge motors are controlled with AC WAVE 2 inverters by REEL supplied by one dc bus allowing to recover the braking energy of one motor to advantage of another one which is accelerating or in working in torque.

The working cycles of the 5 centrifuges were studied and synchronized to exploit the maximum energy between the motors running in torque and the recovering ones; thus the consumption from the mains supply was sensibly reduced.

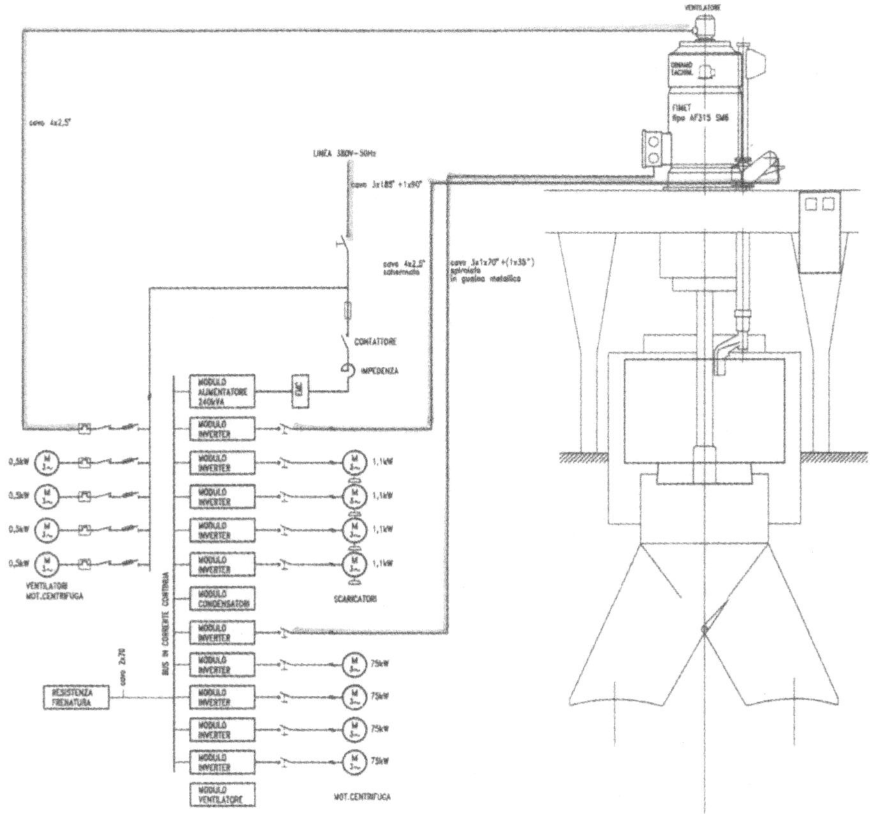

As regards the motor size, the inertia related to the mechanical parts composing the centrifuge had to be considered, especially: rotating shaft, basket, emergency brake, damp material, dry material.

Basing on the working cycle's speed and time values required by the Customer, the related torque values were calculated, needed to carry out the centrifuge cycle:

1. Resisting torque at the load step
2. Acceleration torque to reach the centrifugation speed
3. Resisting torque during centrifugation
4. Braking torque for deceleration
5. Resisting torque at the unload step

The different weights between damp and dry material were obviously taken into account, which means system efficiency and possible value derating due to the heating caused by the inverter supply harmonics.

As shown on the speed chart below, the motor is partially used during the cycle, around the flux reduction zone during the centrifugation step (about .63Hz), when lower torque is required.

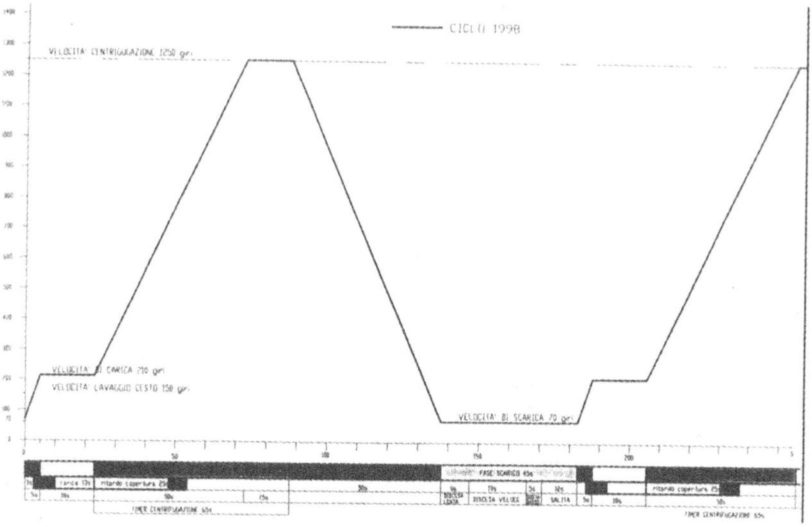

Basing on the required working cycle and the calculated torque values, the power curve of each single machine can be determined; through PLC the start during time is managed in a way to reduce the energy absorbing from the mains to the minimum.

The diagrams show that the average power absorbed by a machine during one cycle is about 30 kW, which means 150 kW with 5 working machines controlled by stand-alone drives.

Thanks to the Wave 2 system having a common dc bus, the average power absorbed by the 5 machines together is about **25-30 kW**.

The data that were kindly given by Eridania regarding the energy consumption of the sugar beet campaign 2000 confirm a value very near to the one calculated above.

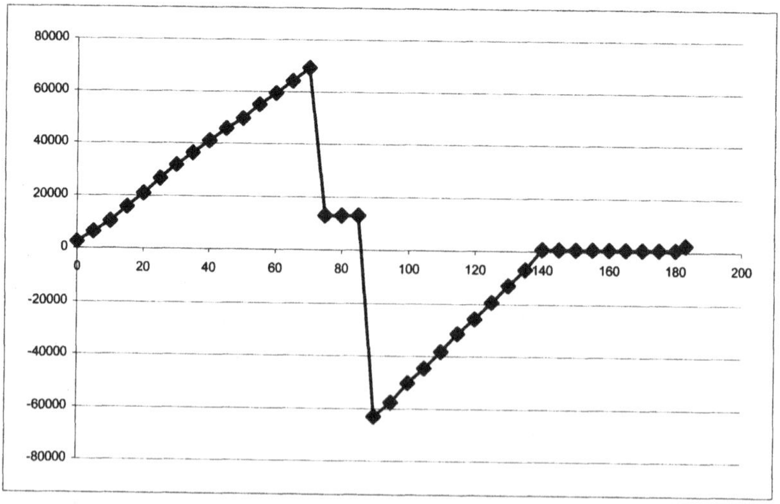

Power referring to a single centrifuge

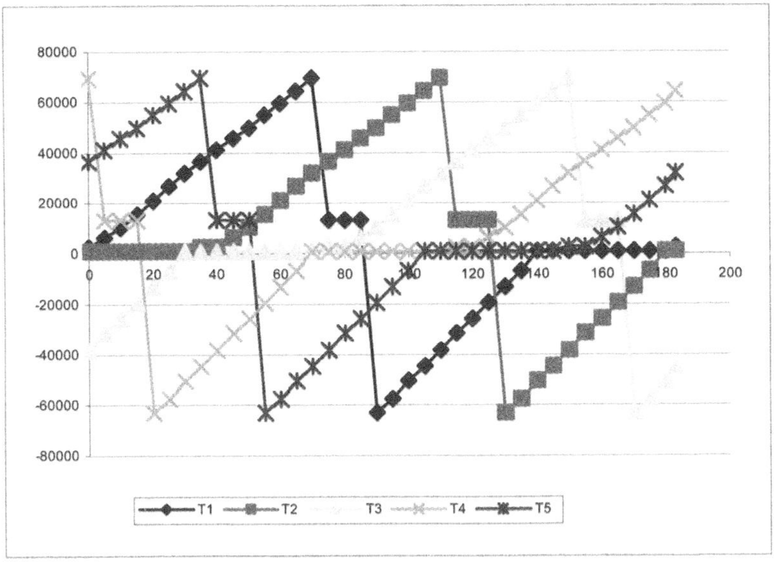

Power referring to the cycle of 5 centrifuges

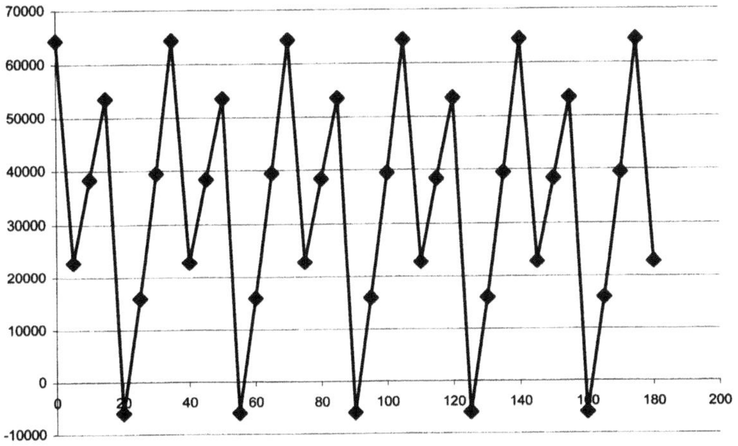

Power resulting from the cycle of 5 machines

The following aspect should be added: In a machine battery where, due to material missing one or more centrifuges were excluded form the working cycle so that the motors turned without respecting the planned sequence, a much higher consumption was found compared to the centrifuge battery that worked continuously 24 hours a day during 105 days.

The energy saving resulting from the REEL system in comparison to a stand-alone solution can be easily seen. Besides, production increases of 30% thanks to the cycle time reduction.

Low-Cost Inverters for Pump and Fan Drives

S. Williamson, S. McDonald, A. Green, M. Barnes

UMIST, PO Box 88, Sackville St, Manchester M60 1QD

Introduction

In a paper originally presented at the First International Conference *Energy Efficiency Improvements in Electric Motors and Drives* [1], Prof T de Almeida examined the potential for energy savings in the European Union by the use of high-efficiency motors and/or drives. His data clearly showed that whilst the energy savings that may be obtained by switching to EEMs are significant, they are not as large as those that may be achieved if VSDs are adopted.

The main barrier that prevents the widespread use of VSDs in flow control applications is the cost of the electronic speed controller or converter. This is particularly true for smaller power ratings, where the cost per kilowatt is greatest, and the speed controller typically costs around three to four times as much as the motor. As a result Almeida concluded that VSDs will not be cost-effective for flow control purposes below 37kW, and will have limited penetration above that size. This caused him to estimate the *likely achievable* energy savings in 2010 as 47.7TWh per year (VSD with standard efficiency motor) or 73.8TWh per year (VSD with EEM). However he suggests that there are an estimated 80TWh per year of energy savings that are technically feasible but economically non-viable because of the converter cost.

In recent years the technology trend in converter development has been towards the enhancement of induction motor performance to achieve the degree of controllability that was previously obtained only with d.c. motors. This has produced ever more sophisticated electronic systems, which are capable of giving fast and precise control of speed and torque. However, speed controllers are usually designed with the standard off-the-shelf industrial induction motor in mind, even though such machines are themselves designed for operation from the three-phase mains. The design of such a motor is a compromise between the requirements of a good starting performance and a good running performance, which often call for diametrically opposed design features. A possible key to reducing drive system cost, therefore, is to invert the standard process and make the controller the primary focus of the design. In order to do this, all assumptions that arise from considerations of the nature of the motor (such as excitation waveform or phase number) have as far as possible been dispensed with. The focus must be solely on the design of an inherently low-cost controller. It is well understood that this will have design and cost implications on the motor but the relative costs are such, that a modest saving in the cost of the controller could more than offset an increase in the cost of the motor, thereby reducing their combined cost. This paper reports the results of such an investigation, based on a 7.5kW bipolar (inverter) drive, as this rating currently has one of the highest installed total capacities per annum in the EU for both motors and inverters. The work was sponsored in the UK by the Energy Efficiency Best Practice Programme [2].

Cost Reduction Opportunities

A detailed cost analysis of the components of a 7.5kW three-phase inverter, using industrial cost data provided by a leading European manufacturer, indicated the most promising avenues for investigating cost reduction to be, output stage power devices (26%), heatsink(s) (8.5%), fan (4.1%). The percentages given in brackets in the above list refer to the proportion of total component cost for a basic inverter. The inverter in question employed a straightforward $v \propto f$ control, and did not incorporate a mains input filter.

Device Rating

The cost of the IGBTs depends on many factors, not the least of which is the economy of scale that can be obtained when production quantities are large. Other things being equal, however, the cost of an IGBT is roughly proportional to its current rating because the cross-sectional area of silicon required increases in proportion to current. For general purpose motor inverters, devices are rated to carry the peak value of the anticipated full-load current, scaled by an overload factor of between 150% and 200%. If the end-use of the inverter is known (eg a fan or a pump) a much lower value may be used. One manufacturer, for example, already uses an overload factor of only 110% for specially designed HVAC inverters.

Phase Number

Conventional inverters produce a three-phase output, largely because they are used to excite three-phase induction motors which can be bought off-the-shelf. However, it is appropriate to investigate the relationship between phase number and cost since an inverter could in principal be designed for any phase number. The breakdown of the cost of a 7.5 kW three-phase power module (derived from the manufacturer's cost for a 750W module) is given in figure 1, which shows that the IGBTs and freewheel diodes account for almost 80% of the module cost. The cost of these two component parts depends to a large extent on the silicon cross-section required. This is largely insensitive to phase number, because the product of the phase number and the rated current per phase remains constant for an inverter of given rating. The costs of the substrate, wire bonding, and gate drive circuits, which together account for 15% of the total cost of the 7.5 kW three-phase module, are assumed to increase linearly with phase number, because of the proportional increases in manufacturing steps and component count, figure 2.

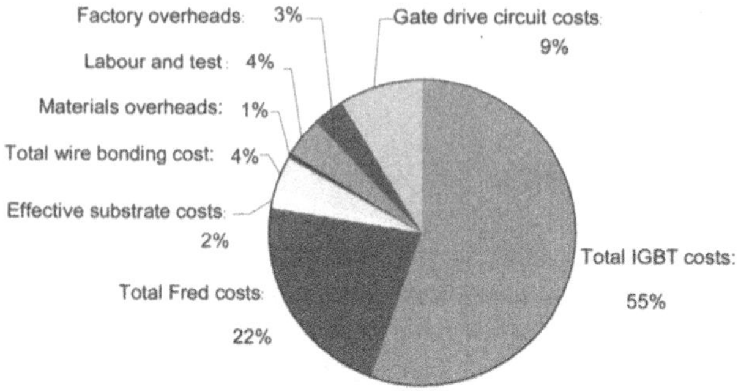

Fig. 1. Cost breakdown for a 7.5kW three-phase power module

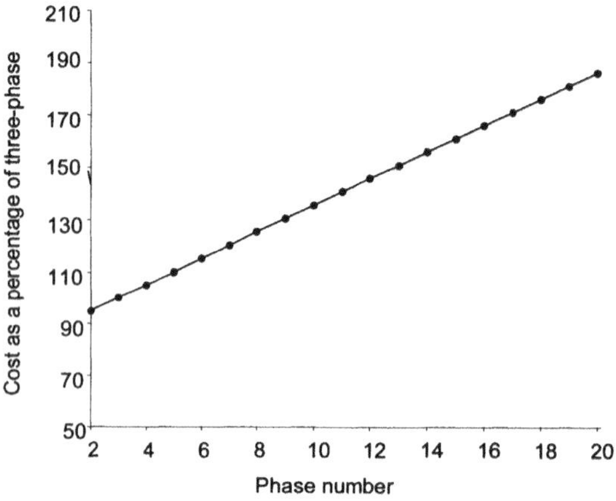

Fig. 2. Variation of power module cost with phase number 7.5kW

Figure 2 needs to be interpreted in the context of the system as a whole. As will be discussed in section 2.3, one method for reducing inverter losses (and thus the heat sink cost) is to use quasi square wave switching. This has the disadvantage of increasing the harmonic content of the current waveform, however, and a multiphase approach may be needed in order to combat this. The use of a phase number smaller than three produces a system in which the line currents do not sum to zero at each instant in time, and so a more expensive inverter topology would have to be adopted. The least expensive inverter drive system might therefore have more than three phases, although the power devices have a higher relative cost.

Heat Sink and Cooling Fan

The cost of the heat sink depends on its physical size, and this in turn depends on the amount of heat it is called upon to handle. Reducing the module switching and conduction loss will therefore reduce the cost of the heat sink and/or enable a smaller cooling fan to be used. The switching frequency of a modern 7.5kW inverter is typically 4kHz, and there is accordingly some saving in loss to be achieved by reducing this frequency.

For the purpose of assessing semiconductor module loss, the load has been represented by the simplified equivalent circuit of a series-connected resistor and inductor, sized to give a power factor of 0.8 at 7.5 kW and rated frequency. The drive manufacturer currently uses two alternative 'space vector' switching strategies, referred to as SVM1 and SVM2. SVM2 is known by the manufacturer to reduce inverter losses. Clearly a strategy that would reduce the number of switching events per cycle to its bare minimum is Quasi Square wave (QS) switching, in which each device is turned on and off only once per output cycle.

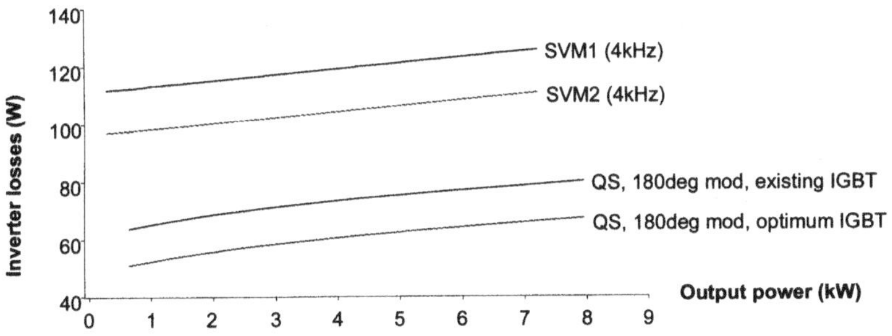

Fig. 3. Inverter loss for different modulation schemes

Figure 3 shows the variation of inverter loss with load for SVM1, SVM2, and QS switching, using the power module currently used in a commercial 7.5kW inverter. Also included is a curve showing the loss using QS switching with power devices in which conduction loss is reduced at the expense of switching loss. The disadvantage of QS switching is that it results in a high harmonic content in the voltage waveform applied to the motor, to which the motor must be designed to be tolerant. It has already been remarked in section 2.2 of this paper, that one possible strategy to achieve this is to adopt a multiphase approach, which could have the effect of increasing the module cost. The results presented in figure 3 for QS switching have assumed a modulation angle of 180^0. The use of such a scheme would require a variable d.c. link voltage, which would add to the cost on the inverter. In practice a fixed d.c. link voltage would be employed, and the magnitude of the voltage applied to the motor varied by varying the conduction angle. However, as both schemes would lead to 180^0 conduction at rated frequency and full load, they would produce the same rated full-load losses, and therefore require the same size of heatsink.

The calculated maximum thermal resistance (based on full-load and maximum ambient temperature) is typically used to determine the required size of the heatsink and fan, with due allowance for a margin of safety. In addition to heatsink and fan type, the other main parameter in making this calculation is the junction-to-case thermal resistance of the semiconductor. This depends on both the current rating of the device (which dictates its physical size) and on the style of packaging. Figure 4, curve (a), shows the effect of device rating on the maximum allowable thermal resistance between the heat sink and the ambient. It is derived from a curve fitting routine employed using manufacturer's data for a large number of devices. For this purpose the motor load was again represented by a simple RL circuit, having a 0.8 power factor (p.f.) at rated frequency.

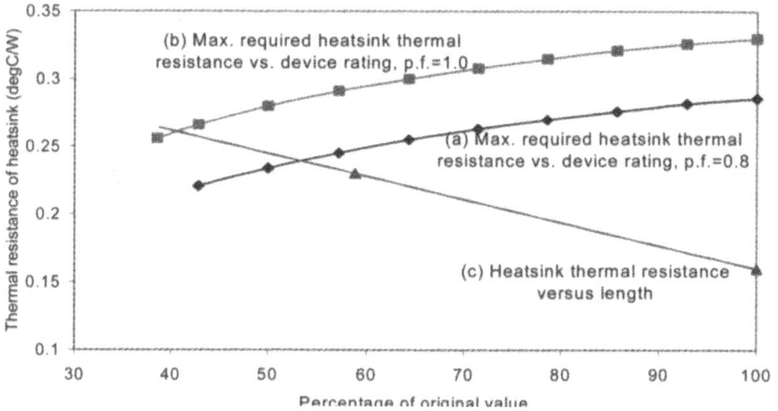

Fig. 4. Effect of device rating on size of heat sink

There are two effects that come into play as the rating of the IGBTs is reduced. The first is an increase in their power loss for a given current waveform, because the cross-sectional area of the silicon is smaller. The second is an increase in the thermal resistance between the junction and the case, for the same reason. Both of these effects conspire to raise the junction temperature, and so the size of the heatsink has to be increased in order to keep that temperature within limits. To put this requirement into context the variation in thermal resistance of the type of heat sink typically used versus size (length) is also shown, curve (c), based on the two data points available and linear interpolation.

Increasing the power factor reduces the current required to produce a given output power, and this in turn reduces the inverter loss. The effect that this has on the heatsink may be gauged from figure 4, curve (b) which shows the variation of the maximum thermal resistance with device rating, for QS switching and unity power factor. Increasing the power factor has the effect of shifting the required thermal resistance versus rating characteristic upwards by approximately 0.045 degC/W, allowing a further reduction in the size of the heat sink.

Estimate of Component Cost Reduction

For a 90% efficient 7.5kW drive with a 0.8 pf load, rating the IGBTs to meet the full load requirement only (that is, no overload capability) would enable the device rating to be reduced from the 35A currently used in commercial inverters to 17.5A if QS switching is used. The maximum acceptable thermal resistance for the heatsink may now be obtained from fig 4 as approximately 0.23 degC/W. Allowing a margin of safety, this suggests that a heatsink of approximately 65% of the length of the one currently used would suffice. The reduction in component cost resulting from these changes may be estimated as

$$\frac{35-17.5}{35}\times 26\% + 0.35\times 8.5\% = 16.0\% \tag{1}$$

Based on information provided by inverter manufacturers, component cost is typically 80% of overall cost. This suggests a reduction of around 13% in the manufacturing cost of the inverter may be obtained. If the power factor were increased to unity, the IGBT device rating could be reduced to 13.5A. Again from fig 4, values for the heatsink thermal resistance (0.25 degC/W) and length (50%) can be found for this case. Using the same method as in (1) this suggest a 20.2% reduction in component cost, i.e. a 16% reduction in converter cost.

Discussion

The cost reduction measures that have been found to show greatest promise call for the power semiconductors to be derated and a simple quasi square wave switching strategy to be adopted. The first of these implies that the motor and the driven load are both well-defined and undemanding, such as fans, pumps, and some types of compressor. The use of quasi square wave switching poses other problems arising from the associated increase in harmonic currents that normally accompany this type of switching. These can cause increased motor heating and noise, increased line filter cost, and problems with conducted EMC as well as a more expensive d.c. link capacitor, or reduced capacitor life. The key here is the motor design. Inverter switching causes a square wave *voltage* to be applied to the motor terminals, and the resulting *current* waveform depends on the harmonic input impedances. It is possible that the harmonic content of the current waveform may be attenuated to lie within acceptable limits by appropriate motor design.

References

[1] A de Almeida, P. Bertoldi and W. Leonhard (eds.), "Energy Improvements in Electric Motors and Drives", Springer-Verlag, Germany, 1997.

[2] S Williamson, M Barnes, and A M Green, "Low Cost Controller for Fans, Pumps and Compressors – Feasibility Study", ETSU agreement E/RD/00913/00/00/4581, 2001.

EcoEfficient Drives – From Product Development to Recycling

Fausto Belotti

ABB Industria SpA
via Luciano Lama,33, I-20099 Sesto San Giovanni, (Milan) Italy
phone + 39 02 2414 3732, fax + 39 02 2414 3979
E-mail: fausto.belotti@it.abb.com

Introduction

In the electrical and electronic equipment sector, as in the rest of industry, companies vary considerably in their rate of progress on environmental issues.

A lot of them are still considering how to build their own Environmental Management System (EMS) for which all the basic policy have to be placed down and followed accordingly.

There are many ways in which the "environmental phenomenon", with EMS and ISO 14001, is analogous to the "quality phenomenon", with QMS (Quality Management System) and ISO 9001. In fact, there are plenty of synergies between environmental and quality management.

Essentially, the two work in parallel: in both processes companies work towards achieving and demonstrating compliance with a set of requirements specified in an international standard.

Organisations that have built an ISO 9001 based on quality management system therefore have a solid foundation on which to build their EMS.

The international organization for standardization (ISO) published ISO 14001 in 1996, three years later the EU published its EMAS (Eco-Management and Audit Scheme) regulation. It provides a framework for integrating environmental factors into industrial operations, and required that the organization demonstrate commitment towards continual improvement and compliance with the relevant legislation.

Variable speed drives market is very competitive and the customers are well aware of different product performance, optional features and cost of different products. Environmental performance of the products is becoming more and more important also in this business.

To make the product comparison easier for the customers, manufacturers have started to publish Environmental Product Declarations (EPD), which provide comparable and verified information about the environmental of products impact and services.

One of the most important properties of a certified environmental product is that it provides comparability between declarations within a given product group or service type based on a common platform reference of understanding.

To achieve this, the basic data must be calculated in the same way and with the same general rules. This applies, for instance, to different forms of assumptions as well as setting of system boundaries and the choices of calculation methods.

Therefore, product-specific requirements (PSR) have been prepared for Variable speed electric drives http://www.environdec.com/eng/.

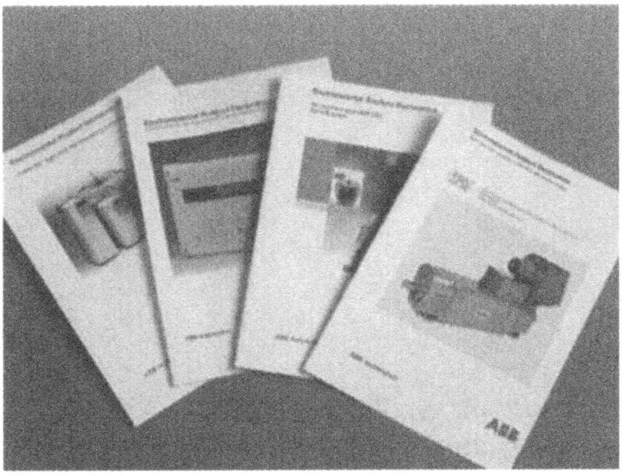

Life Cycle Assessment LCA of a Product

This paper introduces the practical actions included in the Life Cycle Assessment (LCA) of drives and gives the pre-requisites of the entire operational phase from the develop to the waste of the drive itself.

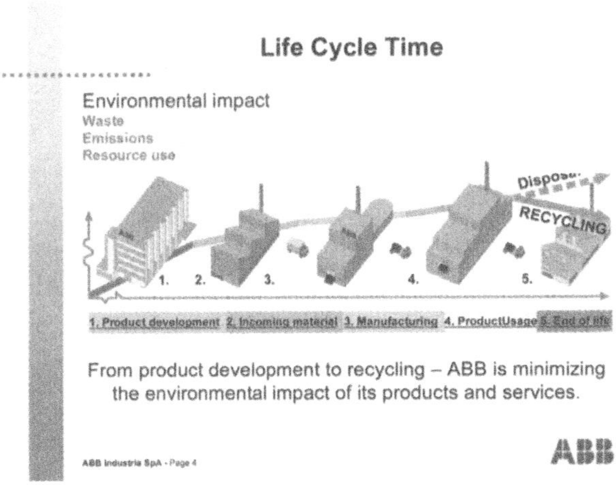

Product Development

Product development is the first phase of the life cycle time where the environmental impact can be reduced with the best results as possible.

Improving the efficiency of electric drives, using Life Cycle Assessment method, to reduce lifetime environmental impact and the Design guidelines have to be able to cover the complete product lifecycle and recycle phases.

Incoming Material

Incoming material has to be controlled so that Suppliers are involved in the development of their products and services according to the LCA policy.

Material & end of life products have to be packaged in reusable boxes and bulk transportation of products are addressed as much as possible to limited number of Customers so that we can optimise the transportation phase themselves.

Manufacturing Process

Manufacturing process has to be dedicated to minimise production waste like smoke; dirty water; solid impact material.

When some industrial waste is inevitable; the process have to be realised in a way that the impact have to be as low as possible with the maximum care in the waste purification phase.

Product Usage

Product usage is a very important part of the lifecycle because 98-99% of environmental impact is caused by the energy consumption during the usage itself and lower energy consumption reduces greenhouse gas (GHG) emissions in electrical power production.

We have here in included a real case of energy saving referred to a screw compressor set controlled by a variable speed drive (VSD) by comparing the energy consumption with the traditional ways of regulation.

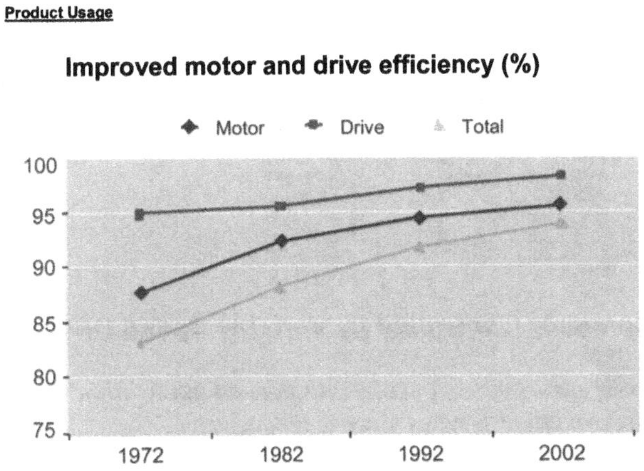

Product End of Life

Product end of life has to include Recycling instructions, with a network of certified recycling partners, in this way is possible to recycle over 90% of the used components minimising the environmental impact and saving money by the re-use of a lot of useful parts.

Most of the decisions for EcoEfficient products, like variable speed drives, are based on a direct payback period or Net Present Value of the investment.

Both energy and investment costs are different in different markets but for the global sustainability it would be more fair to use a method called Return on Natural Capital (RNC) which compare the manufactures by each other and disposal loading the environment to the reduced emission caused by the EcoEfficient products.

Why use variable speed drives

Improved demand-side efficiency saves primary energy

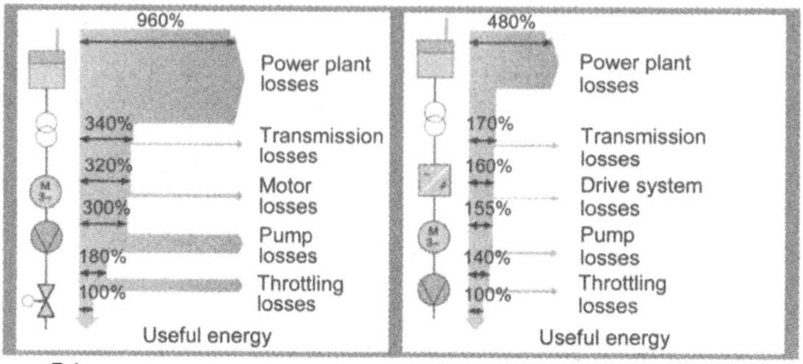

Primary energy consumption with throttling and with variable speed drive. The following efficiency estimations are applied:

Power plant	= 35%	Pump with throttling	= 60%
Transmission	= 95%	Pump with AC drive	= 70%
Motor without AC drive	= 94%	Valve	= 55%
Motor with AC drive	= 93%		

ABB Industria SpA

Real Case: Screw Compressor Controlled by Variable Speed Drive

It has been proven in a real case that in a screw compressor application a variable speed drive (VSD) can save about 50% of electrical energy.

In fact, along with energy savings, there are also other significant advantages, such as suppression of electric absorption peaks during motor start-up in direct on line phase, elimination of wasted compressed air during the on-off regulation (determined by the need to depressurise the air-oil tank each time the machine is emptied) and reduced wear on mechanical parts (compressor screw; motor bearings and outgoing air tubes), with a consequent reduction in maintenance costs.

There are also improvements with regard to power factor: using the on-off command where the value is inevitably 0.8, while with the inverter this value becomes 0.97-0.98.

Based on a real case measurement done in a compressed air station between the on-off regulation solution and the inverter configuration, some amazing results were obtained:

- 32% reduction in maximum absorbed electric power;
- 42% reduction in electric consumption;
- saving of about 115,000 m³ of air per year thanks to elimination of depressurisation of the air-oil tank.

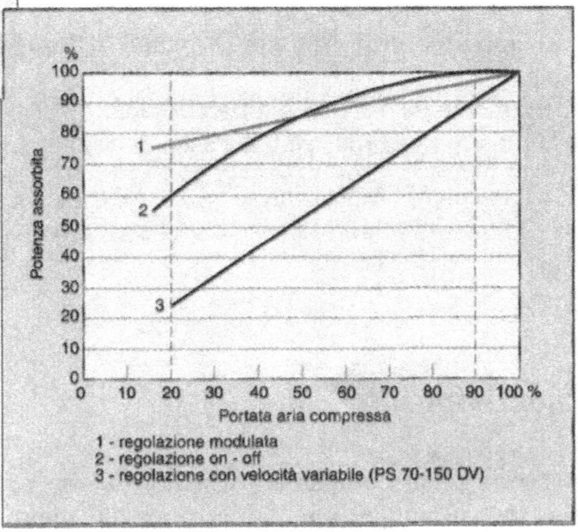

y=f(x) : y is absorbed power % and x is air flow %
1=valve modulation; 2=on-off regulation; 3=VSD variation

Sensorless Control for AC-Motor in Pumping Systems

Pablo Gutiérrez, José Antonio Domínguez, José Miguel Ruiz, Santiago Lorenzo

Departamento de Tecnología Electrónica,
E.T.S. Ingenieros Industriales. Paseo del Cauce s/n., 47011 Valladolid (SPAIN),
Ph.: +34 983 423 340; Fax: +34 983 423 310; E-mail: pablogut@eis.uva.es

Objectives

Several small companies, which work on renewable energy system fittings, made several suggestions concerning the improvement of various Photovoltaic equipment mainly having in view solutions in solar irrigation and solar house isolated from the grid.

The equipment should be able: to extract maximum power, with non stops due to conditions change, to be open to remote communication and to be adaptable to different applications with minimum changes. All these characteristics would turn out the equipment more efficient in operating and with less exploitation costs.

To answer those needs, we proposed to develop the modular system INCA - Innovación en Control y Adaptación, designed for pumping or solar house. In this paper we expose the innovative concepts regarding the pumping case.

Introduction

Today's commercial pumping systems using DC/AC converters are based in the old proportional control volts/hertz (V/w). This kind of systems present a poor dynamic response, needing slow start ramps and not allowing fast changes in the working conditions. This produces discontinuous operation and loss of efficiency. The result of all of them is a lower exploitation of the, still expensive, PV modules.

With the use of Vector Control is possible to achieve a great improvement in the system's performance, obtaining an immediate response of the AC motor to torque steps, but most times using a tachodynamo. The advantage of the control method that we propose is the elimination of the motor mechanical speed measurement, which could be a very important problem in some systems, like for example the pumping systems.

Besides we present the commercial equipment that has been designed and manufactured under the financial support and supervision of the European Union, in a Joule SME Cooperative Research Project. This equipment has been designed in a modular way to increase its applications.

Control Description

The basic scheme of the system is shown in Fig. 1:

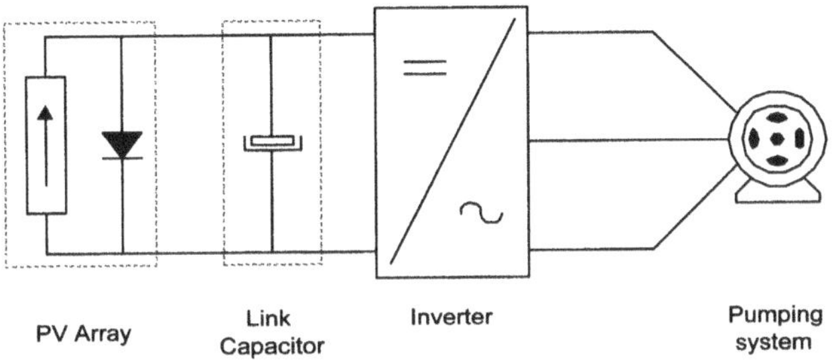

Fig. 1. Block basic scheme of the system.

1. DC subsystem, including the PV array and a link capacitor, besides the measuring probes.
2. AC subsystem, including the DC/AC inverter and the load attached to it, besides the measuring probes.

The inverter's tasks are extracting the maximum power from the PV array, and feeding the load; controlling flux and speed of the AC machine.

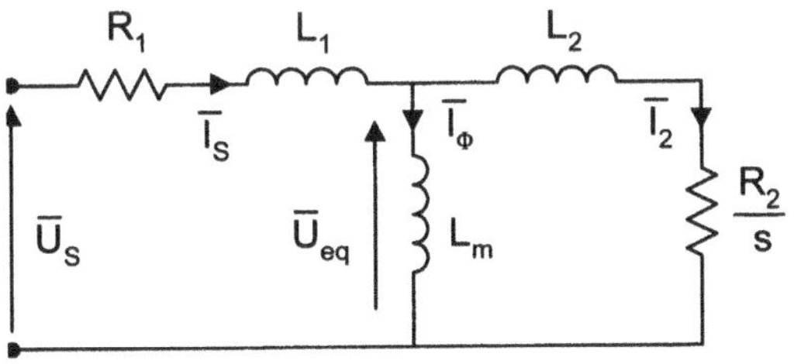

Fig. 2. Asynchronous machine in Field Orientation.

As it's known, applying Field Orientation to Yammamura's asynchronous machine schemes (Fig. 2), it's possible to uncouple the flux and torque magnitudes, regulating both of them independently, because they are projected on different orthogonal revolving axis (d,q).

With the control method shown below, the inverter can work in voltage or current source mode, though the prototype has been developed using a vectorial PWM.

Flux Control Loop

It is known that if current through inductance Lm, is constant (on the revolving d axis), then the AC motor has an immediate response to torque steps. A conventional regulator is employed to do the control as shown in Fig 3.

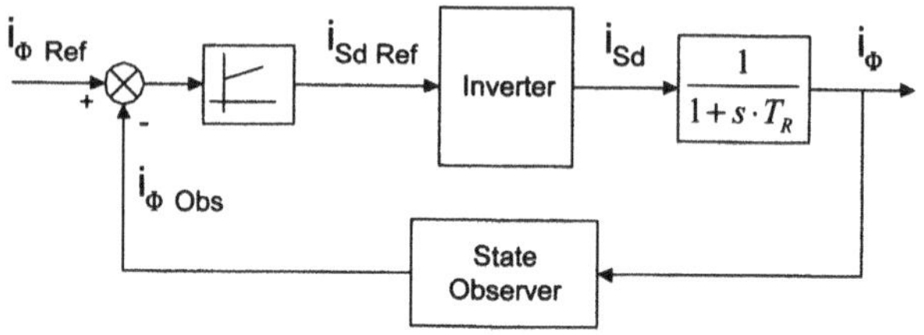

Fig. 3. Flux control loop on d axis.

Power Control Loop

As the PV arrays are current source, it is necessary to include a link capacitor between the array and the DC/AC inverter, to obtain a voltage source in the input of the last one. This voltage has to be maintained as constant as possible, independently of the energy given by the PV array to minimize the level of armonics.

The modelling of the system is as shown in equations 1 and 2 (iron, copper, mechanical and other losses are considered minimal compared to total installed power):

$$P_{PVarray} = P_{capacitor} + P_{engine} \tag{1}$$

$$P_{engine} = T_{engine} \cdot \omega = K \cdot p \cdot I_{\Phi} \cdot I_{sq} \cdot \omega \tag{2}$$

With this control loop we pretend to transfer to the motor all the power given by the PV array. This is performed adjusting the power demanded by the motor (P_{engine}) to the value that keeps constant the DC voltage in the link capacitor ($P_{capacitor}=0$).

The method's innovation is to execute that control (see Ec. 2) shifting the stator current frequency 'ω' and not the torque current 'Isq'. This avoids the need for motor mechanical speed measurement.

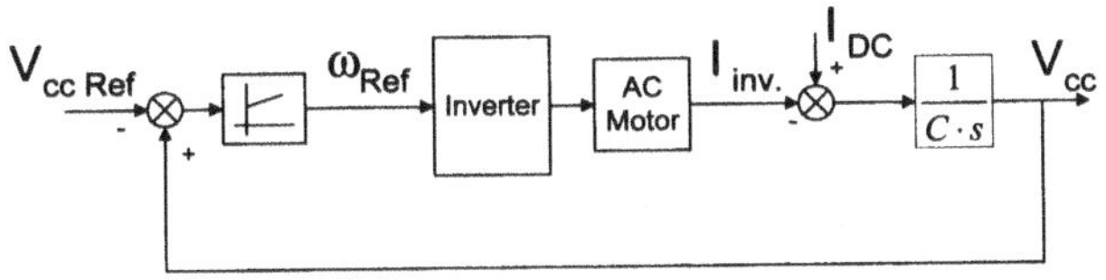

Fig. 4. DC voltage control loop.

If we can control the DC voltage with the inverter and we can calculate, in every instant, the active power developed by the AC engine, then the tracking of the Maximum Power Point (MPP), that's mean the most adequate reference voltage, can be calculated with a very simple algorithm. This avoids the use of a DC/DC converter.

Industrial Equipment

The INCA Power Conditioner has been designed entirely in the Tecnología Electrónica Department of University of Valladolid, Spain.

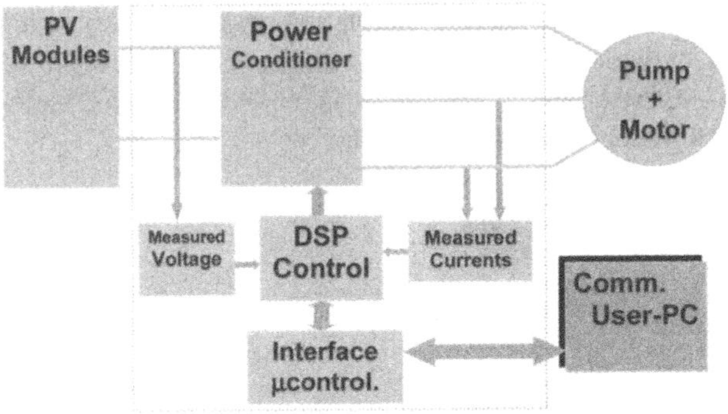

Fig. 5. Scheme of the INCA Pumping equipments.

Several high-tech concepts has been applied to design these equipments:

1. Vector Control is used to drive the AC motor. This control is done with only two direct measures: two-phase input currents.
2. Use of standard asynchronous engines presents lower costs than other commercial systems.
3. Use of DSP allows to complete one regulation cycle in less than 20µseconds. A microcontroller for user interface, PC communication, alarms...etc.
4. This type of control allows DC/AC inverter to impose the DC input voltage, controlling the bias point of the modules without any other device like DC/DC converters.
5. Because of the DSP, the tracking of the Maximum Power Point is done every 8 seconds with great accuracy.

6. The tracking is independent of every external condition. So efficiency is always maximum.

Fig. 6. Control PCB with the DSP and the microcontroller.

7. Above features provide a better efficiency, more simplicity and less power/wire losses.
8. The PV array is arranged to cover the possible applications of the Power Conditioner: Pumping, Isolated Home and Multi-point generation. So as the design of the INCA equipment is modular, several applications can be performed easily. This feature minimizes the manufacturing and setting costs.

Fig. 7. View of the power supply PCB.

9. The PC-communication allows the remote monitoring by mobile phone variables like water counter, electrical conditions, incidents like electrical errors, full water tank, dry well...

Conclusions

A new Sensorless Vector Control has been designed and implemented getting very good results. Its efficiency has been compared versus others AC-drives from General Electric and others, and its performance has been superior, not only in the instant power extracted else in the total energy daily extracted thanks to the robust control gotten.

A successful industrial equipment has been designed. It is modular, so more applications like Isolated home and the Multi-point generation are available with small changes in it.

References

Blaschke, F. "The Principle of Field Orientation as Applied to the New TransVektor Closed Loop Control System for Rotating Field Machines", Siemens Revue, 1972.

Domínguez Vázquez, J.A. "Control de Sistemas Fotovoltaicos para Bombeo y Conexión a Red por Medio de La Teoría de Campo Orientado". Tesis Doctoral, julio 1996. Universidad de Valladolid.

Jansen, L.P. y Lorenz, R.D. "Transducerless Position and Velocity Estimation in Induction and Salient AC Machines", IEEE Transactions on Industry Applications, vol 31, nº2 marzo/abril 1995.

Leonhard, W. "Control of Electrical Drives", Spring Verlag 1985.

Lorenzo, S. y M. Shaker. "Observadores de Estado para Medida de Flujo, Par y Deslizamiento en Motores de CA Insensibles a las Variaciones de la Constante de Tiempo del Rotor", may 1988.

Lobo Salgado, J.I. "Investigación y Desarrollo de Técnicas de Control Basadas en Vectores Espaciales de Campo Orientado para Motores de Inducción", Tesis Doctoral, marzo 1993. Universidad de Valladolid.

Yammamura, S. "AC Motor for High Performance Application Analysis and Control". Marcdel Dekeer Inc. 1988.

S. Lorenzo, J.A. Domínguez, S. de Pablo, S. Cáceres. "New design for photovoltaics systems". PESC'98

Optimal Voltage/Frequency Curve for Inverter-Fed Motor

Sebastião Lauro Nau, Alexandre Postól Sobrinho

WEG Indústrias S.A. – Motors Division, Jaraguá do Sul, SC – Brazil.
E-mail: slnau@weg.com.br, alexandres@weg.com.br

Abstract

This paper presents a technique to optimize the relation torque-temperature in inverter-fed induction motors for speeds below the base speed. As it is very known, the ventilating system is not efficient for low speeds, demanding a torque derating in order to keep the temperature rise unchanged. The aim of this paper is to present a voltage/frequency curve that minimizes total losses of the motor in dependence on the frequency for constant torque.

Introduction

Dealing with losses in PWM inverter-fed three-phase induction motor is a challenge. Boglietti et al have investigated the influence of the inverter characteristics on the iron losses in induction motors and they concluded that the modulation is important but switching frequency is not so important [1]. Otherwise, iron losses change significantly in dependence on the frequency of operation. As frequency decreases, iron losses also decrease. Then, magnetic flux density (induction) could be increased. As torque is a product between current and magnetic flux, for a constant torque current consequently would be reduced. So, it is possible to reduce the voltage proportionally less than the reduction of the frequency in order to find an optimal voltage/frequency (V/f) ratio that minimizes total losses. On the other hand, iron losses and no-load RI^2 losses increase tremendously as V/f ratio increases [2].

A low-loss operation of a converter-fed-machine cannot be achieved by simply obeying to the common V/f=constant control mode [3]. Both iron losses and RI^2 losses are affected by V/f and f. The reduction of the ventilation as frequency decreases is considered by increasing temperature rise of the motor for a constant torque operation. The influence of the voltage drop on the impedance of the stator and the changes in the mechanical losses with the speed are also considered.

Determination of Losses

Many authors [2, 3, 4] have segregated losses of induction motor properly but for this investigation purposes, losses are separated simply in three components:

1. Losses that depend on the current such as losses by Joule effect in the stator windings and the rotor bars (p_i).
2. Losses that depend on the magnetic induction, frequency and the quality of ferromagnetic material such as iron losses (p_{fe}).
3. Losses due to the fan and friction also called mechanical losses (p_m). For a given motor, mechanical losses depend exclusively on the speed.

Total losses can be written as:

$$p = p_i + p_{fe} + p_m \tag{1}$$

Iron losses are usually separated in two components: losses due to hysteresis (p_H) and eddy current (Foucault) (p_F). For a lamination tested according to Epstein Frame Test with a sinusoidal supply, hysteresis losses are directly proportional to the frequency (f) and to the square of the magnetic induction (B^2). Similarly, eddy current losses are proportional to the square of both frequency (f^2) and magnetic induction (B^2), especially for induction above one Tesla (1T). But, in an induction motor, iron losses exhibit a behavior much more dependent on the magnetic induction than shown by Epstein Test. In several three-phase induction motors with full processed steel lamination tested with very different levels of saturation, iron losses presented a dependence on the induction with nearly B^4 for induction above 1.2T that is usual for low voltage induction motors.

On the other hand, torque is directly proportional to the product between magnetic flux and current. To keep the torque constant, as flux increases, current decreases or vice-versa. Therefore, losses due to the current (Joule effect losses) depend inversely on the square of the magnetic flux. By its turn, magnetic flux (or induction) is directly proportional to the ratio V/f (voltage/frequency). Nevertheless, the influence of voltage drop on the primary resistance has to be considered. As input voltage decreases for low speeds, voltage drop on the primary resistance becomes more significant. For the base frequency, the influence of the voltage drop on the primary resistance is minimum but for low frequencies this influence is very important. In this approach, the influence of the primary reactance will be neglected. The electromotive force E is then given by:

$$E = V - R_1 I_1 = V - \Delta V \tag{2}$$

ΔV depends directly on the stator current I_1. In per unit basis, taking rated voltage as reference, E/f can be written as

$$\frac{E}{f} = \frac{V}{f} - \frac{\Delta V_n}{f} \cdot \sqrt{\frac{k_T^{\,2} k_m^{\,2}}{\left(\frac{V}{f}\right)^2} + k_{i0n}^{\,2} \cdot k_{i0}^{\,2}} \; . \tag{3}$$

ΔV_n is the voltage drop at rated frequency and load. All voltages are per unit referred to nominal voltage. Some iterations to calculate *E/f* have to be done until a desirable error is achieved substituting *V/f* by *E/f* in (3). The no-load current factor k_{i0n} is the relation between no-load current and rated load current, both at rated voltage and frequency. k_T, k_m and k_{i0} are described in the paragraphs below.

It is still necessary to take into account that mechanical losses reduce as frequency (and consequently speed) decreases. The reduction of the mechanical losses decreases the current and, consequently, the losses p_i due to the Joule effect. Mechanical losses do not affect iron losses. They act as an additional load to the motor. Therefore, they have to be considered as a torque to be added to the rated torque available on the shaft. The torque due to mechanical losses varies with the cube of the frequency (f^3). The factor due to the mechanical losses k_m is given by (4) where mechanical losses at nominal speed p_{mn} are referred to the rated output power P_n and f to the base (nominal) frequency f_n.

$$k_m = \left(\frac{1 + p_{mn} f^3}{1 + p_{mn}} \right) \tag{4}$$

Due to the poor ventilation of the motor for low speed, the torque has to be reduced in order to keep the temperature rise below the insulation class of the motor or the rated temperature rise for an extended lifetime. The idea presented in this paper is to avoid the torque reduction by minimizing the losses and consequently the temperature rise, despite poor ventilation of the self-ventilated motors (TEFC) for low speed. This is considered by the torque derating factor k_T in (6). The torque influences the losses due to the current only.

It was observed experimentally that the no-load current increases as the ratio *E/f* increases due to the non-linearity of magnetization curve of the lamination according to (5). This is given by the no-load current factor k_{i0}.

$$k_{i0} = \left(\frac{E}{f} \right)^{3.4} \quad for \ \frac{E}{f} \geq 1, \ \ k_{i0} = \left(\frac{E}{f} \right) \quad for \ \frac{E}{f} < 1 \tag{5}$$

So, total losses p (p.u.) in a motor operating with both variable voltage and frequency can be rewritten as follows:

$$p = p_{in} \left[\frac{k_T^{\ 2} \, k_m^{\ 2}}{\left(\frac{E}{f} \right)^2} + k_{i0n}^{\ 2} \cdot k_{i0}^{\ 2} \right] + p_{Hn} \left(\frac{E}{f} \right)^4 f + p_{Fn} \left(\frac{E}{f} \right)^4 f^2 \tag{6}$$

The first term in (6) means losses due to total current of the motor. The second and third terms are respectively iron losses due to hysteresis and eddy current. Induction B was conveniently replaced by *E/f*. The losses p_{in}, p_{Hn} and p_{Fn}, referred to the total losses (except the mechanical losses), for 50Hz, 4-pole motors with low losses full processed lamination (silicon steel) at rated load, voltage and frequency are in average 0,80; 0,12 and 0,08 respectively. All parameters are in per unit (p.u.) referred to the base speed (or base frequency).

The relative minimum losses have to be found for different values of *V/f* in dependence on the frequency. In other words, it is necessary to find out the values of *V/f* that minimizes the losses for each range of frequency. The family of curves shown in Fig.1 was created to determine the *V/f* ratios that minimize total losses. Two corresponding *V/f* curves in dependence on frequency are plotted in Fig.2. Straight line was used in this study.

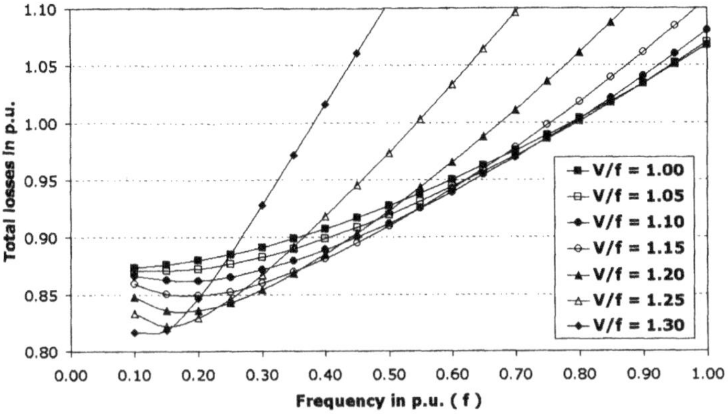

Fig. 1. Total losses x frequency curve for several V/f ratios at rated torque

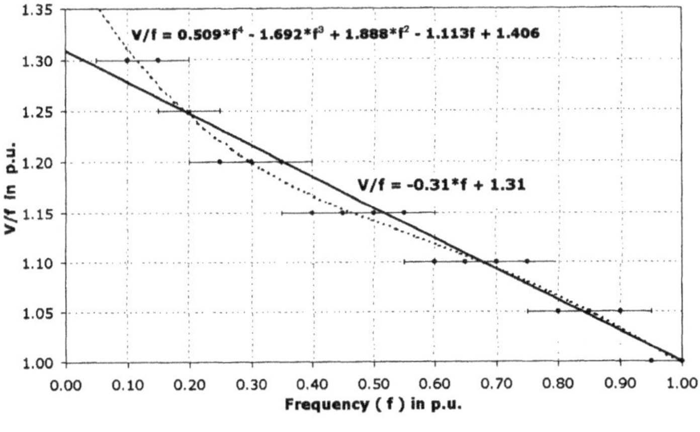

Fig. 2. V/f x frequency curve for minimization of total losses

Influence of the Temperature

After several experiments and tests carried out in motors at rated load, varying separately the speed of the fan from zero to base speed, it was possible to conclude that there is a similar thermal behavior for TEFC three-phase induction motors of very different output power. Fig.3 shows a curve that represents the temperature rise (in p.u.) of 50Hz, 4-pole low voltage cage induction three-phase motors with die cast iron frame in function of the fan speed (also in p.u.).

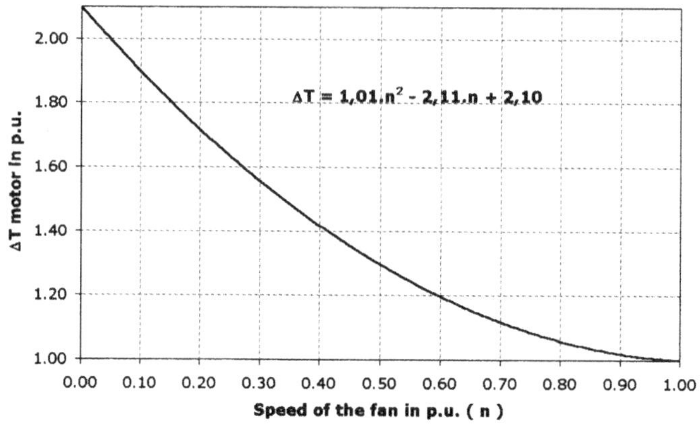

Fig. 3. Temperature rise (p.u.) x speed of the fan at rated load

In this approach, for each value of desired frequency, total losses p are calculated according to (6) for a ratio V/f given by Fig.2. Then, ΔT_{vent} is determined by Fig.3 to consider the influence of the ventilation reduction. A new p has to be calculated for the required temperature rise according to (7). For instance, if the maximum desired temperature rise is ΔT_n, then it is enough to take ΔT=1. Otherwise, if the maximum temperature rise permitted is the insulation class temperature rise, then $\Delta T = \Delta T_{class}$.

$$p = \frac{\Delta T}{\Delta T_{vent}} \cdot p_n \tag{7}$$

With these new total losses p, the torque derating factor k_T is calculated. Separating k_T from (6), it results:

$$k_T = \frac{k_{HVF}}{k_m}\left(\frac{E}{f}\right)\sqrt{\frac{p - p_{Hn}\left(\frac{E}{f}\right)^4 f - p_{Fn}\left(\frac{E}{f}\right)^4 f^2}{p_{in}} - k_{i0n}^2 \cdot k_{i0}^2} \tag{8}$$

k_T can be higher than one if minimized total losses are such that rated temperature rise is reduced even with a poor ventilation. If k_T is calculated for an insulation class temperature rise, it will be normally higher than one if the rated temperature rise is much smaller then the insulation class temperature rise. The harmonic voltage factor k_{HVF} was placed in (8) to consider the influence of the voltage harmonics for a PWM supply according to Figure 30-1 of NEMA Standard MG1-1998, part 30. For most of inverters k_{HVH} is 0.95.

Example

A 4-pole, 50Hz, 30kW three-phase induction motor with constant rated torque for all frequencies was tested and the results are shown in the Fig.4 and Fig.5 below for three different situations: calculation with *V/f* constant, calculation with *V/f* optimized and result of test with *V/f* optimized. Fig.4 shows a comparison between total calculated losses p and Fig.5 shows the torque derating factor k_T.

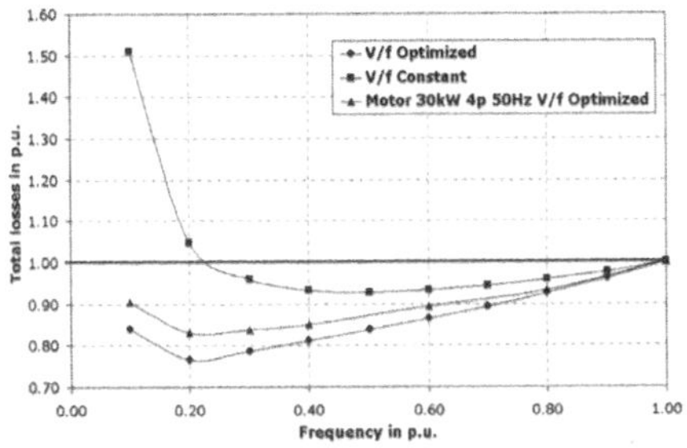

Fig. 4. Total losses x frequency

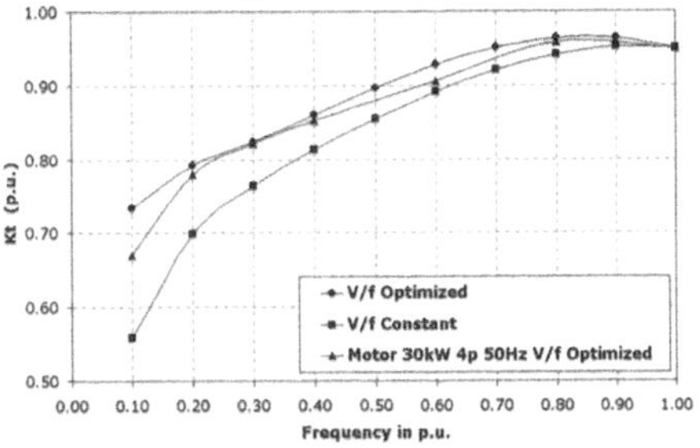

Fig.5. Derating factor x frequency

Conclusion

By analyzing the losses behavior of three-phase induction motor as a function of load, frequency and electromotive force/frequency ratio, an expression for *V/f* that minimizes total losses was created. It has a good agreement with results of tests. The equations shown in Fig.2 can be implemented in the scalar control inverter software in order to select automatically the best ratio *V/f*. This investigation has to be continued. It is necessary to consider the magnetic induction in teeth and yoke of the stator and rotor separately and the influence of the PWM supply on the iron losses more accurately.

References

[1] A. Boglietti, P. Ferraris, M. Lazzari, and M. Pastorelli, “Influence of the inverter characteristics on the iron losses in PWM inverter-fed induction motors”. IEEE Transactions on Industry Applications, Vol. 32. No.5 Sept/Oct 1996, pp. 1190-1194.

[2] S. L. Nau, and S. B. Silveira, “Determination of losses and BxH curve in steel lamination by test in electric motors” (in Portuguese). IV Seminar on Electrical and Magnetic Materials. Rio de Janeiro, 28-29 April, 1994.

[3] H. Röhrdanz, “Losses in PWM converter-fed three phase induction motors and their percentages”. ICEM 98. Istambul, Turkey, 1998, pp. 175-180.

[4] R. Kaczmarek, M. Ama and F. Protat, “Iron losses under PWM voltage supply on Epstein Frame and in induction motor core”. IEEE Transactions on Magnetics, Vol. 32. No.1. January 1996, pp. 189-194.

Maximum Efficiency Control of Brushless Doubly-Fed Reluctance Motors for Large Pump Applications

Milutin G. Jovanović, Jian Yu

Northumbria University, UK

Introduction

The inverter-fed brushless doubly fed reluctance machine (BDFRM) is undoubtedly an attractive solution for medium performance variable speed applications due to its low cost, high reliability and lower harmonic injection into the mains. The cost effectiveness and improved power quality of a BDFRM drive can be attributed to the slip power recovery property of the machine which allows a smaller inverter to be used, and especially in cases when the speed range required is relatively limited (typical examples are pumps [1] and wind turbines [2, 3]). However, because of its modest torque per volume ratio, resulting from the unusual operating principle [4] and unconventional design[1] (Fig. 1), a larger, and therefore more expensive, BD-FRM is needed to achieve the torque output of a synchronous reluctance (Syncrel) or cage induction machine [5]. Nevertheless, despite this deficiency, the total system cost can still be substantially reduced in larger drives (up to about 30%) by significant savings that can be made in the power electronics [6].

The BDFRM shares all the advantages of doubly fed machines over singly excited counterparts – the operational mode flexibility, the greater control freedom and the possibility of sub-synchronous and super-synchronous speed operation in both motoring and generating regimes. It can work as a standard or doubly excited induction machine and as a fixed or adjustable speed synchronous turbo-machine. The latter operating mode means that high speed, field weakened traction applications and high frequency generators become feasible. In the former applications, the BDFRM ability to function as an induction machine in case of the inverter failure is an important "fail-safe" feature. From a control viewpoint, one important merit of the machine is that one can not only control the torque output, but also the power factor [7] or efficiency.

When compared with machines having similar properties, such as the doubly-excited wound rotor induction machine (DEWRIM) or the closely related, brushless doubly-fed induction machine (BDFIM), the BDFRM is superior in many respects. The absence of brush gear means that it is more reliable and with lower maintenance requirements than the DEWRIM. On the other hand, the use of a cageless reluctance rotor (this can be any of the modern Syncrel designs) makes it more efficient [8], more mechanically robust and much easier to control than the BDFIM.

---

[1] The stator windings and the reluctance rotor all have mutually different pole numbers.

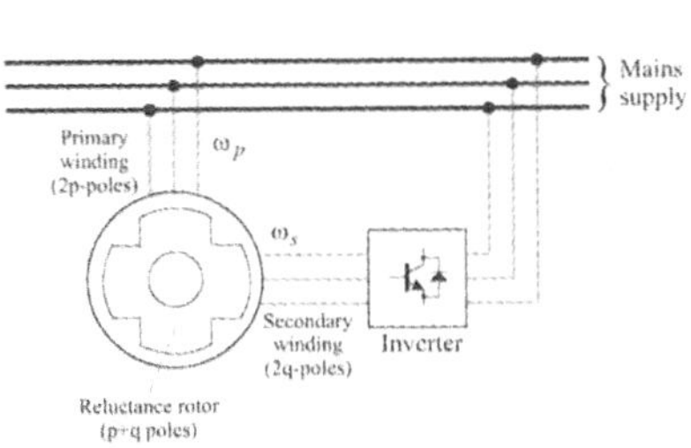

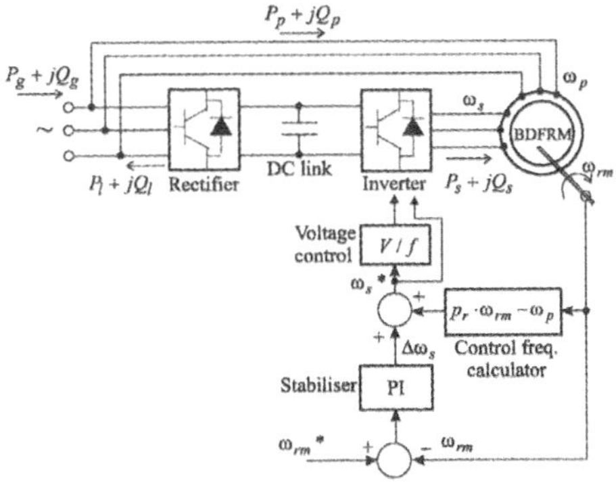

Fig. 1. Simplified diagram of BDFRM drive **Fig. 2.** Scalar control system

The main objective of this paper is to develop a scalar control algorithm (Fig.2) and conditions for higher efficiency vector control of the BDFRM for use in pump-type applications. The maximum torque per inverter ampere (MTPIA) and the minimum copper loss (MCL) control strategies are primarily considered while some attention is also being paid to the power factor control capabilities of the machine (a more detailed study can be found in [7]). Influences of the winding pole-numbers on the machine's copper losses and the inverter rating shall be also addressed.

Steady-State Operation and Control Properties

This section is concerned with the analysis of expressions for different performance parameters of an ideal BDFRM (i.e. no saturation and iron losses) that are suitable for vector control implementation [9]. Using the space-vector model[2] in a primary flux oriented reference frame [3, 5, 7] and assuming motoring convention one can derive the following main relationships:

$$P_{p,s_{3\phi}} = P_{cu_{p,s}} + P_{p,s} = \frac{3}{2}R_{p,s}i_{p,s}^2 + \frac{\omega_{p,s}}{p_r}T_e \tag{1}$$

$$T_e = \frac{P_p + P_s}{\omega_{rm}} = p_r\frac{P_p + P_s}{\omega_p + \omega_s} = \frac{3}{2}p_r\frac{L_{ps}}{L_p}\lambda_p i_{sq} \tag{2}$$

$$Q_p = \frac{3}{2}\frac{\omega_p\lambda_p}{L_p}(\lambda_p - L_{ps}i_{sd}) \tag{3}$$

$$P_{cu} = P_{cu_p} + P_{cu_s} = \frac{3}{2}\frac{R_p}{L_p^2}\lambda_p^2 + \frac{3}{2}\left(R_p\frac{L_{ps}^2}{L_p^2} + R_s\right)i_{sq}^2 + \frac{3}{2}\left(R_p\frac{L_{ps}^2}{L_p^2} + R_s\right)i_{sd}^2 - 3\frac{R_p}{L_p^2}\lambda_p L_{ps}i_{sd} \tag{4}$$

[2] It is interesting that this is almost identical in form to the DEWRIM one although the torque producing mechanism of the two machines is fundamentally different.

where subscripts 'p' and 's' denote the primary (grid-connected) and secondary (inverter-fed) windings respectively, $L_{p,s,ps}$ are the constant 3-phase inductances of the windings, $\omega_{rm} = (\omega_p + \omega_s)/p_r$ is the rotor angular velocity[3] and $\omega_{p,s}$ are the windings applied frequencies. Several important remarks related to the machine operating and control features can be made from the above expressions.

The slip energy recovery nature of the BDFRM's operation immediately follows from (1) as the supply inverter only has to handle a real power proportional to $|\omega_s = -s\omega_p|$ i.e. the degree of slip s. In this respect, the BDFRM behaves as a DEWRIM and this characteristic is the reason that a fractionally rated inverter can be used for restricted speed changes above and below the synchronous speed $(\omega_{syn} = \omega_p/p_r)$ when s is small[4].

In terms of torque control, (2) shows that this can be achieved in an independent manner by the q-axis secondary current as $\lambda_p = \text{const}$ due to the primary winding grid connection. The maximum torque per inverter ampere (MTPIA) i.e. the minimum inverter current for a given torque is obtained if all of the secondary current is torque producing i.e. $i_s = i_{sq} \Leftrightarrow i_{sd} = 0$. The same equation also indicates the general improvement of torque per ampere with increasing the L_{ps}/L_p ratio as this provides better magnetic coupling between the windings.

We can also see from (3) that even though we cannot vary the machine primary flux, we are able to regulate the primary power factor in a decoupled fashion since the primary reactive power (Q_p) is only affected by the d-axis secondary current (i_{sd}). The unity power factor (UPF) is obtained when $Q_p = 0$ and occurs at:

$$i_{sd_{upf}} = \frac{\lambda_p}{L_{ps}} \tag{5}$$

The secondary winding would be carrying all of the magnetising current for the machine and a larger inverter would be consequently required if this control strategy was desired [7].

It is of particular interest for this paper to see how one can minimise the total copper losses in the machine in order to improve its efficiency. The i_{sd} dependent terms in (4) illustrate the possibility of controlling the machine copper losses via i_{sd}. However, unlike the previous case, the control is now not decoupled from torque, due to the presence of the i_{sq} component in (4). It can be shown that the copper losses (MCL) are minimal if:

$$i_{sd_{mcl}} = \frac{\lambda_p L_{ps}}{L_{ps}^2 + \frac{R_s}{R_p} L_p^2} \tag{6}$$

where the resistances of the 2q-pole secondary and 2p-pole primary winding (with conductors of uniform cross section) can be related as [7]:

$$\frac{R_s}{R_p} = \frac{n_s}{n_p} \cdot \frac{q+\pi}{p+\pi} \tag{7}$$

---

[3] It can be shown [4] that under this condition the machine produces usable torque where the number of rotor poles $p_r = p + q$ with p and q $(p \neq q)$ being the windings pole-pairs (Fig. 1).

[4] This property makes the BDFRM a good brushless candidate for pumps where the speed range required is typically 2:1 or less [1].

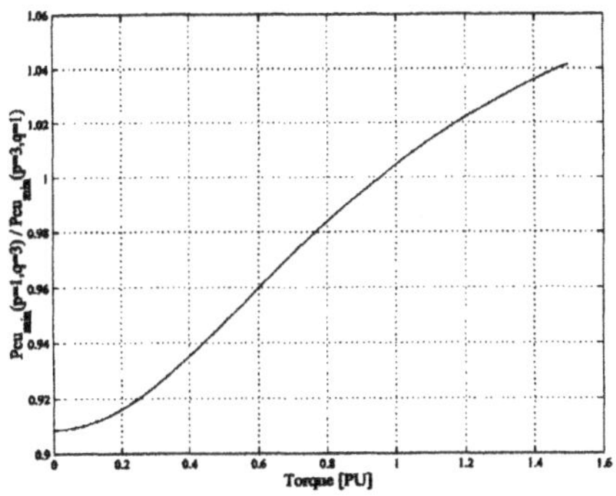

Fig. 3. MCL ratio

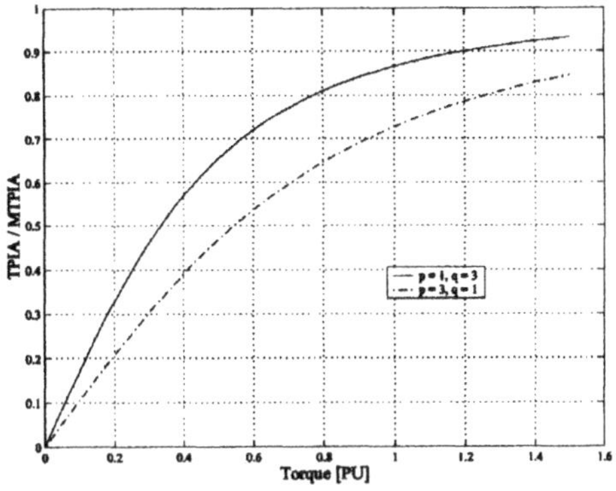

Fig. 4. TPIA performance under MCL conditions

Therefore, to minimise i_{sd} and hence maximise TPIA with MCL control the machine designer needs to increase R_s/R_p ratio which, given (7), means that: $n_s \geq n_p$, and $p < q$. The effective turns/pole relationship ($n_s/n_p \geq 1$) is consistent with improving L_{ps}/L_p this representing a general condition for better torque production as mentioned earlier [7].

Comparative Analysis of MCL and MTPIA Strategies

It is the main purpose of this section to investigate the effects of the windings' functions i.e. which of them should be 'power' winding and which 'control' winding, on the machine copper losses and inverter size. The following plots have been generated for a 6/2-pole machine having the same winding inductances i.e. $L_p = L_s$, the same gauge copper wire and $L_{ps}/L_p = 7/9$ (equivalent to a typical 4-pole Syncrel rotor saliency ratio of 8 [5]). The base value used for torque normalisations is $T_B = \frac{3}{4} p_r \frac{\lambda_p^2}{L_p}$.

From a MCL viewpoint, the $p < q$ case appears to allow lower losses in the torque range up to about 0.95-pu according to Fig. 3. At higher torque values, the machine seems to be more efficient when $p > q$. The differences in losses are more pronounced at very low torques (less than 10%), and hence the overall efficiency may be affected more significantly, as it has inherently decreasing tendency at small output powers. At mid and higher torques, however, these variations are minor (only a few percent), and their impact on the machine efficiency is virtually negligible. Therefore, unless the machine is to be used at low powers relative to its rating, it is irrelevant which winding is grid-connected and which is inverter-fed.

The plots in Fig. 4, that represent the TPIA normalised to its optimum for a machine with minimum copper losses, yet again confirm the previously made conjecture of the values being higher for a 2-pole primary and 6-pole secondary winding combination.

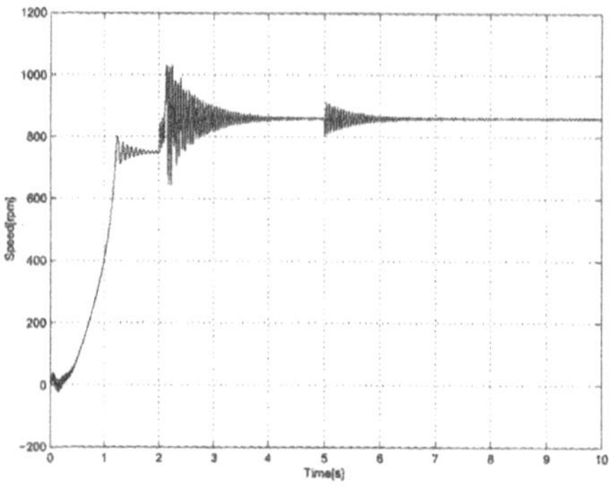

Fig. 5. Open-loop V/f control

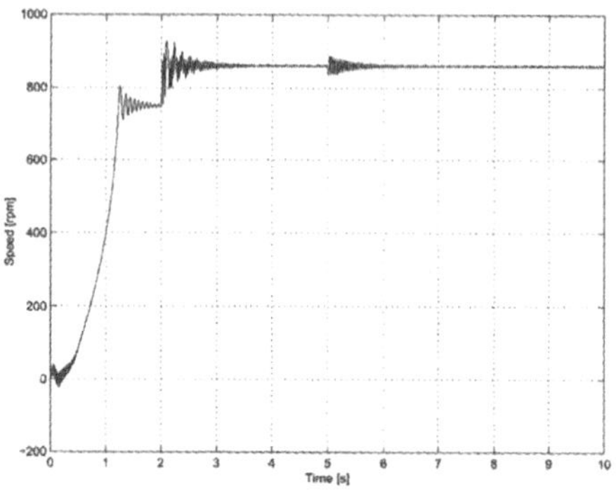

Fig. 6. Closed-loop V/f control

Scalar Control

In pump-type applications simple scalar control is a satisfactory solution as fast dynamics of a machine is not required. Furthermore, speed ranges in these drives are limited which may help alleviate stability problems common with this method. Some preliminary Simulink results generated by executing the V/f=const algorithm in Fig. 2 for a machine data from [9] are shown in Figs. 5 and 6. It can be seen that the machine response to speed and/or load torque step changes (at 2-s and 5-s respectively) is faster and smoother under closed-loop control as expected. The open-loop scheme has experienced stability problems in case of larger step changes of the speed command.

Conclusions

The paper has discussed vector control strategies for improved efficiency of the machine. A V/f scalar control scheme, suitable for general purpose BDFRM drives, has been developed and its good performance demonstrated by computer simulations. The effects of winding pole-numbers on the machine efficiency and especially the inverter size have been closely examined. It has been shown that from this point of view it is generally better to have a two-pole power winding and a multi-pole control winding.

This work should make a significant contribution to fundamental knowledge and general understanding of the BDFRM operation and control. The consideration of the machine performance/inverter rating trade-offs is of particular importance as the future of this interesting and unusual machine will be likely decided by the associated cost benefits in the target applications.

The development of a DSP based experimental test system for a 6/2-pole axially-laminated BDFRM prototype is in the final stage at the Northumbria University, UK. The results of this work will be soon available and will be published in our future papers.

References

[1] B. Gorti, D. Zhou, R.Spée, G. Alexander, and A. Wallace, "Development of a brushless doubly-fed machine for a limited speed pump drive in a waste water treatment plant," *Proc. of the IEEE-IAS Annual Meeting*, pp. 523–529, Denver, Colorado, October 1994.

[2] L. Xu and Y. Tang, "A novel wind-power generating system using field orientation controlled doubly-excited brushless reluctance machine," *Proc. of the IEEE IAS Annual Meeting*, Houston, Texas, October 1992.

[3] M.G. Jovanović, R.E. Betz, and J. Yu, "The use of doubly fed reluctance machines for large pumps and wind turbines," *Proc. of the IEEE-IAS Annual Meet-ing*, Chicago, October 2001.

[4] Y. Liao, L. Xu, and L. Zhen, "Design of a doubly-fed reluctance motor for adjustable speed drives," *IEEE Transactions on Industry Applications*, vol. 32, pp. 1195–1203, Sept/Oct 1996.

[5] R.E. Betz and M.G. Jovanović, "The brushless doubly fed reluctance machine and the synchronous reluctance machine - a comparison," *IEEE Trans. on Industry Applications*, vol. 36, pp. 1103–1110, July/August 2000.

[6] Y. Liao, "Design of a brushless doubly-fed induction motor for adjustable speed drive applications," *Proceedings of the IAS Annual Meeting*, pp. 850–855, San Diego, California, October 1996.

[7] M.G. Jovanović and R.E. Betz, "Power factor control using brushless doubly fed reluctance machines," *Proc. of the IEEE-IAS Annual Meeting*, Rome, Italy, October 2000.

[8] F. Wang, F. Zhang, and L. Xu, "Parameter and performance comparison of doubly-fed brushless machine with cage and reluctance rotors," *Proc. of the IEEE-IAS Annual Meeting*, Rome, Italy, October 2000.

[9] L. Xu, L. Zhen, and E. Kim, "Field-orientation control of a doubly excited brushless reluctance machine," *IEEE Transactions on Industry Applications*, vol. 34, pp. 148–155, Jan/Feb 1998.

Field-Oriented Induction Motor Drive with Efficiency Optimisation

Francesco Parasiliti, Roberto Petrella, Marco Tursini

Department of Electrical Engineering, University of L'Aquila,
I-67040 Monteluco di Roio, L'Aquila-Italy
Fax: +39 0862434403
E-mail: rock@ing.univaq.it, petrella@ing.univaq.it, tursini@ing.univaq.it

Introduction

The application of variable speed Induction Motor (IM) drives based on Field Oriented Control (FOC) is going to be extended from the traditional industrial area to household appliances. A typical example is represented by domestic washing machines, where converter-fed (three-phase) IMs are being taken into account as substitutes for the traditional line-fed commutator motors, due to the featuring variable speed operation, extended speed range, good matching of the torque-to-speed characteristics and fast response to pulsating load torque.

Of course, considering such extensive applications, a particular attention should be paid to the energy consumption and efficiency of the drive systems. The key point is related to variable torque/speed operation. In fact, being the efficiency of power converters close to one, the power consumption of the drive is mainly imputable to the motor. Unfortunately, as known, the efficiency of the induction motor depends on the operating conditions: it is high when the motor works near the rated ("design") conditions, whilst it decreases and can also become poor when the motor is operated at low torque/speed conditions. Clearly, such a behavior depends on the different feeding conditions (voltage, frequency, current) with respect to the "rated" ones. In fact, owing to the feeding through a power converter, not negligible energy efficiency improvements can be obtained by means of a proper dedicated control strategy, a possibility which has attracted the interest of many researchers in the last few years.

This paper reports the authors' experience on a modified controller for induction motor drives, capable to assure both efficiency optimisation at steady-state and maximum torque capability during transient operation, [1][2]. The paper is organized as follows: at first, the drive scheme and the operation principle of the FOC based controller are presented in Section 1; then, the basic relations and the algorithm employed for the on-line efficiency optimisation are explained in Section 2; finally, some laboratory results are shown in Section 3.

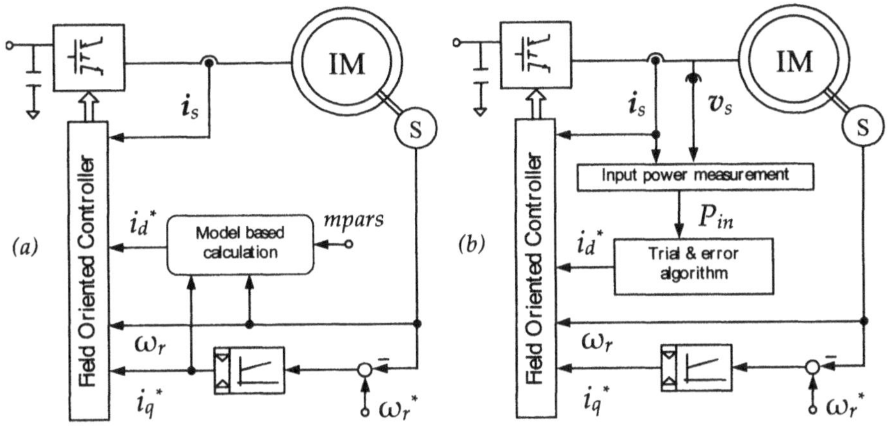

Fig. 1. Model based feed-forward optimisation (a) and performance measurement feed-back optimisation (b).

1. Field oriented controller with efficiency optimisation

The possibility of improving the efficiency of variable speed IM drives has attracted the interest of the researchers in the last few years. Several controllers have been proposed in literature, based on different approaches, such as power factor control, loss model based control or input power minimising (e.g. [3]). They all rely on the consideration that efficiency improvements can be obtained by controlling the balance between the copper and iron losses, i.e. the electromagnetic losses. This balance can be controlled by selecting the flux level in relation to the torque and speed of the machine. Thus, for any specific torque and speed there is a specific flux level which minimizes the total losses and maximizes the efficiency.

As known, in rotor-flux field-oriented control the flux and torque levels correspond to the *direct* (i_d) and *quadrature* (i_q) current components in the synchronously rotating reference frame aligned with the rotor flux, [4]. Thereafter, efficiency optimisation results in the imposition of a proper flux current level for any specific torque/speed operating condition. Basically, two methods have been proposed in literature to achieve this goal: the "model based feed-forward" approach and the "performance measurement feed-back" approach.

The "**model based feed-forward**" approach makes use of proper machine models, including electromagnetic losses, to calculate the flux current command and to impose it in a feed-forward scheme, Fig. 1a). An interesting example of this type of method is given by Garcia in [5], which obtains a simple optimisation formula by considering a simplified motor model where the motor leakage inductances are neglected and the iron loss resistor is supposed constant. This kind of methods can be arranged without the use of additional hardware with respect to the one needed in standard field-oriented drives. They can easily be incorporated in the field-oriented controller and allow fast optimisation transients, but they are affected by model uncertainties and parameters variations with the operating conditions.

These drawbacks are overcome by the "**performance measurement feedback**" approach, in which the input power is constantly measured and the flux current command is updated by a "trial-and-error" algorithm in order to minimize the input power for any steady state condition, Fig. 1b), (e.g. [6]). Such methods require some additional hardware (the voltage signals are needed, in addition to the current, for on-line measurement of the motor input power) and computation but they assure independence with respect to model uncertainties and parameters variations. Nevertheless, the presence of analytical, though approximate, relations which provide the proper flux and torque current levels in a feed-forward scheme allows to speed-up the optimisation transient and to increase the robustness to measurement errors.

As for the authors' experience, this kind of "**mixed model based/performance measurement**" approach seems to give the best performance, [1][2]. The proposed scheme is depicted in Fig. 2

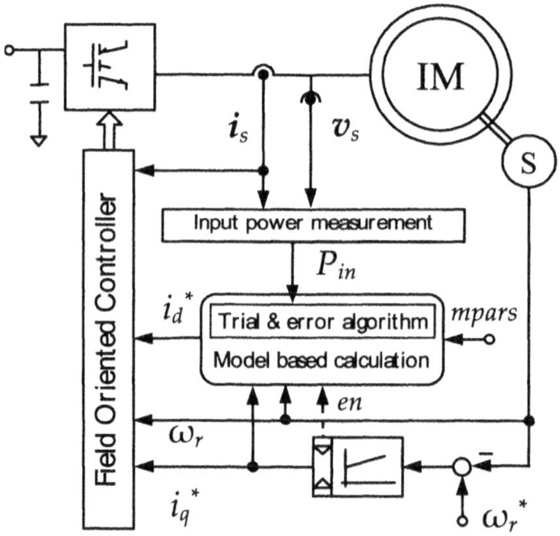

Fig. 2. Proposed mixed model-based/performance-measurement optimisation.

As usual, the flux (i_d^*) and torque (i_q^*) command currents are provided, respectively, by the efficiency optimizer and the speed regulator, but the efficiency optimisation is based on a mixed "analytical/measurement" approach. As a first attempt, once steady-state is attained, the efficiency optimizer calculates i_d^* as a function of i_q^*, according to the value of slip speed which assures maximum efficiency at the given rotor speed. Such a value is obtained by the iterative procedure, described in Section 2, which calculates in real-time the (analytical) solution for the maximum efficiency of the IM. Then, the controller starts to make small step changes in the flux command current (i_d^*) by setting the value which results in minimum input power. As the output power is held constant by the speed regulator (for a fixed load torque), the system will always seek the maximum efficiency.

The changes of the flux current command are allowed at constant speed operation only by the enable signal (*en*). Changes of i_d^* are made slowly with respect to the response time of the drive in order to avoid perturbation of the steady state operation. During transients, i.e. when a large speed error is detected, the enable signal is set to "false" and the efficiency optimisation is disabled. In this case, the flux command current is increased to a level that produces high torque per stator ampere to assist in responding to the speed error.

2. Model based maximum efficiency calculation

As known, in rotor flux field orientation the torque and flux producing current components are related to the slip speed (ω_2) by the following relation, [4]:

$$\omega_2 = \frac{R_r}{L_r}\frac{I_q}{I_d}$$

Assuming to operate at maximum efficiency and field orientation, the same relation subsists in the form:

$$\frac{I_q}{I_d} = \omega_{2\eta}\frac{L_r}{R_r} \tag{1}$$

where ($\omega_{2\eta}$) indicates the slip speed value assumed at maximum efficiency operation. Clearly, if this value can be estimated in the different operating conditions, equation (1) can be used to impose that exact ratio between the flux and torque current components which results in maximum efficiency.

A proposal for the calculation of ($\omega_{2\eta}$) is presented in [1].

Assuming the T-form equivalent circuit to model the steady-state operation of the IM, and supposing:

I. to neglect, as usual in normal operation, the rotor iron losses;
II. to assume for the stator iron losses the expression:

$$P_{loss,iron} = c_h \Phi_m^2 \omega + c_e \Phi_m^2 \omega^2$$

where (c_h) and (c_e) are the loss coefficients due to hysteresis and eddy currents respectively, (Φ_m) the magnetizing flux, and ω the synchronous speed;

III. to assume constant the remaining parameters of equivalent circuit,

the efficiency of the induction motor can be calculated as:

$$\eta = \frac{1-\omega_2/\omega}{1+2R_s/R_o+\alpha\,\omega/\omega_2-\beta\omega_2/\omega} = \eta(\omega_2, \omega, mpars) \tag{2}$$

being the quantities:

$$\alpha = R_r\left(\frac{1}{R_o}+\frac{R_s}{R_o^2}+\frac{R_s}{(\omega M)^2}\right)$$

$$\beta = 1+\frac{R_s}{R_r}\left(1+\frac{L_{\sigma r}}{M}\right)^2+\frac{(\omega L_{\sigma r})^2}{R_o R_r}\left(\frac{R_s}{R_o}+1\right)+\frac{2R_s}{R_o}$$

and the equivalent iron loss resistor

$$R_o = \frac{3\omega}{c_h + c_e\omega}$$

functions of the synchronous speed only ($mpars = R_s, R_r, L_{\sigma s}, L_{\sigma r}, M, c_h, c_e$ is an acronym meaning for the "motor constant parameters set", see Table 1).

The maximum of the function (2) with respect to the slip speed, calculated at constant synchronous speed, yields the analytical solution:

a) $$\omega_{2\eta} = -\frac{\alpha}{\beta}\left(1-\sqrt{1+\frac{\beta}{\alpha}}\right)\cdot\omega \tag{3}$$

Iterative method for maximum efficiency calculation

Due to the dependence on the synchronous speed, equation (3) does not give the value of ($\omega_{2\eta}$) in a closed form at constant rotor speed. Nevertheless it allows the set-up of an iterative process for its calculation, whose principle is illustrated in Fig. 3 In fact, once steady-state operation at commanded rotor speed (ω_r^*) is achieved, the maximum efficiency will be attained for that value of synchronous speed (ω^*) which satisfies both a) and

b) $$\omega_2 = \omega - \omega_r^* \tag{4}$$

This condition is represented by the point P in Fig. 3.

Then, considering a trial solution (ω') for the synchronous speed, it is possible to calculate the difference:

$$\Delta\omega = \omega' - \left(\omega_r^* + \omega'_{2\eta}\right).$$

The sign of this difference defines the position of the trial solution with respect to the exact one according to the following simple reasoning:

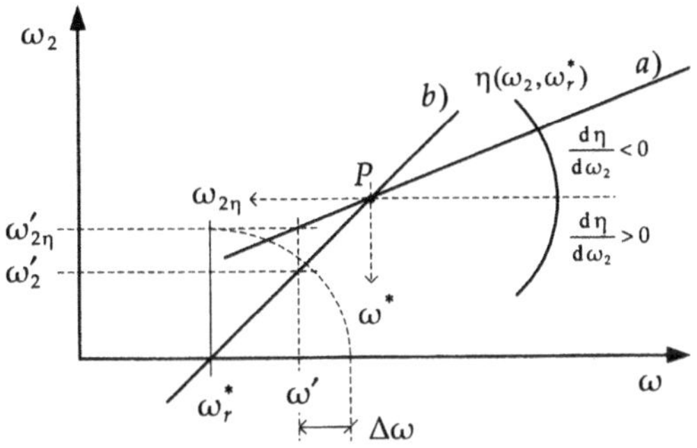

Fig. 3. Iterative calculation of ($\omega_{2\eta}$)

$$\begin{aligned} &\textit{if} \quad \Delta\omega < -\varepsilon \quad \textit{then} \quad \textit{increase} \quad \omega' \\ &\quad \textit{else} \quad \textit{if} \quad \Delta\omega > \varepsilon \quad \textit{then} \quad \textit{decrease} \quad \omega' \\ &\quad\quad \textit{else} \quad \textit{if} \quad |\Delta\omega| < \varepsilon \quad \textit{then} \quad \omega' \quad \textit{is} \quad \textit{ok!} \end{aligned}$$

being $\varepsilon > 0$ the accepted precision range.

The trial and error iteration process has been implemented by using the Binary Search Algorithm, which provides the required rapidity of convergence and robustness in real-time implementations, [7].

3. Results

The rotor flux field oriented controller with efficiency optimisation has been implemented on a drive system using a TMS320F240 DSP controller and an IGBT inverter. It has been tested with two different induction motors:

- a four poles, 500 W motor for washing machine applications, rating 196 V-170 Hz, and
- a two pole, 650W motor for spindle application, rating 220 V-200 Hz.

The parameters of both the motors are resumed in Table 1.

Operations at different torque and speed values over the whole operating range have been carried out, using a commercial brake to impose the load torque and a digital wattmeter for the basic electric measurements.

Fig. 4 shows the efficiency vs. slip speed plots in the case of the **spindle motor**. The maximum locus is also drafted, for the sake of completeness, which shows the dependence, in the actual case, from the speed and the load values.

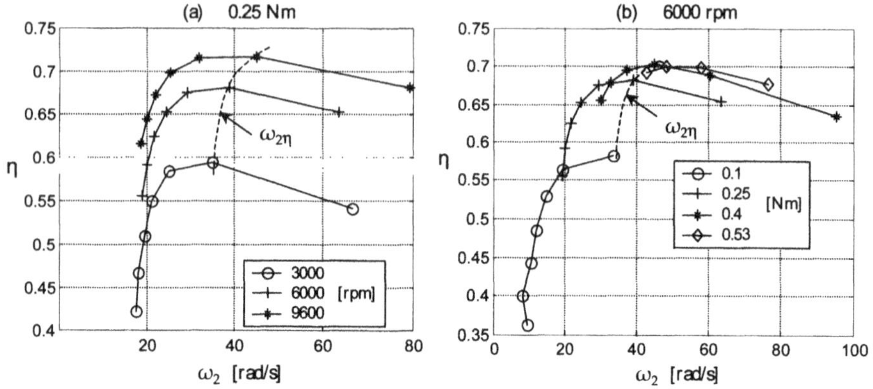

Fig. 4. IM efficiency at different rotor speed (a) and load torque (b) (**spindle motor**).

Fig. 5 shows the efficiency vs. torque plots in the case of the **washing machine motor**. Tests have been done at different speeds in the constant torque region, respectively 550 and 1000 rpm (typical washing speeds) and 4000 rpm (spinning speed region). In order to quantify the gain achieved by the efficiency optimizer, the results are compared with the corresponding ones obtained by setting a fixed value for the flux current command, i.e. without optimisation.

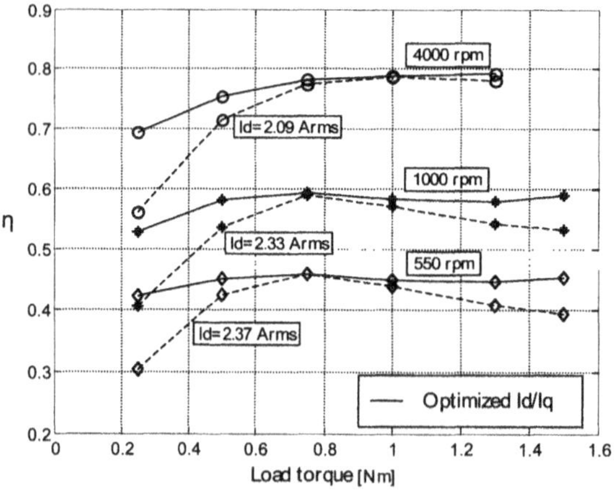

Fig. 5. Efficiency vs. torque with and without optimisation (**washing machine motor**).

Fig. 6 resumes the results in terms of efficiency gain. The maximum efficiency gain is obtained at low torque-low speed operation, where it reaches the 40% (550 rpm, 0.25 Nm). Fig. 7 shows the behaviour of the flux vs. torque current locus achieved by the action of the efficiency optimiser for the operating points reported in the previous figures.

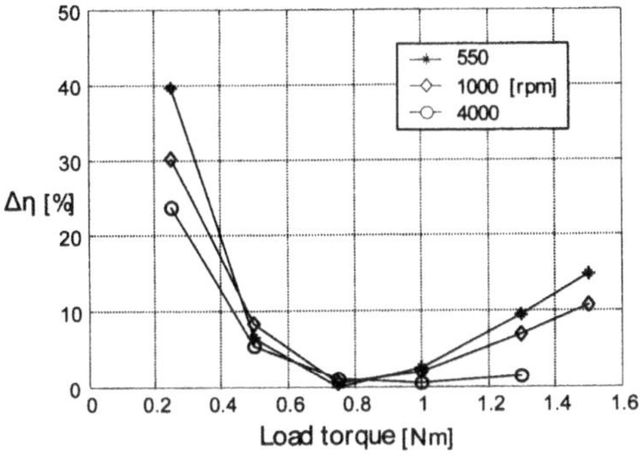

Fig. 6. Efficiency gain with optimisation (**washing machine motor**).

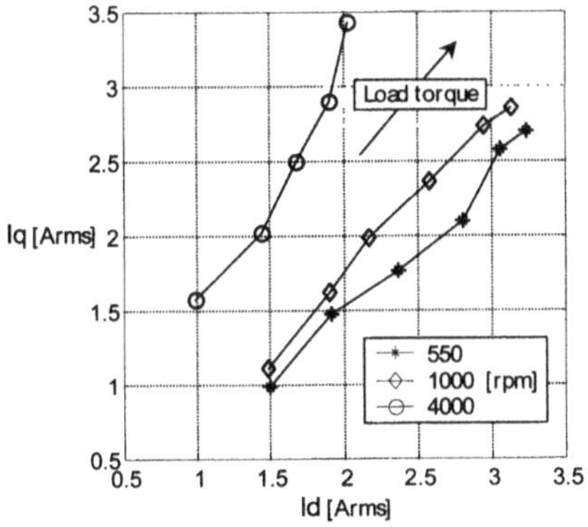

Fig. 7. Flux vs. torque current components locus with maximum efficiency (**washing machine motor**).

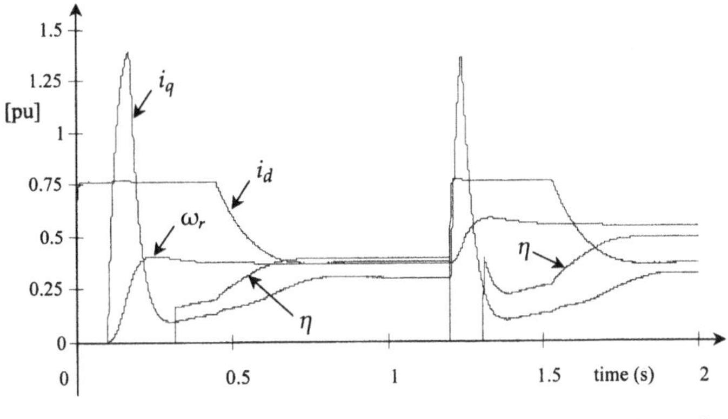

Fig. 8. Main system variables during speed transients and steady-state at 0.25 pu load torque (1pu = 5A, 157rad/s, 2.3Nm) – (**washing machine motor**).

In conclusion, Fig. 8 shows the transient operation of the efficiency optimizer. The figure illustrates the behavior of the flux and torque (actual) currents, the speed, and the efficiency during transients and steady state operations for the considered washing machine motor (simulation). During transients, the flux current is commanded at a constant (relatively high) value in order to give maximum torque. The efficiency optimiser starts to work when the (nearly) steady-state conditions are detected. Its action modifies the commanded flux and torque currents so that the efficiency increases.

Table 1. Induction motors parameters

		washing machine motor	*spindle motor*
p	poles pairs	2	1
R_s	stator resistance, at 20°C	1.415 Ω	2.04 Ω
R_r	rotor resistance, at 20°C	0.815 Ω	1.64 Ω
$L_{\sigma s}$	stator leakage inductance	2.3 mH	2.3 mH
$L_{\sigma r}$	rotor leakage inductance	2.3 mH	1.2 mH
M	magnetizing inductance	29.75 mH	95.3 mH
R_o	equivalent iron losses resistor	171 Ω, (at 50Hz)	1447 Ω, (at 200Hz)

References

[1] S. Frattesi, R. Petrella, M. Tursini, "An Efficient Induction Motor Vector Controller for Washing Machines Applications", *Energy Efficiency in Household Appliances and Lighting*, *Springer*, 2001.

[2] R. Petrella, M. Tursini and M. Villani, "Efficiency Optimisation of Rotor-Flux Field-Oriented Induction Motor Drives", Conf. Record of the 15th International Conference on Electrical Machines (ICEM'02), CD Rom (6 pages), Brugge, Aug. 25-28, 2002.

[3] P. Famouri, J.J. Cathey, "Loss Minimization Control of an Induction Motor Drive," IEEE Trans. on Industry Applications, Vol. 27, No. 1, Jan/Feb. 1991, pp. 32-37.

[4] V. Novotny and T.A. Lipo, "Vector Control and Dynamics of AC Drives," Oxford Science Publications, 1997.

[5] G.O. Garcia, J.C. Mendes Luis, R.M. Stephan and E.H. Watanabe, "An Efficient Controller for an Adjustable Speed Induction Motor Drive," IEEE Trans. on Industrial Electronics, Vol. 41, No. 5, Oct. 1994, pp. 533-539.

[6] S. Kirschen, D.V. Novotny and T.A. Lipo, "Optimal Efficiency Control of an Induction Motor Drive," IEEE Trans. on Energy Conversion, Vol. EC-2, No. 1, Mar.1987, pp. 70-76.

[7] F. Parasiliti, M. Tursini, D.Q. Zhang "Real-time Gain-Tuning of PI Controllers for High Performance PMSM Drives", IEEE Trans. on Industry Applications, Vol. 38, No. 4, Jul/Aug. 2002, pp.1018-1026.

Impact of Cost and Reliability on Energy-Saving for Industrial Electrical Drives

Laszlo Szentirmai, Tivadar Szarka

Department of Electrical and Electronic Engineering, University of Miskolc, H-3515 Miskolc, Hungary, Phone: +36-46-565142; Fax: +36-46-563447; E-mail: elkerika@gold.uni-miskolc.hu

1. Introduction

The total electricity consumption of the world today is about 12,000 TWh (terawatthour) per annum and 80% is dedicated to electrical drives. Thus, energy saving is dominant in electrical drives where each per cent of saving contributes to living standard. The primary energy consumption of the world today is 10 billion tonnes of oil equivalent. For easier conversion 1 tonne of oil equivalent (TOE) = 42.2 x 10^9 joule (J), thus 1 petajoule (PJ) = 10^{15} J = 23,700 TOE.

Energy efficiency is also the key to forestalling the effects of global warming. The less energy used, the less CO_2 is released to the atmosphere. This is a strategy that businesses of all kinds can profitably adopt.

2. General Consideration for Industrial Drives

For most applications, standard or high-efficiency induction motors can be used in drives with variable-voltage, variable-frequency *inverters* with little or no motor de-rating [1].

Typical a.c. *drive performances* taking e.g. the metallurgical industry into consideration are as follows: (a) Needs: Super-high reliability, advanced energy- and labour- saving; (b) Speed accuracy (error):±0.01 to 0.05%; (c) Dynamic response (breakdown frequency): 40 to 60 radians per second; (d) Adjustable speed range: 1:1,000; (e) Torque ripple: ± 2.0%; (f) Current ripple: ±0.5 %; (g) Ratio of response to speed accuracy: 4,000 to 6,000. The requirement for rapid speed response has correspondingly raised the bandwidth to 10 Hz (a rise time of 50 ms) or more with good torsional vibration damping for far better productivity and much higher quality steel products.

The metallurgical industry in the vicinity of the University has more than 10,000 motors and the authors carried out various projects there.

3. Failures

A "failure" is any inability of an item to carry out its specified function. In other words it is defined as the termination of an item's ability to perform its required function.

Three important criteria reflect the time how often the item breaks down:

1. Mean time between failures-*MTBF*: this applies to repairable items, and is given in 1000 hours or months.
2. Mean time to failure-*MTTF*: applies to non-repairable items, and means the average time an item may be expected to function before failure.
3. Mean time to repair-*MTTR*: this applies to *repairable* items, and is usually given in hours indicating the time while the equipment is out of service for maintenance and repair, thus it has also another term "down-time".

For a system with a repair facility, the term "availability" is generally used and defined as

$$Steady\text{ - }state\ availability = \frac{MTBF}{MTBF + MTTR} \tag{1}$$

The failure rate λ is usually expressed per annum, per month or per other time-interval. Roughly 10,000 motor failures have been investigated in England, Finland and the USA and it was found that 25% of the faults have been occurred in motors exceeding 40 kW power each and their repair cost presented 80% of the total repair cost. The tests in Hungarian steel mill proved to be better [2].

The failure distribution found by the investigation are presented as follows: (1) Overheating 30%; (2) Humid, acidic environment 19%; (3) Phase fault 14%; (4) Rotor faults 13%; (5) Insulator ageing 10%; (6) Bearing faults 5%; (7) Others 9%;

The failure rates per annum are ranging between 2.5 to 4.0%. Conclusions from recent studies are that higher failure rates are a function of higher temperatures, only 25% of the failures could have been prevented by preventive maintenance.

The principal electromagnetic faults which may occur within the machine are: (a) Winding open-circuit, (b) Winding short-circuit (phase-to-phase), (c) Winding short-circuit at terminals. Within the power converter the faults under considera-tion are as follows: (a) Power device open-circuit, (b) Power device short-circuit, (c) D.C. link capacitor failure.

It is the *fault tolerant drive* which can continue to operate with any one of these faults [3].

4. Reliability, Availability and Serviceability (RAS)

It is frequently said that the electric motors of one make is more reliable than those of another make [3]. In this case, the *"dependability"* of the motor is generally what is meant, if relative costs are not considered. In the strict sense, dependability is the ability of an entity to comply with one or many required functions under fixed condition. The dependability is composed of four elements:

a) **Reliability** is the characteristic of an item expressed by the probability that it will perform a required function under stated conditions for a stated period of time.

The main characteristics of reliability are (a) the quantitative reliability, (b) the failure rate, (c) the meantime between failure (MTBF), (d) mean time to failure (MTTF) and (e) mean life.

If a unit begins to function at the instant $t = 0$ and the failure could occur at the random instant T, then the reliability function $R(t)$ is defined as the probability of failure-free operation of the unit during the time interval [0, t]:

$$R(t)=Pr(T>t), \qquad R(0)=1, \qquad R(\infty)=0$$

The function

$$f(t) = -\frac{d}{dt}R(t) \tag{2}$$

is called the probability density function (pdf). The product $f(t)\Delta t$ closely approximates the probability of failure for devices under consideration in the time interval $[t, t+\Delta t]$.

b) **Availability** characterises the ability of an entity to be in conditions to realise a required function under fixed conditions at any instant.

1) *Point availability* $A(t)$ is the probability that the system is in an operating state at time t. By contrast, the reliability function $R(t)$ is the probability that the system has operated without failure over the interval $[0, t]$.

The point availability $A(t)$ contains no information on how many failure/repair cycles have occurred on the 100 units prior to, say, 20 years. In general term, $R(t) \le A(t)$.

If system or its components cannot be repaired, then $A(t)=R(t)$.

2) *Interval (mission) availability*

$$A^*(T_1, T_2) = \frac{1}{T_2 - T_1}\int_{T_1}^{T_2} A(t)dt \tag{3}$$

is the expected fractional amount of an interval of specified length T_2-T_1 that the system is in an operable state. This interval may be the design life of the system or the time to accomplish some particular mission.

3) Steady-state (asymptotic, limiting) availability

$$A^*(\infty) = \lim_{T\to\infty}\frac{1}{T}\int_0^T A(t)dt \tag{4}$$

is an interval availability for an infinite interval. If the system under consideration is not repairable, then its steady-state availability is equal to zero because

$$A^*(\infty) = \lim_{T\to\infty}\frac{1}{T}\int_0^T R(t)dt = \lim_{T\to\infty}\frac{MTBF}{T} = 0 \tag{5}$$

c) **Serviceability** or **maintainability** characterises the ability of an entity to be maintained or restored to a state in which it can realise a required function when maintenance is done under fixed conditions with prescribed procedures and means.

d) **Safety** characterises the ability of an entity to avoid doing appear in fixed conditions critical or catastrophic events.

Product reliability means the success of the product from the material stage to its operation in the field.

New results were achieved and published here by the authors concerning RAS as

1. In a production line the highest cost comes from the breakdown of a key-motor, e.g. liquid iron becomes solid, the quality of products is decreasing, thus the replacement cost of the motor is negligible to the losses occurring in the production process.
2. Initial cost should increase reliability rather than reduce investment because running cost during life span of a motor exceeds 80 to 100 times the initial cost.
3. Serviceability contributes to cost-saving in manufacturing process.

5. Energy saving

The provision of variable speed may reduce the energy required by a load by up to 50%. With inverter-fed variable-speed drives, energy saving are possible by matching the voltage to the power demand at a given speed so as to maximize efficiency when under-loaded.

The efficiency, η, of an inverter-fed drive is defined as:

$$\eta = \frac{\omega T}{\omega T + \Delta P_M + \Delta P_{MV} + \Delta P_{inv}} \qquad (6)$$

where T = the torque developed by the motor, ω = the angular speed, ΔP_M = the fundamental loss of the motor, ΔP_{MV} = the harmonic losses of the motor and ΔP_{inv} = the sum of losses of the inverter and converter. The motor internal friction is included in the torque. This improves the efficiency to some extent (1%) and has little effect on determination of the minimal loss point.

Both theoretical computations and measurements show, that with drives operating with the optimum slip – at reduced load – 2-3% of the nominal apparent power can be saved in comparison with drives operating at rated flux. About the same amount of energy can be saved – mainly at low speed – at rated load, but in this case the motor flux must be raised above the rated value.

In practice, energy saving in drives – in the case of inverter supply – can be implemented by voltage control on the motor, while the speed control loop produces the fundamental frequency of the drive. The simplest method of voltage control is to store the value of the optimum slip in memory (as a function of fundamental frequency).

Hungarian Metallurgical Industry Figures	*Power ranges: kW*		
	under 10	10 to 100	over 100
1. Motor ratios with respect to power categories	75%	20%	5%
2. Failure rate in percentage of total motors available..	80%	19%	1%
3. Nature of failures and percentage:			
(a) mechanical faults	10 to 15%	20%	70 to 80%
(b) Winding burnt-out and electric faults....	85 to 90%	80%	20 to 30%
4. MTBF in months	6 to 8	12 to 24	120
5. Repair costs in percentage of initial costs....	80%	40 to 60%	10%
6. Cost of disassembly and transportation to workshop....	10%	5%	1%
7. Expenditure due to replacement of a key-motor (estimation) in percentage of initial cost while the production line is not working...	1,000%	2,000%	1,000%

Note: if the production line is working this percentage exceeds the above values by 10-times.

It is a good assumption that the energy saving potential world-wide is at least 5%, this means that about 600 TWh could be saved annually. Assuming an average 5000 hours duty time per year, the power saving would be 120 GW.

Predicting that the production rate and all other conditions remain the same, 12-16 years would be needed to replace all the existing drives.

6. Manufacturing and Running Costs

Manufacturing cost consists of the cost of stator and rotor iron core, copper for stator winding, aluminium for rotor bars and end rings, and the cost of structural material, insulation and labour [4]. To rewind the stator of a 15 kW-motor costs only 30% of the initial cost while for a 250 kW-motor this ratio is roughly 20%.

A major consideration in drives is the *running* or operational cost. If the overall efficiency can be raised, as a result of reducing the drive losses, then clearly this has a cost saving effect.

The use of *high-efficiency* rather than standard induction motors adds somewhat to the initial cost of the drive. These motors use more conductor material in both stator and rotor to achieve lower winding resistances and losses. For the same rating, these motors maintain the same frame size as standard motors but are made axially longer. The initial cost is increased by up to 25%. For example, a typical increase in efficiency of 6% is obtained in a 10 kW high efficiency motor over that for a standard design. If this motor is operated near rated load 80% of the time, the saving will be 4,200 kWh per year, thus, the added initial investment can be recovered in few months of operation.

7. Reliability Costs

Three separate cost factors are involved - the cost of design (including research and technological development), the initial or production cost and the cost of repair and maintenance. As the reliability of an equipment increases, so will the cost of design and production increase, whereas the cost of repair and maintenance will go down. Design becomes more expensive.

On the production side higher reliability means better quality and therefore more expensive parts. It may be necessary to use costlier materials, to work to finer limits, and to provide additional and more elaborate test and inspection facilities. Usually more skilled and, therefore, more highly paid assemblers must be employed, and the completed drive will in turn have to meet a tight and comprehensive test and inspection schedule.

To obtain a more reliable equipment, maintenance costs can be made lower and the total cost reduced (Fig. 1.)

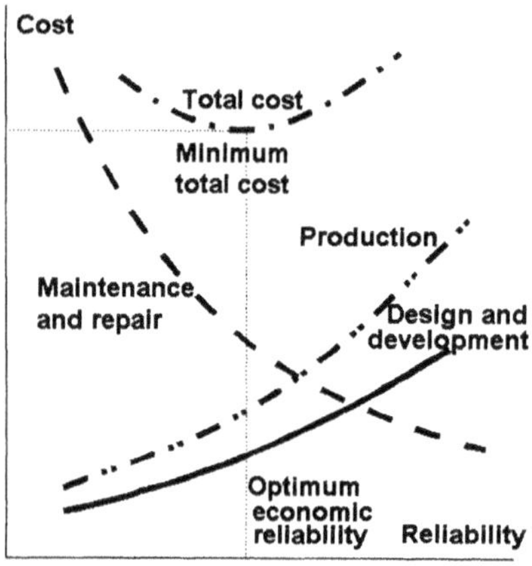

Fig. 1. Variations of costs versus reliability

8. References

[1] Szentirmai, L.: Considerations on the industrial drives. p. 687-722. In: Ertan, H.B. et al.(eds.): Modern Electrical Drives, Kluwer Academic Publishers, The Netherlands, 2000.

[2] Szarka, T.– Szentirmai., L.: Reliability improvement for application-specific industrial drives. Electrical Drives, Power Electronics,, SPEEDAM'96, Capri, Italy. Proceedings, 1996. pp: A2-13-19.

[3] Mecrow, B.C.; Jack, A.G.; Haylock, J.A.; Coles, J.: *Fault tolerant permanent magnet machine drives*. Seventh International Conference on Electrical Machines and Drives, Durham, 1995. IEE Conference Publication No. 412, pp: 433-437.

[4] Shetty, H.V.K.: Concept of built-in reliability in the design of large induction motors. Electric Machines and Electromechanics, Vol. 7. No. 6. Hemisphere Publishing Co., Washington, 1982. pp: 447-461.

Possible Problems and Solutions when Introducing Variable Speed Drives

M. Didden, J. Driesen, R. Belmans

K.U.Leuven, Department of electrical engineering-Div.ESAT/ELECTA
Kasteelpark Arenberg 10, B-3001 Leuven-Heverlee, Belgium
Tel. +32 16 32 10 20, Fax+32 16 32 19 85
E-mail ronnie.belmans@esat.kuleuven.ac.be

Introduction

The number of installed variable speed drives (VSDs) to reduce energy consumption is far below the number that is economically justifiable. One of the main reasons for this is the companies' fear for possible problems resulting in equipment damage or process interruptions that would outweigh benefits of energy savings.

This paper gives a review of possible problems that may arise when using frequency converters to supply induction motors, and indicates precautionary measures that can be taken before installing a VSD and solutions afterwards if problems arise. The following topics are covered: power quality towards the grid, electromechanical vibrations, extra losses, acoustic noise, overvoltages when using long cables, bearing problems and impact of voltage dips on the drive.

Power quality towards the grid

Most VSDs contain a three-phase bridge rectifier with diodes only conducting when the line voltage in their path is higher than the DC bus voltage. While the DC load contains inductances smoothing the load current, the line currents at the AC side consists of current spikes during the conducting period of the diodes. Dependent on the line impedance, these currents can result in considerable harmonic voltages. These harmonics may cause several problems, such as:

- Torque pulsations, reduced efficiency and possible overheating of AC-motors.
- Heating and losses in the transformer core and windings.
- Improper operation of protective relays and breaker failures.

In order to control the harmonics, standards such as IEEE-519-1992 or IEC 1000-3-x specify recommended practices and requirements. To check whether or not problems may occur regarding harmonics, some issues should be addressed, such as capacitor banks without tuning reactors and the short-circuit capability.

A harmonic may be required to determine the interaction of VSD and grid. If it turns out that the harmonic distortion due to the installation of VSDs will rise beyond acceptable limits, one of the following measures can be taken:

- Replacing the 3-phase rectifier with diodes by a 12-pulse rectifier with a 30-degree phase shift or by an active front end, i.e. a fully active PWM-controlled inverter bridge connected to the grid, capable of producing 'clean' current.
- Installing passive filters with one or more tuned resonant circuits (tuned for the 5th, 7th and 11th harmonics).
- Installing active filters having the ability to compensate current harmonics.
- Lowering the impedance of the main distribution transformer.

Electromechanical vibrations

If the rotor of an electrical machine is not at the centre of the stator bore, a resulting force is generated, in general trying to reduce the smallest air-gap even further, generally referred to as unbalanced magnetic pull. The problems linked to it are common in larger machines. The so-called critical speed, where abnormal vibrations occur, for *radial* vibrations is reduced by these magnetic forces. This reduction may be as much as 30%. In speed controlled drives, the reduced critical speed can coincide with the drive speed, leading to large and very dangerous vibrations. Due to flux optimisation of frequency inverters, the critical speed reduction may change. Due to the static eccentricity (caused by the rotor not coinciding with the centre of the stator bare), that often occurs in two-pole induction motors, double supply frequency radial vibrations are generated. If this frequency coincides with the natural frequency for radial vibrations, large vibrations may be expected.

Torsional oscillations due to harmonic components of supply voltages and currents from the inverter generate vibration components of the torque, with frequencies six times the supply frequency and its multiples. The high switching frequencies in modern inverters have made this no longer critical in up-to-date drives.

Additional Losses

Induction motors are manufactured for direct grid supply with a constant voltage and frequency. In order to reduce starting current while increasing the starting torque, special rotor construction have emerged over the years. If a motor is supplied by a frequency converter, the fundamental frequency is no longer constant and both current and voltage may contain harmonics.

As the flux is normally kept at a constant value in the air-gap, the magnetic losses vary with the frequency (hysteresis proportional to the frequency, eddy currents proportional to the frequency squared). Empirically an increase by a factor of 3.25 is found when the frequency goes from 50 to 120 Hz. At constant current an increased stator conductor loss is found due to skin effect. Therefore, it is sometimes advisable to use special types of stator windings when high supply frequencies are used. Mechanical problems should not be overlooked at high speed: increased windage and friction losses, mechanical stresses on the rotor, critical speeds, bearing lubrication, etc. Also at low speeds, problems will occur. Especially at constant torque application, as e.g. compressors, losses at low speed may be considerable. As induction motors in general are equipped with a cooling fan fixed on the rotor shaft, cooling is reduced dramatically at low speed. Therefore, such applications require an independently driven fan. Harmonic voltages produce current components at high frequency. These current harmonic induce relatively high currents in the rotor squirrel cage. In inverters of the "pre-IGBT" type, large additional losses were found. Due to the introduction of high switching frequencies, these problems have been reduced dramatically.

Acoustic noise

Induction motors are most of the time cooled by a fan mounted on the machine shaft, thus having the same speed. Due to the increased speed, not only the fan power consumption, but also its acoustic noise level increases dramatically. The importance of the inverter-motor interactions are illustrated by some tests carried out on a 13.5 kW squirrel cage induction motor. The motor has a single-layer stator winding and a double-cage aluminium cast rotor. Two inverters were used: a transistor based inverter having a switching frequency of 1 kHz while the switching frequency of the second, IGBT inverter can be set in between 1 and 12 kHz. In general, noise decreases as the switching frequency increases. Especially at low speed, acoustic noise reduction due to increased switching frequency is noticeable.

Table 1. Acoustic noise (in dB) of different inverters

50Hz Supply	IGBT-Inverter 1.2 kHz	IGBT-Inverter 12 kHz	Transistor Inverter 1 kHz
68.8	71.5	69.4	74.6

When the different inverters are compared with respect to acoustic noise at 50 Hz, only a 0.5 dB(A) increase is found when compared to the grid supply at a switching frequency of 12 kHz. At 1 kHz the increase is far more pronounced. A switching frequency above 10 kHz does not contribute to a noise reduction. The presence of pure tones is more disturbing to human beings, for a given overall level. Using random sampling of modern inverters, the pure tones may be avoided.

Overvoltages

IGBTs lead to very steep switching fronts, resulting in travelling waves in the electrical connection between motor and inverter. A reflection and subsequent voltage doubling at motor/inverter terminals may occur. The insulation of motor windings may be destroyed, yielding sparks in explosion-endangered zones. Applications in such zones require long distances between inverter and motor, as the inverter always has to be outside ex-zones, while the motor is in it. As the problem is due to the steep front ends, reducing inverter switching frequency is not a solution. Only damping and smoothing voltage waves using chokes or special cable types offer a fundamental way out. However, the introduction of chokes reduces the dynamic drive behaviour and may be inacceptable. Furthermore, they increase losses and introduce a substantial extra cost (up to 20 or 30% of the overall drive cost). A lossy contains several isotropic, absorbing materials (Fig. 1).

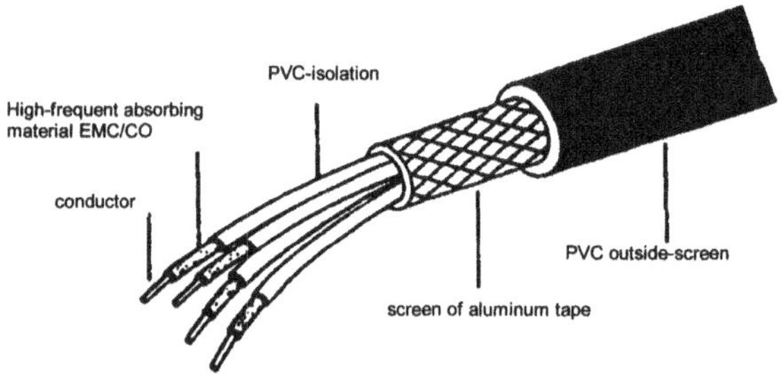

Fig. 1. Low pass cable

Bearing problems

Machine models fail in describing and explaining some parasitic effects in inverter driven machines, especially with respect to bearing currents. The drive (Fig. 2) including supply, DC link, switching elements, cables, motor and load machine is as a complex system. The high-frequency common-mode voltage causes a capacitive current to flow through parasitic capacitances, grounding system, cable shields and through parts of motor and inverter. Fig. 3 shows a sketch of an induction motor. Iinternal distributed capacitances are simplified to concentrated elements: winding-to-frame =C_{WF}, winding-to-rotor C_{WR}, rotor-to-frame C_{RF}.

Several ways of protecting the bearings against premature failure are applicable:

- bearing current interruption using an electrically non-conducting bearing, e.g. ceramic balls or rollers or a ceramic coating on the outer ring. An insulated coupling is necessary if common-mode current flows via the coupling through the load machine, its grounding system and back to the inverter.

- bearing current can be short-circuited by a brush-slipring-construction. This solution may be necessary for capacitive shaft-to-ground voltages.
- bearing currents can be minimized or even avoided by access to the common-mode voltage as source of bearing currents. Filtering at the inverter output is possible but expensive.

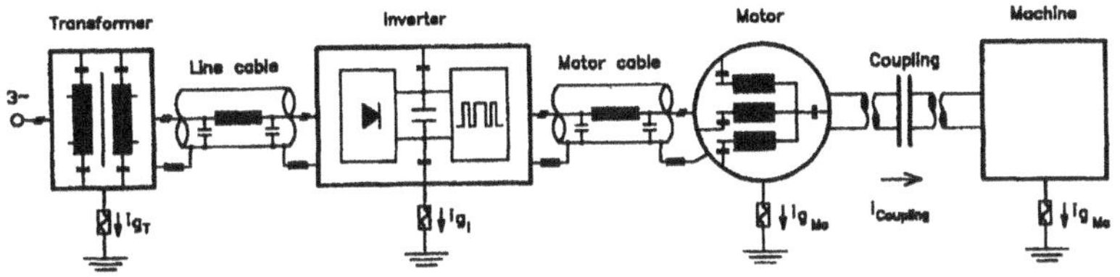

Fig. 2. General configuration of an inverter drive

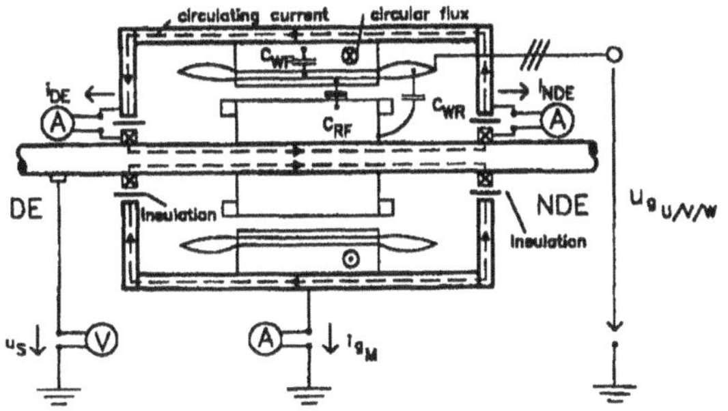

Fig. 3. Parasitic capacitances in an induction motor and test setup

Impact of voltage dips on the drive

Voltage dips are a power quality phenomenon in which the supply voltage is reduced with 10-99% for a short period of time (typically <1 s). Dips are mainly caused by the starting of heavy loads or by short circuits. A voltage dip causes the DC-link not to be loaded to its rated value. Most VSDs are equipped with a minimum voltage protection to prevent damage to the motor or the end product. A three-phase dip causing the voltage to drop by a higher percentage than this threshold causes a motor stop (fig 4a).

Other types of dips, called unbalanced dips, result in a different behaviour of the DC bus voltage. For a two-phase dip at the connection point of the load, the unbalance in supply voltages causes the rectifier to operate in a single-phase mode as one phase remains at its pre-event value (Fig.4b). Whether or not the DC bus voltage will reach the under-voltage protection level U_{min} and consequently trip the drive depends on the load conditions and size of the dc bus capacitor C. In most processes, the energy efficiency gain outweighs the costs of voltage dips since the tripped motor can easily and without any costs be restarted. Alternatively, the 'restart on the fly' is a cheap remedy.

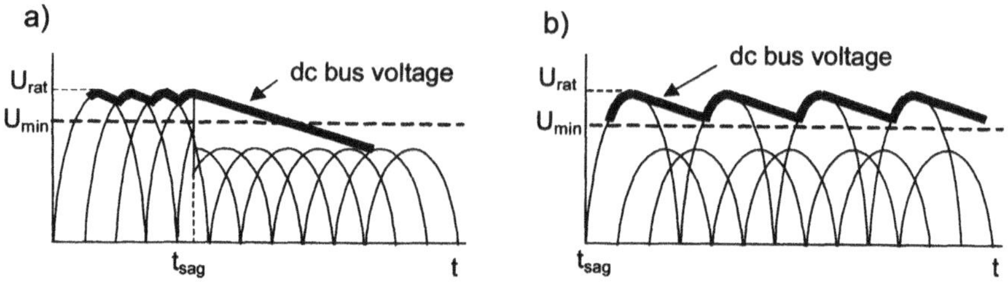

Fig. 4. DC-bus voltage during a) a three-phase dip and b) a two-phase dip

In some processes, such as extrusion where synchronism of a large amount of motors has to be taken into account, the failure of a motor can cause considerable losses. In these cases, installing a boost converter or an active front-end to the DC bus can be an option to mitigate voltage dips. If the DC-bus is not accessible, a Dynamic Voltage Restorer, a flywheel or a DySC is an option.

Conclusions

The introduction of variable speed drives can help in reducing the energy consumption dramatically. However, the problems discussed in this paper have to be avoided from the beginning as otherwise the energy savings benefits are jeopardized by the costs of solving these problems afterwards.

Acknowledgement

The authors are grateful to the Belgian "Fonds voor Wetenschappelijk Onderzoek Vlaanderen", the "Instituut voor de aanmoediging van Innovatie door Wetenschap en Technologie in Vlaanderen", Electrabel and Laborelec for their support of this work and the Research Council of the K.U.Leuven for granting a concerted research action to support this research. J. Driesen holds a postdoctoral research fellowship of the "Fonds voor Wetenschappelijk Onderzoek - Vlaanderen".

References

[1] J.Erdmann, R.Kerkman, D.Schlegel, G.Skibinski: Effect of PWM Inverters on AC Motor Bearing Currents and Shaft Voltages, 10th IEEE Applied Power Electronics Conference, 1995, Vol. 1, pp.24-33

[2] D.Busse, J.Erdmann, R.J.Kerman, D.Schlegel, G.Skibinski: System Electrical Parameters and Their Effect on Bearing Currents, IEEE Applied Power Electronics Conference, 1996, pp.570-578

[3] S.Chen, T. Lipo, D.Fitzgerald: Source of Induction Motor Bearing Currents Caused by PWM Inverters, IEEE Trans. on Energy Conversion, 1996, Vol. 1, No. 1, pp.25-32

[4] S.Chen, T.Lipo, D.Novotny: Circulating Type Motor Bearing Current in Inverter Drives, IEEE-IAS Annual Meeting, 1996, Vol. 1, pp.162-166

[5] T.Green: “The impact of EMC regulations on mains-connected power converters,”IEE Power Engineering Journal, February 1994, pp.35-43

[6] T.Hableter, D.Divan: “Acoustic noise reduction in sinusoidal PWM drives using an randomly modulated carrier,” IEEE Trans. on Power Electronics, 1991, Vol. 6, No. 3.

[7] L.Malesani, P.Tenti: “Three-phase AC/DC converter with sinusoidal AC currents and minimum filter requirements, IEEE Trans. on Ind.Appl., 1987, Vol. 23, No.1, pp.71-77

[8] J. Gray, F. Haydock: "Industrial power quality considerations when installing adjustable speed drive systems", IEEE Trans., 1996, IA-32 (3), pp. 646-652

[9] M. Didden, R. Belmans, W. D'haeseleer: "Lessons learned form a thorough voltage sag case study", PQA North America, June 2001 Pittsburgh

[10] A. Von Jouanne, H. Zhang, A. Wallace: "An evaluation of mitigation techniques for bearing currents, EMI and overvoltages in VSD applications", IEEE Trans., 1996, IA-34 (5) pp.1113-1122

The Successful Retrofit of 5000 Horsepower Boiler Forced-Draft Fans with Perfect Harmony Variable Frequency Drives

Richard Osman[1], Glenn Davis[1], Kevin Kellerman[1], Rick Langley[2]

[1] ASIRobicon, 100 Sagamore Hill Road, Pittsburgh PA 15239 USA
Phone 724 339 9500; e-mail richard.osman@, glenn.davis@us.asirobicon.com

[2] EPRI PEAC Corporation 942 Corridor Park Boulevard, Knoxville TN 37932 USA; Phone 865 218 8016; e-mail rlangley@epri-peac.com

Introduction and Background

One of the missions of EPRI is to encourage the adoption of new electric technology to improve performance and save energy for the producers and consumers of electricity. To this end, EPRI supports the development of new and improved variable-frequency drives, and promotes the use of drives among its power utility members. A new topology of medium-voltage drives, the series-cell multilevel arrangement, had become commercially available and EPRI wanted to install drives of this sort in a generation plant as part of a technology demonstration project.

In the summer of 2000, ASIRobicon engineers met with representatives of EPRI to discuss a technology demonstration project in which state-of-the art VFD's would be installed in a US electric power plant. Later that year, we finalized the application details for the installation at a 500 MW generating station owned by a utility in the Southwestern US. This is a gas-fired power plant that had recently undergone a motor upgrade of its forced-draft fans. Two 5000 Hp motors replaced the older 3000Hp motors. Along with the forced-draft fan upgrades, BACT NOx modifications were also made to the boiler.

The utility uses standby generators to black-start the power plant. Computer simulations revealed that the anticipated voltage drop created by line-starting each of the 5000Hp motors from the standby generators was 21%. This level was unacceptable and could potentially cause nuisance shutdowns of other systems due to undervoltage. VFD's would allow the utility to soft-start the motors without causing large system voltage drops.

The airflow through the fans was originally controlled by inlet guide vanes. This presented an opportunity for energy savings through variable-speed motor control.

So there were several good reasons to install variable frequency drives. As the effect of the VFD's on the source was a major concern, the Perfect Harmony drive was an ideal offering since it had been designed for maximum power quality at both the input and output.

The Perfect Harmony Medium-Voltage Variable Frequency Drive

The development of this VFD dates back to 1993, shortly after IEEE-519 had been revised. Although medium voltage VFD's were reliable and cost-effective at that time, they left something to be desired as far as input and output power quality was concerned. The main concept of the Harmony drive is the use of a multiplicity of power conversion modules connected in series. This approach, which is called distributed power architecture, offers some powerful advantages. Each cell as shown in Fig.1 consists of a 3-phase bridge rectifier, a DC link capacitor, and an H-bridge of IGBT's, plus local control signal electronics. The input is 3-phase 690 volts AC from an isolated secondary on the input transformer. In the cell, no device can be exposed to more than the DC link voltage. Thus, the difficulties of voltage sharing among series semiconductors is avoided. The cells are connected in series to develop the required output voltage, as shown in Fig 2. For the 5000Hp drive, which was rated at 4160V output, there were four identical cells per phase, with a current rating of 660 amps.

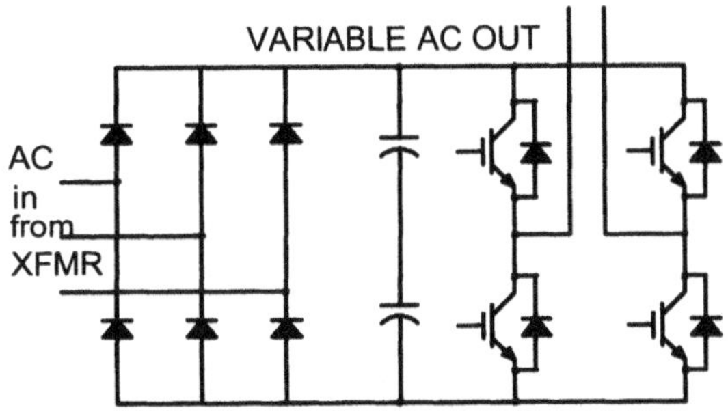

Fig. 1. Power Module (Cell) Schematic

Input Power Quality

The 12 transformer secondaries are phase shifted to present a 24-pulse load to the line. Depending on the output voltage of a particular drive, the cells per phase can vary from 3 to 6, so the input pulse number ranges from 18 to 36. There is a significant advantage in using an uncontrolled bridge as compared to a thyristor controller, since the rate of change of input current during commutation is comparatively low. This minimizes the higher order harmonics, and the phase paratively low. This minimizes the higher order harmonics, and the phase shifting minimizes the lower order harmonics. As the input circuits are diode rectifiers, the input power factor is always above 0.95 regardless of the load. Typically, the input current distortion is around 2%, and always below the most strict IEEE-519 limit of 5%.

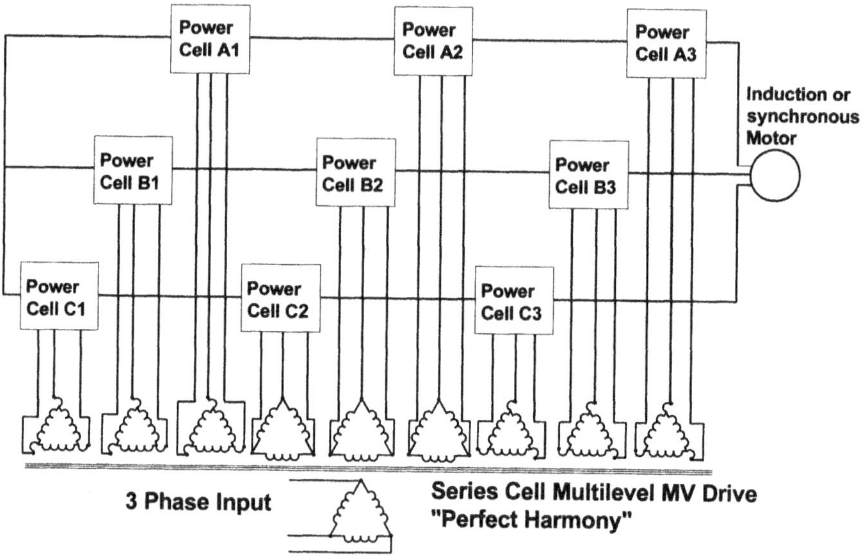

Fig. 2. Complete Schematic of 3 Cell per Phase Perfect Harmony

Output Power Quality

Of course, the user is concerned about the effect of the VFD waveform on the motor. There are three issues to be concerned about, harmonic currents, dv/dt, and common mode voltage. Since there are many cells in series, there are many output levels which can be used to improve the harmonic quality of the output. The switching instants of the cells are interdigitated, so that the effective carrier frequency as seen by the motor is twice the number of cells per phase times the IGBT switching frequency of 600Hz.

Fig. 3 illustrates the output modulation principles. Therefore for a 4 cell VFD, the effective carrier is 4800Hz. This means that the center of the harmonic sidebands is at the 80th harmonic of 60Hz. The low order harmonic current is negligible. The minimization of low order harmonics virtually eliminates torque pulsations. We did review the torsional analysis of the fan as a precaution, however. As only one cell switches at a given time, the step of voltage at the output is equal to the DC link voltage, about 950 volts. Even though this occurs in about one microsecond, it is not harmful to the insulation system of a modern form-wound MV machine. Since there is an input transformer, the machine insulation system does not have to support any common-mode voltage.

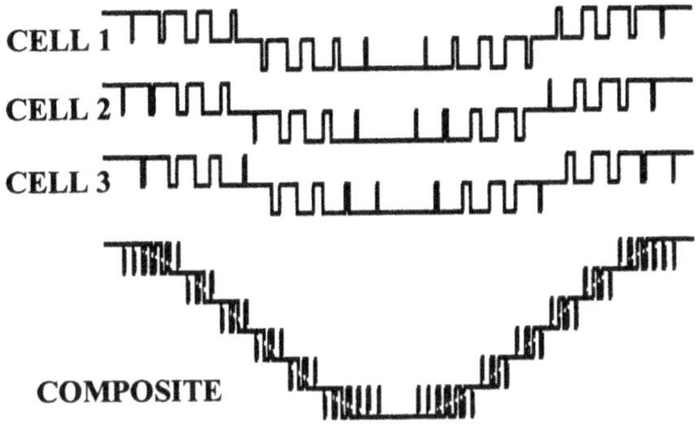

Fig. 3. Modulation Strategy for 3 Cells

Electrical Installation Details at the Utility Station

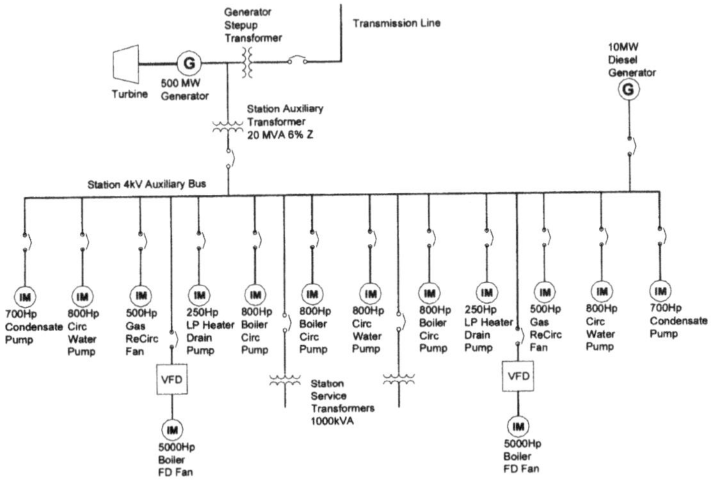

Fig. 4. One-Line Diagram of Power Station

Fig. 4 shows the approximate one-line diagram for the plant showing where the VFD's are tied to the station 4kV auxiliary bus. The 5000Hp motors are by far the largest load on the bus, and therefore the most likely to affect the bus during starting and fault conditions. The benefits of the drives are the elimination of the voltage drop during starting and the elimination of the fault contribution of the 5000Hp motors, as the VFD cannot feed a fault on the bus. There is the inrush current of the input transformer to contend with, but it only lasts a few cycles and does not occur simultaneously with motor starting.

The VFD is equipped with a bypass contactor to permit direct-on-line operation, as shown in Fig. 5.

Both 5000Hp water-cooled drives were packaged in a single control house for outdoor installation. There was an outdoor air-to-water heat exchanger provided to dispose of the VFD losses, which are about 112kW per VFD at rated power. (See Fig. 7) The VFD's had to be integrated with the existing switchgear and control system.

Results

The VFD's were delivered in the fall of 2001 and commissioned shortly thereafter. Although not a primary goal of this project, there has been significant energy savings as a result of fan control with speed rather than guide vanes.

The engineers at the power station collected data on the fan power consumption and current before and after the installation of the VFD's. The results are shown in Fig. 6.

Note that the current drawn by the fan is significantly reduced by the use of the VFD. This is partly because the reactive power requirement of the motor is supplied from the VFD, and does not come from the line. At 450MW, the use of the VFD saves about 700kW, and the power savings increases at lower capacities, because the part load efficiency of the VFD is very high, whereas the guide vanes degrade the fan efficiency at lower flow rates.

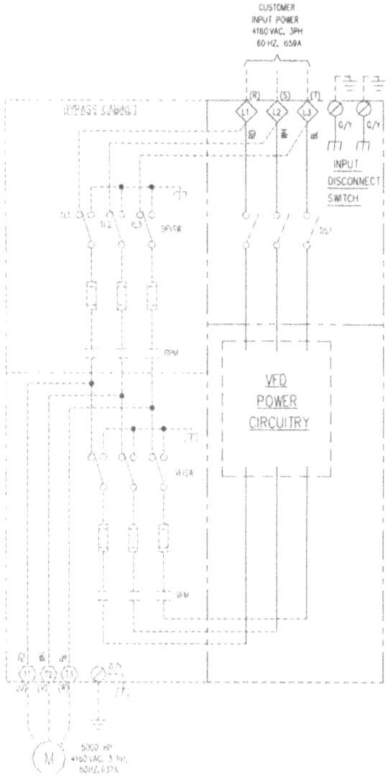

Fig. 5. Details of VFD Circuit

Conclusions and Acknowledgments

This paper has summarized the installation and results of a forced-draft fan VFD project at a power station. The performance of the drives was very satisfactory to the user and the power savings was an extra bonus. This project demonstrates that it is possible to obtain large medium-voltage drives which have excellent input and output power quality.

The authors would like to thank EPRI for its financial support of the project, and the engineers of the Utility whose enthusiastic collaboration was vital to the successful outcome.

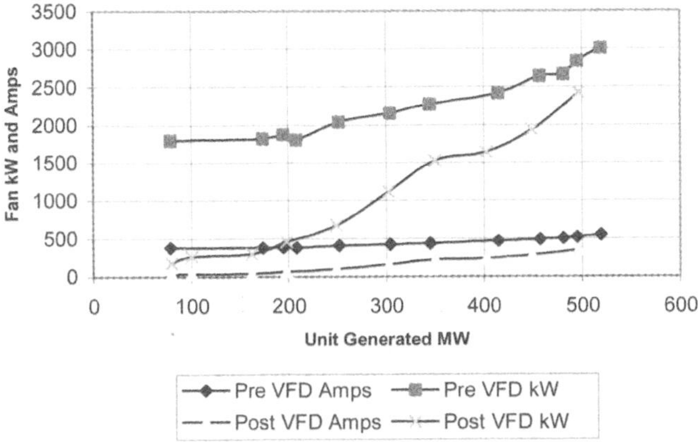

Fig. 6. Power and Current Consumption by a Fan, Pre and Post VFD

Fig. 7. This is a photo of the VFD house (46ft. long by 16ft. deep by 10ft. high) in the foreground, with the heat exchangers in extreme foreground. The fans and one motor are visible behind the VFD house.

Replacement of Old Drives and Motors – Software Tools for Evaluation and Selection

Mauri Peltola

ABB Oy Drives P.O.Box. 184, FIN – 00381 Helsinki, Finland
Tel.: +358 50 332 3647, Fax: +358 10 222 2287 e-mail: mauri.peltola@fi.abb.com

Introduction

A major concern of many drive end users is the availability and reliability of their drives. There are a large number of drive suppliers having models sometimes varying from year to year. Some models get old faster; some can resist the time better. As the drives get older their cost of ownership gets higher. The old products require a lot of expertise and due to the rare need of spare parts the cost of them can be high compared with the mass produced new drives and their spare parts.
Another reason to replace existing motor and drive installations with new ones is the higher efficiencies of the newer products. High efficiency motors and the new drives with higher switching frequency may have several percentage units higher efficiency than the products manufactured some 20 years ago.
The modernization of processes often requires upgrading of automation systems and the older drive systems are seldom equipped with the communication capabilities like fieldbus gateways. The need of new drives can be caused also for this reason.

The History of AC Drives Has Created the Market for Replacement

The real take-off of AC Drives market growth happened in the second half of the 1980's. After the launch of drives with digital control, power transistors, GTO's, field vector control and dramatic cost reduction the average market growth was about 25% per annum. As shown in Table 1. The growth rate has been reduced to 5 – 7% per annum during the second half of 1990's.

Table 1. Estimated AC Drive market in Europe in Million US Dollars

Year	1987	1994	2001
European Market	220	1270	1930

The slower growth of the market today and the relatively high population of drives in use for more than 10 years increases the interest to replace those older drives with new ones. It has been estimated that quite soon this replacement market could be about 30 percent of the total sales of the AC drives.

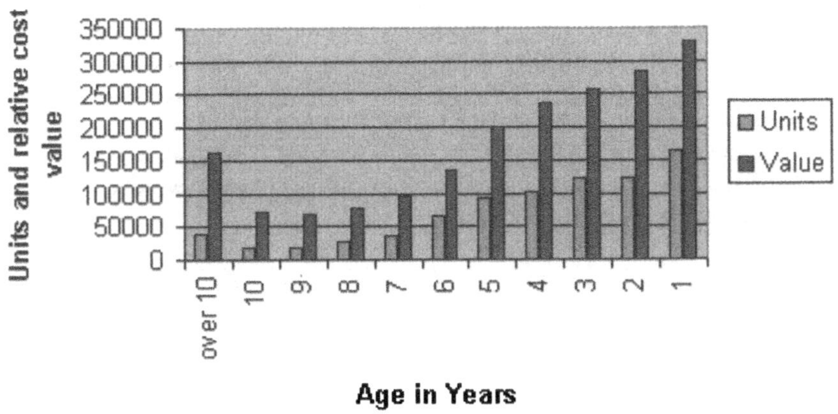

Fig. 1. Potential market for replacement of AC drives delivered by ABB

The market potential can be seen in the Fig. 1. The number of units delivered more than 10 years ago is less than 30 percent of the sales of the new drives, but the value for replacement is actually higher because of higher average size of the drives delivered over 10 years ago.

Why to Replace the Old Drives and Motors ?

The four most common reasons to buy variable speed AC drives and AC motors, instead of some other control, are:

- Improved quality of the end product by better process control accuracy
- Increased profitability by higher process throughput
- Reduced manufacturing cost by energy saving
- Reduced maintenance cost by soft starting

Cost of Maintenance

The above mentioned means that the customer already has quite long and probably good experience with the drives and motors in operation. How to convince him to replace his 10 years old drives or 15 years old AC motors by new equipment? It may be difficult because for the most customers the real cost of ownership can be unknown. As the drives get older their cost of ownership gets higher. This is a common law of nature based on the truth that in nowadays the labor cost is higher than the material cost.
The old products need a lot of expertise and due to the rare need for spare parts the cost for them can be high compared to the cost of mass produced complete drive units and the spare parts for them. Because the technology is developing in faster cycles, the experts who can handle the old technology are more and more rare.
How to maintain the availability of the drives in an acceptable level without investing too much into new drives? The total cost of lifecycle should be known and unexpected system and product failures should be avoided. Therefore a replacement of the old drives is a measure to prevent standstill times of the plant and machines to increase the productivity and the profit.

Cost of Energy Inefficiency

The motors and drives of today are quite a lot more efficient than 20 years ago. This can be seen in the chart shown in Fig. 2. The high efficiency motors are developed by several improvements in stator windings, rotor construction, core iron material, slot geometry etc. Because less heat is produced than in regular motors, a smaller fan can be used, reducing the losses even more.
The efficiency of AC drives has been improved a couple of per cent during the last twenty years mainly because of more efficient power semiconductors. The improvement would be more if there weren't been the demand for more sinusoidal motor voltage and less harmonics content in the line voltage. This has lead to higher switching frequencies and more efficient filtering. These both increase somewhat the losses in the drive itself but they decrease the harmonic losses in the motor. Especially the increased switching frequency reduces the harmonic losses dramatically as shown in Fig. 3.

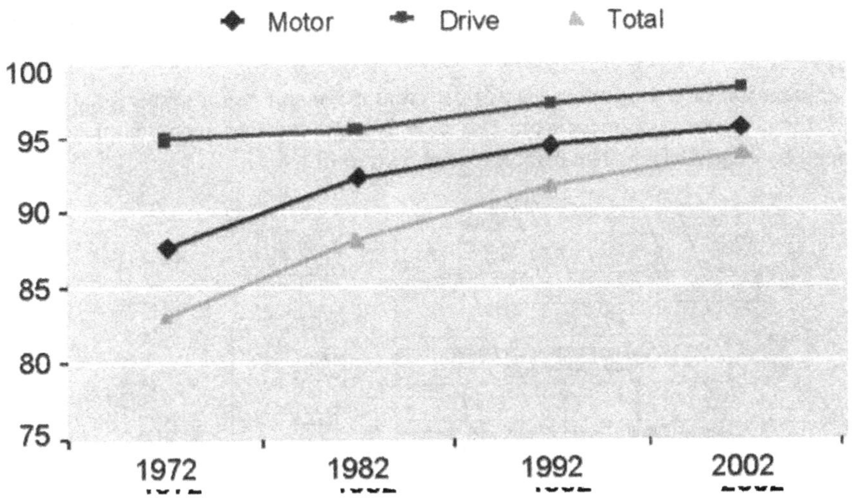

Fig. 2 about 100 kVA AC drive and 75 kW motor efficiency development during last 20 years

For a motor and drive combination, a loss optimum can be found. This means a switching frequency at which the sum of motor and drive losses is at minimum. Depending on the switching components, motor construction and many other factors, it can be said that the optimum generally settles somewhere between 2 and 4 kilohertz.

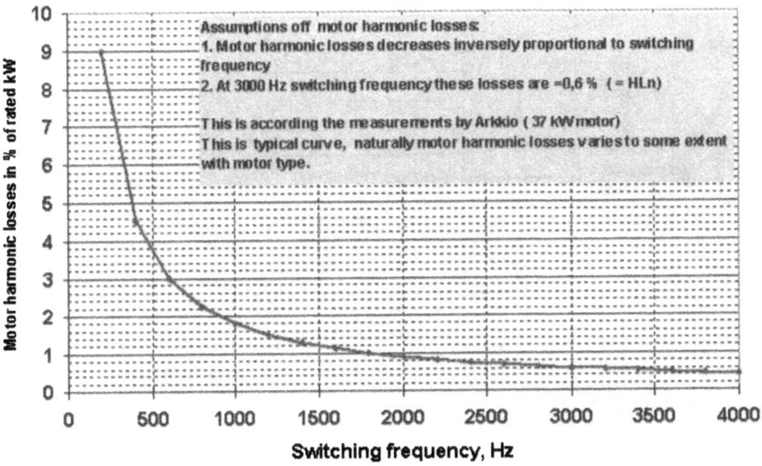

Fig. 3 Harmonic loss with switching frequency for a 37 kW motor

How to Calculate the Economy ?

For calculation of energy efficiency of different drive and motor systems and especially for comparison between old and new systems the Efficiency Tool was developed by ABB Drives. The principle is shown in Fig. 4.

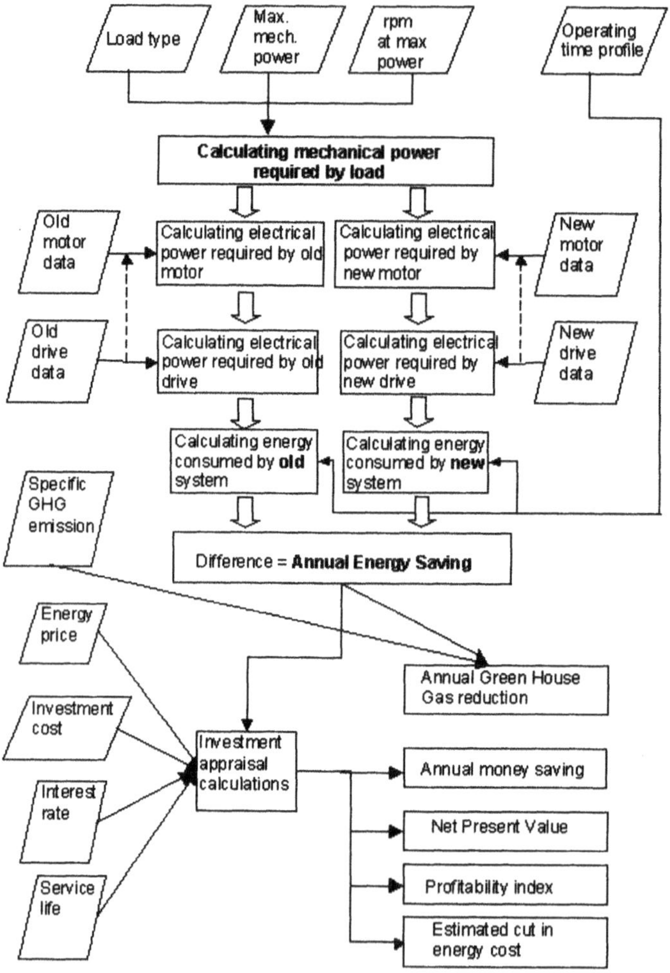

Fig. 4 The calculation path of the Efficiency Tool

How to Find a New Drive for Replacement ?

The DriveUpgrade tool, available on the web can be used for finding a new drive, which is the most suitable to replace the old drive. See Fig. 5.

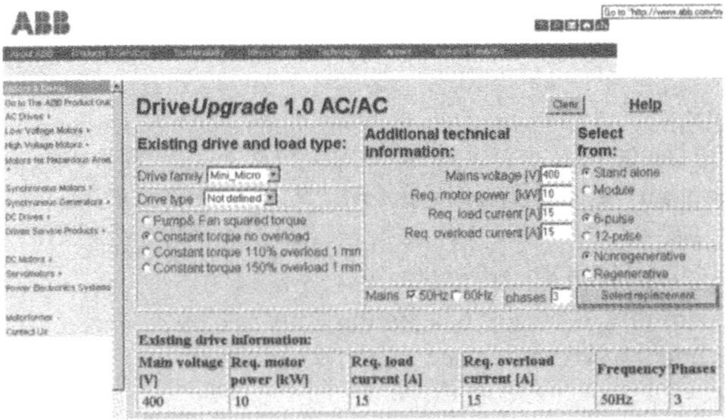

Fig. 5 DriveUpgrade Tool is on the web at http://www.abb.com/motors&drives

Case Raisio Malt Kiln

In April 2000, Raisio Malt replaced the fan drives for two of its kilns with new ABB drives. Drive upgrade reduces energy consumption from 5,245 kWh to 4,755 kWh (9%). Estimated reduction in annual CO_2 emissions for a single kiln is 90 tons. Payback period for investment is only three years as a result of lower electricity costs.

References

[1] Akseli Savolainen. Energy efficiency in industrial motor drive systems and calculation tool. Master's Thesis Helsinki University of Technology, Espoo, October 8, 2001
[2] DeAlmeida, A. et al. VSDs for Electric Motor Systems. Draft EU SAVE Programme Publication No XVII/4.1031/Z/99-04.
[3] Parasiliti, F. et al. Study on technical/economic and cost/benefit analyses on energy efficiency improvements in industrial three-phase induction motors, Final report. University of L'Aquila, Italy, 1999.
[4] Efficiency Tool User's Manual. ABB OY Drives, Helsinki 2001
[5] Type tests of motors and drives. ABB internal documentation
[6] ACS 600 Technical catalogue. ABB Oy Drives. Helsinki, 2001
[7] M3000 combination catalogue. ABB Motors, Vaasa, 2001

Efficiency Improvement of Inverter-Fed Permanent Magnet Synchronous Motors

C. Cavallaro[1], A. O. Di Tommaso[2], R. Miceli[2], A. Raciti[1], G. Ricco Galluzzo[2], M. Trapanese[2]

[1] Dept. of Electrical, Electronic and Systems Engineering, University of Catania, Viale A. Doria, 6 - 95125 Catania, Italy, Tel. +39 0957382323 Fax +39 095339535, E-mail: ccavalla@dees.unict.it - araciti@dees.unict.it

[2] Dept. of Electrical Engineering, University of Palermo, Viale delle Scienze 90128 Palermo, Italy, Tel. +39 0916566285 Fax +39 091488452, E-mail: ditommaso@diepa.unipa.it, miceli@diepa.unipa.it, ricco@diepa.unipa.it, trapanese@diepa.unipa.it

Abstract

In this paper a control algorithm for the efficiency improvement of inverter-fed permanent magnet synchronous motors (PMSMs) is presented. The proposed algorithm allows reducing the losses of the drive without reduction of its dynamic performances. In details, after recalling a dynamic model of the PMSM, which has been purposely modified and that takes into account the iron losses, the basic equations and the constraints to obtain the loss minimization are presented and discussed. Some simulations of a specific PMSM drive employing the proposed algorithm are performed. The results of these simulations show that the dynamic performances are maintained, and enhancement of the efficiency up to 5% can be reached in comparison to a PMSM drive using a more traditional control strategy.

Introduction

Permanent magnet synchronous motors (PMSMs) fed by inverters are widely used in industrial applications for their high performances. The main reasons rely on their optimal characteristics, which are, for example, higher efficiency and higher power-weight ratio than dc and induction motors. PMSMs are convenient because they have a loss-free rotor, and the power losses are mainly related to the stator windings and the stator core. The ratio of the copper and iron power losses is a key issue in determining the maximum efficiency point as function of the mechanical load that is connected to the motor shaft. In the case of constant-speed motors fed by the main, the motor designer performs a trade-off to obtain the maximum of the efficiency at given load conditions, according to the user requirements. Unfortunately, this advantageous condition is lost as long as the motors operate at variable-torque and variable-speed.

This paper deals with a control algorithm, which is able to improve the efficiency of permanent magnet synchronous motor drives by reducing the motor losses (copper and iron losses) through an optimal management of the current space vector in the stator winding, in the case of variable-speed variable-torque applications. In particular, after a brief recall of two loss minimization control strategies [1-5] (the so-called "search control" and "loss-model control" algorithms), both a modified dynamic model of the PMSM, which takes into account the iron losses, and an improved "loss-model" control strategy are presented. The control algorithm here proposed allows determining the optimal current space vector according to the operating speed and the load conditions.

The proposed approach is suitable to be applied to machines with salient or non-salient rotor structure types. The control algorithm is devoted to improve the efficiency in steady-state condition, which is a major opportunity for energy savings. By applying the proposed loss minimization algorithm to a specific drive, the simulation results have shown that its dynamic performances are maintained, and enhancement of the efficiency up to 5% can be reached in comparison to a PMSM drive equipped with a more traditional control strategy (i.e. i_d=0).

Loss Minimization Techniques: State-of-the-Art

Control techniques aiming to obtain the loss minimization have been extensively investigated in literature [4-8]. However, despite the huge of papers they can strictly be summarized into two main categories: papers, which deal with the "loss model control" technique, and papers, which apply the "search control" algorithm. The "loss model control" technique is based on the development of a mathematical model, which allows estimating the energy losses occurring during the running of the motor. Obviously, key issues in this case are the knowledge of a precise system model, an accurate identification of its parameters, and also the variation of the parameters with the temperature, current, etc. By expressing the losses as a function of the control variables of the drive, then it is possible to impose an operating condition to obtain a maximum of the efficiency [1].

The "search control" algorithm, on the contrary, is not based on a model rather on an adaptive routine. The approach mainly consists on changing step by step the value of a control variable, then measuring for each operating point the active power flowing into the motor. Finally, by comparing the measurement result with the previous one at fixed operating conditions, the minimum power consumption of the drive is searched. To this aim, recent works have experimentally demonstrated that the searching procedure successful can individuate a maximum efficiency point [2]. The "search control" algorithm has the advantage that there is no need to know the model of the motor and its parameters. A drawback is that such a technique can originate system oscillation phenomena, thus making unstable the drive. As far as this technique is adopted, a requirement of an additional stabilization network may arise [5].

PMSM Dynamic Model with Iron Losses

The basic hypotheses, which have been used in order to define the proposed dynamic model of a PMSM, are that the spatial distribution of the magnetic flux in the air gap is sinusoidal and the magnetic circuit is linear. Moreover, a dedicated parameter has been considered aiming to account for the loss of the stator iron. In particular, the iron loss is modeled by a resistance R_c, which is inserted in the traditional equivalent circuits of a synchronous machine so that the loss depends on the air-gap linkage flux. According, by considering the two-axes theory of Park and introducing the change in the model as above defined to account for the iron losses, the dynamic d- and q-axis equivalent circuits of the PMSM can be drawn, as it is shown in Fig. 1. With reference to Fig. 1, the state equations of the dynamic model of a PMSM, taking into account also the iron losses, are:

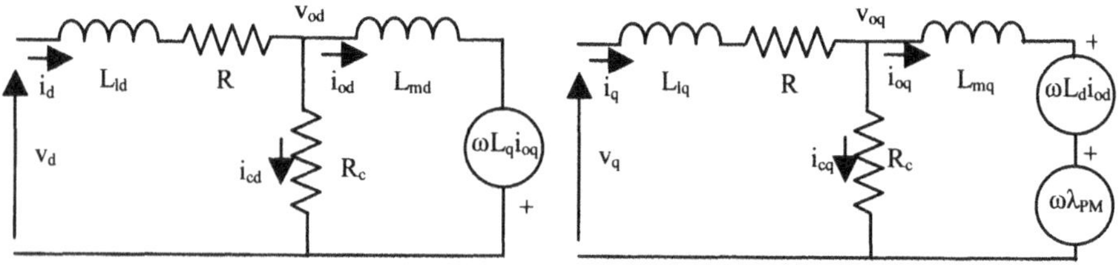

Fig. 1. Dynamic equivalent circuits along the d- and q-axis of a PMSM.

$$\frac{di_d}{dt} = \frac{1}{L_{ld}}\left(v_d - Ri_d - L_{md}\frac{di_{od}}{dt} + L_q P\omega_r i_{oq}\right) \tag{1}$$

$$\frac{di_q}{dt} = \frac{1}{L_{lq}}\left(v_q - Ri_q - L_{mq}\frac{di_{oq}}{dt} - L_d P\omega_r i_{od} - \lambda_{PM} P\omega_r\right) \tag{2}$$

$$\frac{d\omega_r}{dt} = \frac{1}{J}\left[T_e - C\,Sign(\omega_r) - F\omega_r - T_m\right] \tag{3}$$

$$\frac{d\theta_r}{dt} = \omega_r \tag{4}$$

where

$$i_{od} = i_d - i_{cd}\,,\; i_{oq} = i_q - i_{cq} \tag{5}$$

$$i_{cd} = \frac{-\omega k L_d i_{oq} + L_{md}\frac{di_{od}}{dt}}{R_c}\,,\; i_{cq} = \frac{\omega(\lambda_{PM} + L_d i_{od}) + L_{mq}\frac{di_{oq}}{dt}}{R_c} \tag{6}$$

$$T_e = \frac{3}{2}P\left[\lambda_{PM}\, i_{oq} + \left(L_{md} - L_{mq}\right)i_{od} i_{oq}\right] \tag{7}$$

Loss Minimization Algorithm

The power losses in a PMSM are copper and iron losses in the stator, mechanical losses, and additional copper and iron losses. The considered losses can be separated into two categories, namely, controllable and uncontrollable losses. The copper losses, which are caused by the fundamental harmonic component of the stator current, and the iron losses, which are caused by the fundamental harmonic components of the air-gap linkage flux, belong to the first type (in turn, these losses depend on the controllable variables of the motor). Unlike the above quantities, copper and iron losses, which are caused by the higher harmonic components, together with the mechanical losses belong to the uncontrollable ones. For this reason, the mechanical and additional losses are here not considered.

The dynamic model described by relations (1-7) has been prepared aiming to perform both the steady state and transient simulations of a PMSM drive. However, the optimization of the power loss consumption is done by considering an algorithm (see [1]) which refers to a steady-state model. By setting to zero the time derivatives of the d- and q-axis current components in (1), and (2), and also of the rotor angular speed in (3), the steady state model is obtained. Hence, the power losses caused by the fundamental harmonic of the current in the windings (W_{Cu}), and the power losses caused by the fundamental harmonic of the air-gap flux linkage in the iron stack (W_{Fe}), can be expressed as a function of the i_{od} and i_{oq} current components and of the electrical speed, $\omega = P\omega_r$.

$$W_{Cu}(i_{od}, i_{oq}, \omega) = \frac{3}{2}R\,(i_d^2 + i_q^2) = \frac{3}{2}R\left\{\left(i_{od} - \frac{\omega k L_d i_{oq}}{R_c}\right)^2 + \left(i_{oq} + \frac{\omega(\lambda_{PM} + L_d i_{od})}{R_c}\right)^2\right\} \tag{8}$$

$$W_{Fe}(i_{od}, i_{oq}, \omega) = \frac{3}{2}R_c(i_{cd}^2 + i_{cq}^2) = \frac{3}{2}\frac{\omega^2 (kL_d i_{oq})^2}{R_c} + \frac{3}{2}\frac{\omega^2(\lambda_{PM} + L_d i_{od})^2}{R_c} \tag{9}$$

By adding relations (8-9) the total electrical losses are calculated:

$$W_C(i_{od}, i_{oq}, \omega) = W_{Cu} + W_{Fe} \tag{10}$$

By combining relations (7) and (10), a relation that expresses the power losses of the motor as function of the electromagnetic torque T_e, the direct axis current component i_{od}, and the angular speed ω_r is obtained:

$$W_C(i_{od}, T_e, \omega) = W_{Cu}(i_{od}, T_e, \omega) + W_{Fe}(i_{od}, T_e, \omega) \tag{11}$$

Based on inspection of relation (11), a simple consideration can be carried out: at fixed values of both T_e and ω, the total controllable losses depend only on the i_{od} value, then can be minimized by controlling the current space vector. In reference [1] the value of i_{od} that minimizes the electrical losses has been analytically calculated by differentiating expression (11), with respect to the i_{od} variable, and successively equating the resulting expression to zero. The following relations summarize the main results of such a procedure, being the "heart" of a new technique of implementation and solution of the algorithm of the minimum loss condition:

$$AB = T_e^{\,2} C \qquad (12)$$

where

$$A = P^2 [R R_c^{\,2} i_{od} + \omega^2 L_d (R + R_c)(L_d i_{od} + \lambda_{PM})]$$

$$B = [\lambda_{PM} + (1-k) L_d i_{od}]^3, \; C = [R R_c^{\,2} + (R + R_c)(\omega k L_d)^2](1-k) L_d$$

In the case of machines with isotropic rotor structure the parameter of saliency is k=1, and condition (12) simplifies. According, the optimal current i_{od}* can be easily expressed as analytical function [1]. However, for a more general structure of the motor rotor, the problem cannot be easily solved, and a closed solution i_{od}* is not obtainable, because of the non-linear relationship nature of (12). For this reason, in reference [1] the value of i_{od}* which minimizes the losses has been calculated by using an approximate procedure. By the use of a polynomial expression, which is a function of i_{oq}, the solution of the optimal d-axis current is "on line" calculated through the use of an implemented "off line" look-up table previously calculated. This method has the burden relative to the recalculation of a lot of coefficients, in order to update the look-up table, aiming to account for the new parameters in case the motor is changed.

In this paper an alternative solution is proposed. By the use of a subroutine implemented in the simulation tool MATLAB-Simulink expression (12) has been solved. The timing of the procedure is roughly the following. The current i_{oq} is determined by substituting both the calculated optimal current i_{od}* and the desired torque T_e in relation (7). The d- and q-axis components of the current i_d and i_q are then determined based on relations (5-6). The main difference in respect of the procedure used in [1] is the "on line" calculation of the i_d and i_q current components that minimize the power losses. Moreover, any change of the used motor needs the simple introduction of the major motor parameters, thus avoiding the "off line" calculation of many data and the implementation in the look up table.

Simulation Results

Extensive simulation runs have been carried out by implementing the PMSM electrical drive on MATLAB-Simulink environment. Some of the main results are presented in this section. Firstly, a comparison has been made between two different approaches: the loss minimization algorithm (LMA) discussed in the preceding section, and a traditional i_d=0 control. The i_d=0 condition has been reached through a speed-dependant compensation of the rotor instantaneous position. The parameters of the simulated PMSM are listed in Table I.

The efficiency curves, as a function of the mechanical speed, in the case of LMA (conts. curve) and of i_d=0 control (dotted curve) at the rated load of 1.67Nm, are shown in Fig. 2 a). The efficiency of the brushless motor, thanks to the new approach, can increase up to 5% in the high-speed range. The d- and q-axis current components as a function of the speed, at rated load conditions, are shown in

Fig. 2 b). From such a figure, we can see that at high speed the flux weakening significantly reduces the iron losses. The efficiency curves as a function of the load torque, in the case of LMA (conts. curve) and of i_d=0 control (dotted curve) at the rated speed of 2000 rpm, are traced in Fig. 3 a). The d- and q-axis current components, as a function of the load torque at the rated speed (2000 rpm), are shown in Fig. 3 b). As shown in Fig. 3 a) the effectiveness of the LMA grows at increasing load and becomes more significant in proximity of the rated one.

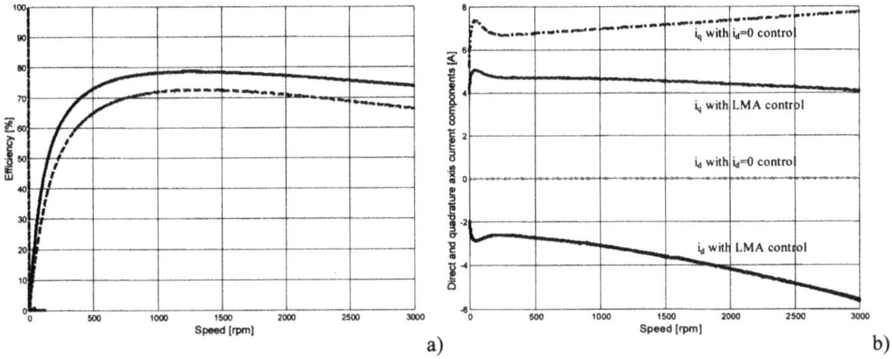

Fig. 2. a) Efficiencies vs. speed with LMA (conts. line) and i_d=0 control (dotted line) at rated load. b) d- and q-axis current components versus speed in the same conditions.

Table 1. Nameplate data and per phase parameters values of a PMSM.

Rated speed [rpm]	2000
Rated current (rms) [A]	5
Rated torque [Nm]	1.67
Number of poles	4
Armature resistance R [Ω]	0.57
Equivalent resistance of the iron losses R_c [Ω]	240
Direct axis inductance [mH]	8.72
Quadrature axis inductance [mH]	22.78
Permanent magnet flux [Wb]	0.088
Mechanical losses (torque) [Nm]	0.058

Finally, in Figs. 4 a) and b) the simulated dynamic response of the PMSM electrical drive to a sudden load change, both with i_d=0 control and LMA at 2000 rpm, is presented. The drive is running at no load condition, then at t=0.5s the rated load is applied (1.67 Nm) on the shaft. From inspection of such Figs. 4 a) and b) we can observe that there is not evidence of appreciable difference in the dynamic response of the speed (Fig. 4 a)), nor in the dynamic of the direct and quadrature current components (Fig. 4 b)), between the traditional control algorithm and the loss minimization one. On the contrary, a great difference is present in the values of the d- and q-axis current components, as it is shown in Fig. 4 b). This figure demonstrates the effectiveness and rapid convergence of the proposed control technique, capable to improve the efficiency subsequently to a transient, without appreciable delay or oscillations, while maintaining the dynamic of the current components. In order to experimentally verify the LMA, an early stage of a PMSM drive has been realized, and actually it is in progress. Preliminary experimental results, during the set up of the drive, encourage more developments in order to implement the proposed algorithm in a commercial DSP.

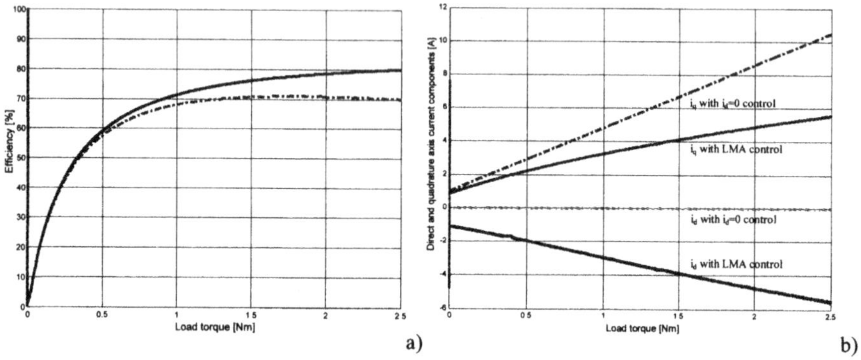

Fig. 3. a) Efficiencies vs. load torque with LMA (conts. line) and i_d=0 control (dotted line) at rated speed. b) d- and q-axis current components vs. load torque in the same conditions.

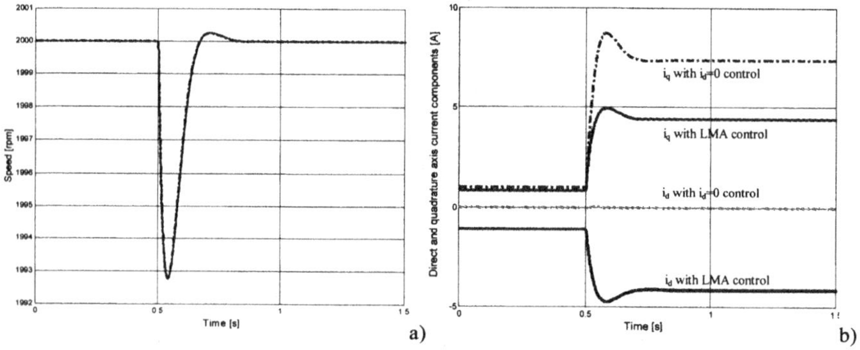

Fig. 4. a) Speed responses of the drive with LMA (cont. line) and i_d=0 control (dotted line). b) d- and q- axis current components with LMA (cont. lines) and i_d=0 control (dotted lines).

Conclusions

In this paper both a new dynamic model of the PMSM, which takes into account the iron losses, and a modified "loss-model" control strategy have been presented. In particular, it has been verified by simulation runs that controlling the stator current space vector can minimize the controllable electrical losses occurring in a brushless motor drive, consisting of the fundamental copper and iron losses. Such a control strategy, accounting for both the instantaneous speed values and the load torque condition, uses the combined effects of the field weakening and the exploitation of the reluctance torque. The loss minimization technique here reported is very flexible and simple to implement because only requires the knowledge of the common motor parameters. The main results of the simulations carried out demonstrated how, in comparison with more traditional control methods, the loss minimization algorithm increments in a significant way the efficiency of a PMSM drive without any reduction on the dynamic performances.

List of Symbols

i_d, i_q	*d- and q-axis current components;*	P	*motor pole pairs;*
i_{cd}, i_{cq}	*d- and q-axis iron loss current components;*	ω	*electrical angular speed;*
		ω_r	*rotor mechanical angular speed;*
v_d, v_q	*d- and q-axis voltage components;*	T_e	*electromagnetic torque;*
L_{md}, L_{mq}	*d- and q-axis mutual inductances;*	T_m	*load torque;*
L_{ld}, L_{lq}	*d- and q-axis leakage inductances;*	J	*rotor inertia;*
k	*saliency ratio ($k=L_q/L_d$);*	C	*Coulomb friction factor;*
R	*stator resistance;*	F	*viscous friction factor;*
R_c	*transversal resistance;*	θ_r	*instantaneous rotor position.*
λ_{PM}	*permanent magnet rotor flux;*		

References

[1] S. Morimoto, Y. Tong, Y. Takeda, T. Hirasa, *"Loss Minimization Control of Permanent Magnet Synchronous Motor Drives"* – IEEE Transaction on Industrial Electronics, Vol.41, No. 5, October, 1994.

[2] S. Vaez, M. A. Rahaman, *"Adaptive Loss Minimization Control of Inverter-Fed IMP Motor Drives"* – IEEE, Vol. 5, pp. 861-868, 1997.

[3] T. M. Jahns, G. B. Kliman, T. W. Neumann, *"Interior Permanent-Magnet Synchronous Motors for Adjustable-Speed Drives"* – IEEE Transaction on Industry Applications, Vol. IA-22, No. 4, pp. 738-747, July/August 1986.

[4] Roy S. Colby, Donald W. Novotny, *"Efficient Operation on Surface-Mounted PM Synchronous Motors"* – IEEE Transaction on Industry Applications, Vol. IA-23, No. 6, pp. 1048-1054, November/December 1987.

[5] Roy S. Colby, Donald W. Novotny, *"An Efficiency-Optimizing Permanent Magnet Synchronous Motor Drive"* – IEEE Transaction on Industry Applications, Vol. 24, No. 3, pp. 462-469, May/June 1988.

[6] S. Morimoto, Y. Takeda, T. Hirasa, *"Current Phase Control Methods for Permanent Magnet Synchronous Motors"* – IEEE Transaction on Industrial Electronics, Vol. 5, No. 2, pp. 133-139, April 1990.

[7] C. Mademlis, J. Xypteras, N. Margaris, *"Loss Minimization in Surface Permanent-Magnet Synchronous Motor Drives"* - IEEE Transaction on Industrial Electronics, VOL. 47, No. 1, pp. 115-122, February 2000.

[8] F. Fernandèz-Bernal, A. Garcìa Cerrada, R. Faure, *"Model-Based Loss Minimization for DC and AC Vector Controlled Motors Including Core Saturation"* – IEEE, vol. X, pp. 1608-1615, 1999.

High Energy Efficiency in Continuous Cycle Production Lines with Drives

Eugenio Santacatterina, Sergio Vellante

Control Techniques SpA, Via Brodolini 7, 20089, Rozzano (MI), Italy
Ph. +39 02 57 57 5 1; Fax +39 02 57 51 2858

Introduction

The scope of this brief work is to show the evolutions of technologies of power feeding of continuous and semi-continuous production lines , with the target of increasing overall energy efficiency.

With the terms “continuous and semi-continuous lines”, here are intended medium size production plants, such as Paper Making Lines, Steel & Aluminum Treatment Lines (Pickling, Annealing, Painting, Galvanizing,…) and Cold Rolling Mills. Common to all these kind of plants is the continuous load duty and the use of a consistent quantity of Electrical Energy (installed motor power is, typically greater than 1 MW): the major costs for material production, in facts, are due to raw materials and power consumptions due to motorization.

Another important characteristic is the plant complexity: a deep analysis of the plant functioning features and of the mechanical architecture, allows to design a Power Supply System which is specifically “tailored” on the application.

This important f

act can lead to a power consumption optimization and to wide benefits of the impacts on Mains supply.

Migration from DC to AC technology? The “Cold Rolling Mill” case

Today we can say that the power converter technology, either DC and AC are in a simple and well-established technique. The challenge of today is to install the most suitable and reliable system architecture into the plant environment.

An electrical equipment for an industrial plant must comply with a dual set of requirements: one is related to the “network side”, the other one to the “process side”.

“Network side” design criteria are:

- Voltage level/voltage variation

- Frequency/frequency variation
- Available network short circuit capability
- Allowed generated power factor
- Allowed generated harmonic distortion

The "Process Side" selection criteria are much more related to production efficiency and quality and can be summarized as follow:

- System efficiency
- System reliability
- Reparability and diagnostic capabilities
- Long term engineering, service and spare parts support
- Investment costs/installed costs/life-cycle cost
- Production load schedule criteria
- Plant start-up

The job of finding the right system topology is very complex and very critic, but most high power End Users are very experienced in this analysis and know what the process and plant requirements exactly are.

When analyzing the energy efficiency of a Drive System, the Power Factor (relationship between active and reactive power) is a major issue.

Ideal value for Power Factor is 1, which means only Active Power consumption and no bad consequences on the Mains Supply.

The Power Factor of a thyristor rectifier (DC drive system) depends on the firing angle of the system (which depends by the motor load) : when operating with a low firing angle, it is in the range of 0,93 to 0,90 near rated output, with larger fire angles the power factor is reduced. Normally, it is below 0,77.

The typical PF of an AC system is approximately between 0,92 and 0,95 and it is not dependant by the process load (i.e. Line Speed).

So, on the PF basis, AC technology seems to be preferred, but let's look a specific case: the Power Electrical Equipment of a Cold Rolling Mill.

A simple structure of a Cold Rolling Mill is shown in fig. 1.

This kind of plant is dedicated to thickness reduction of metal strip, typically for steel, aluminum and copper. Peculiar of this plant is the wide range of production (i.e. Thickness range from 1 to 30) which means a range in the motors speed that could reach 1:100. It is clear that the Power Factor is, actually, the main issue in the choice of a Drive System.

DC Solution

All the Reels and Mill DC Motors are driven independently by DC thyristor drives. Strip tension control and speed control guaranties, both during acceleration and steady state, are reached by using the technology of Full-Digital DC drives, but the typical Bandwidth of thyristor technology, is limiting more and more the future process performances.

The fully digital microprocessor based thyristor controllers are equipped an on-board powerful second processor (MD29) dedicated to the real time processing of application programs such as coiler center-winding, mill motors S-ramps,….

All of the DC Drives on the Mill are Four Quadrant Drives - 6 pulse or 12 pulse type with back to back thyristor sets for full regeneration capability.

The reels which acts as the uncoiler during a rolling pass, will be generating energy into the mains supply.

This also ensures that for any part or continuous regenerative applications (eg. Rolling Mill line tension control or emergency / fast stop conditions), the actual braking energy is directed straight back into the Mains Supply.

This is a real energy and cost saving.

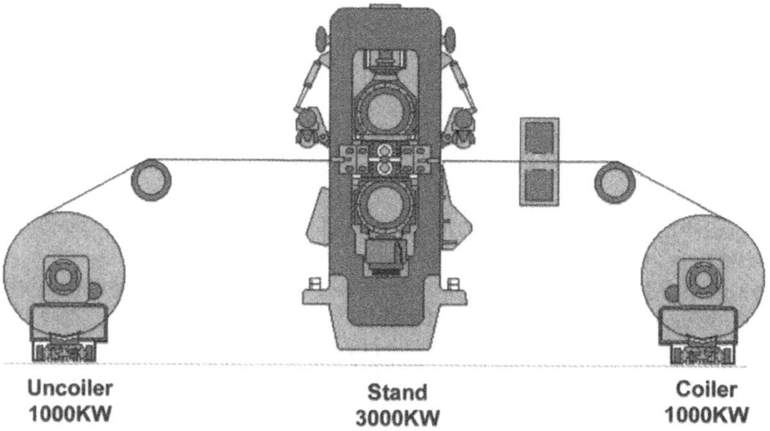

Fig. 1. Cold Rolling Mill

By the "energetic point of view", the ideal balance would be the compensation between the uncoiler energy (generating) and the coiler energy (motoring): this ideal solution will permit to feed only the power due to motor and mechanical cinematic friction and losses and the power for metal thickness reduction (Stand Rolling Load).

Because of the wide speed and load working range (Speed = 1:30 – Diameter build-up = 1:4 – Material tension 1:10), practically Reactive Power is very high during every working phase (even at highest speeds) and only Active Power given by Uncoiler can be used by Coiler and Stand.

The feeding MV/LV Transformers, their auxiliaries and power cables (MV cable from the Power station to the transformer and LV cables from Transformers to Power cubicles) must be rated for the total energy, not only for the energy used for the thickness reduction.

This means that a very efficient Power Factor Corrector System must be designed in order to reduce the very low power efficiency of the DC Drive System.

In order to compensate the very variable load, PF Corrector must be of the "static type" with SCR sets dedicated to the dynamic insertion of the various chokes and capacitors groups. Such kind of PFC implies system complexity, potential unreliability and an overall increase in the investment costs.

AC Solution

Reactive power elimination is the target of AC approach.

Individual AC motors cannot be fed by individual Inverters, which are directly powered by the Mains Supply, because of the reversible functioning of the plant.

In fact, a common AC Vector Drive, is comparable to a single-quadrant DC Drive, without any regenerating feature.

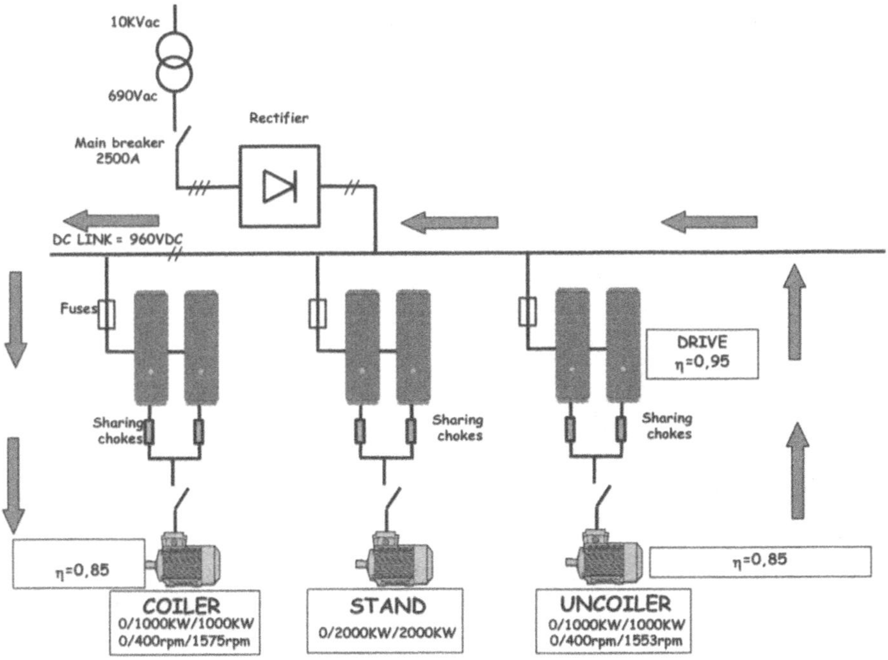

Fig. 2. Cold Rolling Mill: DC Bus and AC Drives

To allow regeneration of the Uncoiler motor, the DC Bus architecture must be adopted. The use of a common DC Bus to supply Vector Inverters gives the best performance in terms of sharing of energy between the AC drives and of space saving inside the electrical cubicle. (it is not necessary the use of the RFI filter for each Drive)

The common DC bus reduces also the cost of electrical power supply, because it works with the automatic compensation of the energy required from the drive working in power mode and the energy gave from the drive working in regenerating mode.

The DC drive works in order to maintain constant the voltage value in the DC link (540 Vdc or 960Vdc – depending by the Motor Voltage) giving energy to the system when the drives need.

With this configuration the DC drive works practically with a very high power factor, maintaining one of the most important advantage using the AC drives: the exclusive use of the Active Power from the line.

Instead of a DC drive, a simple AC to DC rectifier can be used, even increasing the global Power Factor.

With this solution, an external, dedicated circuit for inverters capacitor charging, must be designed.

The electrical design leads to a great Apparent Power reduction if AC solution is adopted and the DC Bus (Rectifier or DC drive) can be rated on the 60% of the total motor installed power (fig. 3).

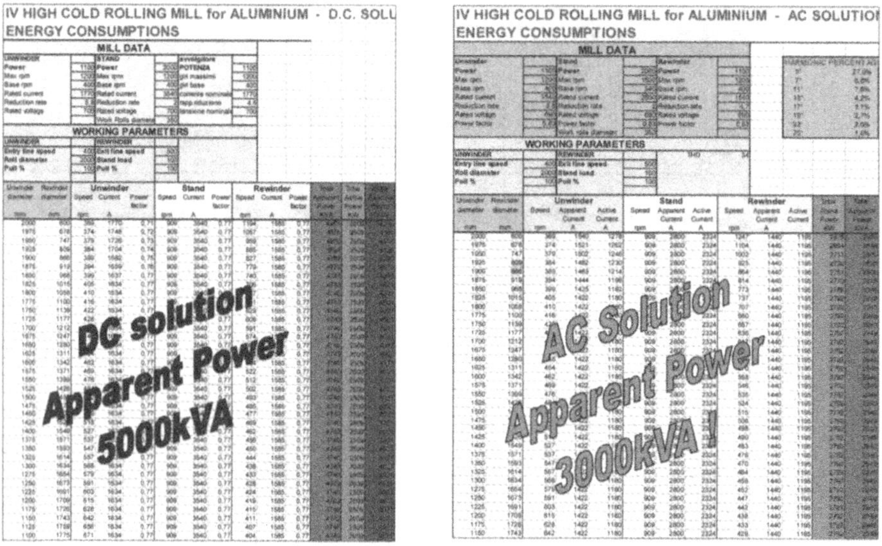

Fig. 3. Cold Rolling Mill: AC vs. DC design calculations

The same "design rating/cost reduction" approach can be transferred to Transformers, cables and auxiliary devices, with clear cost benefits.

In spite of the above advantage, the adoption of the AC technology in Cold Rolling Mills meets today some obstacles : against it, in fact, there are some technical and some major economical factors.

Main technical limitations are the following:

- It is not possible to regenerate energy into the Mains supply.
  If the whole mechanical system is regenerating (i.e. Emergency stop, Fast stop, Strip break…), the generated energy must be dissipated by braking resistor packs. A real regeneration to the Mains can be obtained with complex and expensive architectures like the double DC converter (one for feeding

and the other one for regenerating provided with Voltage Adapting Transformer – Fig.4) or the Active Front End topology.

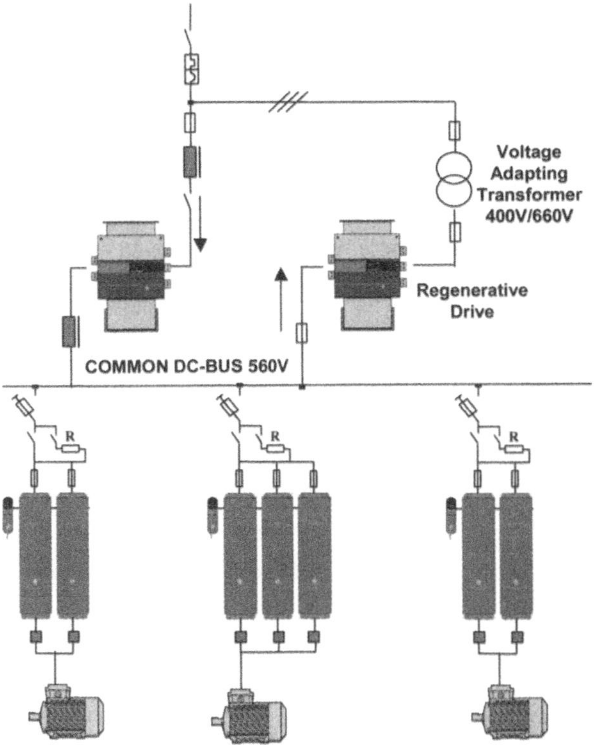

Fig. 4. Cold Rolling Mill: regenerative DCbus

- It is not possible to brake easily a motor in case of Mains Loss or Drive Fault.
  Mainly for Aluminum Rolling Mills it is absolutely necessary to stop the uncoiler rotation in order to avoid the intrusion of the metal strip into the mill, thus causing a destructive burn of the mill (presence of kerosene cooling). With DC drives, by holding the field supply, it is possible to stop the motor under any condition. With the AC Inverter, there is are no simple techniques to control the motor if the drive fails or if the power supply is down – only mechanical braking seems to be practicable.

Finally, the investment costs for large AC Drives is still very high, as well as installation (wiring) costs. It is evaluated that the break-even-point for an investment for an Aluminum rolling mill, total installed power 5MW, AC vs. DC technology is reached not earlier than 5 years. The long payback period is discouraging many EU in investing in large AC drives in rolling mills.

Energy efficiency with AC technology. The "Pickling Line" case

If it is proven that AC technology is not costly efficient for some systems with low number of motors and in presence of big individual motor power, it can be demonstrated that in those plants driven by several motors and low individual power (approx. less than 400kW), both the energy efficiency and the overall cost reduction are a real benefit of the today's AC technology.

A typical example is the Electrical Equipment for Pickling Line. The Pickling line is dedicated to clean (acid solutions) the surface of Steel Strip coming from a hot rolling mill and to bright it, through a surface rolling action (Skinpass Rolling Mill). The system is supplied by 32 AC motors fed by AC vector drives for a total of 2.1 MW power installed (Fig. 5).

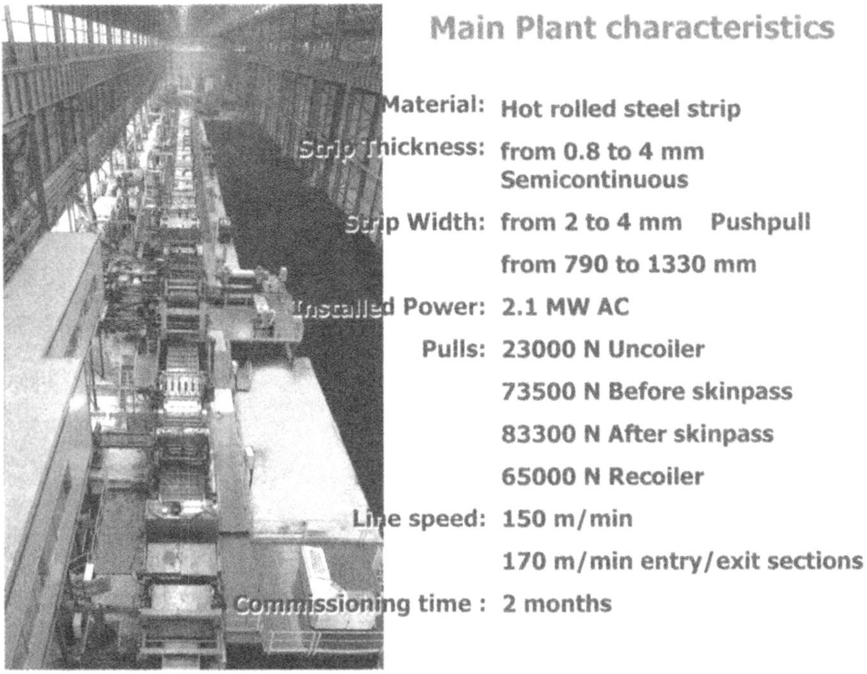

Fig. 5. Pickling line: Main characteristics

Because of its particular mechanical lay-out, the Drive Power System can be sectioned into N. 3 independent "energetic zones": Entry Zone (Fig. 6) - Process&Skinpass Zone (Fig.7) – Exit Zone (Fig. 8). All the drives of an "energetic zone" share energy between them, but not with the others, so it is possible to create N. 3 separated DC Bus supplies. Inside each "energetic zone/DC Bus" there is a great degree of energy regeneration with the presence of many motors that have the only charge to keep the pre-set tension on the material and not to perform a work on the material itself. At the end, only the "flattener" and the "skin pass rolling mill" motors are dedicated to convert electrical energy into a strip deformation job.

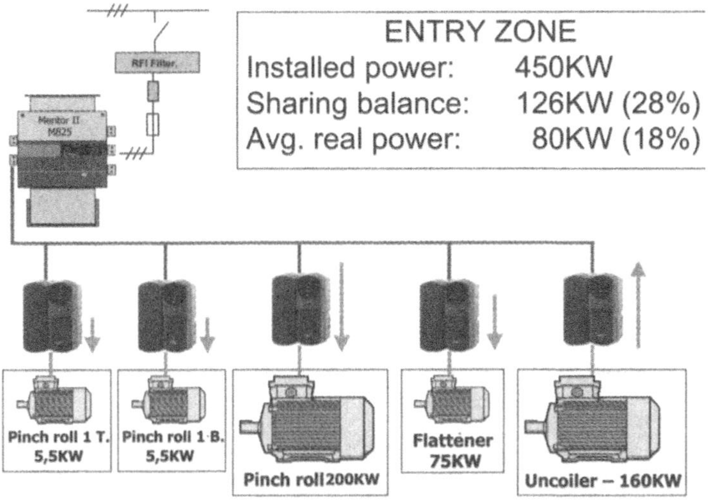

Fig. 6. Pickling line: Entry zone DC Bus

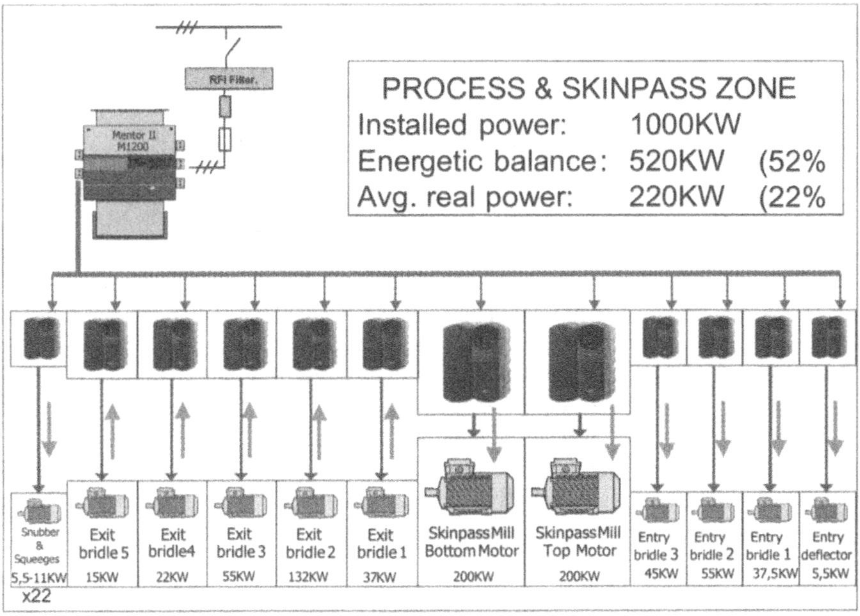

Fig. 7. Pickling line: Process&Skin-pass zone DC Bus

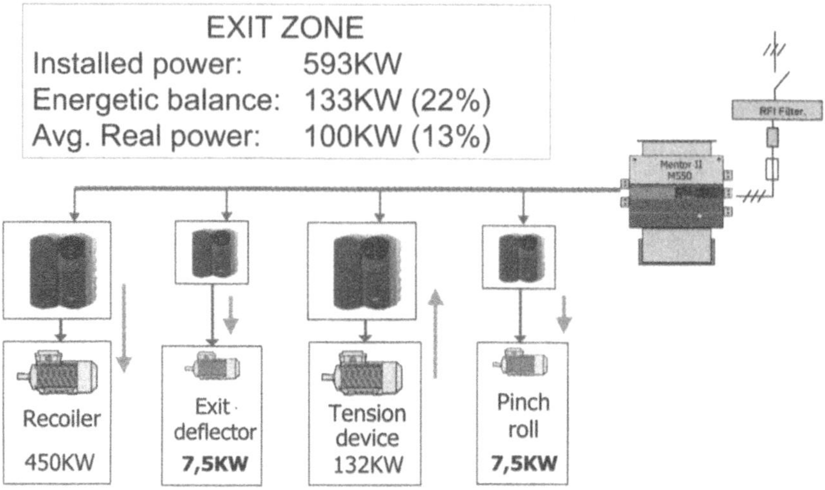

Fig. 8. Pickling line: Exit zone DC Bus

In fig. 6-7-8, it is reported that, in spite of a total mechanical installed power of 2.053KW , the calculated effective total power needed by the motor system is about 800KW, by taking in account the energy balance between all units.

Effectively, the total installed power of the three DC bus is 1200KW, because of the necessary system overload, the possible future power enhancements and the "good habitude" to provide the 60% of the total power.

The really important goal of the application came to evidence during the day-bay-day plant management: the actual average used power is about 400KW, which is the 19% of the mechanical installed power.

This final result is due to the following main factors:

- Proven efficiency in the motor/drive system architecture.
- Mechanical layout with many regenerating motors.
- Wide line speed range, due to the 0,8:4mm thickness range.

In its normal functioning, the plant doesn't need to regenerate energy to the Mains and even during Fast Stop or Emergency Stop, the friction of the steel strip with the feeding rolls avoids regeneration. This is the reason why the three DC Bus are not regenerative.

If the Drive System were based on DC technology, the strong presence of Reactive Power, and the difficulty in calculating it dynamically, would have lead to a very inefficient drive system. Furthermore, the Investment cost for AC solution in this kind of plant is comparable with DC, whilst AC offers the best payback in terms of maintenance costs and energy consumptions.

Harmonic distortion : invisible pollution

When designing a Power & Control system, the optimisation of power consumption is only a partial care, because also harmonic distortion is a real kind of environmental pollution. The impact on the Mains are invisible, can be very damaging and yet, even when the symptoms are clearly seen, can be hard to diagnose. The symptoms of harmonics can be severe: transformers may overheat causing damages to insulation, cables may get too hot, motors can have overheating, noisy effects, rotor torque oscillations with destructive mechanical resonance and vibration.

Also, capacitors can overheat and explode, displays and electronic equipments may fail or have false functioning. Filtering can be an effective cure, but choosing the right type of drive is a better approach. The magnitude of supply harmonics generated is largely unaffected by the type of output Inverter circuit used in the drive. Instead, the size of the motor load and (again) the configuration of the drive's input converter and DC link circuits mainly affect the level of harmonics.

A comparison of the THD values of different types of input circuits, follows:

Solution	Front end circuit	THD current
1	Thyristor rectifier	50%-150%
2	Six-pulse diode rectifier without inductor	130%
3	Six-pulse diode rectifier with small inductor	70%
4	Six-pulse diode rectifier with large inductor	45%
5	Twelve-pulse diode rectifier with large inductor	10%
6	Drive with IGBT inverter front end	Less than 10%

In plants like the "Rolling Mill" or the "Pickling Line", a good solution for Harmonics reduction and Investment cost control is to provide solutions 4 or 5.

Low effect solutions 1-2-3 are today unacceptable and architecture 6 is too expensive with the actual technology (2 times the actual cost).

Conclusion

The concept of "energy efficiency" incorporates both the power consumption and the impact on the electrical supply. Furthermore, the Process to be controlled must not be affected by the choice, but must have technological benefits itself.

By our experience (and matching the needs of many OEMs and EUs in metal and paper sectors), also investments constraints are major decision criteria.

That's the reason why there is not a "best solution" for every plant, but only a "best approach", that is the deep knowledge of Drives Technology and of Plant Management.

Literature

Prof. W. Drury, "Power Electronics" – 04/99
AA.VV., "AC to DC Power Conversion Now and in the Future" – 05/01
AA.VV., "Active Front End Inverters in Rolling Mill Drives" – 01/99
CT, "Harmonics Guide" – 08/00
Geff Brown, "Harmonics is better than cure"

Optimized Control of Active Front-Ends to Improve Efficiency and Power Quality in Systems with Closed-Loop Controlled Electrical Drives

G. Brando, A. Coccia, A. Del Pizzo

Dept. of Electrical Engineering - University of Napoli Federico II
Via Claudio 21 I80125 Napoli (Italy). tel. +39 0817683144; fax +39 0817683223;
e-mail: brando@unina.it; delpizzo@unina.it; ancoccia@unina.it

Abstract

With reference to the traditional full-bridge converter topology, a very efficacious control technique of active front-ends (A.F.E.) is proposed. Centered on very fast dynamic predictive evaluation of some electrical quantities, this method allows a quick selection of the voltage space-vectors useful for the V.S. Rectifier modulation and the evaluation of the modulation intervals by means of very simple analytical expressions. The paper underlines the positive contribution of A.F.E. to the efficiency and power quality of systems feeding electrical drives.

1. Introduction

The growing request of high-performance drives has been satisfied by technological progresses in power electronics devices and in digital microcontrollers. However, electrical drives introduce not negligible problems, mainly producing distortion in voltage and current waveforms both on motor-side and on line-side. At the same time they are also victims of any kind of disturbances, like non-sinusoidal, asymmetrical or non-stationary feeding networks. Active filtering is now becoming one of the most followed way to improve features of the electric feeding networks, ensuring high levels of power quality.

For electrical drives having two-stages converters, these considerations have led to use PWM rectifiers, in order to reduce the impact of the drive both towards and from the network. Combining the actions of a traditional diode-bridge rectifier and of an active filter, these active front-ends are able to shorten distortion and eventual unsymmetries of line currents, increasing the power factor and reducing ripples in the d.c.-link voltage. Moreover, all these characteristics can positively influence the efficiency either of the drive containing the A.F.E. or of other loads connected to the same network, due to the limitation of additional losses.

Performances of A.F.E. are strictly dependent on the operating switching frequency, on the control algorithm and on the features of current controllers. In order to contribute to improve some performances of A.F.E., in the following a very efficacious control technique is discussed with reference to traditional full-bridge converter topology.

2. Basics and Control Technique

For a voltage source rectifier (V.S.R.) used as first stage of a converter-fed electrical drive (fig.1), the electric mathematical model is expressed by:

$$L\frac{\mathrm{d}\mathbf{i}_s}{\mathrm{d}t} = -R\,\mathbf{i}_s + \mathbf{v}_s - \mathbf{s}\,v_0 \tag{1}$$

where the space-vectors $\mathbf{i}_s$, $\mathbf{v}_s$ and $\mathbf{s}$ are defined by:

$$\mathbf{i}_s = \sum_{k=1}^{3} i_{s,k}\,\mathrm{e}^{\mathrm{j}\,(k-1)2\pi/3} \;;\mathbf{v}_s = \sum_{k=1}^{3} v_{s,k}\,\mathrm{e}^{\mathrm{j}\,(k-1)2\pi/3} \;;\mathbf{s} = \sum_{k=1}^{3} s_k\,\mathrm{e}^{\mathrm{j}\,(k-1)2\pi/3} \tag{2}$$

and where s_k, is the conduction mode of the k-th phase of the V.S.R..

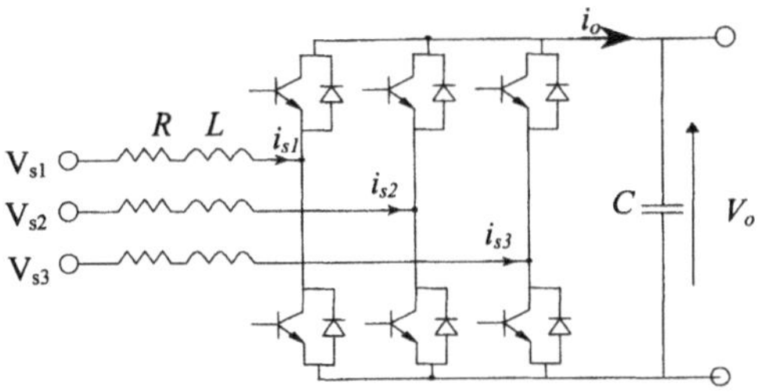

Fig. 1. Schematic representation of a V.S.R. as a first stage of a converter-fed drive

If we consider a time interval (t_{in}, t_{fin}) where the quantities $\mathbf{v}_s$, $\mathbf{s}$ and v_0 can be assumed constant, the simple analytical solution of eq. (1) is given by:

$$\mathbf{i}_s(t) = \mathbf{i}_s(t_{in})\,\mathrm{e}^{-(R/L)t} + \frac{1}{R}\left[\mathbf{v}_s(t_{in}) - \mathbf{s}\,v_0(t_{in})\right]\left(1-\mathrm{e}^{-(R/L)t}\right) \text{ with } t\in(t_{in},t_{fin}) \tag{3}$$

If a digital control of A.F.E. is provided, in a generic sampling interval (0, T_s) we can force the actual current $\mathbf{i}_s$ to reach at the instant $t=T_s$ the value of the reference current $\mathbf{i}_s^*$ at the initial instant $t=0$; i.e.:

$$\mathbf{i}_s(T_s) = \mathbf{i}_s^*(0) \tag{4}$$

The reference currents $\mathbf{i}_s^*$ are evaluated on the basis of the actual values of both supplying and d.c.-link voltages, as illustrated in fig.2.

After the classic decomposition of the line voltages system in a direct ($\mathbf{v}_s^d$) and an inverse ($\mathbf{v}_s^i$) sequence, each $i_{s,k}^*$ current is imposed in phase with the corresponding component of the line-voltage direct sequence $v_{s,k}^d$.

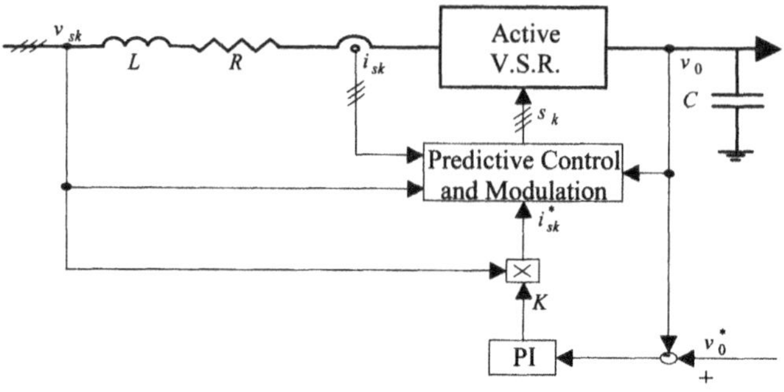

Fig. 2. Control block diagram

The instantaneous values of $i_{s,k}^*$ are a fraction K of $v_{s,k}^d$, where K is a value given by a P.I. controller of v_0 d.c. voltage, which is forced to follow a prefixed value v_0^*. Together with the values $\mathbf{i}_{s,0}$, $\mathbf{v}_{s,0}$ and $v_{0,0}$ directly measured at the initial instant $t_{in} = 0$ of a generic sampling period, the condition (4) allows the evaluation of the inverter conduction state **s** to have during the same T_s. If a modulation of the V.S.- rectifier is carried out dividing T_s in four different subintervals (see tab. I) where two consequent non-zero voltage vectors are applied (intervals t_1 and t_2) together with a zero-voltage (interval t_0), the sequence of the corresponding inverter states can be correctly evaluated by recursively applying the condition (2). We obtain (see [4]):

$$\mathbf{s}_1\, t_1 + \mathbf{s}_2\, t_2 = \frac{1}{v_{0,0}} \left[(T_s - KL)\, \mathbf{v}_{s,0} + (L - RT_s)\, \mathbf{i}_{s,0} \right] \tag{5}$$

The unknown quantities of this complex equation are the two-non zero conduction states $\mathbf{s}_1$ and $\mathbf{s}_2$, and their corresponding application time-intervals t_1 and t_2. The two apparent freedom degrees of (5) can be eliminated by adding some considerations: *a*) the converter states $\mathbf{s}_1$, $\mathbf{s}_2$ are adjacent in the space-vector star; *b*) among the different possibilities only a pair of $\mathbf{s}_1$, $\mathbf{s}_2$ allows to yield positive values of both the times t_1 and t_2 in (5).

This last condition can be exploited in order to rapidly solve eq.(5) in two steps; t_1 and t_2 are evaluated on the basis of very simple computations, that is no divisions or trigonometric functions are required (see [4]).

Table 1. Inverter states and modulation intervals in a sampling period

$t_0/2$	t_1	t_2	$t_0/2$
000	100 010 001	110 011 101	111

$t_0/2$	t_1	t_2	$t_0/2$
111	110 011 101	100 010 001	000

3. Numerical Results

Validity of the proposed control algorithm of the A.F.E. has been tested in [4] with reference to low-power applications for which very reduced influence on the feeding network is obtained also during havy transient operations of the drive. In order to show how an A.F.E. can really improve efficiency and power quality in a *medium-power* system containing electrical drives, some cases are numerically investigated. A closed-loop vector-controlled induction motor drive (with an A.F.E. as a first conversion stage) of 100 *kW* rated power is supplied by a transformer of 400 *kVA,* 20 *kV/400V* together with only another direct-connected asynchronous motor of about 100 *kW* (case ***a*** in fig. 3) or together also a thyristor-controlled dc-drive of about 110 *kW* (case ***b***). An average switching frequency of 5 *kHZ* is used in V.S.R. control. In fig.4 (case ***a***), in fig.5 (case ***b***) and in the table II, waveforms and numerical results are compared to the ones obtained by using a classic diode-bridge+passive filters (instead of VSR).

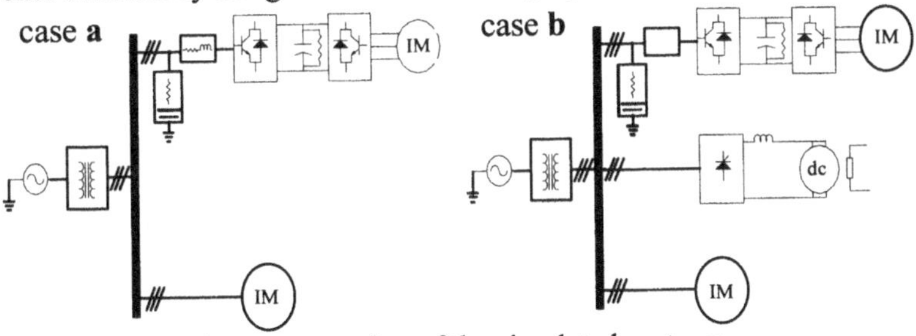

Fig. 3. Schematic representation of the simulated systems

In the case ***a*** we can deduce from fig. 4 evident improvements in line voltage and current when V.S.R. is used. Also the electromagnetic torque of the direct-connected induction motor is not affected by disturbances. Analogous considerations can be made in case ***b*** (fig.5), from which we can also deduce that the thyristor-controlled drive is able to compensate the feeding distorted voltage. Finally, for the considered systems ***a*** and ***b***, Table II compares numerical values of global efficiency , line current THD and line voltage THD in correspondence of rated steady-state operating conditions.

4. References

[1] R.Wu, S.B.Dewan and G.R.Slemon, "A PWM AC-to-DC converter with fixed switching frequency", *Industry Applications, IEEE Transactions on*, Sept.-Oct. 1990, vol. 26, no. 5, pages 880–885.

[2] S.Fukuda, "LQ control of sinusoidal current PWM rectifiers", *Electric Power Applications, IEE Proceedings-*, March 1997, vol. 144, no. 2, pages 95–100.

[3] T.Kataoka, Y.Fuse, D.Nakajima and S.Nishikata, "A three-phase voltage-type PWM rectifier with the function of an active power filter ", in *Power Electronics and Variable Speed Drives, 2000. Eighth International Conference on*, 2000, IEE Conf. Publ. No. 475, pages 386–391.

[4] G.Brando, A.Del Pizzo, "A predictive control technique for PWM-Rectifier in vector controlled induction motor driver", *ISIE 2002 L'Aquila, July 2002*

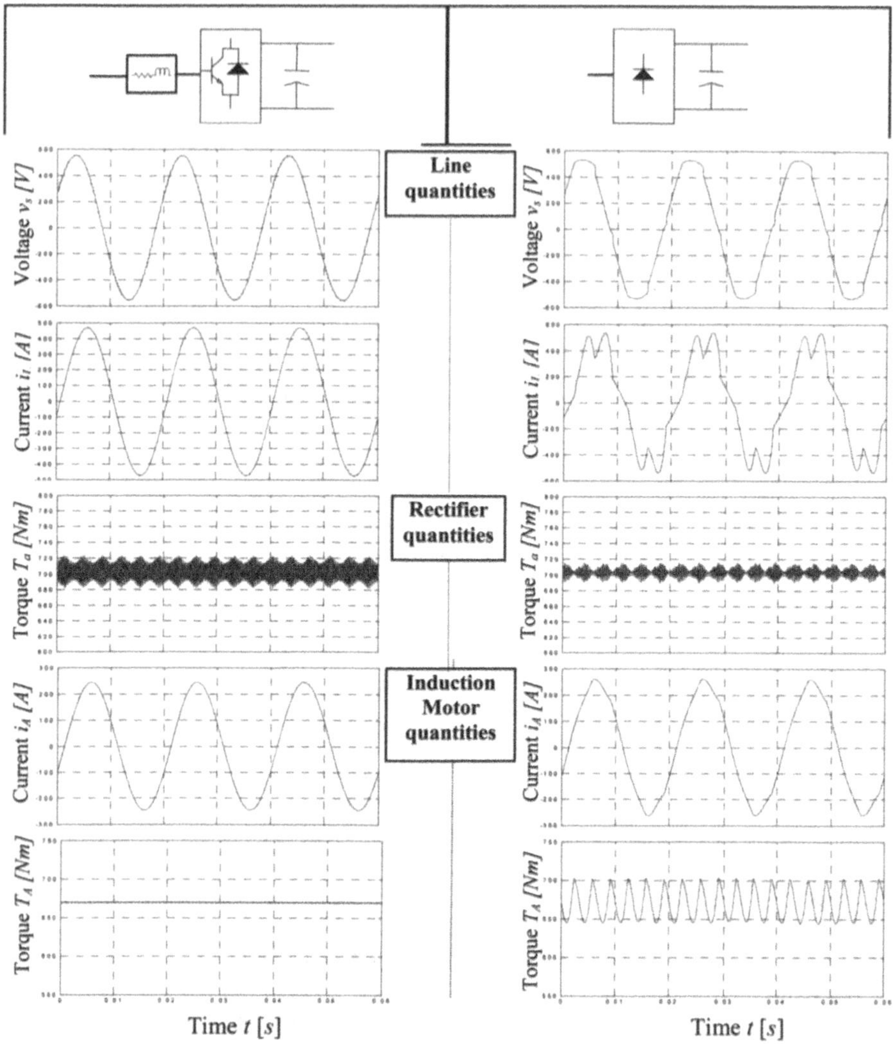

Fig. 4 Time behavior of some load and line quantities at steady state for case ***a*** f fig.3 with VSR and Diode bridge

Table 2. Total line power, efficiency and THD of line currents and line to line voltage of the simulated systems

	Case a		Case b	
	VSR	**Diode Bridge**	**VSR**	**Diode Bridge**
$\mathbf{P_L}$	227.3 *kW*	229.1 *kW*	341 *kW*	341.3 *kW*
η	88.3 %	87.4 %	88.6 %	88.3 %
THD$_v$	1.2 %	4.4 %	4 %	4.8 %
THDi	0.4 %	19.3 %	9.8 %	17 %

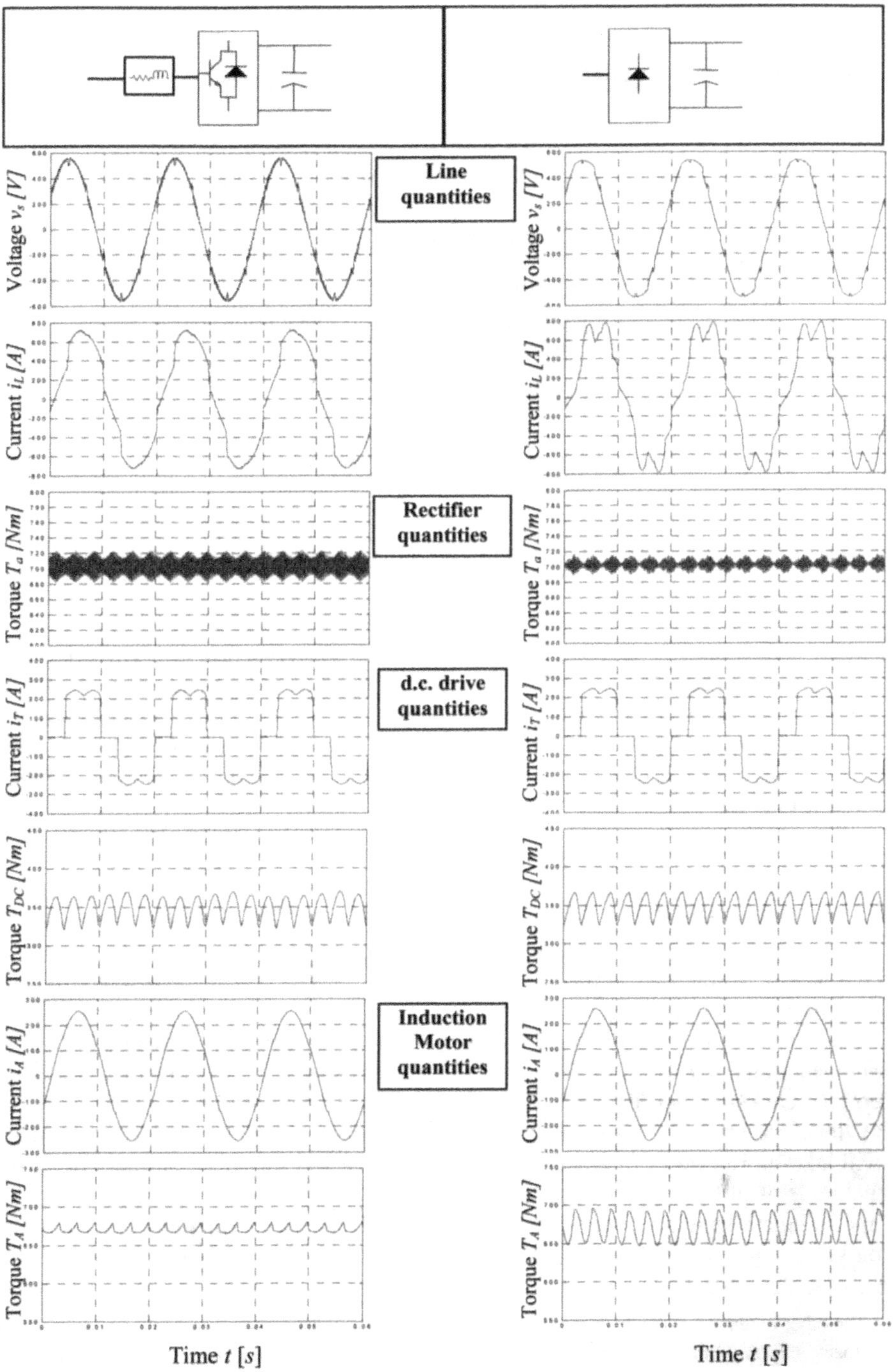

Fig. 5 Time behavior of some load and line quantities at steady state for case ***b*** of fig.3 with VSR and Diode bridge

European Commission – Europump Guidance on Efficiencies for Pumps

Paolo Bertoldi[1], Hugh Falkner[2], David Reeves[3] [(1)]

[1] European Commission Joint Research Centre, Ispra (VA), Italia.

[2] AEA Technology PLC, Harwell, Didcot, Oxon, OX11 OQJ, United Kingdom.

[3] Independent Pump Consultant, Mulberry House, 4 Oaklands Close, Collingham, Notts, NG23 7RQ, United Kingdom.

Abstract

With the exception of a small group of experienced engineers, most buyers of pumps find it hard to know how the efficiency of a pump that they are being offered compares to the best available. To encourage buyers to search for more efficient pumps, the European Commission funded a 2-Year SAVE study to provide clear guidance to buyers. This paper presents the results of this study and the new scheme for giving guidance on available efficiencies of single stage centrifugal pumps for clean cold water pumping duty.

Why the Efficiency Guidance Scheme Approach

It was immediately clear to the study group that because the efficiency of any specific pump varies so greatly with actual flow and head, categorising pump efficiency by a simple labelling scheme would be very misleading. Instead, based on published characteristics of pumps from a wide range of international manufacturers, typical efficiencies for a pump for any desired head and flow within the limits of the work are given. It is hoped that by giving specifiers this information, they will take more care to look around for a more efficient pump, which may often be available at little or no extra cost. Also through seeing more clearly the large loss in efficiency that can occur through just a small mis-calculation in estimated pump duty, it is hoped that specifiers will be encouraged to devote more attention to

(1) The other members of the project team were CETIM, France, Darmstadt Technical University, Germany, NESA – Energy Technology, Denmark. In addition the team was very grateful for the comments and suggestions of both members of Europump and end users who either attended meetings or commented on the proposals.

more accurate calculation of pump duty. Indeed, the study team acknowledge the much bigger energy savings that are possible through much better system design.[2] It should also be noted that not only will operating a pump a long way from its design point lead to much reduced efficiency, it is also likely to severely limit the life of the pump.

In any efficiency classification scheme, it is always important that products are only compared on a "like for like" basis, and so this scheme can only be used for three types of pumps, which between them account for about 15% of EC pumping energy consumption. These are:

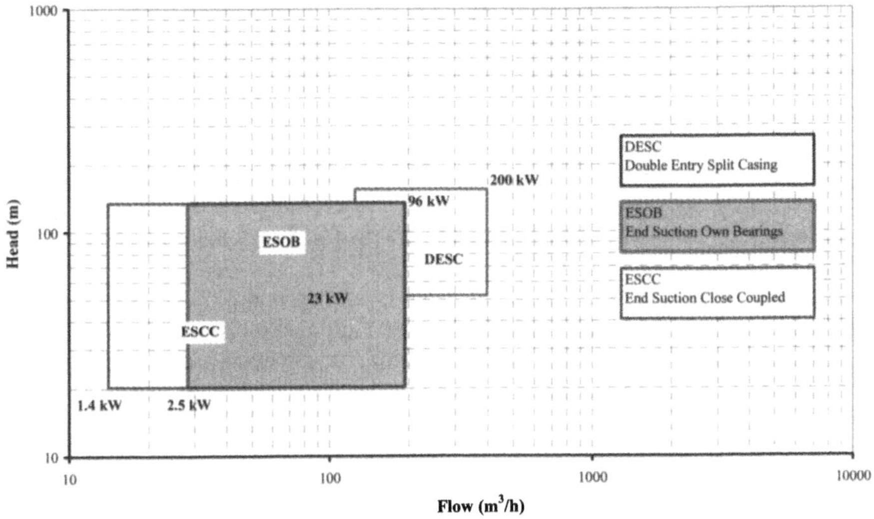

Fig. 1. Range coverage at best-efficiency duties at 2900 rev/min

- Close coupled end suction pumps, with the impeller on the motor shaft.
- End suction pumps having their own bearings.
- Double entry pumps having an axially split casing.

The applicable ranges of flow and head are shown in figure 1 for pumps running at nominally 2900 rev/min, with corresponding ranges for pumps running at 1450 rev/min. In addition, the scheme is only applicable to pumps in clean cold water duties.

Collection and Presentation of Data

For the types of pump listed, manufacturers were asked to submit data on the pumps that they would offer for the duties shown in the above ranges. From this

2 In particular, the *'guide to life cycle cost analysis for pumping systems', Europump and Hydraulic Institute, 2000,* identifies the many savings that can be made through better attention to system design.

data, the mean of the Best Efficiency Points was calculated and plotted. Head correction curves were also added to enable the quick translation of the resultant efficiency for different heads. The wide scatter of actual BEPs is shown in the sample graph in figure 2, showing clearly the very wide range of efficiencies likely to be offered for any one duty. A useful spin-off from these results is that it enables manufacturers to compare the efficiencies of their own pumps with the range of efficiencies available, hopefully encouraging them to improve future designs.

Derivation of the Specific Speed Correction Factor

Correction 'C', as shown on figure 2, is based on pump flow, head and speed. It is actually a correction for pump Specific Speed. Pumps of relatively low head (high Specific Speed) and high head (low Specific Speed) lose efficiency due to unavoidable secondary hydraulic losses. Ref 3 shows a curve to allow efficiency correction for Specific Speed. This is actually the correction proposed by Anderson in Ref 4. Alternative curves to correct for Specific Speed can be derived from Ref 5 and Ref 6.

None of the these curves appear to provide an 'optimum' correction for the published manufacturers' data analysed for this Guide. We have therefore chosen to use an intermediate Specific Speed correction curve which produces the minimum scatter of the points plotted. There is no strict theoretical justification for this approach but we believe that it serves well for the purpose for which the Guide is intended. (Further information can be found in Ref 1.)

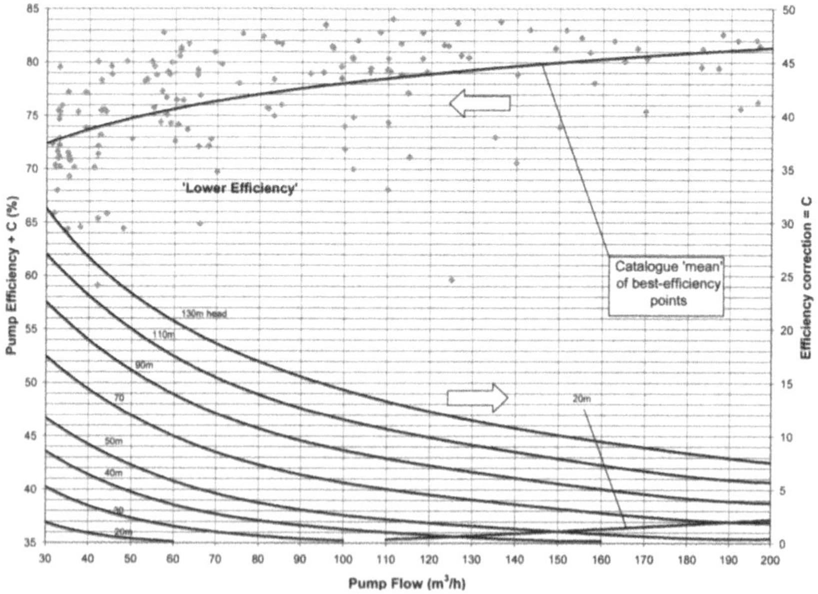

Fig. 2. Catalogue mean of best efficiency points and the specific speed correction factor

Efficiency Classification Scheme

Figure 3 shows an extension of figure 2 with a lower line added based on manufacturer's quotes for duties within the range of figure 1. The 3 bands have been labeled to give an indication of how the efficiency of a chosen pump compares to the optimum selection for a particular duty.

Selection class 1: High efficiency pumps operating at or close to the best efficiency point.

Selection class 2. Pumps operating at intermediate efficiency points. Selections in this area will be for pumps which probably have reasonably good best efficiencies, but have been quoted away from the best efficiency point.

Selection class 3. Pumps operating at lower efficiency. Selection in this area shall be avoided, unless other parameters override efficiency.

Although a quoted low efficiency may be due to poor pump quality, it is more likely to be due to the chosen duty not coinciding with that pump's best-efficiency point. For instance the flow may be below or above the optimum for that pump.

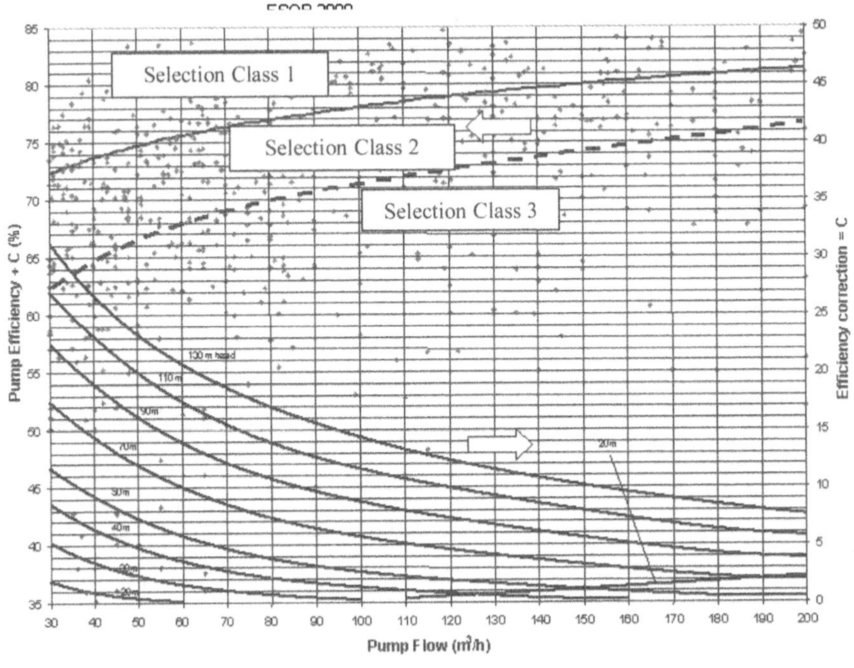

Fig. 3. The Pump selection efficiency classification scheme

Using the Plots

To use the plots:

Decide which type of pump you want.
Choose the flow and head at which you would like maximum efficiency.
Check that the chosen flow and head are within the range of figure 1..
Get efficiency quotes from manufacturers (or use published information).
Enter your chosen flow and head on the plot that suits the quotes (figure 3).
Read correction factor 'C' on the right-hand axis.
Add 'C' to the efficiency that has been quoted.
Plot (quoted pump efficiency + correction 'C') at your chosen flow.
If the point lies in the Selection class 3 or 2 areas, then seek higher efficiency quotes.

Worked Example:

Chosen pump type? End suction with own bearings.
Chosen duty for maximum efficiency? 80 m^3/h at 110 m.
Quoted pump performance? 60% efficiency at 2900 rev/min.
(Check materials, suction performance, etc, are satisfactory)
Is chosen duty within ranges covered? From Fig 1, yes.
From Fig 3: 'C' = 14.
Plot on Fig 3: 'Pump Efficiency + C' = 60 + 14 = 74%.
Is this satisfactory? Fig 3 suggests that an additional 4 to 8 points is possible.
Action: Seek further quotes.

Comparison of Results with Other Sources of Data

For practical reasons it was only possible to source a limited amount of data to produce the plots in this Guide. To assess how meaningful the results are, it is useful to compare them with other sources.

The best efficiencies at optimum Specific Speed for end suction pumps having their own bearings at 2900 rev/min from Fig 3 are replotted below in Fig 4 and marked 'Catalogue 'mean''. Additional curves (all at optimum Specific Speed) are derived from:

- Ref 3, marked 'Hydraulic Institute'.
- Ref 4, marked 'Anderson'.
- Ref 5, marked 'Maximum practically attainable'.
- Ref 6, marked 'Theoretically attainable'

From Fig 4 we deduce that the Catalogue 'mean' curve is suitable for the guidance purposes intended. The 'Hydraulic Institute 'ANSI/API'' curve is low, particularly at low flows. This is probably mainly due to the use of relatively large wear ring clearances, as required for pumps in special materials or to meet the American Petroleum Institute (API) Standard. Comparing the catalogue mean results with those published by other sources shows excellent agreement with that of Anderson. It is salutary to see that pump efficiency has not improved significantly in the time since this earlier work.

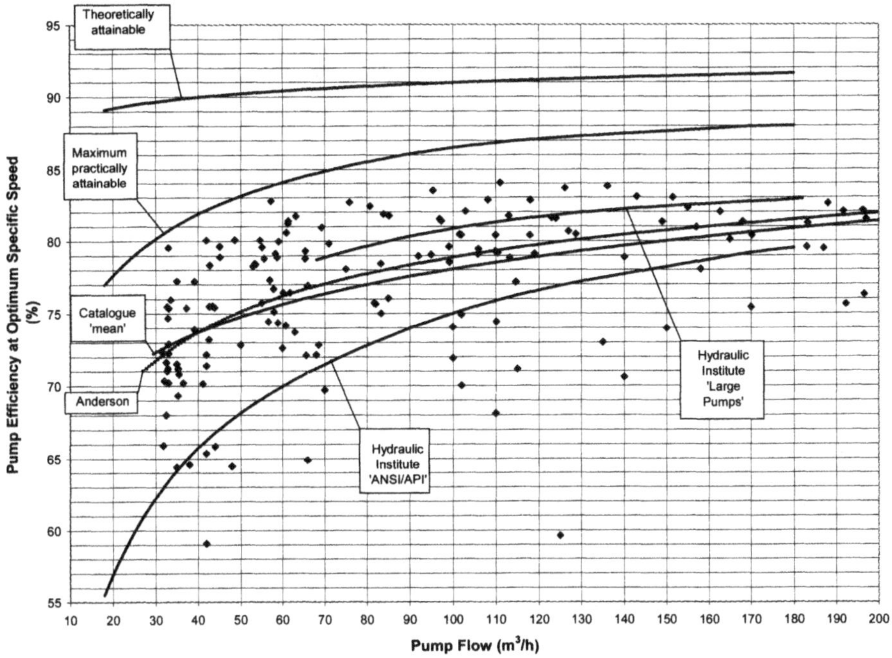

Fig. 4. Comparison of efficiencies of end suction pumps from various sources

Conclusions

This new scheme for the first time gives purchasers of pumps a clear idea of the sort of efficiency they can expect for their particular duty. It is therefore hoped to stimulate the better selection of pumps through purchasers taking more time to consider alternatives. A brochure aimed at pump specifiers will be published early 2003. Once the scheme has been evaluated, it is hoped to extend it to other types of pump.

Unlike other component efficiency classification schemes, this scheme focuses on the duty rather than the component. It therefore emphasises the need for Users to accurately estimate the actual duty, which is essential if the system is to have lowest energy costs. Manufacturers may also use the scheme to be able to point out in quotes when a pump is in the Class 1 selection category.

This methodology could be successfully used for other types of pumps and even other components such as fans.

References

[1] 'Improving the energy efficiency of pumps', Report produced for the European Commission (DG Energy – SAVE), 2000.

[2] 'A guide to life cycle cost analysis for pumping systems', Europump and Hydraulic Institute, 2000.

[3] 'Efficiency prediction method for centrifugal pumps', Hydraulic Institute, 1994.

[4] Anderson, H.H., 'Centrifugal pumps and allied machinery', Elsevier Advanced Technology, 1994.

[5] 'Attainable efficiencies of volute casing pumps', Europump Guide No. 2, 1999.

[6] Stoffel, B. and Lauer, J., Summary of the final report on the research project for VDMA 'Theoretically attainable efficiency of centrifugal pumps', Technical University of Darmstadt, 1994.

Evaluation of Pump System Efficiency and Performance

Maurice A. Yates

The Energy Centre, Finnimore Industrial Estate, Ottery St Mary,
Devon EX11 1NR UK

Introduction

Pumps represent 20% of the world's electricity demand and, whilst the pumps themselves may have a good efficiency, the application of a pump into an operating system will considerably reduce the pumping system efficiency.

In addition to the energy losses inherent in pumps, motors, valves, pipework and controllers the energy losses in the total system will significantly increase with time.

The increase in energy losses is largely due to the deterioration of both the pump and the hydraulic system. Therefore, even though the pumping system may be well designed it is essential that the system efficiency is reviewed on a regular basis, say every 10,000 hours.

World Electricity Consumption

Europe with 6% of the world's population consumes 20 % of the world's energy and the United States with 5% of the world's population consumes 26%. Therefore on a per capita basis the USA consumes 56% more energy than Europe.

Pumping Systems

The following components are common to all pumping systems – Pump, Motor (Driver), Pipework, Valves and Control System.

At the design stage each element of the system must be considered, both on an individual basis and on its integration in the system.

In selecting the pump the duty required by the pump should be close to the best efficiency point of the pump. The main problem in selecting the pump is the pump will only operate where the pump head curve meets the system head curve.

This process becomes more complex when there are two or more pumps operating in parallel.

The recently published Pump Life Cycle Costs by the American Hydraulic Institute and Europump is an excellent guide for pumping system design. (Ref 1)

Existing Installations

Moving now to looking at existing stations. Various studies have found that existing installations are a good source of energy savings.

Source	Year	Savings %	Considered Saving Opportunities(*)
Norgaard et.al. (DK)	1983	55	(1),(2),(3),(4),D
Fichtner (D)	1988	25	(5)
KWW (NL)	1985	"large"	(3),(4),(5),D,M
Lovins e.a. (US)	1989	28-60	(1),(2),(5),D,M
Nadel e.a. (US)	1991	16-40	(1),(2),(5),D,M
Larson and Nilsson (S)	1990	50-75	(1),(2),(3),(4),(5),D,M

(*)(1) = Motor, (2) = Transmission, (3) = pump, (4) = piping and fittings, (5) = control, D = Dimensioning, M = Maintenance

(Ref 2 – 7)

The UK experience is that savings of up to 30% are readily achievable with some systems having a potential savings in excess of 50%. The areas of savings are:

Pump Condition

Here we consider the actual condition of the pump. The rate of deterioration of the pump is dependent upon several factors. Firstly the fluid being pumped. Fluids containing abrasive material will wear faster, although it should be remembered that even water supply pumps will show significant wear in a period as short as one to two years. The second factor is the nett positive suction head available at the pump suction. Low $NPSH_a$ will lead to cavitation which in turn will lead to severe damage to the impeller and reduced life of the pump.

Another significant area of accelerated wear is where a pump is running away from its best efficiency point. Pumps are designed to run at this best efficiency point and working outside the range – 30% - + 10% of BEP flow rate will reduce the life of the pump and increase its energy consumption. There is an optimal point of refurbishment where it is cost effective to carry out refurbishment. By carrying out economic refurbishment the owner will only have a limited amount of expenditure. The work will mainly be to the wearing rings and the bearings, whereas if the pump is left until it is say 15% down then, in addition to the above items, both the shaft and impeller may require replacing at considerable expense.

Also in addition to saving repair costs early refurbishment will also save on energy as the pump will be maintained at a higher efficiency.

Economic refurbishments can be carried out with pump efficiency shortfalls as low as 5%. Before the advent of thermodynamic testing it was very difficult to determine such shortfalls in efficiency, however, particularly in the UK Water Industry, thermodynamic testing is a standard process and has lead to very significant energy reductions in water supply pumping.

At this point it should be noted that up to 20% of a water utility's operating cost will be the purchase of electricity to drive the pumps. (Ref 8)A Paper on the pump efficiency project in Thames Water which used the thermodynamic technique (Ref 9) gives in great detail the measures taken which have lead to E1m/year energy savings in Thames Water.

Scheduling and Optimization of Pumping Systems

This is very applicable to pumping systems which consist of several parallel pumps and is imperative when variable drives are involved system. Optimization of pumping systems reduces the energy used in the system by selecting the best pumping combination to achieve the desired flow rate. This combination may be a mixture of large and small pumps at various variable speed combinations.

Two large pumping systems in the UK have shown energy reductions of 14% and 18% respectively. This has been achieved by optimizing the pumping schedules against the diurnal flow rates required. The combined saving of these two stations is in excess of E100,000/year.

Tariff

This area is very regionalised and whilst it does not save energy for the pump owner his benefit is reduced cost. The electricity utility does save energy because it is able to optimize its electricity generating schedules. In particular the water industry does have the potential to reduce costs by increasing night and weekend pumping.

Variable Speed Drives

The good the bad and the ugly!

Variable speed drives are a real benefit to the pumping business, particularly when there is a single pump operating say under pressure or flow control or when a number of parallel pumps can be replaced by a variable speed driven pump. However, when variable speed drives are involved in parallel pumping great care has to be taken. Often variable speed drives will be used in conjunction with fixed speed drives and it is very easy to effectively drive either the fixed speed drive or the variable speed towards closed valve, hence reducing the operational efficiency of the system. Plant operators have a real financial responsibility. It is not unusual for a pumping installation to consume electricity to the value of E1m/year. Unfortunately very few installations have sufficient instrumentation to enable operators to be aware of the energy losses associated with pumping systems.

Installation Effects

Here uprating of pumping stations may cause significant changes to the pipework which in turn can lead to pump suction problems and increased losses due to high fluid velocites in the pipework.

One point to note is that there is always a temptation to make the main pipework and valve sizes match the pump suction and discharge branch size. This is very false economy, particularly when the pumps are running for long periods. Pumping velocities should rarely exceed 2m/sec whereas it is not unusual for pump discharge to exceed 5m/sec.

Remember good suction means easy delivery. Any had loss in the pump system can be catastrophic to the pump. Look carefully at all those bends and tees. Right angle tee junctions should be avoided, swept tees and Y pieces are then order of the day. Typically in the UK Water Industry saving 1m of pumping head will reduce the total energy consumption by 1%.

Decisions made at the design stage will influence the energy consumption of the pumping system for life. Great care must be taken to ensure that the pipework is at optimum size considering both the capital cost and the energy cost over the life of the system.

Pump Selection

The initial pump selection is critical to reducing energy costs. Low cost pumps often lead to lower efficiency and again reference should be made to the life cycle costing manual.

In addition water pumping stations have a very long life and whilst the main civil works should be designed for all known eventualities the initial pump installation should be for a period of probably less that ten years.

The Prize

The prize is to reduce CO_2 emissions. Every 1kW of hydraulic energy saved will reduce the generating requirement by a factor of 10 and the annual CO_2 emission by 40 Tonnes.

References

[1] Pump Life Cycle Costs by the American Hydraulic Institute and Europump

[2] Norgaard JS, Holck J and K Mehlsen, Long Range Technical Potential for Electricity Conservation, Physics Lab III, Danish Institute of Technology, Lyngby 1983 (in Danish)

[3] Fichtner Beratende Ingenieure, The Potential for Energy Savings in the Application of Electrical Energy – Study for the Commission of the European Communities, DG for Energy, Stuttgart 1988

[4] KWW The application of Electric motors in industry and related energy saving possibilities, Krekel van der Woerd Wouterse by Rotterdam 1986 (in Dutch)

[5] Lovins AB, J Neymark, T Flanigan, PB Kiernan, B Bancroft and m Shepard, The State fo the Art: Drivepower, Rocky Mountains Institute, Snowmass (Aspen) CO 1989

[6] Nadel S, Shepard M, G Katz, S Greenberg, A de Almeida, Energy Efficient Motor Systems: A Handbook on Technology, Programs and Policy Opportunities, American Council for an Energy Efficient Economy, Washington DC 1991

[7] Larson E, Nilsson, L A System-Oriented Assessment of Electricity Use and Efficiency in Pumping and Air Handling, Dept of Energy and Environmental Systems Studies, University of Lund 1990

[8] Pump Efficiency Project in Thames Water J C Drake and M Sims Thames Water Utilities Limited Reading UK

[9] Thermodynamic/Conventional Tests on Two 4MW Pumps Maurice A Yates, A Kumar 7th European Congress on Fluid Machinery 15-16 April 1999. The Hague, Netherlands. Published by Professional Engineering Ltd Bury St Edmunds IP32 6BW UK

Thermodynamic On-Site Measurement of the Efficiency of Centrifugal Water Supply Pumps in the Frame of Condition Orientated Maintenance

Bernd Stoffel[1], Stephan Meschkat[1], Gerhard Mollenkopf[2], Klaus-Joachim Prien[3]

[1] Prof. Dr.-Ing. Bernd Stoffel, Dipl.-Ing. Stephan Meschkat
Institute for Turbomachinery and Fluid Power, Darmstadt University of Technology, Magdalenenstrasse 4, D-64289 Darmstadt, Germany

[2] Dr.-Ing. Gerhard Mollenkopf, KSB AG, 67227 Frankenthal

[3] Dipl.-Ing. Klaus-Joachim Prien, Wasserverband Aabach-Talsperre, 33181 Wünnenberg

Summary

The thermodynamic method of efficiency measurement is based on the law of conservation of energy. In a simplified description, the raise of usable energy in the form of pressure, fluid velocity and elevation represents the pump benefit and the raise of fluid-temperature represents the internal losses by friction and leakage. Therefore, the thermodynamic method requires the experimental determination of the difference of suction- and pressure-side temperatures (which is normally in the order of 10^{-3} to 10^{-2} Kelvin), pressures, averaged fluid velocities and elevations. By appropriate measuring devices and skilful application, this method provides the feasibility to determine the efficiency of continuously operating pumps without operation time-out and with low effort of staff and time.

The Institute for Turbomachinery and Fluid Power of the Darmstadt University of Technology was assigned by the German Association for Gas and Water Distribution (Deutsche Vereinigung für das Gas- und Wasserfach, DVGW) to carry out a research project on the applicability and accuracy of the thermodynamic method for the purpose of efficiency monitoring of centrifugal pumps in drinking water supply plants in respect to condition orientated maintenance.

A theoretical part of this project aimed at the investigation and quantification of influences of various measuring errors on the determined value of efficiency and of external effects on the quantities to be measured.

In the frame of this project, also on-site efficiency measurements were carried in several drinking water plants. In this paper, it is reported on measurements on centrifugal water supply pumps installed at the drinking water plant of the Wasserverband Aabach-Talsperre in Germany. Using the thermodynamic method, efficiencies were determined before and after the maintenance of the pumps by the pump manufacturer.

Theoretical Background

For the thermodynamic determination of efficiency of centrifugal pumps, the principle of conservation of energy is used as it is stated in the first law of thermodynamics. As a result, by neglecting peripheral heat transfer, that part of the pump power input which is not transformed into the usable power output is transformed into an increase of the inner energy of the fluid which results in a raise of the fluid temperature from pump inlet to outlet.

It is assumed that the friction losses of the bearings and shaft seals have little influence on the temperature increase of the delivered liquid and that they are not detected by the temperature probes. Thus the thermodynamically determined efficiency closely corresponds to the inner efficiency $\eta_{i,Pump}$ of the pump.

$$\eta_{i,Pump} = \frac{P_{deliver}}{P_{deliver} + P_{loss}} = \frac{\Delta h_{t,deliver}}{\Delta h_{t,deliver} + \Delta h_{t,loss}} \tag{1.1}$$

$$= \frac{\frac{1}{\rho}(p_O - p_I) + \frac{1}{2}(c_O^2 - c_I^2) + g(z_O - z_I)}{\frac{1}{\rho}(p_O - p_I) + \frac{1}{2}(c_O^2 - c_I^2) + g(z_O - z_I) \; + \; \Delta h_{t,loss}}$$

Herein, Δh_t is the increase in total enthalpy. In order to calculate the efficiency, the temperatures T_I, T_O and the pressures p_I, p_O at suction- and pressure-side, as well as the electric power input are measured. Implying the state laws of water, the specific losses $\Delta h_{t,loss}$ can be calculated directly from T_I, T_O and p_I, p_O in a first step. To determine the velocities c_I and c_O at suction- and pressure-side of the pump, the rate of flow is calculated from the conventional equation for the efficiency including the shaft-power P_S :

$$\eta_{Pump} = \frac{\dot{V} \cdot (p_O - p_I + \frac{\rho}{2}(c_O^2 - c_I^2) + g(z_O - z_I))}{P_S} \tag{1.2}$$

$$\Leftrightarrow \quad \dot{V} = \eta_{Pump} \cdot \frac{P_S}{p_O - p_I + \frac{\rho}{2}(c_O^2 - c_I^2) + g(z_O - z_I)}$$

The shaft-power P_S is quantified as the product of the efficiency of the electric driver and the electric power input. For the efficiency η_{Pump} the efficiency $\eta_{i,Pump}$ is inserted by neglecting the mechanical losses. With the rate of flow the velocities c_I and c_O are determined by the pipe diameters D_I, D_O at suction- and pressure-side as:

$$c_O = \frac{\dot{V}}{\frac{\pi}{4} D_O^2}, \quad c_I = \frac{\dot{V}}{\frac{\pi}{4} D_I^2} \tag{1.3}$$

Because of the appearance of the velocities in equ. (1.2) where they are determined from themselves, the calculation has to be repeated iteratively setting the difference of the squares of velocity to zero in the first step.

By setting $\eta_{i,Pump}$ for η_{Pump} an error in η_{Pump} occurs which is of little relevance; the same holds true for a possible error in measuring of power input, because it changes the numerator and the denominator of equ. (1.) for the efficiency by errors in the difference of velocity square by nearly the same magnitude. In case of equal pipe diameters at suction- and pressure-side, the calculation is exact because the difference of velocity disappears. From this determination, the rate of flow arises as an additional information. As a consequence of its kind of determination, it is most sensitive to measuring errors.

The Influence of Errors

To clarify the influence of measurement errors for different pumps, the resulting error of efficiency in percent-points is plotted against the pump total head in Fig. 1. for a constant error of the measured quantities, but for different values of pump efficiency.

In addition to the larger errors in efficiency at smaller pump head, the graphs show the increase of errors for increasing pump efficiency. The reason is the smaller temperature rise as a consequence of smaller losses at higher efficiency where the same absolute error of the measured temperature difference leads to a larger relative error. The assumed magnitude of an error of 2 Millikelvin in the measured temperature difference directly refers to specifications of suppliers of special measuring equipment and was also determined by comparison of the graphs shown above with graphs published by measuring equipment suppliers. In the frame of our research work, it could be shown that the accuracy of commercial equipment is better than some tenths of a Millikelvin. Larger errors must be ascribed to external influences like erratic temperature distribution in the measurement cross section, erratic velocity distribution in the measurement cross section, heat transfer from the pump to the ambient (of small amount) and heat transfer from the fluid to the pump parts (disappears by timer averaging over sufficiently long periods).

As is shown in the left hand graph of Fig. 1., an error in efficiency of more than 6 percent-points results for pump heads below 10m for an actual efficiency of 90% which is an unacceptable error. For pumps of medium efficiency even at pump heads as low as 20m the thermodynamic method can serve to gain rough information on the pump efficiency. In the right hand graph the additional influence of an error in the measured pressure difference is included. The assumed error of 0.05 bar is larger than equipment suppliers specifications for such pump heads. At the lowest pump heads the additional error in efficiency amounts to another percent-point. With increasing pump head the additional error drops far below one tenth of a percent-point.

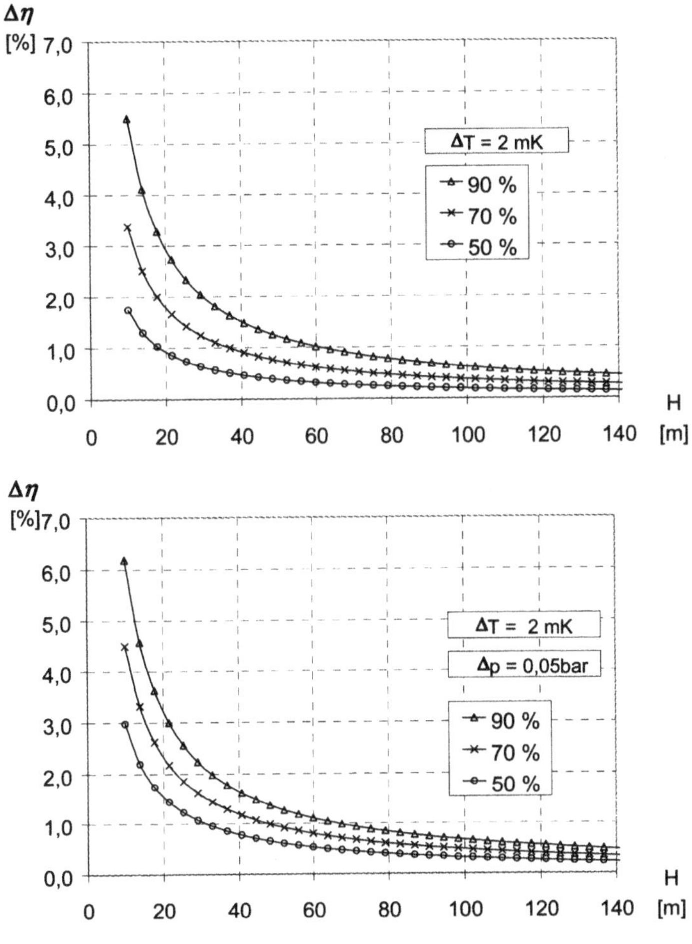

Fig. 1. Influence of errors at differing head and efficiency

Although the error in determination of the absolut value of efficiency can be relatively high for pumps of good efficiency and low head, the method works well even for those pumps in respect to monitoring of relative variations in efficiency (i.e. in comparison to an initial value) as they arise during life cycle wear or before and after overhaul.

Results from Practical Applications

To study the practical applicability and accuracy of the thermodynamic method, on-site efficiency measurements were carried out at several drinking water plants in Germany. Especially, measurements at the drinking water plant of the Wasserverband Aabach-Talsperre have shown the usability of the method for the purpose of efficiency monitoring in respect to condition orientated maintenance. Fig. 2. shows the four pumps of the Wasserverband Aabach-Talsperre for the untreated water on some of which measurements were carried out before and after an overhaul.

The maintenance work comprised the replacement of sealing gap rings at the impeller (to reduce volumetric losses) and smoothing of internal surfaces (to reduce hydraulic losses).

Fig. 2. Pumps at the Wasserverband Aabach-Talsperre

Fig. 3. shows efficiencies measured on-site by the thermodynamic method. The measurements were performed on two of the four identical pumps before overhaul and on three ones after overhaul for various flow rates. The flow rate was varied by variation of the rotational speed in steps (from right to left 1350, 1310, 1280, 1250, 1230 min^{-1}). The pump head as determined by the plant head characteristics at the time of the measurements was about 70m during the tests before and 66m during the tests after overhaul. The resulting improvement of the efficiency was clearly proven by the results of the measurement.

Additionally to the replacement of the worn out gap rings, at pump1 impeller and volute were coated with a special ceramic material, at pump2 the surface of the volute was smoothed and at pump3 the volute was coated with the ceramic material. The resulting improvements were detected by the thermodynamic method as an increase in efficiency of about 1.5 percent-points for pump2, about 3 percent-points for pump3 and 4.5 percent-points for pump3.

As the thermodynamic method was well investigated in respect to precision and external influences during the mentioned research project the results can be considered reliable. Thus the results are showing the valuable effects of smoothing and covering of inner surfaces in volute casing pumps on the one hand and the usability of the thermodynamic method on the other hand.

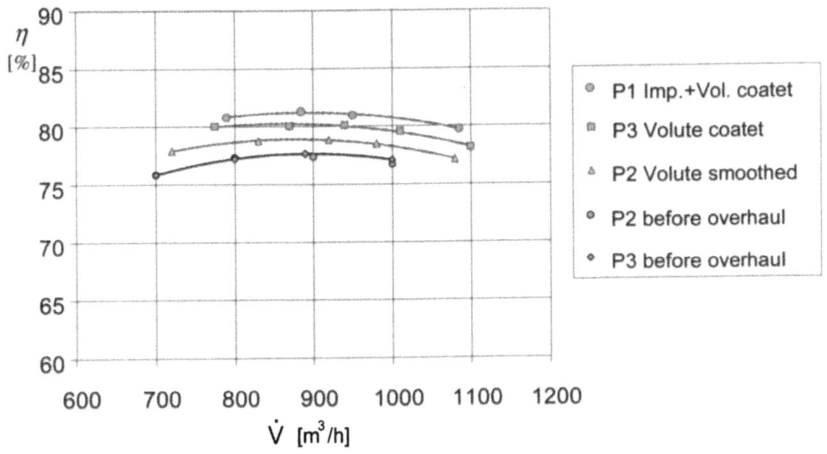

Fig. 3. Measured efficiency before and after overhaul

Conclusion

The thermodynamic efficiency determination is especially useful for detection of relative variations. For measurements of absolute values of efficiency external influences on the measurement of the temperature difference have to be comprised.

On-site measurements can be performed with low effort of staff and time because no measurements of flow rate via tank method are necessary. For the execution of measurements no operation time-out is necessary.

Thus the method is well suited for life cycle efficiency monitoring and condition orientated maintenance.

Design Factors Affecting Pump Efficiency

Gerhard Ludwig, Stephan Meschkat, Bernd Stoffel

Chair of Turbomachinery and Fluid Power, Darmstadt University of Technology
Magdalenenstraße 4, D-64289 Darmstadt, Germany
E-mail: ludwig@tfa.maschinenbau.tu-darmstadt.de

Introduction

The investigations presented as follows were carried out in the frame of an EU funded SAVE project headed *"Study on Improving the Energy Efficiency of Pumps"*. One aim of this study was to quantify the effects of several design factors affecting pump efficiency of single-stage centrifugal pumps. The theoretical investigations that were carried out at the chair of Turbomachinery and Fluid Power at Darmstadt University of Technology where mainly focused on the influence of surface roughness, the aspect of smoothing several parts of pumps and the internal leakage flows corresponding to the sized gap clearances. The specific speeds of the considered pumps covered the range from $n_s = 10\ \text{min}^{-1}$ up to $n_s = 100\ \text{min}^{-1}$ which represents the typical field of application of standard centrifugal pumps. The theoretical estimations were obtained by the aid of a special (non CFD) software tool, which was developed within the scope of a former research project named *"Attainable Efficiencies of Volute Casing Pumps"* sponsored by the Research Fund of the German Pump Manufacturer Association (VDMA).

Theoretical Investigations

Loss-Causing Components of a Centrifugal Pump

To determine the friction losses for the parts shown in fig. 1 respectively the leakage flow rates through the sealing gaps the program uses differential equations as well as simplified mathematically loss approaches. All calculations are carried out on the base of a hydraulic design process considering common industrial design standards in respect to the geometrical settings.

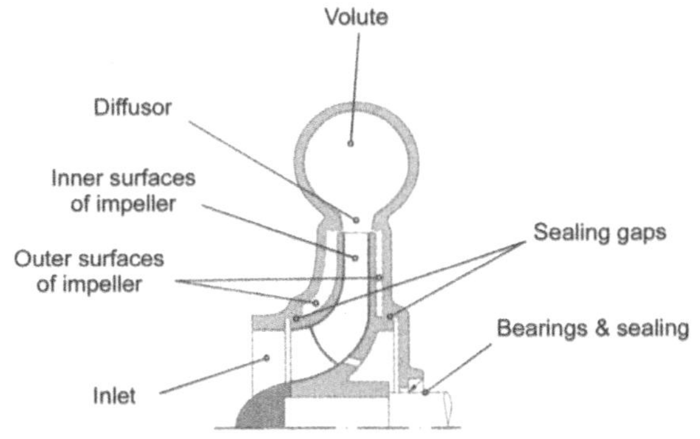

Considered Losses

- hydraulic losses:

  friction losses
  deceleration losses
  wake losses
  mixing losses (Volute)

- volumetric losses (gap flow)

- mechanical losses

Fig. 1. Loss-causing components of a centrifugal pump

All the following figures which demonstrate the influence of the parameters surface roughness as well as gap clearance show efficiency values η respectively differences of efficiency values $\Delta\eta$ in per cent points and are plotted versus the value of specific speed n_s as defined in fig. 2 (where n is the speed of rotation, Q the rate of flow and H the pump head).

Fig. 2 exemplarily shows the dependence of the overall efficiency η on the rate of flow respectively pump size at constant speed of rotation ($n = 1450\ \text{min}^{-1}$). It can be stated that an increase of the rate of flow leads to higher values of the overall pump efficiency, which is the effect of an increasing Reynolds number *Re*. The figure also shows that the efficiency values for very low specific speeds are definitely smaller than for higher ones, which is due to the well known geometric as well as hydraulic attributes of such types of pumps.

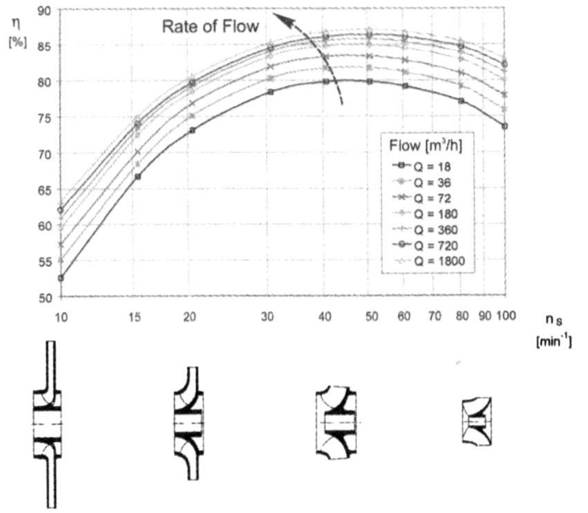

Specific speed:

$$n_s = n \cdot \frac{Q^{1/2}}{H^{3/4}}$$

n_s [min^{-1}], n [min^{-1}], Q[$\frac{\text{m}^3}{s}$], H [m]

$$n_{s,US-units} = 51.64 \cdot n_s$$

In the following:

$$n = 1450\,\text{min}^{-1}$$
$$Q = 180\,m^3/h$$

Fig. 2. The influence of rate of flow

In respect to a better comparability all results of the investigations shown in the following diagrams were generated for operating conditions characterised by a flow rate Q of 180 m^3/h and a speed of rotation n = 1450 min^{-1}.

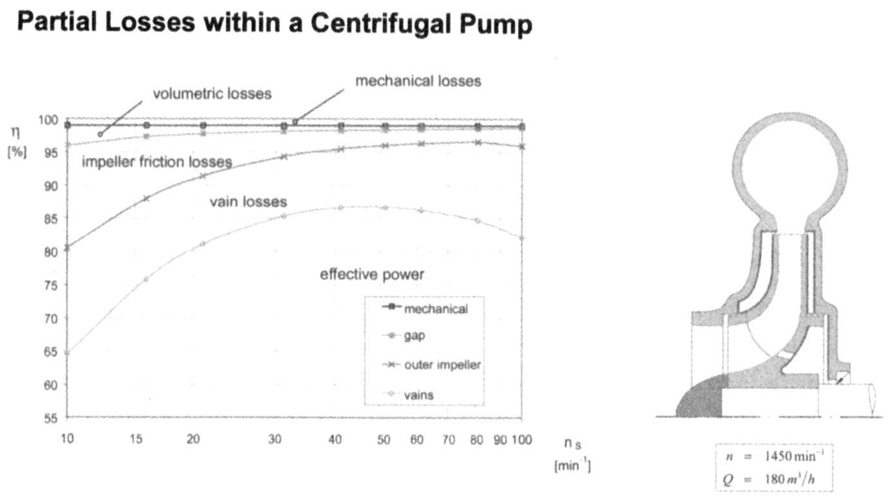

Fig. 3. Partial losses within a centrifugal pump

Depending on the value of specific speed which directly corresponds to the shape of the impeller different influences on the losses caused by several pump components can be mentioned (cp. fig. 3). For pumps with lower specific speeds volumetric losses as well as losses due to disk friction at the back and front shroud of the impeller are very significant. This also applies to the losses within in the volute casing. For higher specific speeds the influence of blade friction losses within the impeller dominates and mainly determines the level of the overall efficiency η.

Influence of Different Values of Surface Roughness

Fig. 4. demonstrates the general influence of different values of surface roughness. It is remarkable that the gain of efficiency due to smoothing the inner surfaces of a centrifugal pump is estimated more than 5 per cent points compared to pumps showing top quality sand-cast-rough surfaces (surface roughness $k_s \approx 0.024$ mm). Compared to pumps showing a very low surface quality (e.g. due to low quality of manufacturing, corrosion or incrustation which can result in a value for the surface roughness up to $k_s = 0.4$ mm) a theoretical efficiency improvement of even more than 20 per cent points could be estimated for pumps of very low specific speed (n_s = 10 min^{-1}).

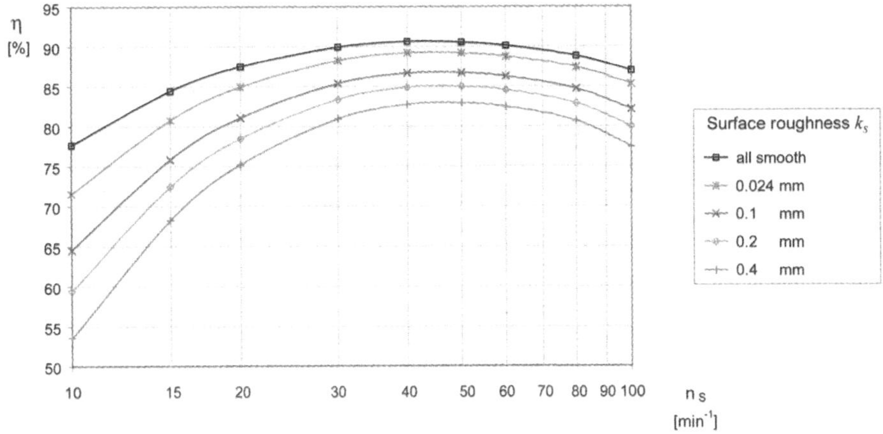

Fig. 4. The influence of surface roughness

The Influence of Smoothing Several Parts of Pumps

Since smoothing a whole pump is a very cost intensive manufacturing process (especially for small and medium sized pumps produced by a normal sand cast method) the influence of smoothing only several parts of the pump (i.e. volute, casing, outer surface of impeller, inner surface of impeller) was investigated theoretically by the aid of the described software.

The result of this parameter study shows, that also in case of partial smoothing the maximum efficiency improvement is to be expected for pumps with low specific speeds and can reach values of roughly 6.5 per cent points (e.q. in case of smoothing the outer surfaces of a radial impeller showing a origin surface roughness $k_s = 0.2$ mm). With respect to an overall surface treatment of the impeller the investigations showed also, that a smoothing of the inner surfaces is primarily favourable for pumps with higher specific speeds ($n_s > 30$ min^{-1}), where the hydraulic losses were mainly quantified by the flow velocity within the impeller.

The Influence of Only Partly Smoothing the Outer Surface of the Impeller

Due to the fact that especially for radial pump impellers ($n_s < 30$ min^{-1}) smoothing the outer surfaces of the impeller front and back shroud (by turning) is a very efficient and less costly procedure to reduce the losses, i.e. improvement of pump efficiency, the effect of smoothing the impeller front and back shroud only partly was also investigated by an appropriate theoretical parameter study.

As a result of this study it could be estimated, that smoothing only 40 % of the outer surface of the back and front shroud (starting the turning process at the impeller outlet diameter D_2) an efficiency improvement of roughly 5.5 per cent points still can be estimated. This value decreases to 3.5 per cent points in the case of smoothing only 20 % of the outer surface of the back and front shroud. Due to this fact there is no need to smooth the impeller at smaller diameters where turning gets more difficult because of the more complicated impeller contour. Fig. 5 summarises the theoretical estimation results obtained by the several parameter studies.

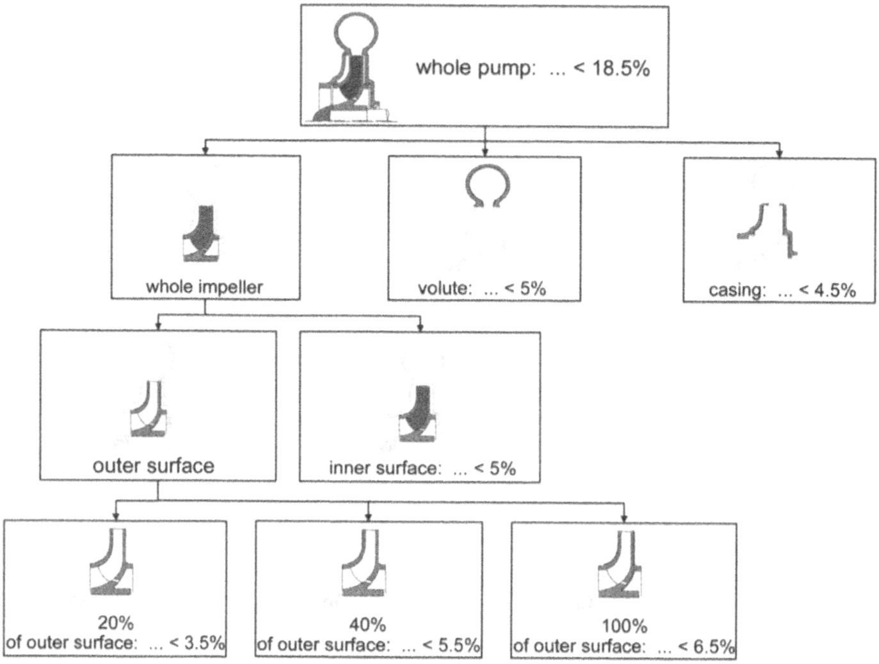

Fig. 5. Maximum improvement of efficiency for several smoothing steps (estimated by theoretical calculations for a medium size pump, with BEP at 180 m^3/h)

The labelled efficiency values roughly quantify the maximum gain of efficiency (in per cent points) that can be expected by smoothing the wetted surfaces of a centrifugal pump showing a surface roughness equal to a k_s value of 0.2 mm.

Influence of Different Gap Clearances on the Internal Leakage Flow

Volumetric losses are mainly caused by the existence of a suction-sided sealing gap which serves as a throttle in order to reduce the secondary flow from the impeller out- to inlet as well as a an additional pressure-sided sealing gap which usually belongs to the axial thrust balancing system of a single-stage centrifugal pump. This internal leakage flows strongly depend on the clearance of the sealing gaps. Fig. 6 shows the change in efficiency due to a variation of the gap clearance (the change in efficiency refers to a smooth gap with a radial gap clearance of 0.6 mm).

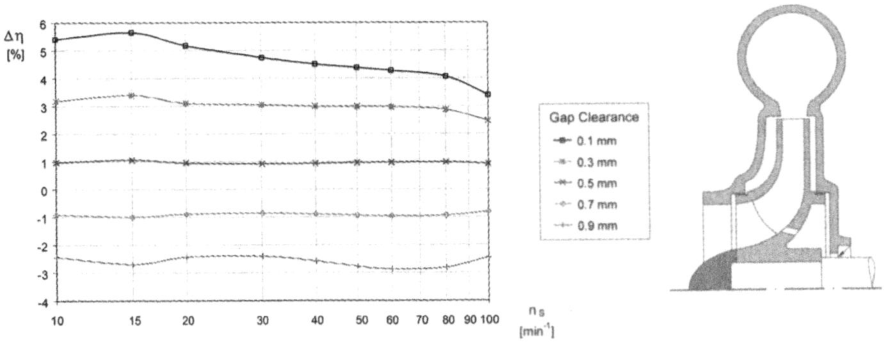

Fig. 6. The influence of secondary flow through the sealing gaps

The above diagrammed estimation results demonstrate that a reduction of gap clearance for instance from 0.6 mm to 0.3 mm can improve the pump efficiency about 3 per cent points. A possible additional treatment to reduce the internal leakage flows is to furnish one or both gap surfaces with circumferential notches (cp. fig. 7).

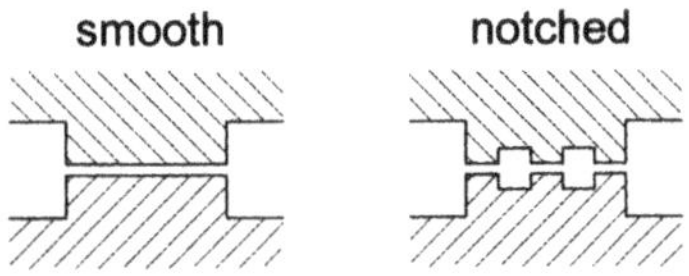

Fig. 7. Different types of sealing gaps

The theoretically carried out parameter study showed, that notching gaps of the type usually used in standard centrifugal pumps (cylindrical gaps with a relatively short gap length), only leads to a slight improvement of the pump efficiency.

Accompanying Experimental Investigations

In order to evaluate the efficiency values estimated by the computer program additional experimental investigations, funded by the German Federation of Industrial Cooperative Research Associations (AiF), were carried out at two centrifugal pumps one with a specific speed of $n_s = 12\ \text{min}^{-1}$ and another with a specific speed of $n_s = 20\ \text{min}^{-1}$ These special test pumps, designed according to usual industrial standards were equipped with very narrow sealing gaps (radial gap clearance equal to 0.1 mm) as well as hydraulic smooth surfaces. For these optimised pumps the value of the inner efficiency η_i was measured at a special high precision test rig and compared to the theoretical values obtained by the computer program. The comparison of both values (for pump operation at the point of best efficiency) results in a very good agreement.

Conclusions

The presented results show, that for single-stage standard centrifugal pumps within a range of specific speed n_s = 10 min^{-1} up to n_s = 100 min^{-1} (corresponding to values from 520 min^{-1} up to 5200 min^{-1} in US-Units), the highest potential for an efficiency improvement can generally be found in the region of low specific speeds. All efforts aimed at an improvement of the surface quality of several parts of the pump which are in contact with the flow cause a gain of efficiency. With regard to the manufacturing costs which result from such additional surface treatments the smoothing of the outer front and back shroud of the impeller can be proposed as a cost-efficient procedure to improve the efficiency (cp. fig. 8). Furthermore it could be shown that it is recommendable to reduce the clearance of the sealing gaps to the smallest possible value in order to increase the volumetric efficiency.

Example:
Pumps showing a low surface quality
(k_S = 0,2 mm)
Q = 180 m^3/h
n = 1450 min^{-1}

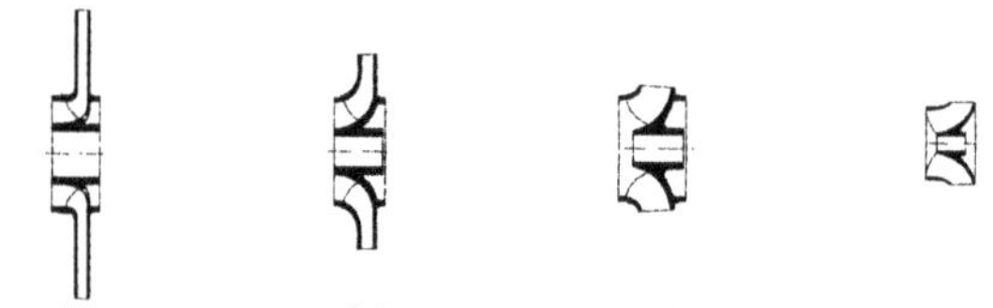

n_s :	10	... 30	... 60	... 100
Smoothing	Improvement of efficiency η in per cent points			
Whole pump :	18,5	7	5,5	7
Volute :	5	1,5	0,5	0,2
Casing :	4,5	1,5	0,5	0,4
Impeller inner surface :	1,5	2,5	3,5	5
Impeller outer surface :	6,5	2	1	1
Reducing gap clearance :	1-3	1-3	1-3	1-3

Effort of production High — Low

Fig. 8. Effort of manufacturing versus the theoretically estimated gain of efficiency

It should also to be mentioned, that the conditions of the surfaces as well as the sealing gaps within a centrifugal pump normally depend on the time of operation, which means that there is a strong necessity to check these parameters at reasonable intervals during the life time of a pump.

Screening Pumping Systems for Energy Savings Opportunities

Vestal Tutterow[1], Don Casada[2], Aimee McKane[1]

[1] Lawrence Berkeley National Laboratory

[2] Diagnostic Solutions, LLC

Abstract

In most industrial settings, energy consumed by pumping systems is responsible for a major part of the overall electricity bill. In some cases, the energy is used quite efficiently; in others, it is not. Facility operators may be very familiar with pumping system equipment controllability, reliability, and availability, but only marginally aware of system efficiency.

The cost of energy consumed by pumps usually dominates the pump life cycle cost. But many end users, already stretched to support day-to-day facility operations, lack the time and resources to perform a methodical engineering study of, in some cases, hundreds of pumps within their facilities to understand the energy costs and the potential opportunity for reduction.

Under the auspices of the Department of Energy's (DOE) Best Practices Program, prescreening guidance documents and a computer program called PSAT (Pumping System Assessment Tool) have been developed to help end users, consultants, and equipment distributors recognize, both qualitatively and quantitatively, pumping system efficiency improvement opportunities.

This paper describes the general methodologies employed and shows case study examples of the prescreening and software application.

Background

Industrial electrical motors account for two-thirds of the US industrial electricity. Pumping systems account for an estimated 25% of this electrical motor consumption. A recent study funded by the US Department of Energy estimates potential energy savings of approximately 20%, representing over 20,000 GWh/year, through industrial pumping systems optimization using existing, proven techniques and technologies (Ref. 1). This energy savings potential represents significant cost savings potential for industrial facilities. For example, an average paper mill can save up to $200,000/year through pumping systems optimization alone, based on the DOE study. Additionally, it has been shown that energy efficiency improvements to industrial systems usually provide improved reliability, improved productivity, and reduced environmental costs.

Many end-users are consumed by the day-to-day activities required to support facility operations, and lack the time and resources to perform a methodical engineering study of what can be hundreds of pumps at their facility. The following discussions will provide a background on pumping systems efficiency and introduce some tools available to help end-users quickly recognize opportunities to profit from their pumping systems.

System Optimization

Generally speaking, the best strategy for pumping systems optimization is to begin at the end of the line and work backwards through the pump, motor, and back to the transformer. The elements and the order of review contemplated is:

1. Ultimate goal or purpose
2. Piping system
3. Pump
4. Gear or coupling
5. Motor
6. Adjustable speed drive (if applicable)
7. Motor starter
8. Transformer

There are several reasons for this reverse sequence. In many cases, the greatest efficiency improvement opportunities are found in situations where the fluid system is simply doing a lot more work (e.g., delivering a higher flow rate or head) than is truly needed to support the ultimate goal of the system. Another important reason is that energy savings identified at the end of the energy transfer path are multiplied at the power line (the billing point) because of the inefficiencies of the upstream elements.

Some of these elements, such as the motor starter and shaft-to-shaft coupling, are important from a reliability standpoint, but are inconsequential from an energy efficiency perspective.

For those elements that are more important, how would one go about analyzing the individual components? There are methods for estimating efficiencies for the individual components, such as the motor), in the field (Ref. 2). But as noted previously, the BestPractices program encourages taking a systems approach. The general thesis of this approach is that it is more important to gain a measure of the overall system efficiency or effectiveness rather than dwell on individual components (Ref. 3). In common terms, the idea is to see the forest, not just the trees.

Pump efficiency is defined as the pump's fluid power divided by the input shaft power. The efficiency of any pump is influenced by hydraulic effects, mechanical losses and internal leakage. Pump manufacturers have many ways to improve pump efficiencies. For example, the pump surface finish can be made smoother by polishing to reduce hydraulic losses, but the additional first cost must be weighed against the energy savings. A "good" efficiency for a pump will vary depending on the type of pump. Pumps with special characteristics, such as canned motor pumps, self-priming pumps, pumps for solids handling, and low flow/high head duty pumps will have low efficiency ratings.

A more useful efficiency term is the wire-to-water efficiency, which is the product of the pump and motor efficiency. An even better measure of efficiency for analysis purposes is the system efficiency, which is defined as the combined efficiency of the pump, motor, and distribution system. These efficiency measures are illustrated in the following figures (Ref. 4).

In the simple pumping system shown in Fig. 1, fluid is drawn from a tank and pumped through a piping system which includes one flow control valve, to an elevated tank. A recirculation line with another control valve is also included. The pump is driven by an electric motor fed from a motor control center (MCC) which is, in turn, fed by a station transformer. How would one measure efficiency in such a system? Consider a series of boxes drawn at different levels around the system. At each level, the power input and useful output are measured, and the ratio defines the efficiency of the process inside the box.

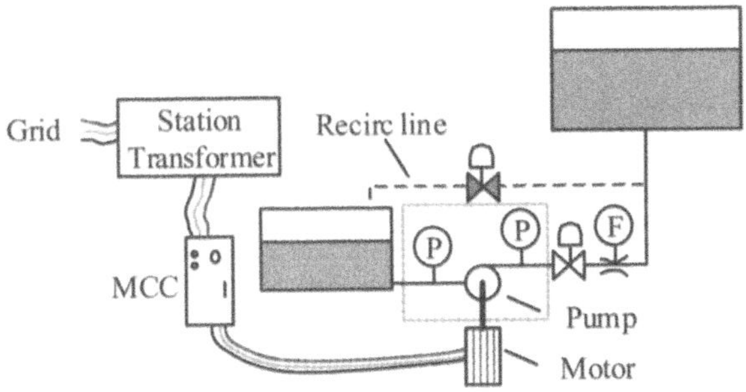

Fig. 1. The pump efficiency

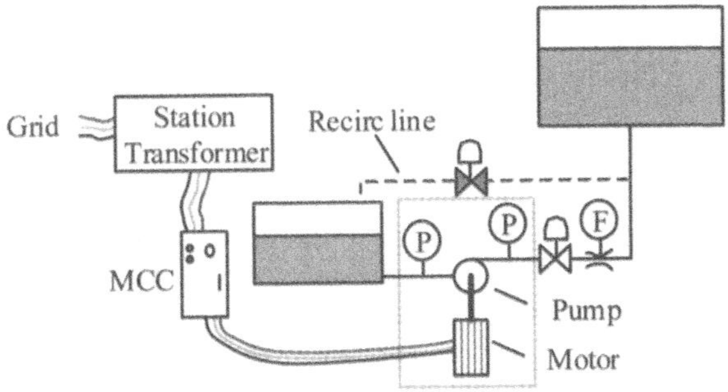

Fig. 2. Combined motor and pump efficiency

In Fig. 1, a box is drawn around the pump. The input power is the shaft mechanical power; the output power is the hydraulic power delivered to the system. The ratio is the *pump* efficiency.

In Fig. 2, a box is drawn around both the pump and the motor. The input power is the electrical power supplied to the motor; the useful output power is the hydraulic power delivered to the system. The ratio is the combined motor and pump, or *wire-to-water* efficiency.

In Fig. 3, a box is drawn around the pump, the motor, and the entire normal distribution piping network, from the source to the discharge tank. The input power is, again, the electrical power supplied to the motor; the useful output power is the net hydraulic power delivered *across* the fluid system.

The *system* efficiency, as defined by the power transfer in and out of the box in Fig. 3 is:

$$\eta_{sys} = \frac{P_f}{P_e} = \frac{QH_s\gamma}{P_e}$$

where:

H_s = static head (includes elevation & pressure head)
P_e = motor input power
P_f = fluid power
P_s = shaft power
Q = volumetric flow rate delivered to the tank
η_{sys} = overall system efficiency
γ = fluid specific weight

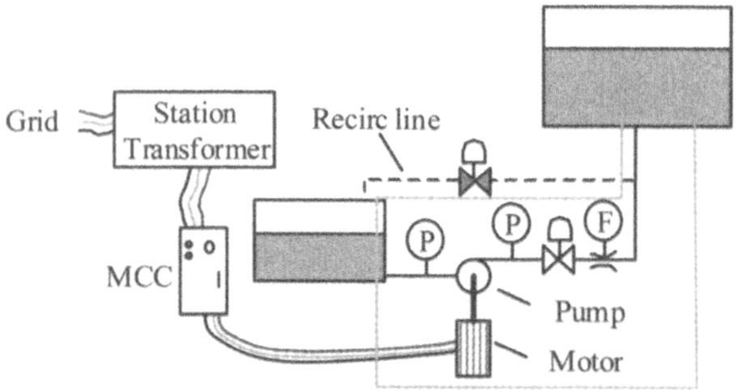

Fig. 3. System efficiency

The flow rate in the above equation is the net flow between tanks (ignoring recirculation flow). The head is the elevation difference between the tanks, or static head (implicitly ignoring friction losses). This is a true *system* efficiency – it overlooks the details and sees only the big picture. This approach can be quite useful. However, it does not work for all situations. For example, Equation 1 will produce a system efficiency of zero for a closed-cycle circulating system (with no static head).

Analysis Tools

Prescreening Guideline

The US Department of Energy has developed a guideline for prescreening pumping systems for potential energy savings. The guideline provides a methodology that can help to identify and prioritize candidate systems for optimization. This prescreening guideline includes sample data collection forms, and can be downloaded at **www.ornl.gov/etd-equip/Prescreen/Prescreening.pdf**.

Pumping System Assessment Tool

Once the prescreening process has identified pumping systems with potential cost savings opportunities, the Pumping System Assessment Tool (PSAT) can be used to further screen systems and quantify the potential savings. The PSAT software was developed for DOE as a tool to assist end users (and others) in assessing the overall effectiveness of pumping systems (Ref. 4). PSAT is available at no cost through the DOE BestPractices program web site (**www.oit.doe.gov/bestpractices/software_tools.shtml/**). The prescreening guide noted above is included in the PSAT installation. PSAT applies to centrifugal pumps directly coupled to 3-phase induction motors. The minimum size motor is 5 hp; there is no maximum.

PSAT estimates the existing motor and pump efficiency using field measurements and nameplate type motor and pump information. It also estimates achievable efficiencies if the motor and pump were optimally selected to meet the specified flow and head requirements. The "existing" and "optimal" results are compared, and potential power savings are determined. Finally, potential cost and energy savings are estimated based on user-specified power cost rates and operating times.

Display layout

The primary or front panel of the software is shown in Fig. 4. The "inputs" are all located in the three boxes that cover the left one-third of the panel.

The *Pump, motor, system information* in the upper left corner is general design information:

- Pump style, nameplate speed, and number of stages
- Fluid viscosity, specific gravity
- Motor class, nameplate hp, rpm, and voltage.

The *Operating parameters* for the system are at the middle left:

- Operating fraction (fraction of time the pump is operated at the specified conditions)

- Cost of electricity (cents/kwh).

The *Measured or required conditions*, at the middle to lower left are:

- Measured (or required) flow rate and head
- Measured motor power or current (depending on the selected load estimation method)
- Measured bus voltage.

The *Calculated Results* are displayed in the upper right portion. In the next sections, the sources of motor and pump data, and how they are used in PSAT to achieve these results will be discussed.

Finally, *Log, summary file controls* are just below the *Calculated Results*. The log controls enable the user to store and retrieve data for subsequent display and comparison. Data can be stored by plant, process, or any other structure the user elects. Tab-delimited text summary files that save the essential inputs and results can be created or existing summary files appended. The summary files can be opened and manipulated in other applications. For example, a summary file with the analysis results for all of a particular facility's pumps can be created, allowing further analysis in a spreadsheet program.

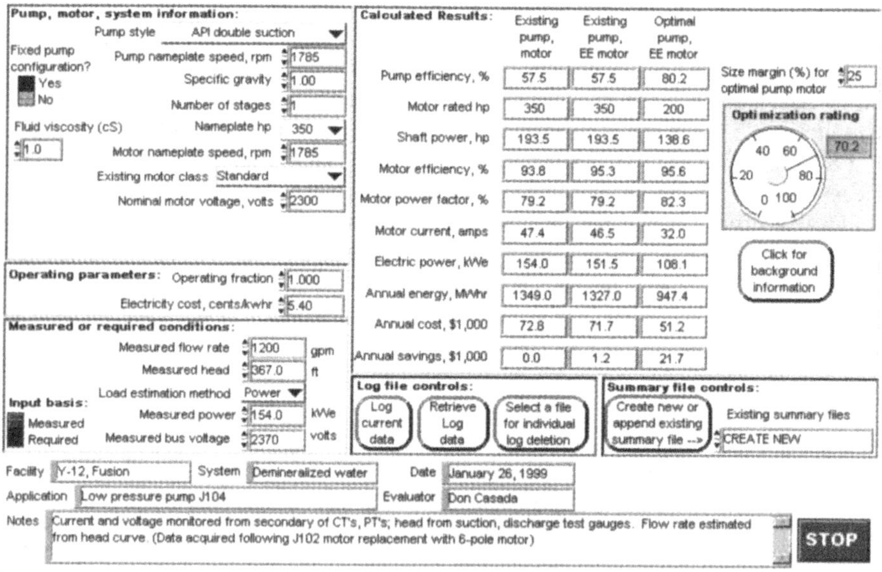

Fig. 4. PSAT primary panel

Motor performance characteristics used by PSAT

The BestPractices Program distributes the MotorMaster+ (MM+) software, a motor selection, comparison, and management tool (Ref. 5). Part of the underlying supporting structure for the MM+ package is an extensive database of motors. The database, constructed using motor manufacturer-supplied data, includes a fairly comprehensive list of parameters such as motor rated power, efficiency, power factor, speed, full load current, enclosure style, NEMA design type, rated voltage, and price.

This motor database was used to develop algorithms used in PSAT. The motor population was thus categorized by size, speed and efficiency class, and average performance characteristics (current, power factor, and efficiency vs. load) were established. Using these average values, curve fits of the performance characteristics were developed.

The curve fits developed from the average performance characteristics of the MM+ database allow motor efficiency to be estimated based on motor size, speed, and measurement of either motor input power or current. If power is measured, PSAT determines the shaft power and efficiency that is consistent with the specified motor size and speed. If current is measured, the power is estimated from the current vs. load profiles in PSAT. A full set of motor characteristics (shaft power, current, power factor, and electrical power) can be established, regardless of whether current or power is measured.

Although the motor characteristics used in PSAT were derived exclusively from 460-V motors, the user can select from other nominal voltages, such as 230, 2300, or 4160 Volts. The current data is linearly adjusted for nominal voltage.

The user also selects from one of three motor efficiency classes – energy efficient, standard efficiency, and average. The average selection simply calculates motor performance characteristics based on the average of the standard efficiency and energy efficient motor values.

Pump capability estimation used by PSAT

There are many different pump designs applied to the broad spectrum of pumping applications. As mentioned previously, certain applications, such as sewage or stock pumping, have service reliability considerations that prevent the use of more efficient designs that are used in clean water pumping.

Fortunately, the Hydraulic Institute (HI) has published a standard that provides guidance on achievable efficiencies (Ref. 6). The standard addresses the effects of general pump style, capacity, specific speed, and variability in achievable efficiency from miscellaneous other factors such as surface roughness and internal clearances.

Putting the pieces together

Based on the input data, PSAT first estimates the existing shaft power from the motor data measurements. It then calculates fluid power from the specified flow rate, head, and specific gravity. At this point, the motor input power, the shaft power, and the fluid power are known, as are the existing motor and pump efficiencies. Given the fraction of time the pump is operated and the electricity cost rate, the annual energy and energy cost is also calculated.

Two actions are considered for comparison:

1. Apply an energy efficient motor in lieu of the existing one. The middle results column shows the potential savings associated with just motor replacement.

2. Employ an optimal pump for the application, with the pump driven by an energy efficient motor.

Of course if the existing motor is energy efficient, no improvement in motor efficiency would be seen.

PSAT results are useful in identifying the approximate energy and cost savings that could be achieved if the existing pump system was optimized. PSAT does *not* identify how the savings can be achieved; in other words, it is not a solution provider, but rather an opportunity identifier.

Example using PSAT

DOE representatives and personnel from the US Steel Edgar Thompson Works in Pittsburgh, Pennsylvania, USA recently used the pumping systems prescreening checklist and the PSAT software to identify opportunities at the US Steel facility. The process began with discussions between DOE and US Steel personnel. The prescreening checklist was employed to identify several pumping systems that were likely candidates for energy reduction. Several symptoms (cavitation noise and damage, continuously open bypass flow valves, and constant pump operation under varying load) from the checklist that are generally indicative of energy (and reliability) suggested the pumps used for the basic oxygen furnace hood spray were good candidates for further analysis using PSAT. Also, through the earlier discussions, it was known that these pumping systems were high maintenance systems.

Next, field measurements were taken to gather input necessary for the PSAT software tool. Data on flow rate, head, speed, power, and process duration time were recorded. Calculations performed by PSAT estimate that pumping energy costs could be reduced by up to 87%. Using PSAT and the experience of the DOE and US Steel personnel, a project was developed that including replacing a pump with a smaller pump, isolating the normally open bypass line, and installing a soft starter for the motor. A simple payback of less than one year has been calculated, with projected annual energy savings exceeding $40,000.

References

[1] U.S. Department of Energy, United States Industrial Motor Systems Market Opportunities Assessment, December 1998.

[2] Oak Ridge National Laboratory, Assessment of Methods for Estimating Motor Efficiency and Load Under Field Conditions, ORNL/TM-13165, January 1996.

[3] US Department of Energy, Improving Pumping System Performance: A Sourcebook for Industry, January 1999.

[4] Casada, Don, "Screening Pumping Systems for Energy Savings Opportunities", Proceedings from 1999 Industrial Energy Technology Conference, May 1999, pp. 261-270.

[5] MotorMaster+ software program, Web page: http://www.oit.doe.gov/bestpractices/.

[6] Hydraulic Institute, Efficiency Prediction Method for Centrifugal Pumps, 1994.

Complex Automation of Pump Stations Based on Energy Saving Drives

I. Ya. Braslavsky[1], Z. Sh. Ishmatov[1], I. A. Averbakh[2], E. I. Barats[2]

[1] Ural State Technical University, Electric Drive Department,
Mira st. 19, 620002 Yekaterinburg, Russia
E-mail: braslav@ep.etf.ustu.ru

[2] Tyazhpromelectromet
Lenin st. 24/8–602, 620014 Yekaterinburg, Russia
E-mail: tech@elmet.epn.ru

The economic efficiency of frequency-controlled induction motor drives used at water supply stations is proved well as theoretically as practically. The application of these drives for solving energy saving and automation tasks at pump stations provides a number of advantages in compare with constant speed ones:

- electric energy saving (up to 50%);
- water saving (up to 25%);
- reducing the risk of plumbing faults due to motor soft starting;
- automation of pressure (flow, level e t.c.) control.

The energy saving feature is determined by motor load depending on rotating speed and elimination of shutter control while the water saving occurs mainly in low quality systems with increased water losses. This condition also makes the soft-start feature very attractive because the smooth pressure control significantly reduces the repair charges.

Authors obtained the practical experience of frequency-controlled drives application during taking participation in upgrading a number of pump stations of Yekaterinburg and Sverdlovsk region (Russia). Several frequency converters (15, 15, 37 and 200 kW) were implemented. The measuring of energy and water consumption before and after upgrade gave the following results:

- 40–50% energy consumption reducing;
- 20–25% water consumption reducing;
- short period of repayment (6–8 months).

Fig. 1 shows the daily diagrams of input and output pressure of one station. The automatic stabilization of output pressure when the water consumption varies is evident in case of frequency converter application.

As a rule, modern pump stations are the complex objects and consist of not only main pumps but also of a number of auxiliary equipment such as reserve pumps, fire pumps, drain pumps, shutters, gates, filters, reservoirs e t.c. In many cases the combining of all these devices in one automation system with use of PLC is preferable. Moreover, PLC can provide the automatic control of output pressure or other technological parameter. Interaction with up-level system is built if required.

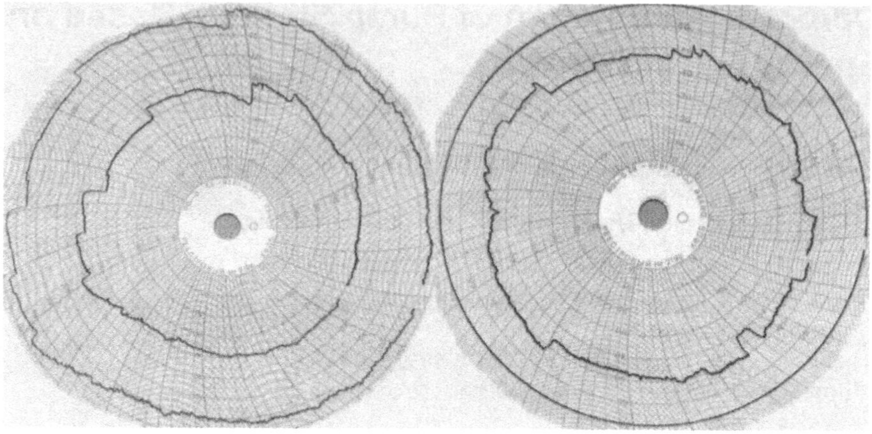

Fig. 1. Diagrams of output and input pressure of a pump station: *left* – before upgrade; *right* – after upgrade

Pump units pose rather low requirements to speed control quality. This allows application of relatively simple frequency converters without vector control, high overload capability, integrated chopper and other features specific for converters used in complicated drives. This results in reducing of converter cost. Besides, many of modern converters developed for HVAC applications have a number of useful specific features:

- motor staging (control of additional drives via output relay);
- belt failure detection (when pump has run dry);
- bypass function;
- energy saving mode e t.c.

As a rule, there is no necessity to equip all the pumps by frequency converters. When several pumps work in parallel a single converter could be enough. Constant speed pumps are started and stopped automatically when it's required. This allows the smooth control of output pressure in all range of station productivity. To avoid water-hammers the application of soft-starters for additional pumps is advisable.

The example of automatic control of pumps group can be seen in project of step-up pump station developed by "Tyazhpromelectromet" Company. There are seven pumps equipped by 75 and 55 kW induction motors and connected in parallel. Some of them are the main ones and others are the reserve and fire pumps. Dispatcher configures the set of main and reserve pumps with use of control switches. Pressure control is provided mainly by single frequency converter "Midimaster Eco" ("Siemens"). The automatic pumps selection is implemented by Simatic S7-200 PLC. At present, the "Midimaster Eco" series is replaced by "Micromaster 430" series that has a lot of new functions.

The task becomes more difficult when large pump stations with medium-voltage motors are upgraded. Low-voltage converters, as a rule, have traditional topology (diode rectifier – PWM inverter with IGBT) and have approximately close prices. Unlike them, medium-voltage ones could have different power circuit topology and consequently different prices. The converter selection in this case is determined by motor type, it's design, supply performances, speed range, available place for equipment. There are several basic topologies for medium-voltage converters:

1. Topology close to low-voltage converters but implemented with use of medium-voltage devices (diode rectifier – voltage source PWM inverter with IGBT). The "Simovert MV" converter ("Siemens") is an example of such topology. Elimination of supply distortion and output current curve close to sine are the advantages of this topology. However, these converters are expensive and require filtering of output voltage to protect motor isolation.
2. Topology based on current source PWM inverter with GTO or IGCT devices and thyristor rectifier. The "Powerflex7000" converter ("Rockwell Automation") is an example of this type of power circuit. These converters have an advantage in possibility of working with any motors without output filtering. Producing the high harmonics and reactive power to supply is a disadvantage. To increase the compatibility with supply one may use 18-pulse rectification instead of standard 6-pulse scheme.
3. Multi-level inverter topology. In this case each motor phase is fed from several single-phase low-voltage IGBT-inverters (cells) connected in series. The diode rectifier of each cell is connected to individual winding of input transformer. This multi-pulse topology provides maximum compatibility as with supply as with motor with respect to quality of current and voltage form but has a disadvantage in rather high cost and increased dimensions of converters. Examples: "MV-GP Type H" ("General Electric") and "MELTRAC-F500HV" ("Mitsubishi Electric").
4. Medium-voltage converters based on low-voltage converters with input step-down and output step-up transformers. Reasonable price and elimination of negative influence on supply are advantages of these converters. Reduced efficiency due to output transformer losses and increased dimensions are disadvantages. Converters based on "Simovert Masterdrives" ("Siemens") and "ACS-600" ("ABB") are examples of such topology.
5. Converters based on thyristor rectifier and inverter with simple SCR thyristors and amplitude-pulse modulation. Such converters have significantly reduced price but have worst performances with respect to influence on supply and output current quality resulting in increased losses in motor.

In conclusion we note that frequency-controlled drive implementation on pump station should follow the feasibility study of upgrade expedience. In some cases such upgrade may not give expectable effect or results in too long period of repayment comparable with equipment life. Our experience of study, design and implementation of frequency control drive allows concluding that maximum effect can be reached on stations with small amount of pumps connected in parallel, equipped by low-voltage motors. This is especially topical for systems with wide range of water consumption and high level of water losses.

References

[1] Braslavsky I.Ya. Energy saving with use of adjustable induction motor drives. Electrotechnika, 1998. No 8.

[2] Braslavsky I., Ishmatov Z., Shilin S., Haehle W. Der geregelte Asynchronantrieb – eine Moeglichkeit zur Energiesparung. Beitrage fuer Lehre und Forschung. Heft 1/98. Hochschule Technik, Wirtschaft, Kultur Leipzig, Deutschland.

[3] Braslavsky I. Energy saving in electric drive. Proceeding of the 5th international conference on unconventional electromechanical and electrical systems. Vol. 2, p.235-238. Poland, 2001.

Factors Influencing Energy Consumption of Waste Water Pumps

Thomas Pensler

Dipl. Ing. T. Pensler, KSB Aktiengesellschaft., Johann Klein Str. 9,
D-67227 Frankenthal

The Judgement of Energetic Quality of a Pump Cannot Be Done Without Examination of Real Operating Point

Pumps have considerable differences in efficiency between optimum and partload and overload conditions in opposit to electrical drives. For example a pump reaches on the best point a very good efficiency of $\eta = 81$ % (1) but on part load conditions only $\eta = 68$ % (2).

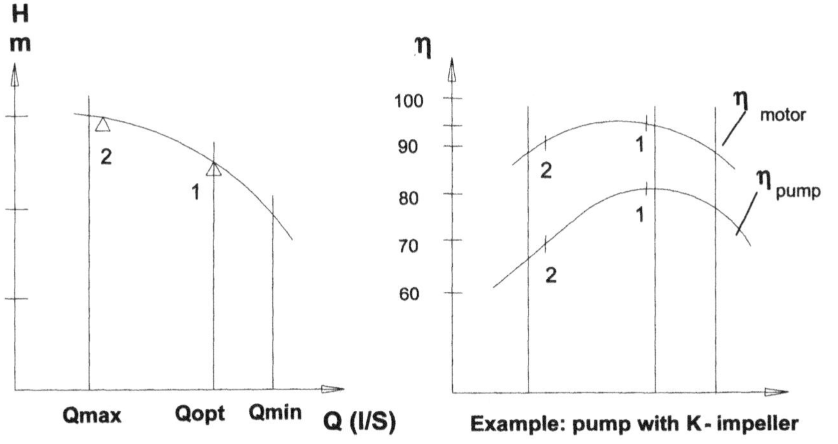

Why Is the Duty Point Not Always Near the Best Point of the Pump?

- We get the definition of duty point of pumps from Head (H)and Flowrate (Q). A series pumps has limited number of versions. Every pump can be used for a range of Q-H.

- In the case of bigger pump stations the amount of waste water is depend from day-time and season. It is necessary to use more than one pump in parallel commision. With increase of capacity the resistance of tube rises and the head too.
- In waste water nets the pumpstations have to pump into one common maintube. In result of the discontinous pump-process the duty points are different.

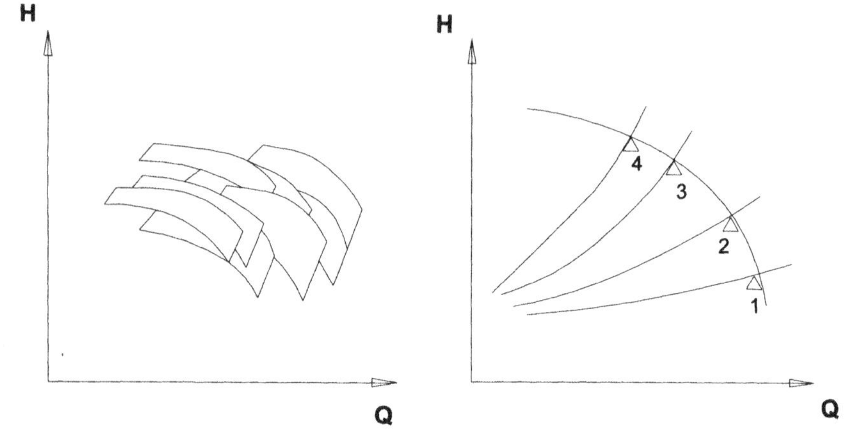

- The waste water is coming to the pump in chanels, pits and other basins. The different water level has a strong influence on the duty point, especially on applications in waste water treatment plants with low total head. Sometimes is it possible to bring pumps into optimal operating conditions with speed control. But frequency converter itself consume also a small part of energy.

- The deposit of solids inside the tube increases the energy loss. Installation of new pump stations in old nets can lead to selection of wrong pumps. Assumption of tube conditions can be wrong caused by calcination.

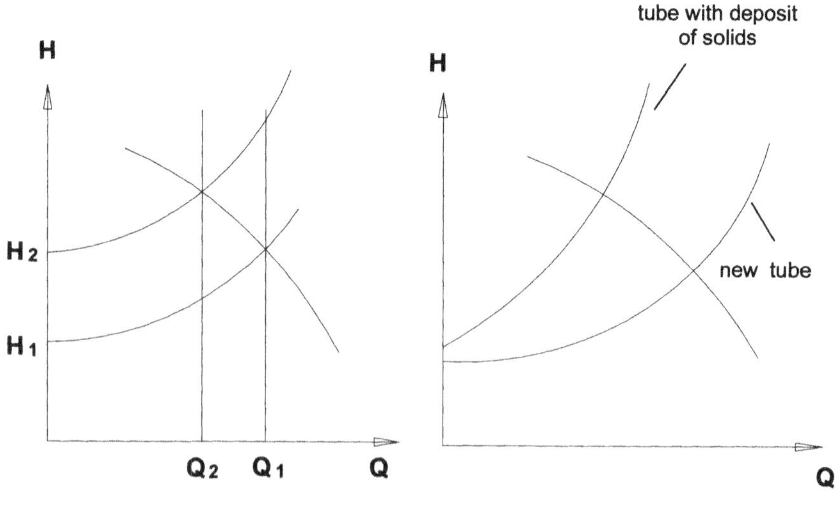

Pumped Media in Sewage Technology

Table 1. Essential parameters to characterize sewage

			gas content (%)	fibre content	solids size	TSR (%)	sand content (g/l)
rain and surface water			-	low	small	-	0 – 3
waste water	municipal WW	household WW	0 - 2	med.	med.	-	0 – 3
		trade WW	0 - 2	large	large	-	0 – 3
	industrial WW		-	low	small	-	0 – 3
	agricultural WW		0 - 2	large	large	0 – 5	0 – 3
sand water			-	-	-	-	8 – 10
sludges	activated sludge		2 - 4	low	small	1 - 2	-
	primary sludge		2 – 4	low	small	2 – 6	-
	secondary sludge		2 – 4	low	small	2 – 3,5	-
	thickener sludge		3 – 6	low	small	2 – 5	0 – 2
	stabilised sludge		-	low	small	5 – 10	-
	dehydrated sludge		-	low	small	20 – 30	-
	dried sludge		-	low	small	30 - 50	-

Comparison of Efficiency Curves for Different Impeller Design Concepts

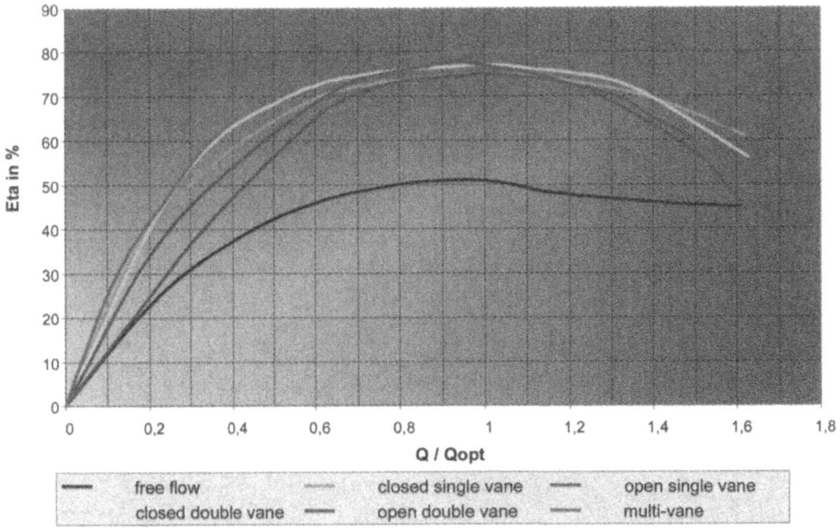

open single vane impeller

closed single vane impeller

open double vane impeller

free flow impeller

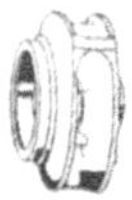

closed double vane impeller

Decrease in Efficiency Due to Wear in Relation to Sealing Clearance Enlargement

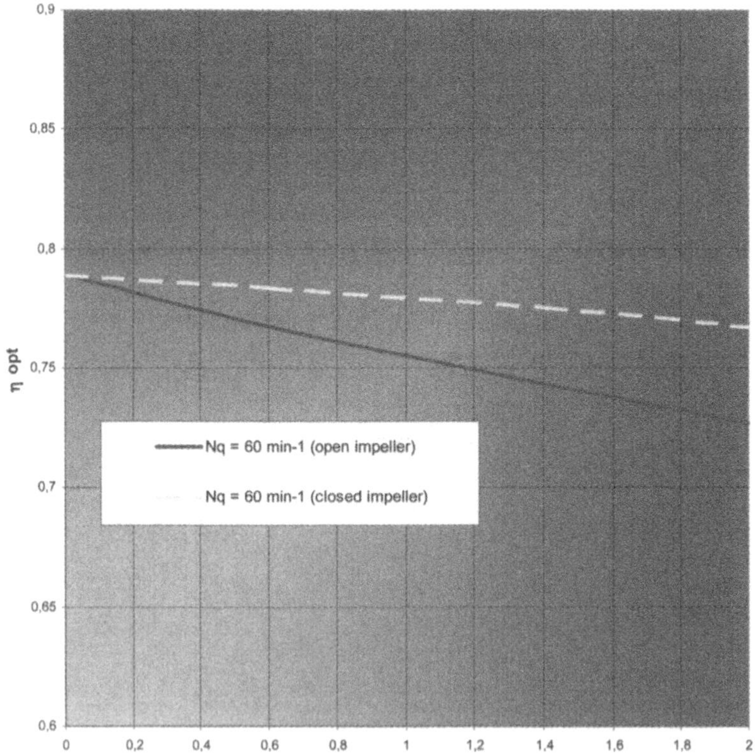

New Intelligent Pump Status Monitoring Reduces LCC

Gerhard Berge[1], Norbert Kastrup[2]

[1] Dipl. Ing. Gerhard Berge, KSB Aktiengesellschaft,
Johann Klein Str. 9, D – 67227 Frankenthal

[2] Dr. Norbert Kastrup, KSB Aktiengesellschaft,
Bahnhofplatz 1, D – 91253 Pegnitz

Pumps take care of the transport of liquids in plants and processes. If they stand still, the result can be loss of production or even failure of entire hydraulic systems. The more expensive a pump is or as more valuable a process is, the more important it is to regularly record and control the condition of the pumps. As pumps also react to changes in a plant, they are also indicators of the conditions prevailing there. In order to achieve real process improvement, it is important to find out the cause of failures and disturbances. With respect to centrifugal pumps, this could be the following:

- defective mechanical seals point to dry-running or partial-/ over-load
- damaged rolling element bearings are the consequence of gas content of the pumped medium
- eroded impellers and casings are the result of cavitation.

This shows that the determination of the pump condition allows for the drawing of conclusions regarding the condition of the process.

In order to increase the availability of plants, this knowledge constitutes an important approach and starting basis for a new, intelligent development in the area of monitoring the condition of centrifugal pumps. The goal was to be achieved with a minimum of sensory devices. An analysis of failure statistics of service units and operators (Fig. 1) showed that 2 to 3 sensors suffice to detect the most important faults of a pump.

- Dry-run recognition
- Bearing temperature
- Vibrations

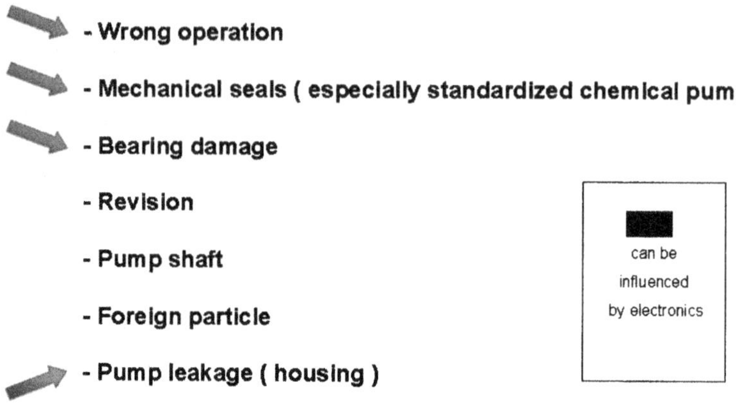

Fig. 1. Failure Reasons of Circulating Pumps in Chemical Industry and Process Engineering

It is important to the user of such a system to have a simple, clear and easy to understand presentation of the pump condition. This eliminates a solution where the measurement values are connected each to a separate conditioning device (Fig.2) and each value having to be interpreted separately by experts. In this case the user would have to draw his own conclusions from the combination of reports.

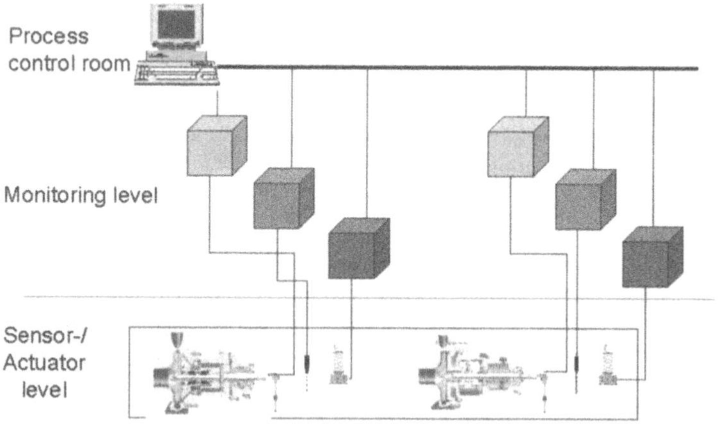

Fig. 2. Standard monitoring solutions need seperate signal - wiring and – conditioning per each sensor

A first step in the direction of benefit from the knowledge gained resulted in the definition of the threshold value for early warning signals. However, this solution requires a well founded pump knowledge on the part of the user in order to define the appropriate magnitude of the threshold value. If however the goal is preventive maintenance, the solution only has limited significance with respect to the general pump condition. It does not at all allow conclusions regarding plant-induced influences. When identifying impermissible operating conditions, the question also is which pump components could be damaged. Therefore, the analyses may not be limited to a simple, singular recording of threshold values.

An intelligent combination of information which each sensor supplies, makes it possible to draw conclusions with regard to pump and plant which clearly exceed the interpretation of individual values. The utilization of micro-processors allows the linking of individual data and signals (Fig. 3). The application of "fault tree analysis" is thus possible in principle. However, it requires high computer capacities in order to generate really predicative information. For this reason a simpler solution was sought. The result was the use of the fault table which is derived from the fault tree. The fault table links/compares several data/signals according to the principle " if... then" and deduces from this accordingly predicative information. This generated information can be transmitted very easily instead of the original large data quantity.

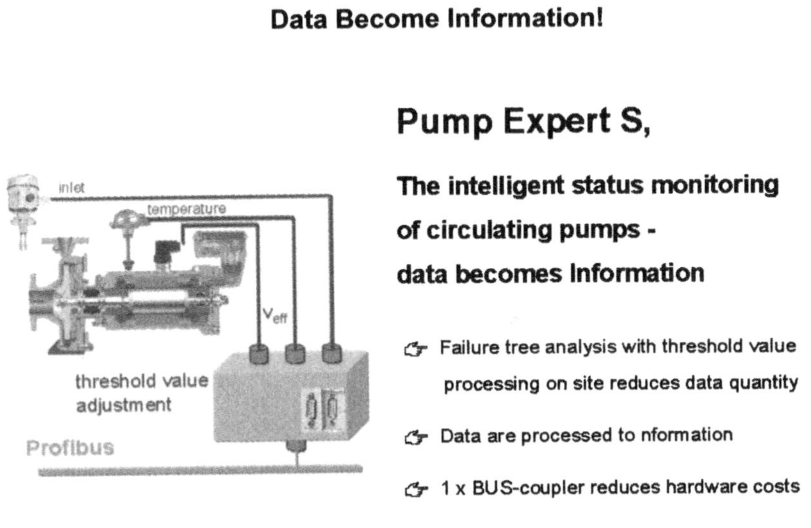

Fig. 3. The intelligent status monitoring of circulating pumps

For this purpose, a programme was developed which can be stored in a micro-processor and which must be adapted specially to every pump type series. Practice has shown that individual values can be in an “acceptable range", the combination however may well signalize a critical condition. Example: Bearing temperature warning “yellow” plus vibration warning “yellow” at the same time can cause a "red" traffic light! With the use of this method of fault table (Fig. 4), this can also be recognized.

method of fault table

fault information \ measured value	Value 1a	Value 1b	Value 2a	Value 2b	Value 3a	Value 3b	
cavitation		X					
plain bearing damage	X			X		X	
wear out of impeller		X	X				
. . .							

Fig. 4. Method of fault table

Thus pump components at risk can be localized in time and in part even the cause of the problem can be determined. To assure easy handling despite the existing complexity and large data quantity, a simple traffic light information – green, yellow, red – is made available. If needed, e.g. depending on the condition, further details can be requested and presented to the pump user such as for example the following messages:

- Dry-running
- Rolling element bearing defect
- Impeller blocked
- Viscosity of pumped medium too high
- Alignment mistake
- Plant not filled
- Medium contains too much air or gas
- Flow cut off
- Cooling of motor bearing bracket disturbed by cooling air of motor

Additional information like operation time counter, start/ stop counter etc. is available to get a general impression of the pump status.

Besides the overall information by the traffic light and the precise diagnosis messages supplied the user retains full access to the data collected from the sensors if requested. Thus the expert on site can use the general pump knowledge at his hand to analyse and improve the individual process. An additional specialty of the system is that it not only creates precise diagnosis messages but also instructions for adaequate actions. Such detailed information can only be generated by the one who best knows the behaviour of the pump and that is the pump manufacturer!

These findings constitute great progress in the improvement of pump availability and even of the entire process. They facilitate service and maintenance and enable the user to take first steps in the direction of preventive maintenance. A digital, bi-directional data transfer, e.g. on Profibus-DP basis, transmits electronically the information calculated out of a multitude of collected data to the control station. Thus details on the pump condition can be requested from a distance. When realizing a product, recourse should be had to less expensive components which nevertheless are flexible in their usage. Furthermore, the components should be introduced on the market in order to assure high reliability and stable functionality as well as a trouble-free acceptance by the users. An explosion protected version for installation in Zone 1 should be relatively simple to realize.

The decision, when selecting the electronics, fell to an intelligent entry/exit modular system supplemented by a mini-PLC (programmable logic controller) with different fieldbus-interfaces, e. g. Profibus-DP (Fig. 5).

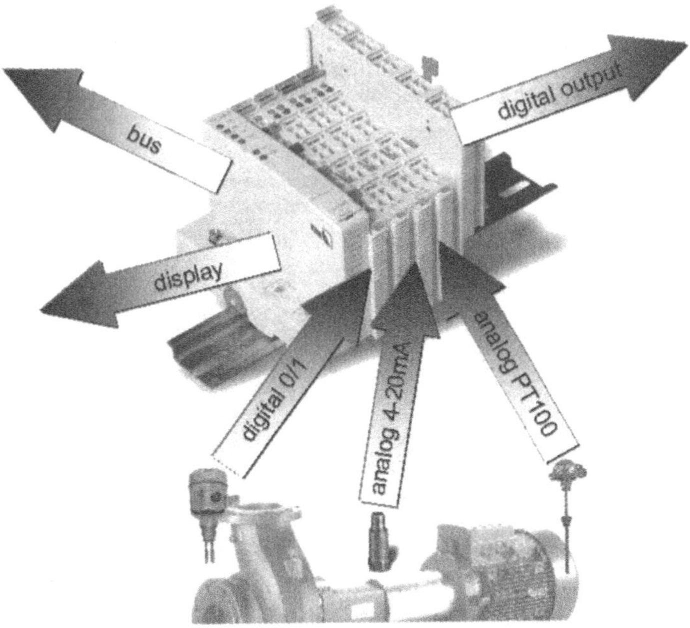

Fig. 5. Hardware components of Pump Expert S

In the mini-PLC, the software of the pump manufacturer is stored which does the evaluation of the data by means of fault table analysis. Thanks to the open structure, different types (Fig. 6) and also other pump makes can be equipped with the new "Pump Expert S".

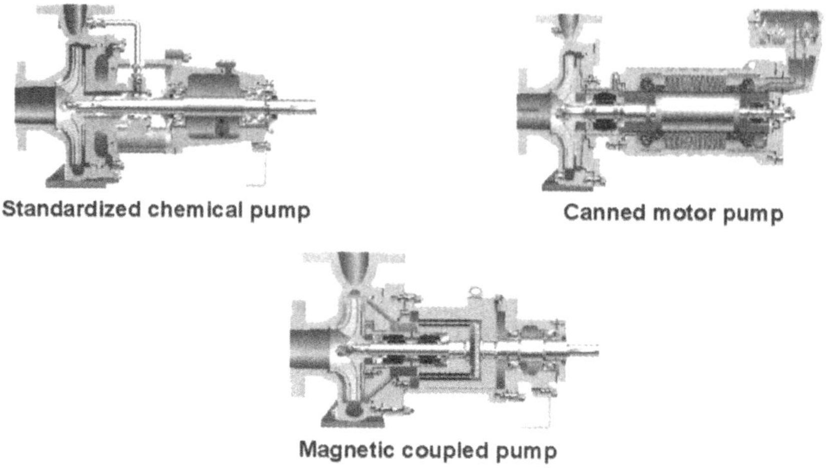

Fig. 6. Types of circulating pumps

Depending on the pump design, sensors will be mounted to the unit upon delivery and thus delivered in cabled condition. Even already on site mounted pumps can be equipped. With the Profibus, detailed data can be called for when needed and presented (Fig. 7).

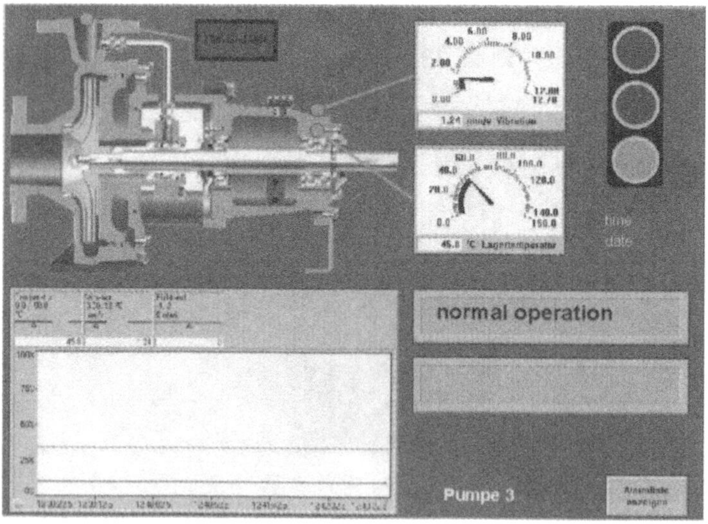

Fig. 7. Visualisation (example)

All components are mounted as a device in an enclosure of class IP 54 on top hat rails (Fig. 8). This robust casing enables the mounting on site at the pump and thus the cabling costs are reduced drastically. As sensors, all standard models on the market can be used, even those already available in the process. For the explosion protected design, intrinsically safe sensors are used, the remaining electronics will then be mounted in a pressure-proof enclosure.

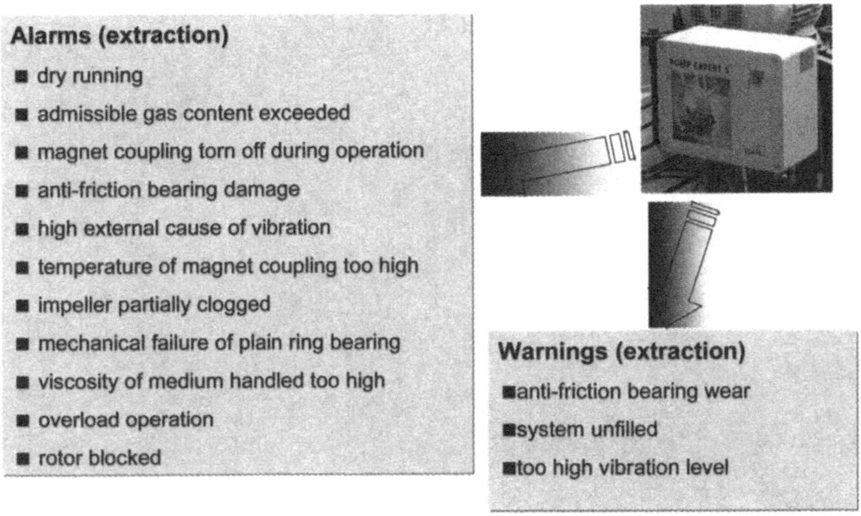

Fig. 8. Alarms and warnings

With the new concept, combining a minimal number of sensors to an intelligent unit with the aid of a programmable logic controller, the process quality can clearly be improved in an economic manner especially for critical plants and expensive pumps. Installation, commissioning and maintenance costs are reduced because alignment mistakes or dry-running are discovered immediately. With the use of advance alarms, standstill periods can be kept very short or even be completely avoided. Environmental impacts and leaks are prevented by timely recognition. The availability of the plant is thus significantly increased. That means Pump Expert S makes an important contribution to the reduction of life-cycle costs of the pump as well as of the plant.

Index

GPSR Compliance

The European Union's (EU) General Product Safety Regulation (GPSR) is a set of rules that requires consumer products to be safe and our obligations to ensure this.

If you have any concerns about our products, you can contact us on ProductSafety@springernature.com

In case Publisher is established outside the EU, the EU authorized representative is:

Springer Nature Customer Service Center GmbH
Europaplatz 3
69115 Heidelberg, Germany

Batch number: 10761599

Printed by Printforce, the Netherlands